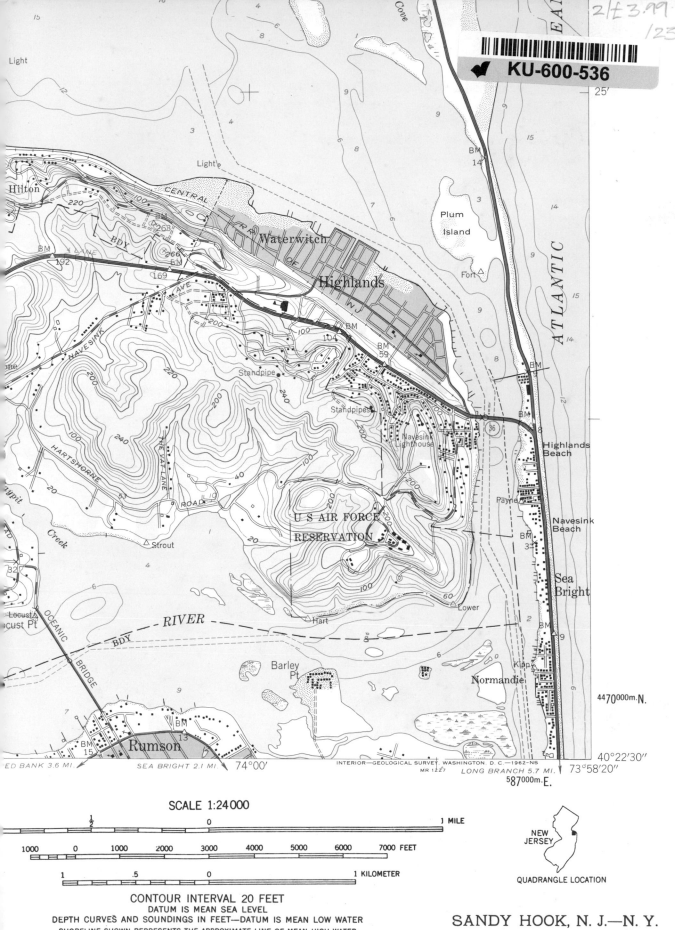

2/£3.99
/23

SCALE 1:24 000

1 MILE

1000   0   1000   2000   3000   4000   5000   6000   7000 FEET

1   .5   0   1 KILOMETER

CONTOUR INTERVAL 20 FEET
DATUM IS MEAN SEA LEVEL
DEPTH CURVES AND SOUNDINGS IN FEET—DATUM IS MEAN LOW WATER
SHORELINE SHOWN REPRESENTS THE APPROXIMATE LINE OF MEAN HIGH WATER
THE MEAN RANGE OF TIDE IS 3·8 FEET

NEW JERSEY

QUADRANGLE LOCATION

SANDY HOOK, N. J.—N. Y.
N4022.5—W7358.33/7.5x9.17

1954

INTERIOR—GEOLOGICAL SURVEY, WASHINGTON, D. C.—1962-NS
MR 1227

INTRODUCTION TO PHYSICAL GEOGRAPHY

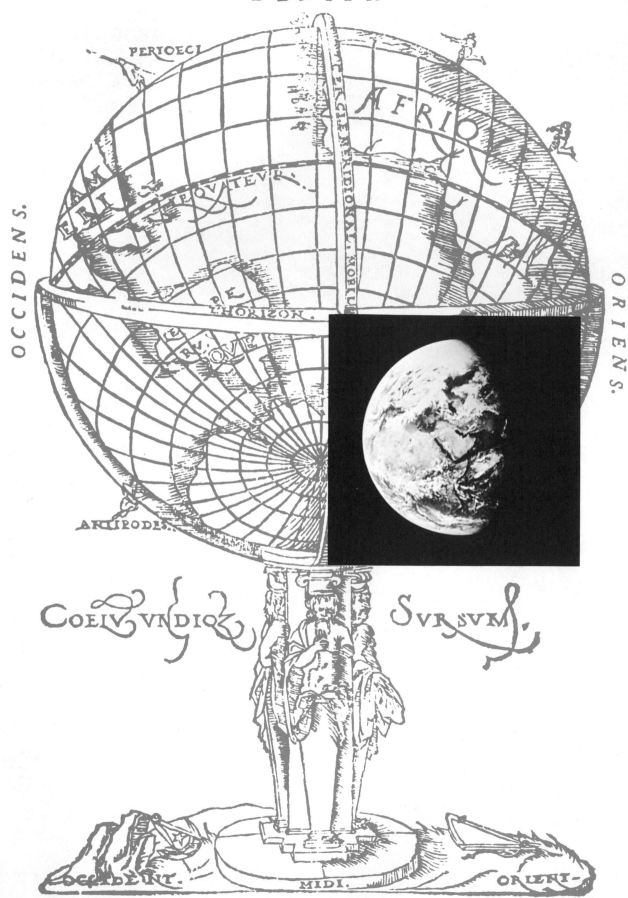

ARTHUR N. STRAHLER

INTRODUCTION TO PHYSICAL GEOGRAPHY

SECOND EDITION

JOHN WILEY & SONS, INC.

New York · London · Sydney · Toronto

Library of Congress Catalogue Card Number: 73-91648

SBN 471   83168   9

Printed in the United States of America

10   9   8   7   6   5   4   3

Preface

FIRST published in 1965, *Introduction to Physical Geography* met the need for a shortened version of its parent work, *Physical Geography*, 1960, which was then in its second edition. The requisite abbreviation was accomplished by deletion of entire topics and chapters dealing with certain aspects of astronomy, geophysics, geodesy, and cartography. At the same time, new material was added in fields of plant ecology, distribution of natural vegetation, climatology, and soil hydrology. Climatic and pedogenic regimes were introduced and related to the soil-water budget.

Since the publication of *Introduction to Physical Geography*, the parent volume has itself been revised. The appearance of new material on the atmosphere and its heat budget, soils, and world landform classification in the third edition of *Physical Geography*, 1969, as well as the introduction of color in the line drawings, made imperative this second edition of *Introduction to Physical Geography*.

We have taken as a basic premise that the aim of physical geography is an understanding of man's physical environment. Of primary importance is the thin planetary zone in which most human beings have lived and have developed their economic, political, and cultural structures. Thus the near-surface layer of the lands has first priority; it requires concentration of study upon climates, landforms, surface waters, soils, and vegetation. First attention has been given to those topics which will prove most directly enlightening to the student when he proceeds to a course in human geography. With regret, much natural science of the earth's crust and interior, the oceans and their basins, the upper and outer atmosphere, and planetary space is omitted from this abbreviated textbook.

New material includes separate chapters for a survey of the earth's atmosphere as a layered structure and the heating and cooling of the atmosphere as viewed from the standpoint of a global radiation budget. New world maps of surface air temperatures, pressures, and winds are a major improvement. A full-color plate of representative soil profiles greatly improves the appreciation of great soil groups, while a brief review of the *Seventh Approximation* system of soil classification, written by Dr. Roy W. Simonson, allows the student to learn what changes are in store in that field.

Introduction of Dr. Richard E. Murphy's worldwide system of landform classification, together with his map, *Landforms of the World* (reproduced in full color as Plate 5), fills a gap that has long existed in the world geography of landforms.

To compensate for the added pages required by the new material, the lengthy appendix of climatic data has been deleted. However, the appendix dealing with topographic map reading has been retained and strengthened.

Over a period of many years, many helpful suggestions have been sent in by colleagues and students who have used the foregoing editions of my physical geography textbooks. It is a source of regret that I cannot acknowledge individually the aid of these many contributors. Their total effort has played a major role in strengthening this newest work and making it more useful in the teaching of physical geography.

Columbia University
November, 1969   *Arthur N. Strahler*

Contents

Color Plates Bound at End of Book

Plate 1 Mean Annual Precipitation of the World

Plate 2 Climates of the World

Plate 3 Great Soil Groups of the World

Plate 4 Natural Vegetation Regions of the World

Plate 5 Landforms of the World

World Population Distribution

World Political Division

End Papers

Front: U.S.G.S. Topographic Map and Legend

Back: Weather Map Symbols

INTRODUCTION TO PHYSICAL GEOGRAPHY

Introduction

WHAT is physical geography? As a first step in understanding this term we might well expand it to read "the physical basis of geography," for physical geography is simply the study and unification of a number of earth sciences which give general insight into the nature of man's environment. Not in itself a distinct branch of science, physical geography is a body of basic principles of natural science selected with a view to including primarily the environmental influences that vary from place to place over the earth's surface.

What, then, are the individual sciences that comprise physical geography, and why are they selected? First and most fundamental are the form of the earth, a concern of the science of *geodesy*, and the relationship between earth and sun, a part of *astronomy*. Much of astronomy is beyond the concern of the geographer, for only two bodies, the sun and the moon, appreciably affect life on earth. Because all energy for sustaining life, all motive power for streams, winds, and ocean currents, comes by radiant emanation from the sun, and because the intensity of this energy changes through daily and annual cycles, an understanding of the motions of the earth in its orbit about the sun is a prime essential. The moon, as the body that controls ocean tides, enters into physical geography only in a minor way.

Because data of the earth sciences are often best represented by maps, and perhaps many are impossible to describe without maps, the science of maps, *cartography*, is an essential ingredient of physical geography. True, cartography is really a science of technique, rather than a basic earth science, but it deserves a place early in the list of topics so that it may provide a means for representing the information to follow.

Man, though he lives on the earth's solid surface, is an air-breather *in* the atmosphere and owes his very survival to favorable conditions of weather and climate. The sciences of *meteorology* and *climatology*, which treat these topics, are thus a major concern of the physical geographer. Lying between the atmosphere and the land masses of the earth is a thin layer, the soil, which reflects the influence of both climate and topography. Soil science, or *pedology*, is thus another constituent of physical geography. The structure and distribution of natural vegetation types, though a botanical

rather than a physical-science subject, must be included for consideration in physical geography because plants are important features of the landscape and are remarkably consistent indicators of climate, soils, and landforms. Thus the science of *plant geography* is taken into the fold of physical geography.

In our concern for the lands of the earth we should not overlook the oceans of the globe. *Physical oceanography*, which includes a study of ocean waves, currents, and tides, finds its way into physical geography inevitably because man uses the oceans for intercontinental communication, for naval and air operations, and as a source of food as well.

The topographic features, or landforms, of the earth's surface are of prime concern to man because they influence the placement of his agricultural lands, his cities, his lines of communication. The science of *geomorphology* treats the origin and systematic development of all types of landforms and is a major part of physical geography. Very often landforms express the varieties and structures of rocks under the surface, so that a certain minimum knowledge of the principles of *geology* is included. An understanding of geological principles will have the further value of explaining the origin and distribution of the principal types of mineral deposits—coal, petroleum, natural gas, metallic ores, building stones, and many others. Closely involved with geomorphology is the science of *hydrology*, which treats the earth's surface and underground waters, including rivers, lakes, springs, and marshes. Fresh water, a basic essential for man's survival, thus looms as an important element of the physical basis of geography.

The professional physical geographer will usually be a specialist in only one of the several fields involved, such as climatology, geomorphology, or soil science. Besides carrying out original research in his chosen specialty, to which he may be making important scientific contributions, the physical geographer attempts to keep informed on important developments as they occur in the other fields of specialization. He is thus able to assemble and integrate pertinent fragments of knowledge into a unified picture of the natural environment of man at any place on the globe at any season of year.

CHAPTER 1
The Geographic Grid and Its Projections

THE spherical form of the earth is one of the facts of our physical environment, which children learn at an early age, but probably few people give much thought to some of the simple proofs of the earth's sphericity. For example, the evidence that people have repeatedly sailed or flown completely around the globe is tacitly assumed to prove the sphericity of the earth, whereas it means only that the earth is a solid body. Circumnavigation could also be performed on a cubical or cylindrical earth.

Even without optical instruments we can get some inkling that the earth's surface curves downward away from us through the observation that sunlight illuminates the tops of high clouds after sunset and before sunrise.

One proof of the earth's sphericity may be had from observations at sea. As a ship recedes farther and farther into the distance, it appears to sink slowly beneath the water level (Figure 1.1). Seen through binoculars or a telescope the sea surface will appear to rise until the decks are awash, then gradually to submerge the funnel and the masts, leaving finally only smoke visible above the horizon. The explanation obviously lies in the fact that the sea surface curves downward away from us. To prove that this curvature is spherical would require numerous observations in which measurements were made of the amount of apparent sinking of a vessel per unit of distance in many different directions away from the observing point.

A second proof is found in the observation that in all lunar eclipses, at which time the earth's shadow falls on the moon, the edge of the shadow appears as an arc of a circle. It can be shown by geometrical proof that a sphere is the only body that will always cast a circular shadow upon another sphere. Because at the time of these eclipses the earth is rarely turned in just the same position, we may conclude that, no matter what earth profile is cast on the moon, the circular shadows are all alike and the earth must be spherical.

Photographs taken from rockets and earth satellites (such as the *Tiros* weather satellites) at extremely high altitudes show the horizon as a curved line (Figure 1.2). Because the curvature appears to be the same in many widely separated parts of the earth, a series of such photographs provides us with a third proof of the earth's spherical form.

A fourth proof may be had from observation of the position of Polaris, the north star. To the observer at the equator Polaris is on the horizon, but as he travels toward the north pole the star seems to be located higher and higher in the sky, until, at the north pole, it is directly overhead in the sky. It would be found that Polaris rises 1° higher in the sky for every 69 mi (111 km) of northward travel by the observer. A similar condition would hold for travel from the equator to the south pole if a star nearly in line with the earth's axis in the southern sky could be observed. Thus we can prove that all north-south lines drawn from pole to pole (e.g., meridians) are arcs of circles, and that the earth is spherical.

A fifth proof of the earth's sphericity comes from the observation that an object near sea level will weigh very nearly the same amount on a spring type of scales at any place on the globe. Knowing that the weight depends upon the pull of gravity, we conclude that the object weighs the same everywhere because all points on the earth's sur-

Figure 1.1   Seen through a telescope, a distant ship seems to be partly submerged.

Figure 1.2 Curvature of the earth's horizon shows clearly on this photograph of the south-western United States and northern Mexico, taken from a Navy Viking-12 rocket at an altitude of 143 mi (230 km). On the left are Lower California and the Gulf of California. To the right the view extends as far as the Los Angeles area. (Official U.S. Navy Photograph.)

face are equidistant from its center, hence, the earth is a sphere.

Now, a pendulum clock will also serve to measure the force of gravity. If the pendulum is kept of exactly constant length the clock will keep constant time so long as it is acted upon by a constant gravity. Thus, if a pendulum clock is found to keep good time at all points at sea level over the earth, the spherical form is proved. Extremely precise calculations based upon this principle, however, show slight variations in gravity, which, as will be explained later, led to the discovery that the earth's true form is not a perfect sphere.

As a last proof it may be noted that modern navigation methods are based on the assumption that the earth is a sphere. When we consider that, for more than a century, positions of vessels have been correctly determined innumerable times by these methods, it becomes obvious that the correctness of the assumption has been established many times over.

Although the ancient Greeks, among them Pythagoras (540 B.C.) and associates of Aristotle (384–322 B.C.), believed the earth to be spherical and had speculated upon its circumference, it remained until 200 B.C. for Eratosthenes, librarian at Alexandria, to perform a direct measurement of the earth's circumference. He obtained a value believed to be equivalent to about 26,660 statute miles (42,900 km), somewhat larger than the true value, which is not far from 25,000 mi (40,000 km). Arabs of the ninth century, using measure-

ments of the positions of stars combined with direct ground-distance measurements, also arrived at estimates of the earth's size. Their measurements were probably much more accurate than those of Eratosthenes, but because the units of measure are not known in modern equivalents, their work cannot be checked.

The earth as an oblate ellipsoid

In 1671 a French astronomer, Jean Richer, was sent by Louis XIV to the Island of Cayenne, French Guiana, to make certain astronomical observations. His clock had been so adjusted that its pendulum, slightly over 39 in (99 cm) long, beat the exact seconds in Paris. Upon arriving in Cayenne, which is near the equator, Richer found the clock to be losing about two and one-half minutes per day. As soon as Newton's laws of gravitation and motion were published (1686) it became possible to attribute the slowing of the clock at Cayenne to a somewhat reduced force of gravity near the equator. It was soon realized that this phenomenon could be accounted for by supposing that the equatorial portions of the earth's surface lie further from the earth's center than do more northerly places.

Refined measurements have since revealed that the true form of the earth is more like a sphere which has been compressed along the polar axis and made to bulge slightly around the equator (Figure 1.3). This form is known as an *oblate ellipsoid.* A cross section through the poles gives an *ellipse* rather than a circle. The equator remains a circle and is the largest possible circumference on the ellipsoid. The earth's oblateness is attributed to the centrifugal force of the earth's rotation, which deforms the somewhat plastic earth into a form in equilibrium with respect to the forces of gravity and rotation.

Rounding off the earth's dimensions to the nearest whole mile, the equatorial diameter is 7927 mi (12,757 km), whereas the length of polar axis is 7900 mi (12,714 km), a difference of about 27 mi (43 km). The *oblateness* of the earth ellipsoid, or *flattening of the poles*, is the ratio of this difference to the equatorial diameter, or roughly 27/7927, which reduces to a fraction only slightly larger than 1/300. Thus we can say that the earth's polar axis is about 1/300 shorter than the equatorial diameter. Using these figures the earth's equatorial circumference is about 24,900 mi (40,075 km). For rough calculations the figure of 25,000 mi (40,000 km) is close enough.

Great and small circles

For many purposes of physical geography the earth can be treated as if it were a true sphere. Generally, flattening of the poles can be disregarded in order to simplify an understanding of the important concepts of the earth as an object turning under the sun's rays.

If a sphere is divided exactly in half by a plane passed through the center, the intersection of the plane with the sphere is the largest circle that can be drawn on the sphere and is known as a *great circle* (Figure 1.4). Circles produced by planes passing through a sphere anywhere except through the center are smaller than great circles and are designated *small circles*.

Become thoroughly familiar with the following properties of great circles, because they frequently enter into such global subjects as meridians, navigation, illumination of the globe, and map projections:

1. A great circle always results when a plane passes through the center of a sphere, regardless of the attitude of the plane.

2. A great circle is the largest possible circle that can be drawn on the surface of a sphere.

3. An infinite number of great circles can be drawn on a sphere.

4. One and only one great circle can be found that will pass through two given points on the surface of the sphere (unless the two points are at the extremities of the same diameter, in which case an infinite number of great circles can be drawn through them).

5. An arc of a great circle is the shortest surface distance between any two points on a sphere.

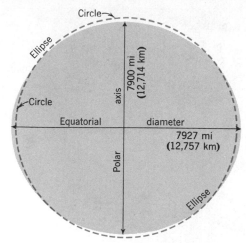

Figure 1.3 Dimensions of the earth according to the International Ellipsoid of Reference (Hayford, 1909).

6. Intersecting great circles always bisect each other.

One use of great circles that may be elaborated on here is in navigation. Wherever ships must travel over vast expanses of open ocean between distant ports, or planes must make long flights, it is desirable in the interests of saving fuel and time to follow the great-circle arc between the two points, provided, of course, that there are no obstacles or other deterring factors preventing the use of the great-circle path. Navigators use special types of maps which have the property of always showing great-circle arcs as straight lines. These are known as *great-circle sailing charts* and are discussed more fully under the subject of map projections. To plot the shortest course between any two points it is necessary only to draw a straight line between the two points on such a chart.

Great-circle courses may easily be found on a small globe by using only a piece of thin string or a rubber band (Figure 1.5). The string can be held in such a way that it is stretched tightly against the surface of the globe between the two thumbnails, each of which is on one of the points

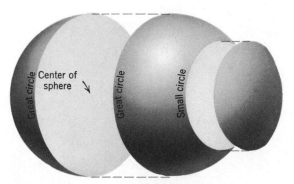

Figure 1.4 Great circle and small circle.

Figure 1.5 A great-circle course. (Charles Phelps Cushing.)

lines running parallel with the equator—the *parallels* (Figure 1.6).

All meridians are halves of great circles, whose ends coincide with the earth's north and south poles. Though it is true that opposite meridians taken together comprise a complete great circle, it is well to remember that a single meridian is only half of a great circle and contains 180° of arc. Additional characteristics of meridians are:

1. All meridians run in a true north-south direction.

2. Meridians are spaced farthest apart at the equator and converge to common points at the poles.

3. An infinite number of meridians may be drawn on a globe. Thus a meridian exists for any point selected on the globe. For representation on maps and globes, however, meridians are selected at suitable equal distances apart.

Parallels are entire small circles, produced by passing planes through the earth parallel to the plane of the equator. They possess the following characteristics:

1. Parallels are always parallel to one another. Although they are circular lines, they always remain equal distances apart.

2. All parallels represent true east-west lines.

3. Parallels intersect meridians at right angles. This fact holds true for any place on the globe, except the two poles, despite the fact that the parallels are strongly curved near the poles.

4. All parallels except the equator are small circles; the equator is a complete great circle.

5. An infinite number of parallels may be drawn on the globe. Therefore, every point on the globe, except the north or south pole, lies on a parallel.

between which the great-circle course is desired. If a rubber band is used, a complete great circle can be shown; it is of special value for points on opposite sides of the globe. Many globes show great-circle routes between distant ports on the Pacific, Atlantic, or Indian oceans. These may readily be checked by the stretched piece of string.

Meridians and parallels

The spinning of the earth on its axis provides two natural points—the poles—upon which to base the *geographic grid*, a network of intersecting lines inscribed upon the globe for purposes of fixing the location of surface features. The grid consists of a set of north-south lines connecting the poles—the *meridians*—and a set of east-west

Longitude

The location of points on the earth's surface follows a system in which lengths of arc are meas-

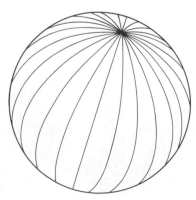

Figure 1.6 *A*, Meridians.

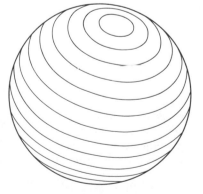

B, Parallels.

ured along meridians and parallels (Figure 1.7). Taking the equator as the starting line, arcs are measured north or south to the desired points. Taking a selected meridian, or *prime meridian,* as a reference line, arcs are measured eastward or westward to the desired points.

Longitude of a place may be defined as the arc, measured in degrees, of a parallel between that place and the prime meridian (Figure 1.7). The prime meridian is almost universally accepted as that which passes through the Royal Observatory at Greenwich, near London, England, and is often referred to as the *meridian of Greenwich.* This meridian has the value 0° longitude. The longitude of any given point on the globe is measured eastward or westward from this meridian, whichever is the shorter arc. Longitude may thus range from 0° to 180°, either east or west. It is commonly written in the following form: *long. 77° 03′ 41″ W,* which may be read "longitude 77 degrees, 3 minutes, 41 seconds west of Greenwich."

If only the longitude of a point is stated we cannot tell its precise location because the same arc of measure applies to an entire meridian. For this reason, a meridian might be defined as a line representing all points having the same longitude. This definition explains why the expression "a meridian of longitude" is often used. Confusion may arise because of the statement that longitude is measured along a parallel of latitude, but this may be clarified by the realization that in order to measure the arc between a point and the prime meridian it is necessary to follow eastward or westward along one of the parallels (Figure 1.7).

The actual length, in miles or kilometers, of a degree of longitude will depend upon where it is measured. At the equator this distance may be computed by dividing the earth's circumference by 360°:

$$\frac{24,900 \text{ mi}}{360°} = 69 \text{ statute miles (approx.)}$$

$$\frac{40,075 \text{ km}}{360°} = 111 \text{ kilometers (approx.)}$$

Because of the rapid convergence of the meridians northward or southward, care should be taken not to employ this equivalent inadvertently except close to the equator. It is a further useful item of knowledge that the length of 1° of longitude is reduced to about one-half as much at the 60th parallels, or about $34\frac{1}{2}$ mi ($55\frac{1}{2}$ km).

Latitude

Latitude of a place may be defined as the arc, measured in degrees, of a meridian between that place and the equator (Figure 1.7). Latitude may thus range from 0° at the equator to 90° north

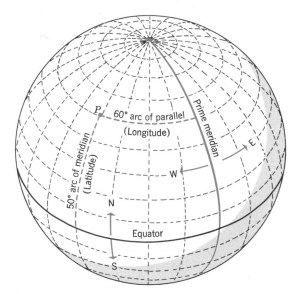

Figure 1.7 The point *P* has a latitude of 50° N, a longitude of 60° W.

or south at the poles. The latitude of a place, written as *lat. 34° 10′ 31″ N,* may be read "latitude 34 degrees, 10 minutes, 31 seconds north." When both the latitude and longitude of a place are given, it is accurately and precisely located with respect to the geographic grid.

For most purposes of physical geography we consider the earth to be a sphere, and therefore the parallels of latitude are taken to be exactly equidistantly spaced if they are drawn on a globe for unit amounts of arc, as, for example, every 10 degrees. The length of a degree of latitude is almost the same as the length of a degree of longitude at the equator, so that the value of 69 mi (111 km) per degree may be used for ordinary purposes.

To be very precise, and take into account the oblateness of the earth, it must be recognized that a degree of latitude changes slightly in length from equator to poles. Using figures of the Clarke ellipsoid of 1866, the length of 1° of latitude at the equator is 68.704 statute miles (110.569 km); at the poles it is 69.407 mi (111.700 km), or 0.7 mi (1.1 km) longer. One degree at the poles is 1 percent longer than at the equator. The difference is by no means trivial and must be taken into account in construction of large-scale maps.

The explanation of this variation in length of a degree of latitude may be had from a diagram showing how degrees of latitude are determined (Figure 1.8). Because of the earth's oblateness, the surface curvature is less strong near the poles than at the equator. That is to say, a smaller circle can be fitted to the curvature near the equator than at the poles, as shown in Figure 1.8. A single degree

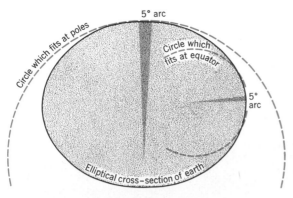

Figure 1.8 The length of a degree of latitude is very slightly greater at the poles than at the equator.

on the largest circle has a greater length of arc than a degree on the smallest circle. Hence the length of a single degree of latitude will be greatest near the poles and least near the equator. In order to obtain the correct values for specific latitudes, it is necessary to consult prepared tables. Table 1.1 gives the lengths of single degrees of both latitude and longitude for various latitudes.

Statute mile and nautical mile

Both marine and air navigation use the *nautical mile* as the unit of length or distance. Meteorology (weather science) of the upper atmosphere has also adopted as the unit of wind speed the mariner's *knot*, which is a velocity of one nautical mile per hour. It is therefore worthwhile for the geographer to understand the nautical mile.

On July 1, 1954, the U.S. Department of Defense adopted the *international nautical mile*, defined as exactly equivalent to 1852 international meters, or 6076.103333 feet (the digit 3 is repeated indefinitely). Therefore, dividing this value in feet by 5280, the number of feet per *statute mile*, we arrive at the equivalent: 1 international nautical mile = 1.150777 statute miles. For ordinary calculations, then, the value of 1.15 statute miles (1.85 km) per nautical mile is quite satisfactory.

At what place on the earth does the international nautical mile equal the length of one minute of arc of the earth ellipsoid? This can be computed by first multiplying 1.150777 by 60 to give 69.04663 statute miles per degree of arc. Next, consulting Table 1.1, this figure is seen to be very close to the length of one degree of latitude at 45°, which is given as 69.054 miles according to the Clarke ellipsoid of reference. Furthermore, if all of the values of the middle column of Table 1.1 are added and the average value computed, it will be found to be 69.055. This leads to the conclusion that the international nautical mile very closely approximates the average length of one

minute of latitude, or that it is the $\frac{1}{5400}$ part of the length of a meridian between equator and pole.

Map projections

A map projection is an orderly system of parallels and meridians used as a basis for drawing a map on a flat surface. The fundamental problem is to transfer the geographic grid from its actual spherical form to a flat surface in such a way as to present the earth's surface or some part of it in the most advantageous way possible for the purposes desired.

One way to avoid the map-projection problem is to use only a globe. Unfortunately a globe has shortcomings. First, we can see only one side of a globe at a time. Second, a globe is on too small a scale for many purposes. On globes ranging from a few inches to two or three feet in diameter, only the barest essentials of geography can be shown. The few large globes in existence, those several feet in diameter, may show considerable detail, but they serve also to accentuate a third shortcoming of globes—their lack of portability. Flat maps printed on paper can be folded compactly so that many may be carried in a small pocket, whereas even the smallest globe is a cumbersome and delicate object. Ease of reproduction greatly favors maps over globes. Making a quality globe requires not only that a map be printed but also that the map be trimmed and carefully pasted onto a spherical shell.

The problem of map projection must therefore be squarely faced in an endeavor to learn what types of networks of parallels and meridians are best suited to illustration of various portions of the earth's surface. It is well to point out, however, that no map projection will ever substitute fully for a globe to show general world relations, and use of a globe is to be recommended in conjunction with flat maps.

Developable geometric surfaces

Certain geometric surfaces are said to be *developable* because by cutting along certain lines they can be made to unroll or unfold to make a flat sheet. Two such forms are the *cone* and the *cylinder* (Figure 1.9). Were the earth conical or cylindrical, the map-projection problem would be solved once and for all by using the developed surface. No distortion of surface shapes or areas would occur, although it is true that the surface would be cut apart along certain lines. The earth belongs to a group of geometric forms said to be *undevelopable*, because, no matter how they are cut, they cannot be unrolled or unfolded to lie flat. It is possible to draw a true straight line in

TABLE 1.1[a]

Latitude, Degrees	Length of 1° of Latitude		Length of 1° of Longitude	
	Statute Miles	Kilometers	Statute Miles	Kilometers
0	68.704	110.569	69.172	111.322
5	68.710	110.578	68.911	110.902
10	68.725	110.603	68.129	109.643
15	68.751	110.644	66.830	107.553
20	68.786	110.701	65.026	104.650
25	68.829	110.770	62.729	100.953
30	68.879	110.850	59.956	96.490
35	68.935	110.941	56.725	91.290
40	68.993	111.034	53.063	85.397
45	69.054	111.132	48.995	78.850
50	69.115	111.230	44.552	71.700
55	69.175	111.327	39.766	63.997
60	69.230	111.415	34.674	55.803
65	69.281	111.497	29.315	47.178
70	69.324	111.567	23.729	38.188
75	69.360	111.625	17.960	28.904
80	69.386	111.666	12.051	19.394
85	69.402	111.692	6.049	9.735
90	69.407	111.700	0.000	0.000

[a]Based on Clarke ellipsoid of 1866, from *U.S. Geological Survey Bulletin* 650, "Geographic Tables and Formulas," by S. S. Gannett, 1916, pp. 36–37.

one or more directions on the surface of a developable solid, but nowhere can this be done on an undevelopable form such as a spherical surface. In order to make the parts of a spherical surface lie perfectly flat, it must be stretched—more in some places than in others. Thus, it is impossible to make a perfect map projection.

When a map is made of a very small part of the earth's surface, for example, an area four miles across, the map-projection problem can be ignored. If the meridians and parallels are drawn as straight lines, intersecting at right angles and correctly spaced apart, the actual error present is probably so small as to fall within the width of the lines drawn and is not worth correcting. As the area included on the map is increased, however, the problem gains in importance. When an attempt is made to show the whole globe, very serious trouble develops. Only by some compromise can the distortion be reduced to a reasonable degree over important parts of the earth's surface.

Although for purposes of simplicity this chapter treats projections of a spherical globe upon a flat map, it is well to point out that in precise plotting of a map projection the earth ellipsoid is the geometrical form that is actually used. But because the earth's oblateness is slight, the earth can be assumed a true sphere for an elementary and descriptive study of map projections.

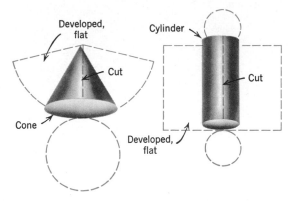

Figure 1.9 The cone and cylinder are developable geometric forms.

Map scale

All globes and maps depict the earth's features in much smaller size than the true features which they attempt to represent. Globes are intended in principle to be perfect models of the earth itself, differing from the earth only in size, but not in shape. The *scale* of a globe is therefore the ratio between the size of the globe and the size of the earth, where size is expressed by some measure of length or distance (but not area or volume). Take, for example, a globe 10 inches in diameter representing the earth, whose diameter is about 8000 miles. The scale of the globe is therefore the ratio between 10 inches and 8000 miles. Dividing both figures by 10, this reduces to a scale stated as: *1 inch represents 800 miles*, a relationship that holds true for distances between any two points on the globe. (Stated in metric units, the scale of this globe would be the ratio of 25 cm to 12,900 km, or *1 centimeter represents 516 kilometers*.)

Scale is more usefully stated as a simple fraction, termed the *fractional scale*, or *representative fraction (R.F.)*, which can be obtained by reducing both map and globe distances to the same unit of measure, thus:

$$\frac{1 \text{ in. on globe}}{800 \text{ mi on earth}} = \frac{1 \text{ in.}}{800 \times 63{,}360 \text{ in. (per mile)}}$$

$$= \frac{1 \text{ in.}}{50{,}688{,}000 \text{ in.}} = \frac{1}{50{,}688{,}000}$$

This fraction may be written as 1 : 50,688,000 for convenience in printing. The advantage of the representative fraction is that it is entirely free of any specified units of measure, such as the foot, mile, meter, or kilometer. Persons of any nationality understand the fraction, regardless of the language or units of measure used in their nation, provided only that the arabic numerals are understood.

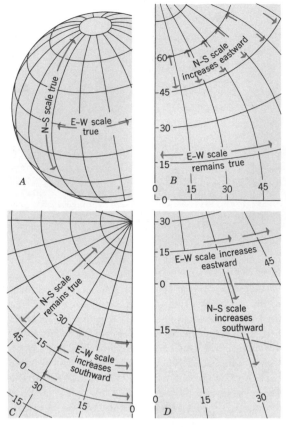

Figure 1.10 *A*, Scale true in all directions on a globe. *B*, Scale true along all parallels but not along all meridians. *C*, Scale true along all meridians but not along all parallels. *D*, Scale changes along both parallels and meridians.

A globe is a *true-scale model* of the earth, in that the same scale of miles applies to any distances on the globe, regardless of the latitude or longitude, and regardless of the compass direction of the line whose distance is being considered (Figure 1.10*A*). This is to say that the scale remains constant over the entire globe. Map projections, however, cannot have the uniform-scale property of a globe, no matter how cleverly devised. In flattening the curved surface of the sphere to conform to a flat plane, all map projections stretch the earth's surface in a nonuniform manner, so that the scale changes from place to place. Thus, we cannot say about a map of the world: "the scale of this map is 1:50,000,000," for the statement is false for any form of projection.

It is quite possible, however, to have the scale of a flat map remain true, or constant, in certain specified directions. For example, one type of projection preserves constant scale along all parallels, but not along the meridians. This condition is illustrated in Figure 1.10*B*. Another type of projection keeps scale constant along all meridians,

but not along parallels, as shown in Figure 1.10*C*. Still other projections have changing scale along both meridians and parallels, as illustrated in Figure 1.10*D*.

Preserving areas on map projections

Because a globe is a true-scale model of the earth, given areas of the earth's surface are shown to correct relative size everywhere over its surface. If we should take a small wire ring, say one inch in diameter, and place it anywhere on the surface of the ten-inch globe, the area enclosed will represent an equal amount of area of the earth's surface. But a similar procedure would not enclose constant areas on all parts of most map projections, only on those having the special property of being *equal-area* projections.

At this point a good question arises. If, as stated above, no projection preserves a true, or constant, scale of distances in all directions over the projection, how can circles of equal diameter placed on the map enclose equal amounts of earth area? The answer is suggested in Figure 1.11. The square, one mile on a side, encloses one square mile between two meridians and two parallels. The square can be deformed into rectangles of different shapes, but if the dimensions are changed in an inverse manner, each will still enclose one square mile. The scale has been changed in one direction to compensate for change in another in just the right way to preserve equal areas of map between corresponding parts of intersecting meridians and parallels. Hence, any small square or circle moved about over the map surface will enclose a piece of the map representing a constant quantity of area of the earth's surface. Projections shown in Figure 1.23 and 1.24 have the equal-area property, but it is also obvious that these networks have had distortions of shape, particularly near the outer edges of the map.

Preserving shapes on map projections

A map projection is said to be *conformal* when any small piece of the earth's surface has the same

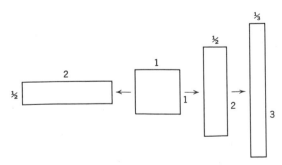

Figure 1.11 Areas can be preserved even though scales and shapes change radically.

shape on the map as it does on a globe. Thus, the appearance of small islands or countries is faithfully preserved by a conformal map. One characteristic of a conformal projection is that parallels and meridians cross each other at right angles everywhere on the map, just as they do on the globe. However, not all projections whose parallels and meridians cross at right angles are conformal.

Another way of saying that parallels and meridians intersect at right angles is that *shearing* of areas does not occur. Figure 1.12 illustrates the meaning of shearing. For projections consisting of straight parallels and meridians, shearing gives parallelograms formed of acute and obtuse angles. For projections with curved meridians and parallels, straight lines are drawn tangent to the curves at the point of intersection. If these tangent lines cross at right angles, the projection is not sheared; but if the tangents form obtuse and acute angles, shearing is present. Conformal maps are not sheared, but not all maps without shearing are conformal. A conformal map cannot have equal-area properties besides, so that some areas are greatly enlarged at the expense of others. Generally speaking, areas near the margin of a conformal map have a much larger scale than central ones.

Whether a conformal or equal-area projection is to be selected depends on what is to be shown. Where the surface extent of something, such as grain crops, or forest-covered lands, is to be shown, an equal-area projection is needed. For most general purposes a conformal type is preferable because physical features most nearly resemble their true shapes on the globe. Many map projections are neither perfectly conformal nor equal area, but represent a compromise between the two. This compromise may be desired either to achieve a map of more all-around usefulness or because the projection has some other very special property that makes its use essential for certain purposes.

Classification of map projections

Map projections may be classified according to the following groups: (1) *zenithal* (*azimuthal*), (2) *conic*, (3) *cylindric*, and (4) individual, or unique types.

The *zenithal*, or *azimuthal*, group of projections includes all types that are centered about a point and have a radial, or wheel-like, symmetry. Some zenithal projections can actually be demonstrated in the laboratory by the following method. (See Figure 1.13.) A wire replica of the earth, in which the wires represent parallels and meridians, is used. A tiny light source, such as a flashlight bulb or an arc light, is placed at the center of the wire globe (or at any one of several prescribed positions). In a darkened room the shadow of the wire

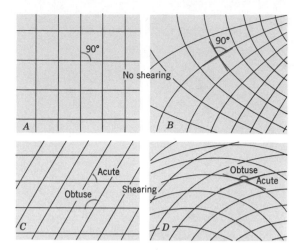

Figure 1.12 Shearing of areas.

globe is cast upon a screen or upon the wall or ceiling. This shadow is a true geometric projection. All projections made with this apparatus are of the zenithal type, which are characterized by the following properties (Figure 1.14).

1. A line drawn from center point of the map to any other point gives the true compass direction taken by a great circle as it leaves the center point, headed for the outer point.

2. When a complete globe or hemisphere is shown, the map is circular in outline. Because any map can be trimmed down to have a circular outline, this feature is not a reliable criterion of the zenithal class.

3. The map possesses a center point around which all its properties are grouped. All changes

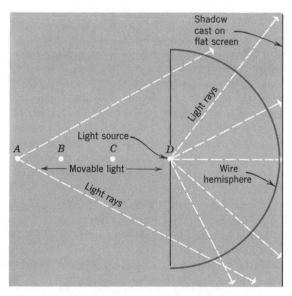

Figure 1.13 Principle of zenithal projections.

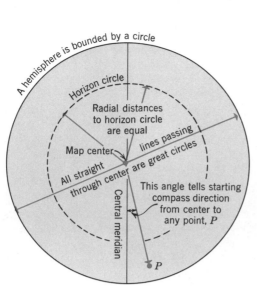

Figure 1.14 Properties of zenithal projections.

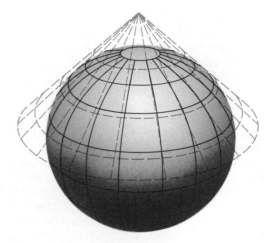

Figure 1.15 Principle of the conic projections.

of scale and distortion of shapes occur uniformly (concentrically) outward from this center.

4. All points equidistant from the center lie on a circle, known as the *horizon circle*. When the entire globe is shown by a zenithal map, the circular edge of the map represents the opposite point, or *antipode*, on the globe. When a hemisphere is shown, the outer edge of the map represents a great circle, everywhere equidistant from the point on which the projection is centered.

5. All great circles which pass through the center point of the projection appear as straight lines on the map. Likewise, all straight lines drawn through the center point of the map are true great circles.

Zenithal projections appear in three positions, or orientations: (1) *polar*, (2) *equatorial*, and (3) *oblique* or *tilted*, illustrated in Figures 1.17, 1.18, and 1.19. In the polar position, the center of the projection coincides with the north or south pole; in the equatorial position, the center is somewhere on the equator; in the oblique position, the center is at any desired point intermediate between the equator and poles. Although the equatorial and oblique types may not seem to be radially symmetrical, they nevertheless possess, just as truly as the polar type, the five characteristics described above.

The *conic* group of projections is based on the principle of transferring the geographic grid from a globe to a cone, then developing the cone to a flat map. This principle, too, can be demonstrated in the laboratory with the wire globe and a point source of light (Figure 1.15). Instead of a vertical flat screen, however, a translucent cone of stiff paper is seated on the wire globe, much as a

lampshade is seated on a lamp. The shadow of the wires cast upon the conical shade gives a conic projection. If this shadow were traced in pencil or ink and the cone unrolled, a true conic projection would result. Simple conic projections possess the following features (Figure 1.21). All meridians are straight lines, converging to a common point

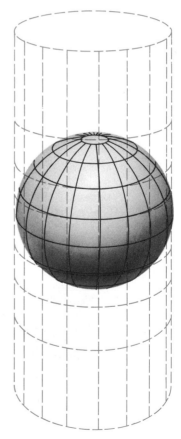

Figure 1.16 Principle of the cylindric projections.

at the north (or south) pole. All parallels are arcs of concentric circles, whose common center lies at the north (or south) pole. A complete conic projection is a sector of a circle, never a complete circle. A conic projection cannot show the whole globe and usually shows little more than the northern (or southern) hemisphere.

Cylindric projections are based on the principle of transferring the geographic grid first onto a cylinder wrapped about the earth, then unrolling the cylinder to make a flat map (Figure 1.16). Simple cylindric projections are easy to draw because they consist of intersecting horizontal and vertical lines. (See Figure 1.22.) The completed map is rectangular in outline, and the whole circumference of the globe can be shown. When the cylinder is tangent to the equator, meridians are equally spaced vertical lines. Parallels are spaced in various ways, according to the particular projection desired.

Many other kinds of map projections exist, each based upon some unique principle.

Several important map projections have been selected to illustrate the wide range of qualities available. Included are conformal and equal-area types, as well as those with unique properties.

1. *Orthographic projection.* The *orthographic* projection belongs to the zenithal class and employs a principle of construction illustrated in Figure 1.17. Parallel rays, or lines, are used to

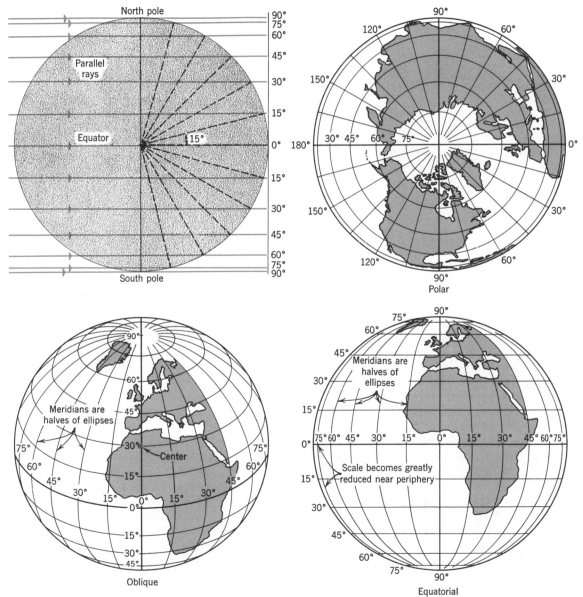

Figure 1.17 The orthographic projection.

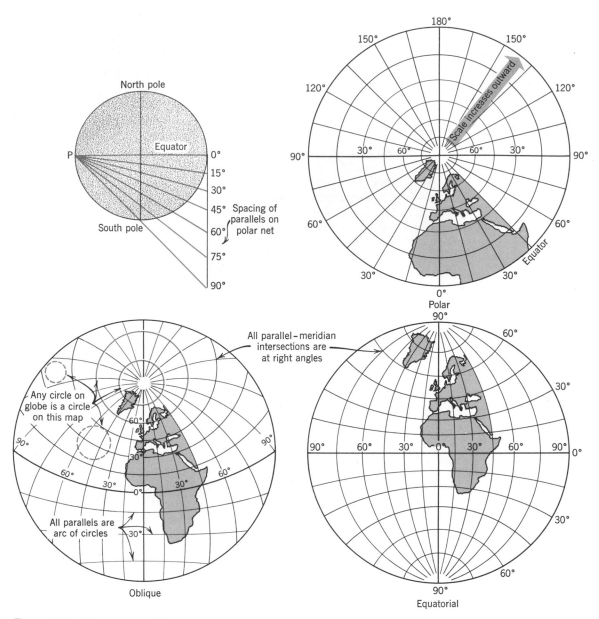

Figure 1.18 The stereographic projection.

project the geographic grid of one hemisphere on a plane touching the globe at pole, equator, or at some intermediate point. The projection can also be imagined as resulting when the shadow of a hemispherical wire globe is cast on a screen by light rays coming from a very distant source, such as the sun. In the polar position the parallels of latitude crowd close together near the outer margin. In the equatorial position, the meridians are parts of true ellipses and show close crowding near the outer margin, whereas the parallels are straight, horizontal lines, spaced more closely near the poles. In the oblique position, the closer crowding of meridians and parallels near the outer margin

is also noticeable.

The largest possible portion of a globe that can be shown on the orthographic projection is one hemisphere. The projection is neither equal area nor conformal. The scale of miles is much larger near the center than near the outer edges. Use of this projection is quite limited, but it gives a visual effect of a globe in three dimensions and is very similar to a photograph taken of a globe. For this reason it often appears to illustrate articles or books on global political or military strategic problems. It gives a true picture of relations between countries or continents which are located near the central point of the projection. A little

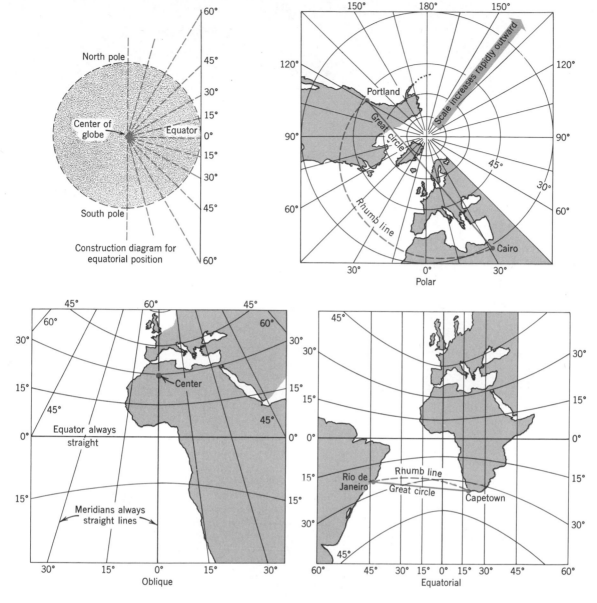

Figure 1.19 The gnomonic projection.

shading added to the map accentuates the perspective effect.

2. Stereographic projection. In the *stereographic* projection, a zenithal type, the point from which construction lines, or rays, emanate is located on the globe at a point diametrically opposite to the point where a plane touches the globe (Figure 1.18). Whereas the orthographic projection gives a map of exactly the same diameter as the original globe used, the stereographic projection gives a much larger map than the original globe. Furthermore, the stereographic net can show much more than one hemisphere, although it cannot show the whole globe. The principal

distinguishing characteristic of this projection is evident in all three of its positions: parallels and meridians show closer spacing near the map center and increasingly wider spacing toward the outer margins. On any stereographic projection the parallels and meridians are either straight lines or arcs of circles. No other kinds of curved lines occur. The reason for this is that the stereographic projection is truly conformal. All lines that are circles on the globe are shown as circles on the map. The scale of miles, however, always increases from the map center toward the periphery.

With the enormous growth in importance of polar regions in the age of scientific discovery and

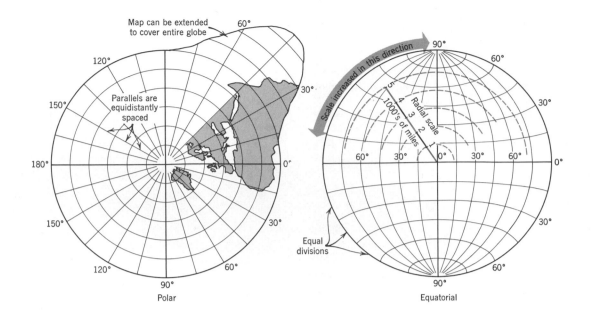

Polar

Equatorial

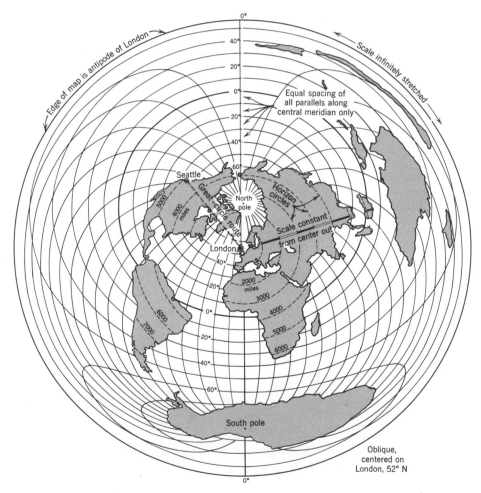

Figure 1.20 The equidistant projection.

of long-range missile and aircraft operation, the polar stereographic projection has assumed great importance.

World Aeronautical Charts issued by the U.S. Coast and Geodetic Survey are based on a polar stereographic projection for latitudes 80° to 90°. The data of scientific observations of the Arctic Ocean and Antarctic continent are usually shown on this projection. The U.S. Weather Bureau's daily weather map is printed on a polar stereographic projection.

3. *Gnomonic projection (great-circle sailing chart)*. The *gnomonic* projection, a zenithal type, is made by drawing rays from a point at the center of the globe, as illustrated in Figure 1.19. The resulting network is easily recognized because the spacing of meridians and parallels increases enormously outward from the map center and results in great distortion of shapes of land areas in the outer part of the map. For geometrical reasons evident from the construction lines in Figure 1.19, it is impossible to show a complete hemisphere; it is even impractical to include the greater part of a hemisphere. For this reason, a gnomonic map is usually trimmed to a rectangular shape.

The gnomonic projection, with its grotesque distortions of both scale and shapes, would find little use were it not for one unique and important property: On a gnomonic map all straight lines are great circles. Conversely, all great circles appear as straight lines. Note that on all three of the projections illustrated in Figure 1.19, all meridians and the equator are straight lines, regardless of where they are located on the map. For navigational purposes the plotting of great-circle courses is accomplished by merely connecting with a straight line any two desired points. For this reason the gnomonic projection goes by the name *great-circle sailing chart* when adapted to navigational uses. Illustrations of great-circle routes plotted on a gnomonic net are shown in Figure 1.19 on the polar and equatorial maps. (Refer to Figure 1.22 to see these same routes plotted on the Mercator projection.)

4. *Equidistant projection*. Fourth of the zenithal types is the *equidistant* projection (Figure 1.20). As its name implies, this network is made by spacing the meridians (or parallels) equidistantly outward from the map center. Moreover, there is nothing to stop the cartographer from extending the map to include the whole globe. The opposite pole, or *antipode*, is then shown as a circle surrounding the map. Thus constructed the scale of miles remains constant along all radial straight lines emanating from the map center. This gives the map a specialized use for air navigation. When centered on a particular city or airport,

great-circle routes can easily be laid off and measured by drawing a line from the central point to any desired point on the map and finding the distance on a graphic scale having equally spaced units.

The equidistant net is often used for small-scale hemispherical maps. The polar position makes a pleasing map to show grouping of the world's principal land areas about the Arctic Ocean.

5. *Conic projection*. The conic projection is based on the principle that a cone can be placed over a globe in such a way as to have its apex directly over the north pole and to touch the globe along a single parallel (Figure 1.15). If the parallels and meridians are then projected upon the cone by drawing rays from the globe's center, and the cone is unrolled to a flat surface, a conic map results (Figure 1.21). Meridians are straight lines radiating from the pole; parallels are arcs of concentric circles, centered on the pole. The conic projection makes a fairly good general-purpose map, without shearing of areas but can effectively show little more than the northern hemisphere. Certain improved types, notably the Lambert conformal conic, are in wide use.

6. *Mercator projection*. Perhaps the best known of all map projections is the *Mercator* net, devised by Gerardus Mercator in 1569 (Figure 1.22). It is based upon a mathematical formula. The principle, however, can be explained without mathematical expression, as follows. On any cylindrical projection in which the meridians are straight vertical lines, equidistantly spaced, the meridians have had to be spread apart. (See right side of Figure 1.22.) Only along the equator are they the same distance apart as on a globe of the same equatorial scale. In order to maintain them as parallel lines, the normally converging meridians have had to be spread apart in a greater and greater ratio as the poles are approached. At 60° N and S lat., the meridians are spread apart twice as far as originally, because at that place a degree of longitude is only half what it is at the equator. At the poles the spreading is infinitely greater, because the poles themselves are infinitely tiny points. Now, to maintain the map as a truly conformal map, we must space the parallels increasingly far apart toward the poles, using the same ratio of increase that resulted when the meridians were spread to make vertical lines. For example, near the 60th parallel north, parallels must be spread twice as far apart as on the globe because, as explained above, the meridians here are also spread twice as far apart. At 80° latitude the scale is enlarged almost six times. Near the poles the spacing of parallels increases enormously and rapidly approaches infinity. Because an enormous

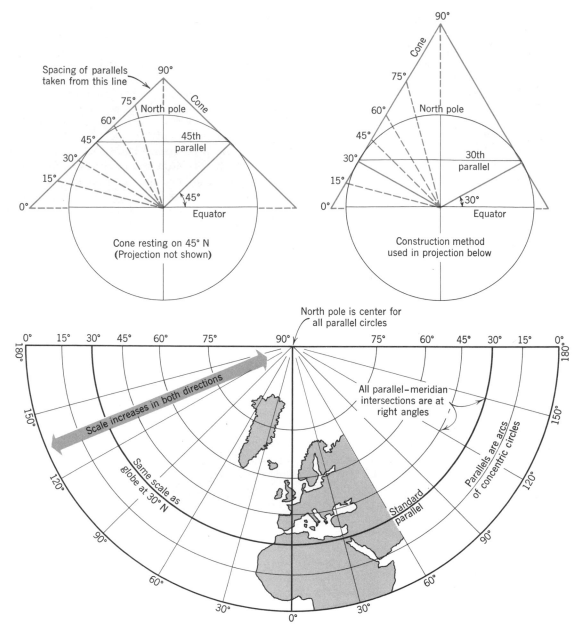

Figure 1.21 The conic projection.

sheet of paper would be needed to show extreme polar regions, the Mercator map is usually cut off about 80° or 85° N and S lat. The poles can never be shown.

The Mercator chart is a true conformal projection. Any small island or country is shown in its true shape. The scale of the map, however, becomes enormously greater toward the poles.

The really important, unique feature of an equatorial Mercator projection is that a straight line drawn anywhere on the map, in any direction desired, is a line of constant compass bearing. Such a line is known to navigators as a *rhumb line*, or

loxodrome (Figure 1.22). If this line is followed, the ship's (or plane's) compass will show that the course is always at a constant angle with respect to geographic north. Once the proper compass bearing is determined, the ship is kept on the same bearing throughout the voyage, if the rhumb line is to be followed. The Mercator chart is the only one of all known projections on which all rhumb lines are true straight lines, and vice versa. A protractor can be used with reference to any meridian on the map, and the compass bearing of any straight line can be measured off directly.

The relation of great-circle routes to rhumb lines

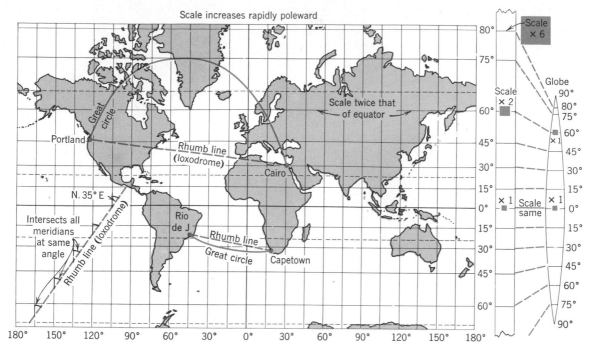

Figure 1.22 The Mercator projection.

is shown by two examples on Figures 1.19 and 1.22. Note that on the gnomonic map great circles are straight and rhumb lines curved, whereas on the Mercator chart rhumb lines are straight and great circles curved. Along the equator and all meridians (but only on these lines) rhumb lines and great circles are identical and are straight lines on both charts.

Although indispensable for navigational uses,

the equatorial Mercator projection has serious shortcomings for use as a world map to show geographical information dealing with areas of distribution. Except for equatorial regions, for which it provides an excellent grid, distortions of scale are very serious. Because of infinite stretching toward the poles, this map fails completely to show how the land areas of North America, Asia, and Europe are grouped around the polar sea. In the

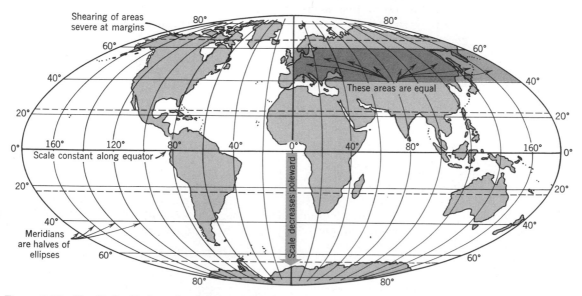

Figure 1.23 The Mollweide homolographic projection.

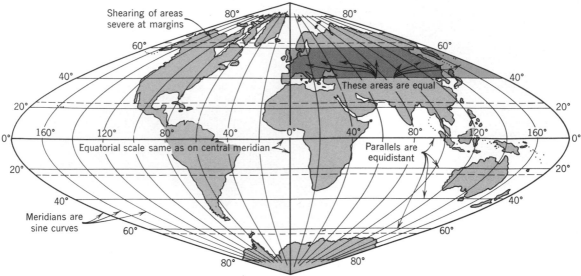

Figure 1.24 The sinusoidal projection.

mind of an inexperienced user it may enhance a false sense of isolation between inhabitants of these lands.

On the other hand, certain forms of geographical information are best shown on the Mercator projection. Because of its accurate depiction of the compass directions of lines, the Mercator net is preferred for maps of direction of flow of ocean currents and winds, direction of pointing of the compass needle, or lines of equal value of air pressure and air temperature. Examples of such uses of the Mercator projection will be seen in later chapters.

7. *Homolographic projection.* One projection rather widely used by geographers to show the entire globe is the *homolographic* projection (Figure 1.23). "Homolographic" is a word often used to mean "equal area," a property this projection possesses. One hemisphere is outlined by a circle; the other hemisphere is divided into two parts and added with an elliptical outline to either side of the circle. All other meridians, except the straight central meridian, are halves of ellipses. The equator is twice as long as the central meridian, which is also true on a globe. Parallels are straight, horizontal lines, becoming more closely spaced toward the poles.

The homolographic projection has distinct advantages and disadvantages. Its equal-area property makes it valuable for showing the global distribution of geographical properties that cover areas, such as crops, soil types, or political units. Severe distortion in the polar regions, however, has hindered its wider use.

8. *Sinusoidal projection.* In some ways the *sinusoidal* projection is similar to the homolographic projection. It is an equal-area projection

with straight central meridian and horizontal straight parallels (Figure 1.24). The difference lies in the type of curve used in meridians. Whereas the homolographic net uses ellipses, the sinusoidal net uses a type of line known as a *sine curve.*

The same advantages and disadvantages apply to the sinusoidal projection. Distortion in polar areas is not quite so great in the sinusoidal net but is nevertheless offensive.

9. *Homolosine projection.* The *homolosine* projection, invented by Dr. Paul Goode in 1923, is a combination of the homolographic and sinusoidal types (Figure 1.25). The sinusoidal projection is used between 40° N and S lat., the homolographic for the remaining poleward parts.

In both homolographic and sinusoidal projections the shearing of polar areas is especially marked to the extreme right and left of the map. This distortion can be reduced by centering each important land area on its own straight central meridian and fitting together the parts. The *interrupted* form of the homolosine projection, shown in Figure 1.25, has North America, Eurasia, South America, Africa, and Australia, each based on the best-suited meridian. Because the map cannot thus fit together between the land areas, except along the equator, large gaps occur. If our interest is in land areas only, as, for example, if we wish to show the areas under wheat cultivation, the interruption is not serious. To show oceans of the world we may center the oceans on central meridians, making the interruptions occur on land areas.

Summary of projections

Nine projections have been described and illustrated. This should serve to give the geography student a good start into map-projection principles.

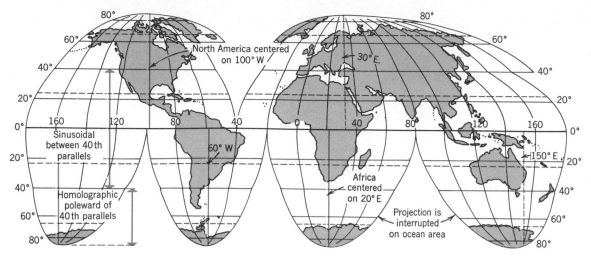

Figure 1.25   Goode's interrupted homolosine projection. (Based on Goode Base Map. Copyright by the University of Chicago. Used by permission of the University of Chicago Press.)

But, because more than 200 projections have been invented, unfamiliar types will frequently be encountered.

The following admonitions may serve to point out some of the principles of selecting a projection:

Where a map is required, select the projection best suited to the need. For example, use an equal-area map to show the areal distribution of things. Always choose a conformal projection to show lines whose compass directions are important and cannot be distorted. Use pole-centered maps for high latitudes.

Certain properties of interest to geographers change systematically from the equator toward either pole. That is to say, the properties depend on latitude. Examples are world temperatures, climates, soils, or vegetation types. To show such properties, choose a map with horizontal, straight parallels so that the eye can easily follow a given latitude zone.

Before you begin to draw conclusions and analyze geographical factors from a map, be sure that you know the qualities of the projection, whether equal-area, conformal, or neither. If in doubt, compare two or more projections of the same areas

and see what possible erroneous concepts each might give. The more of the earth's surface a map shows, the greater should your caution become.

Consult a good globe whenever points or areas are to be related in distance and direction. There is no satisfactory substitute for a true-scale model of the earth's surface features as a study aid to understanding the problems and principles of geography.

REFERENCES FOR FURTHER STUDY

Deetz, C. H. and O. S. Adams (1945), Elements of map projection, *Special Publ.* 68, U.S. Dept. Commerce, U.S. Govt. Printing Office, Washington, D.C., 226 pp.

Chamberlin, W. (1947), *The round earth on flat paper*, Nat. Geog. Soc., Washington, D.C., 126 pp.

Raisz, E. (1948), *General cartography*, McGraw-Hill Book Co., New York, 354 pp.

Robinson, A. H. and R. D. Sale (1969), *Elements of cartography*, third edition, John Wiley and Sons, New York, 343 pp.

Strahler, A. N. (1969), *Physical geography*, third edition, John Wiley and Sons, New York, 733 pp. See Chapters 1 and 2.

REVIEW QUESTIONS

1. Offer at least five proofs of the earth's approximate sphericity.

2. How did Richer contribute evidence that the earth is an oblate ellipsoid rather than a true sphere?

3. What geometrical form has a cross section of the earth cutting through the poles? What is the oblateness of the earth? What fraction approximately expresses the oblateness?

4. What is a great circle? How is it formed? How many great circles can be drawn upon the surface of a sphere? What is a small circle?

5. List six properties of great circles. Of what practical importance are great circles?

6. What is a meridian? How are meridians formed on a globe? List the characteristics of meridians.

7. What is a parallel? How are parallels formed? List the characteristics of parallels.

8. Define and explain longitude. How is longitude written? Give an example. What is a prime meridian? Where is the Greenwich meridian? How long is a degree of longitude, in miles, at the equator? at 60° latitude? at the poles?

9. Define and explain latitude. How is latitude written? Give an example. How long is a degree of latitude, in miles? Does it vary from equator to poles? How much? Why?

10. What is a nautical mile? A knot? At what latitude is the international nautical mile most nearly equal to the length of one minute of arc of the ellipsoid?

11. What is a map projection? What is the basic problem of map projection?

12. For what purposes is a globe inferior to a flat map? For what purposes is a globe preferable to a flat map?

13. Explain the concept of map scale. What is the representative fraction? Do any map projections preserve constant scale in all places and directions over the map?

14. What is an equal-area map projection? Name two equal-area types.

15. What is a conformal map projection? Name two conformal types.

16. Are projections which are neither equal-area nor conformal of any value? Illustrate with a specific example.

17. For what purposes might you select the orthographic projection?

18. Why is the stereographic projection favored for polar regions? Why is it a good base for the U.S. Weather Bureau daily weather maps?

19. Explain how a navigator uses the gnomonic projection. Why is it often called a great-circle sailing chart?

20. How would you go about constructing an equidistant projection? For what purpose might it be used? What properties has the simple conic projection?

21. Explain the principle of the Mercator projection. Is this an equal-area or conformal projection? What shape has a complete Mercator map showing the entire globe? How does the map scale at 60° latitude compare with the equatorial scale?

22. What is a rhumb line? What is a loxodrome? What projection shows all rhumb lines as straight lines? What use is made of rhumb lines and great circles in navigation?

23. Compare the homolographic and sinusoidal projections in appearance. Are these projections equal-area, conformal, or neither?

24. What advantages has Goode's interrupted homolosine projection over the homolographic and sinusoidal projections?

CHAPTER 2
Seasons and Time

OF the utmost importance to man is the relation between the earth and the sun's rays. The angles at which rays strike the earth at different latitudes and at different times of day and year determine the apparent path of the sun in the sky, the lengths of day and night, and the occurrence of seasons. Rhythmic variations in the rates of receipt of solar energy by different parts of the earth at different times act as fundamental controls of atmospheric temperatures, which in turn have a major effect on air pressure, winds, precipitation, storms, and oceanic circulation—all of which taken together make up the earth's varied climates. That is why one must thoroughly master earth-sun relationships before going ahead to the subjects of weather and climate. Because the earth is turning on its axis at the same time that it is moving in a path about the sun, and because the earth's axis is tilted with respect to the plane of its orbit, these relationships are often difficult to understand. We must learn to think in terms of three dimensions, to imagine ourselves viewing the earth from various vantage points in space; then imagine the same situations as they would appear to an observer standing at various points on the earth.

Rotation of the earth

The spinning of the earth on its polar axis is termed *rotation*. In this study of earth-sun relationships we use the period of rotation, the *mean solar day*, consisting of 24 mean solar hours. This day is the average time required for the earth to make one complete turn in respect to the sun.

Direction of earth rotation can be determined by applying one of the following rules. (*a*) If we imagine ourselves to be looking down upon the north pole of the earth, the direction of turning is counterclockwise. (*b*) If we place a finger upon a point on a globe near the equator and push the finger eastward, it will cause the globe to rotate in the correct direction (Figure 2.1). This demonstrates a common expression, "eastward rotation of the earth." (*c*) Direction of earth rotation is opposite that of the apparent motion of the sun, moon, and stars. Because these bodies appear to travel westward across the sky the earth must be turning in an eastward direction.

The velocity of earth rotation, defined as rate of travel of a point on the earth's surface in a circular path due to rotation alone, may easily be computed by dividing the length of parallel at the latitude of the point in question by 24, the approximate period of rotation. Thus at the equator, where the circumference is about 25,000 mi (40,000 km), the velocity of an object on the surface is about 1050 mi (1700 km) per hour. At the 60th parallel the velocity is half this amount, or about 525 mi (850 km) per hour. At the poles it is, of course, zero. We are unaware of this motion because the rotation is at an almost perfectly constant rate.

Two important physical phenomena result from the decrease in rotational velocity with increase in latitude. First, there is a *centrifugal force* generated by the earth's turning which gives surface objects a faint tendency to fly off into space. Because the force of gravity is 289 times greater than this centrifugal force at the equator, objects cannot leave the surface, but the practical effect is to reduce the weight of objects slightly. Near the equator, where the centrifugal force is strongest, this effect is most marked. For example, an object which would weigh 289 pounds at the equator if the earth were not turning actually weighs 1 pound less.

Another effect of the decreasing rotational velocity with increasing latitude is to cause objects in motion to be deflected slightly to the right or left of their paths. This effect will be more fully treated under the subject of the earth's wind systems.

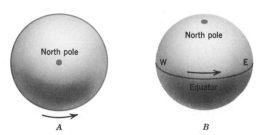

Figure 2.1 Direction of rotation is (*A*) counterclockwise at the north pole, and (*B*) eastward at the equator.

Proof of the earth's rotation

The several kinds of proofs offered for the earth's rotation are beyond the needs of the student of physical geography but one proof in particular, the *Foucault experiment*, is so outstanding as to warrant study. It has been described as follows.[1]

In 1851, the French physicist, M. Leon Foucault, suspended from the dome of the Pantheon, in Paris, a heavy iron ball by wire 200 feet long. A pin was fastened to the lowest side of the ball so that when swinging it traced a slight mark in a layer of sand placed beneath it. Carefully the long pendulum was set swinging. It was found that the path gradually moved around toward the right. [see Figure 2.2.] Now either the pendulum changed its plane or the building was gradually turned around. By experimenting with a ball suspended from a ruler one can readily see that turning the ruler will not change the plane of the swinging pendulum. If the pendulum swings back and forth in a north and south direction, the ruler can be entirely turned around without changing the direction of the pendulum's swing. If at the north pole a pendulum was set swinging toward a fixed star, say Arcturus, it would continue swinging toward the same star and the earth would thus be seen to turn around in a day. The earth would not seem to turn but the pendulum would seem to deviate toward the right, or clockwise.

At first thought it might seem as though the floor would turn completely around under the pendulum in a day, regardless of the latitude. It will be readily seen, however, that it is only at the pole that the earth would make one complete rotation under the pendulum in one day, or show a deviation of 15° in an hour. At the equator the pendulum will show no deviation, and at intermediate latitudes the rate of deviation varies.

Table 2.1 gives the hourly change in direction of the pendulum's swing and the total time re-

[1] W. E. Johnson, *Mathematical geography*, American Book Co., New York, 1907, pp. 54–57.

TABLE 2.1

Latitude	Hourly Change in Pendulum Direction, Degrees	Total Time for 360° Change in Direction, Hours
0°	None	None
5	1.31	275
10	2.60	138
15	3.88	93
20	5.13	70
25	6.34	57
30	7.50	48
35	8.60	42
40	9.64	37
45	10.61	34
50	11.49	31
55	12.29	29
60	12.99	28
65	13.59	26.5
70	14.10	25.6
75	14.49	24.9
80	14.77	24.5
85	14.94	24.1
90	15.00	24.0

quired for the direction to change through 360°.

Another proof of the earth's rotation lies in the oblate ellipsoidal form of the earth. In order to explain the bulging at the equator and shortening of the polar axis, centrifugal force due to rotation on an axis is required.

Revolution of the earth

The motion of the earth in its orbit around the sun is termed *revolution*. The period of revolution, or *year*, is the time required for the earth to complete one circuit around the sun. The year is, however, defined in different ways. For example, the time required for the earth to return to a given point in its orbit with reference to the fixed stars is called the *sidereal year*.

For earth-sun relationships we use the *tropical year*, which is the period of time from one vernal equinox to the next, and has a length of approximately 365¼ days. Every four years the extra one-fourth day difference between the tropical year and the calendar year of 365 days accumulates to nearly one whole day. By inserting a 29th day in February every leap year we are able to correct the calendar with respect to the tropical year. Further minor corrections are necessary to perfect this system.

In its orbit the earth moves in such a direction that if we imagine ourselves in space, looking down upon the earth and sun so as to see the north pole of the earth, the earth is traveling counterclockwise around the sun (Figure 2.3). This is the same direction of turning as the earth's rotation.

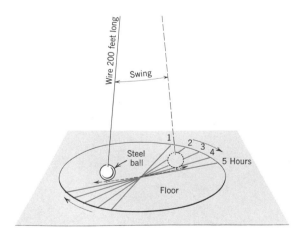

Figure 2.2   The Foucault pendulum.

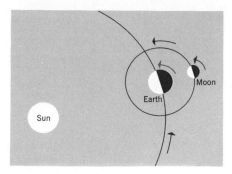

Figure 2.3   Direction of earth rotation and revolution.

Earth's orbit

The earth's orbit is an ellipse, rather than circle, although the *ellipticity*, or degree of flattening of the ellipse, is very slight. The sun occupies one *focus* of the ellipse. Ellipses of various shapes are easily constructed using only a drawing board, two pins or thumb tacks, a piece of thread or thin string, and a pencil, as illustrated in Figure 2.4. The thread is made into a loop which passes around the two pins and serves to guide the pencil point. By this device we maintain the sum of the distances from pencil point to each pin always the same. Each pin is located at one *focus* of the ellipse. The two *foci* lie on a line which is the maximum diameter of the ellipse and is called the *major axis*. The shortest diameter, drawn at right angles to the major axis, is known as the *minor axis*. The size of the ellipse may be controlled by the length of the loop of thread, whereas the ellipticity may be controlled by changing the distance between the two foci.

Perihelion and aphelion

The average distance between earth and sun is about 93 million miles (150 million km), but

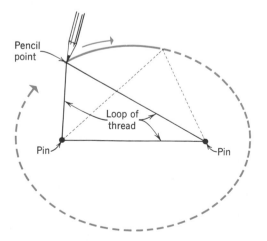

Figure 2.4   Construction of an ellipse.

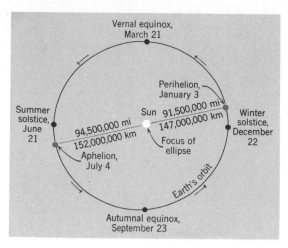

Figure 2.5   Earth's orbit and seasons.

because of the ellipticity of the orbit the distance may be 1½ million miles (2.5 million km) greater or less than this figure (Figure 2.5). The distance is least, or about 91½ million miles (147 million km), on about January 3, at which time the earth is said to be in *perihelion* (from the Greek *peri*, around or near; and *helios*, the sun). On about July 4 the earth is at its farthest point from the sun, or in *aphelion* (from the Greek *ap*, away from; *helios*, sun), at a distance of 94½ million miles (152 million km).

These differences in distance do cause some differences in the amount of solar energy received by the earth, but they are not the cause of summer and winter seasons. This is obvious because perihelion, when the earth should receive most heat, falls at the coldest time of year in the northern hemisphere. Moreover, opposite seasons are present simultaneously in the northern and southern hemispheres, proving that another cause exists. In theory, however, summers and winters should be slightly intensified in the southern hemisphere and slightly moderated in the northern hemisphere as a result of the relation of the dates of perihelion and aphelion to the summer and winter seasons.

Inclination of the earth's axis

Most globes that are made to rotate on the polar axis are fixed in a tilted position. So accustomed are we to seeing the earth represented this way that a globe with its axis vertical seems unnatural. In this position the globe may be used to illustrate the fact that the plane of the equator is inclined $23\frac{1}{2}°$ with the plane of the orbit. The earth's axis makes an angle of $66\frac{1}{2}°$ with plane of the orbit, and is tilted $23\frac{1}{2}°$ from a line perpendicular to that plane (Figure 2.6). No other single fact connected with earth-sun relationships is so important as the inclination of the earth's axis.

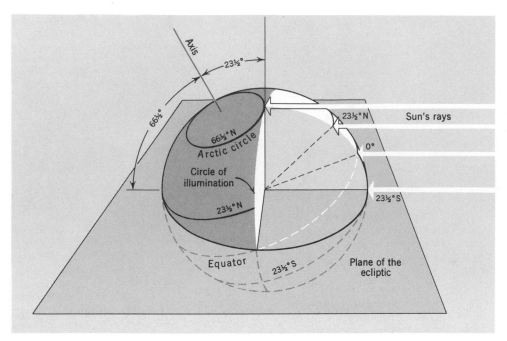

Figure 2.6 Inclination of the earth's axis. (After A. N. Strahler, *The Earth Sciences*, Harper and Row, New York.)

The earth's axis, although always making the angle $66\frac{1}{2}°$ with the plane of the orbit, maintains a fixed orientation with respect to the stars. That is to say, the earth's axis constantly points to the same spot in the heavens as it makes its yearly circuit around the sun. To help in visualizing this movement hold a globe so as always to keep the axis tilted at $66\frac{1}{2}°$ with horizontal. Move the globe in a small horizontal circle, representing the orbit, at the same time keeping the axis pointed at the same point on the ceiling (Figure 2.7).

As a direct consequence of the facts (1) that the earth's axis keeps a fixed angle with the plane of the orbit and (2) that the axis always points to the same place among the stars, it will be seen that at one point in its orbit the earth's axis leans toward the sun, that at an opposite point in the orbit the axis leans away from the sun, and at the two intermediate points the axis leans neither toward, nor away from, the sun (Figure 2.7). Here again, a globe, or, better still, four globes, may be used to aid in visualizing the facts.

Solstice and equinox

On June 21 or 22 the earth is so located in its orbit that the north polar end of its axis leans at the full $23\frac{1}{2}°$ angle toward the sun. The northern hemisphere is tipped toward the sun, and the southern hemisphere is tipped away from the sun. This condition is named the *summer solstice*. Six months later, on December 21 or 22, the earth is in an equivalent position on the opposite point

in its orbit. At this time, known as the *winter solstice*, the axis again leans the full $23\frac{1}{2}°$ directly toward the sun, but now it is the southern hemisphere that is tipped toward the sun.

Midway between the dates of the solstices occur the *equinoxes*, at which time the earth's axis makes a right angle with a line drawn to the sun, and neither the north nor south pole has any inclination toward the sun. The *vernal equinox* occurs on March 20 or 21; the *autumnal equinox* on September 22 or 23.[2] Conditions are identical on the two equinoxes as far as earth-sun relationships are concerned, whereas on the two solstices, the conditions of one hemisphere are the exact reverse of the other. For this reason it is necessary to consider each solstice separately, whereas the equinoxes can be treated together.

Winter solstice

Conditions at the winter solstice, December 21 or 22, are best studied with the aid of diagrams. Figure 2.8 is a cross-sectional representation to show the angles at which the sun's rays strike the earth. Keep in mind that "winter" applies only to the northern hemisphere and that the southern hemisphere is experiencing its summer season.

The great circle which at all times marks the

[2] The exact time of solstices and equinoxes ranges into two calendar days because of the one-fourth day difference in the tropical year and the calendar year, which builds up to one whole day every fourth year and is corrected by adding one day in leap year.

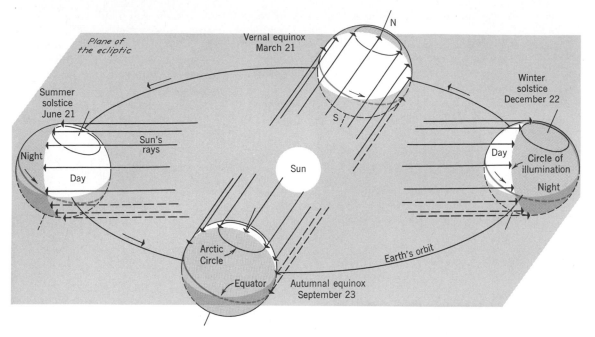

Figure 2.7   The seasons.

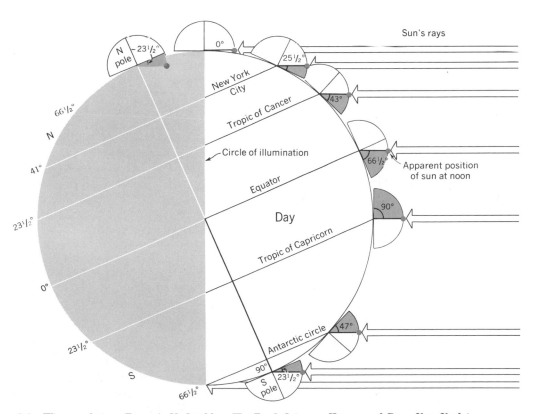

Figure 2.8   Winter solstice. (From A. N. Strahler, *The Earth Sciences*, Harper and Row, New York.)

boundary between sunlit and shadowed halves of the earth is called the *circle of illumination*. At the winter solstice it divides all parallels of latitude which it crosses except the equator into unequal parts. The circle of illumination is tangent to the *Arctic Circle* (66½° N lat.) and the *Antarctic Circle* (66½° S lat.). This occurrence explains why the two parallels are given special designation on the globe. The circle of illumination bisects the equator

in accordance with the law that any two intersecting great circles bisect each other.

Because of the position of the circle of illumination at the winter solstice, day and night are unequal in length over most of the globe. This inequality may be estimated from Figure 2.8 by noting what proportions of a given parallel lie on either side of the circle of illumination. The following facts are evident.

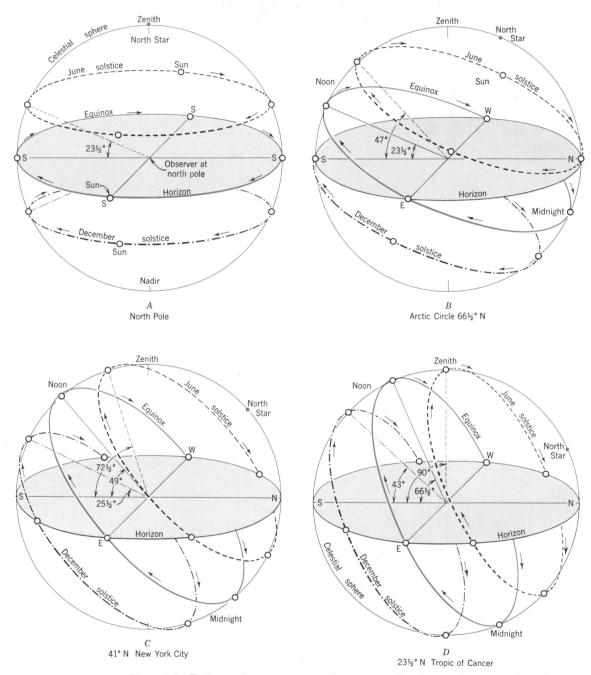

Figure 2.9   To the earth-bound observer the earth's surface is a flat, horizontal disc. The sun, moon, and stars seem to travel on the inner surface of a hemispherical dome above him. The path of the sun in the sky at various latitudes is shown here for the equinox and both solstices.

(*a*) Night is longer than day in the northern hemisphere.

(*b*) Day is longer than night in the southern hemisphere.

(*c*) The inequality between day and night increases from the equator poleward.

(*d*) At corresponding latitudes north and south of the equator the relative lengths of day and night are in exact opposite relation.

(*e*) Between the Arctic Circle, 66½° N, and the north pole, night lasts the entire 24 hours.[3] This fact is evident from Figure 2.8 because the entire polar area north of the Arctic Circle lies on the shaded side of the circle of illumination and hence will not come into the sun's rays even though the earth turns through 360°.

[3] This statement does not take into account *twilight*, which provides considerable light near the Arctic Circle.

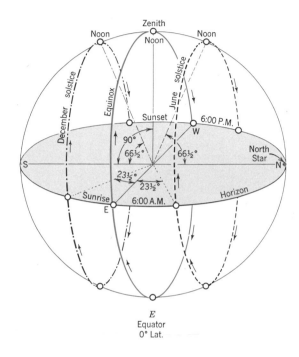

E
Equator
0° Lat.

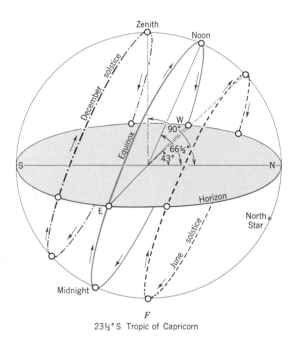

F
23½° S Tropic of Capricorn

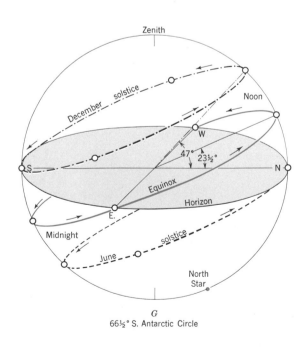

G
66½° S. Antarctic Circle

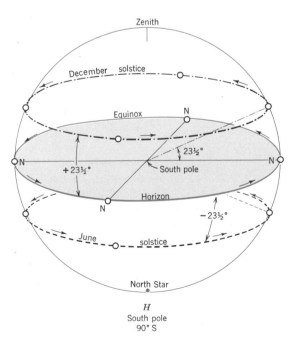

H
South pole
90° S

(*f*) Between the Antarctic Circle, 66½° S, and the south pole day lasts the entire 24 hours, because the earth's turning fails to bring any part of this area into the zone of darkness.

Solar noon occurs when the sun reaches its highest point in the sky. Noon occurs simultaneously at all points having the same longitude, that is, for all points lying on a single meridian. The vertical angle of the sun above the horizon at noon, designated the *sun's noon altitude*, may be determined from Figure 2.8 by measuring the angle between a ray from the sun and a line tangent to the globe at a selected latitude. Sun's noon altitude in degrees is given for various latitudes on the right-hand side of Figure 2.8. Although the earth's surface is curved, the apparent world in which we, as tiny individuals, live is a flat world. Within the limits of vision the horizon appears to make a circle on a flat plane. This explains why a straight tangent line may be used for measuring the altitude angle on the diagram.

At lat. 23½° S the sun's rays at noon strike the earth at an angle of 90° above the horizon. Thus the sun is exactly in the center of the sky, or *zenith*. The parallel 23½° S has therefore been designated the *Tropic of Capricorn*. At the Arctic Circle, 66½° N, the sun at noon is exactly on the horizon. At the south pole, the noon sun has an altitude of 23½° above the horizon, and maintains this angle throughout the full 24 hours.

The path of the sun in the sky at the winter solstice is illustrated for various latitudes in Figure 2.9. The horizon is drawn as a circle lying in a horizontal plane, and the sky is visualized as a hemispherical celestial dome. Below the horizon is the opposite celestial hemisphere, the two hemispheres together comprising the *celestial sphere*. The sun daily completes an entire circle inscribed on the celestial sphere. On certain occasions, as on the equinoxes, these circular paths are great circles on the celestial sphere; otherwise they are small circles.

Direction and time of sunrise and sunset at solstice

The compass direction of sunrise and sunset points on the horizon varies greatly with latitude, as shown in Figure 2.9. At the Antarctic Circle on the December solstice, sunrise and sunset occur at the same instant—midnight—at a point due south on the horizon. At all places between the Antarctic Circle and the Arctic Circle, the sun rises at some point between south and east, and sets at a point between south and west.

To determine approximately the length of day and night and times of sunrise and sunset at solstice for various latitudes, a small globe and a rubber

band can be used (Figure 2.10). Proceed as follows:

Winter solstice. Place the rubber band so as to represent the circle of illumination (great circle) crossing the equator at the 90° meridians west and east, and just tangent to the Arctic and Antarctic Circles where they cross the Greenwich and 180th meridians. Now select a particular latitude, for example, 40° N. Taking the Greenwich meridian as noon, count 15° of longitude to the hour, westward along the 40th parallel, until the rubber band is reached. This number of hours subtracted from 12:00 noon gives the approximate time of sunrise (local apparent solar time), and added to 12:00 noon gives the time of sunset. The lengths of day and night can be obtained by simple calculation. The shorter arc in this case gives the length of day; the longer arc, by way of the Pacific, gives length of night.

Summer solstice. Turn the globe around and use the Pacific side, leaving the rubber band in the same position as before. Be sure that the rubber band maintains a great circle and cuts the equator at the 90th meridians west and east.

Solstice conditions at the poles

At the poles the path of the sun in the sky is the most extraordinary of all places on the earth (Figure 2.9*A,H*). Here the sun does not rise and sink in a slanting path with respect to the horizon, as at other latitudes. Instead it follows a horizontal circle, remaining parallel with the horizon throughout the day. (In actuality this path is spiral, but so low a spiral that it cannot be detected by

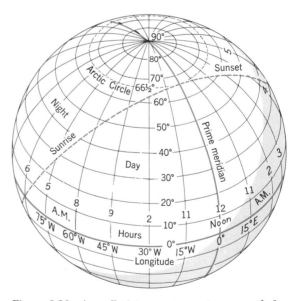

Figure 2.10 A small globe can be used as a graph for determining the lengths of day and night at any latitude.

Figure 2.11 This photograph of the sun was taken at midnight at Hammerfest, Norway, 70°40′ N. lat., during the summer solstice period. The sun has reached its lowest point in the sky. (Photograph by A. Kalland, Hammerfest.)

ordinary observation.) At the December solstice the sun at the north pole remains $23\frac{1}{2}°$ below the horizon throughout the day, while at the south pole it is constantly $23\frac{1}{2}°$ above the horizon. At the south pole we would have no natural way of determining when noon would occur because the sun's altitude remains constant. Moreover, all meridians converge to a point at the pole so that we could not refer the time to any local meridian.

Summer solstice

In almost every way the conditions at summer solstice, June 21 or 22, are the exact reverse of winter solstice conditions. At this time the north polar end of the earth's axis is inclined directly toward the sun and the northern hemisphere enjoys the same conditions of increased sunshine that the southern hemisphere had during the winter solstice (Figure 2.11). Instead of a special diagram to show the relation of sun's rays to the earth, it is suggested that Figure 2.8 be turned upside down, exchanging "north" for "south," "Arctic" for "Antarctic," and "Tropic of Cancer" for "Tropic of Capricorn." The various statements made in the previous pages concerning circle of illumination, length of day and night, and the sun's noon altitude at the winter solstice may be reread, with suitable changes to fit the reversed conditions of the summer solstice.

The path of the sun in the sky on June 21 is

shown in Figure 2.9. For all latitudes between the Arctic and Antarctic Circles the sun rises on the northeastern horizon, and sets on the northwestern horizon.

The equinoxes

On March 20 or 21 and September 22 or 23, the vernal and autumnal equinoxes, respectively, the relation of earth to sun's rays is identical and the two dates may be treated jointly. Figure 2.7 shows the general situation. Although the earth's axis is, as always, inclined $66\frac{1}{2}°$ to the plane of the orbit, the inclination is so oriented that it is neither toward nor away from the sun. The sun's rays make an angle of 90° with the earth's axis. Further details of the equinoctial conditions are shown in Figure 2.12.

The circle of illumination at the equinoxes passes through the poles and hence coincides with the meridians as the earth turns.

As evident from Figure 2.12, the parallels are divided into equal halves by the circle of illumination. Hence day and night are of exactly equal length, twelve hours each, at all latitudes.[4] Conditions are the same for both northern and southern hemispheres. Sunrise occurs at 6:00 A.M. (local

[4]This explains the word *equinox*, from the Latin *aequus*, equal; and *nox*, night. We are not taking into account twilight, which extends the period of daylight before sunrise and after sunset.

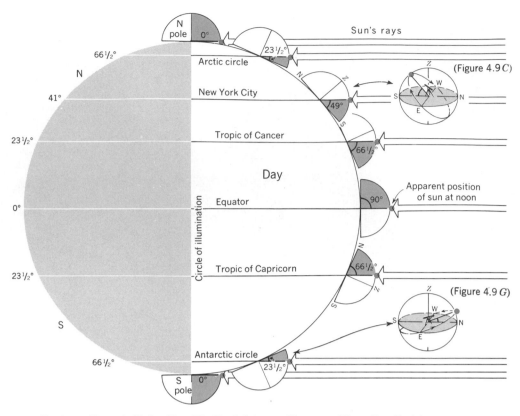

Figure 2.12   Equinox. (From A. N. Strahler, *The Earth Sciences*, Harper and Row, New York.)

solar time), sunset at 6:00 P.M., at all places on the globe, except at the poles, where special conditions prevail.

The sun's noon altitude is found by direct measurement of the angle between the sun's parallel rays and tangent lines drawn at selected points (Figure 2.12). A few such measurements should reveal that the altitude is always the *colatitude*, or 90° minus the latitude. Thus, on the equinoxes, but not for any other time of year, the sun's noon altitude may be computed by a single simple subtraction, if only the latitude is given. Although the altitude is the same for similar latitudes both north and south of the equator, it should be remembered that the angle is measured from the southern horizon in the northern hemisphere, from the northern horizon in the southern hemisphere.

The path of the sun in the sky at the equinoxes is illustrated in Figure 2.9. In each case the path is midway between the paths at the solstices. The sun rises at a point due east on the horizon and sets at a point due west on the horizon at all latitudes except at the two poles. At the poles, the sun remains on the horizon all day long, traveling one complete circuit of the horizon in 24 hours. Note, however, that the direction of apparent movement of the sun is opposite for the two poles.

At the equator the sun has an altitude of 90° at noon; moreover, its path is in a plane perpendicular to the horizon plane. For this reason the sun changes altitude 15° per hour throughout the day at the equator on the equinoxes. Furthermore, at the equator on these dates the shadow of any vertical rod will point due west from 6:00 A.M. to noon, will disappear precisely at noon, and will point due east from noon until 6:00 P.M.

Time

The need for understanding global time relationships and the various kinds of time in use scarcely needs emphasis in this modern day of instantaneous communication and high-speed travel. Before the coming of the telegraph, problems of time differences were of little or no concern to people who lived most of their lives in one community. Even the traveler was caused only the inconvenience of resetting his watch to the time used by local communities. The amount of time consumed in getting from one place to another was so much more than the difference in watch time between the two places that the difference was of little practical consequence.

When it became possible to transmit messages

instantaneously by telegraph, differences in local time resulting from differences in longitudinal position were immediately apparent. With the development of rapid means of travel it became important to correct schedules for the gain or loss of time incurred by passage across the meridians. East-to-west flight in the middle latitudes can now approximate the speed necessary to keep pace with the sun. For example, a plane leaving New York at 12:00 noon, Eastern Standard time, can by traveling about 800 mph (1300 km per hr) arrive in San Francisco at 12:00 noon, Pacific Standard time.

Longitude and time

To avoid confusion and make the study of time relations as simple as possible it is necessary to think of the earth as standing still and of the sun as completing one circuit about the earth every 24 hours. This is a perfectly permissible concept inasmuch as earth-sun relationships are purely relative. Imagine that a meridian is free to sweep westward around the globe and that it constantly maintains such a speed as to be located always where the sun's rays strike the earth's surface at the highest possible angle. This line we shall call the *noon meridian* (Figure 2.13). Directly opposite this meridian, on the other side of the globe, is the *midnight meridian*. It, too, sweeps westward over the globe and remains constantly 180° of longitude apart from the noon meridian. Whereas the noon meridian separates the forenoon and afternoon of the same calendar day, the midnight meridian is the dividing line between one calendar

day and the next.

Because the noon meridian sweeps over 360° of longitude every 24 hours, it must cover 15° of longitude every hour, or 1° of longitude every 4 minutes. We therefore find it convenient to state that one hour of time is the equivalent of 15° of longitude. This equality forms the basis for all calculations concerning time belts of the globe. For example, if the noon meridian reaches one place on the globe 4 hours after it leaves another place, the two places are separated by 60° of longitude.

Enlarging this concept of time meridians still further, let it be imagined that in addition to the noon and midnight meridians there are 22 hour circles, each one a half great circle, 15° of longitude apart from its neighbor. The hour circles are equidistantly spaced between the noon and midnight meridians (Figure 2.13). Each hour circle will then represent a given hour of the day and can be labeled with a specific hour number which it keeps permanently. Together with the noon and midnight meridians, the hour circles can be imagined to form a birdcagelike net enclosing the globe and attached only at the north and south poles. As a further convenience in analyzing global time relations, it is most helpful if the globe used has meridians drawn for every 15°. If so, wherever the noon meridian of the time net coincides with the Greenwich meridian, all other hour circles coincide with true meridians on the globe.

A working model of global time relations is illustrated in Figure 2.14. If a small globe is available, make a cardboard girdle to fit about the equator and mark upon the cardboard the posi-

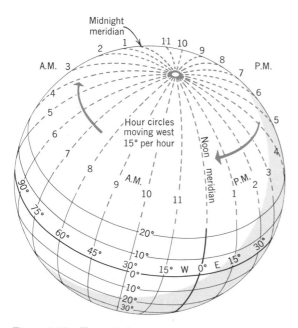

Figure 2.13 Hour circles.

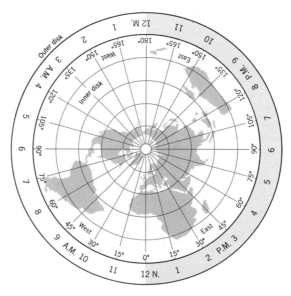

Figure 2.14 A working model of global time.

tions of the time meridians. If no globe is available, make two cardboard discs of different radius and attach them at their centers in such a way that one disc can be turned while the other remains still. On the inner disc draw radii to represent 15° meridians of a globe seen from a point above the north pole. On the outer disc draw similar radii, but label them in hours to represent the time net. As a further refinement, the inner disc may consist of a hemispherical world map.

People sometimes become confused when trying to decide whether the watch time of places to the east (or west) of them is ahead or behind their own watch time. This is especially evident where the problem is to calculate when a radio or television program will be on the air in different parts of the country, or to decide whether to set one's watch ahead or behind one hour when traveling from one standard time zone to another. To avoid confusion let the time meridians be visualized as moving westward around the globe. Then consider, for example, that you are in New York City and the time there is 12:00 noon. The noon meridian which is at New York left Greenwich, England, five hours earlier. Therefore, in England five hours must have elapsed since noon and it must be 5 P.M. in that country. From this we get the general rule: *places that lie east of you have a later hour.* Again, consider that the noon meridian is at New York City. It will take that meridian about three hours to travel westward to reach San Francisco. Hence it must be 9:00 A.M. in that city. This gives a counterpart of the rule stated above, namely: *places that lie west of you have an earlier hour.* Both rules are subject to qualifications where the International Date Line lies between the places.

Local time

One means of establishing a time system for a small community is to take the meridian of longitude that passes through some central point in the town or city, as for example the courthouse or a cathedral. All clocks of the community are set to read 12:00 noon when the sun is directly over that meridian, that is to say, when a shadow cast by a vertical rod points due north. The sun is an erratic timekeeper, and it would be necessary to take the mean value of the sun's noon position, but this need not concern us now. The time system thus derived is called *local time* and may be defined as mean solar time based on the local meridian. All places located on the same meridian, regardless of how far apart they may be, have the same local time, whereas all places located on different meridians have unlike local times, differing by four minutes for every degree of longitude between them.

Standard time

The undesirability of using local time systems in each community of a highly populated country in our modern day is obvious. American railroads, about 1870, introduced a standardized system covering considerable belts of territory, but this system was developed by the railroad companies for their own convenience. Consequently, if several railroads met or passed in a single town, the inhabitants might have had to contend with several different kinds of railroad time in addition to their own local time. It is said that before 1883 as many as five different time systems were used in a single town and that the railroads of the United States altogether followed 53 different systems of time.

The obvious solution to such problems is *standard time*, based on a *standard meridian*, whose local time is arbitrarily given to wide strips of country on both sides. Thus, all clocks within the belt are set to a single time. It is evident that if standard meridians are 15° apart, adjacent zones will have standard times differing by exactly one hour. Furthermore, if these meridians represent longitudes which are multiples of 15, for example, 60°, 75°, 90°, or 105°, each successive standard time zone will differ from the standard time of Greenwich, England, by whole hour units. This is the system of standard time zones employed over most of the globe.

Standard time in the United States

The present system of standard time in the United States was placed in operation on November 18, 1883, but it was not until March 19, 1918, that Congress passed legislation directing the Interstate Commerce Commission to determine time-zone boundaries. The standard meridians and boundaries are shown on Figure 2.15. The six time zones and their meridians are as follows:

Eastern standard time	75th meridian
Central standard time	90th meridian
Mountain standard time	105th meridian
Pacific standard time	120th meridian
Alaskan standard time	150th meridian
Hawaiian standard time	150th meridian

If it had been carried out precisely, the system would have resulted in belts extending exactly $7\frac{1}{2}°$ east and west of each standard meridian, but a glance at the map shows that great liberties have been taken in locating the boundaries. Wherever the time-zone boundary could conveniently be located along some already existing and widely recognized line, this was done. Natural physiographic boundaries have been used. For example, the Eastern time-Central time boundary line follows Lake Michigan down its center, and the Mountain

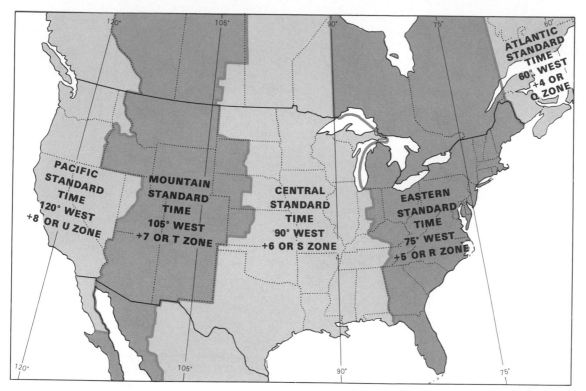

Figure 2.15 Time-zone map of the United States and southern Canada. (Based on data by Interstate Commerce Commission and Department of Transportation.)

time-Pacific time boundary follows a ridge-crest line also used by the Idaho-Montana state boundary. Most frequently, the time-zone boundary follows state and county boundaries. For example, before 1941, the Eastern time-Central time boundary passed north to south through Georgia, but in that year the ICC held hearings in Georgia and officially moved the time-zone boundary westward to the Alabama-Georgia state line, thus bringing all of Georgia into the Eastern standard zone.

The time belts are by no means equally distributed on both sides of the standard meridians, as a glance at the map will show. An extreme example is found in western Texas, where the Central time-Mountain time boundary follows the western boundary of Texas, even crossing west of the standard meridian (105° W) of the Mountain time zone. Although such deviations may seem odd, they cause little difficulty when the boundaries are once well established. The advantage gained by extreme deviations of the boundaries is in permitting entire states to operate under one kind of time. In some places the determining factor was a railroad division point junction or terminal. Where the line was drawn through such a point, as for example at Ogden or Salt Lake City, divisions east and west of the city each ran under a single time.

Daylight saving time and War time

Because many human activities, especially in cities and manufacturing areas, start well after sunrise but continue long after sunset, it would seem desirable to set forward the hours of daylight so as to utilize them to best advantage. A considerable saving in electric power would be made if the early morning daylight period, wasted while people are still in bed and offices and factories are closed, were transferred to the early evening when the large majority of persons are awake and busy. The adjusted time system is known as *daylight saving time* and is obtained by setting ahead all timepieces by one hour. Thus, when the sun is over the standard meridian (i.e., noon by the mean sun), all clocks in that time zone read 1:00 P.M. Sunrise and sunset at the equinoxes or equator, instead of occurring at 6:00 A.M. and 6:00 P.M., would occur at 7:00 A.M. and 7:00 P.M., respectively.

In terms of the standard time system, we may describe daylight saving time as based on the standard meridian lying 15° of longitude east of the standard meridian normally giving the standard time to that zone. For example, Eastern daylight saving time is the same as standard time of the meridian 60° W long.; Central daylight saving time is the same as Eastern standard time, both being based on the meridian 75° W long.

Daylight saving time was adopted in the United States during the First World War, and by act of Congress was put into effect from the last Sunday in April to the last Sunday in September of 1918. After that war it was used locally throughout the United States where authorized by local legislation. During the Second World War, daylight saving time was used nationally throughout the entire period of February 1942, to October 1945, and was known as War time. England during the same war period employed a double daylight saving time in which clocks were running two hours ahead of Greenwich civil time. This was desirable because of the unusually long summer days which England enjoys as a result of her relatively far northerly latitude. Many European countries normally use daylight saving time, which goes under the name *summer time*, during a part of the year. Nations whose time is advanced by an hour throughout the entire year are Great Britain, Ireland, Spain, France, the Netherlands, Belgium, and the USSR.

In April 1966, The United States Congress passed the *Uniform Time Act*, which requires that daylight saving time be applied uniformly throughout each state, unless the legislature of that state has voted to remain instead on standard time. In the latter case, standard time must be applied uniformly throughout the entire state. Daylight time goes into effect at 2:00 A.M. of the last Sunday of April and continues until 2:00 A.M. of the last Sunday of October. The Uniform Time Act is administered by the Department of Transportation, which also has authority to adjust time-zone boundaries.

World time zones

In 1884 an international congress was held in Washington to consider the subject of world standard time. As a result, standard times of countries throughout the world are based on standard meridians, which are multiples of the unit 15° and thus differ from one another by whole hourly amounts. In all global time calculations, the prime meridian of Greenwich, England, is taken as the reference meridian. All time zones of the globe are described in terms of the number of hours' difference between the standard meridian of that zone and the Greenwich meridian. In order to distinguish whether the time zones lie east or west of the Greenwich meridian the time is designated *fast* for all places east of Greenwich (east longitude), and *slow* for all places west of Greenwich (west longitude). U.S. Eastern Standard time, for example, is said to be "5 hours slow."

An alternative system of designating world time zones uses letters of the alphabet, as shown in Figures 2.15 and 2.16.

Figure 2.16 is a world map on which the 24 principal standard time zones of the world are shown. Fifteen-degree meridians are in black lines; $7\frac{1}{2}$° meridians, which form large elements of the zone boundaries, are indicated in color lines. Within each time zone is inscribed the number of hours' difference between the zone time and Greenwich time. Careful examination of this map brings out a number of interesting facts. Some small countries or islands lie about midway between 15° meridians. Under such circumstances a standard meridian is chosen which lies halfway between the two and is thus a multiple of $7\frac{1}{2}$°. The standard time of the country is therefore fast or slow by some multiple of a half hour. Iran ($3\frac{1}{2}$ hours fast) and Surinam ($3\frac{1}{2}$ hours slow) illustrate this point. India, for a large country, is unusual in having the time $5\frac{1}{2}$ hours fast. The country having the greatest east-west extent is the Soviet Union, with eleven standard time zones, but these are all advanced by one hour with respect to the standard time meridians in each zone, to give a perpetual daylight saving time. Canada occupies six time zones. Notice that Newfoundland and Labrador use the time $3\frac{1}{2}$ hours fast.

Although virtually all countries today have standard times related by whole hours or half hours to Greenwich Civil time, it was not so long ago that many countries determined their standard times by meridians of the capital cities or those passing through local observatories. For example, in 1905, all of France used the local time of the meridian passing through the Paris Observatory. This gave a time which was $0^h 9^m 20.9^s$ fast. India in 1905 used the local time of the Madras Observatory, which is $5^h 20^m 59.1^s$ fast of Greenwich Civil time. Ireland in 1905 was using the time $0^h 25^m 21.1^s$ slow, which is the local time of the Dublin meridian.

By 1970, only a few small nations or colonies persisted in the use of odd local meridians for their standard time. These were Liberia ($0^h 44\frac{1}{2}^m$ slow), Guyana ($3^h 45^m$ slow), Chatham Island ($12^h 45^m$ fast), Saudi Arabia, and Mongolia. An up-to-date list of world standard times will be found in *The Air Almanac*.

The International Date Line

If we were to take a world map or globe on which 15° meridians are drawn and count them in an eastward direction, starting with the Greenwich meridian as 0, we would find that the 180th meridian is number 12, and that the time of this meridian is, therefore, 12 hours fast. Counting in a similar manner westward from the Greenwich

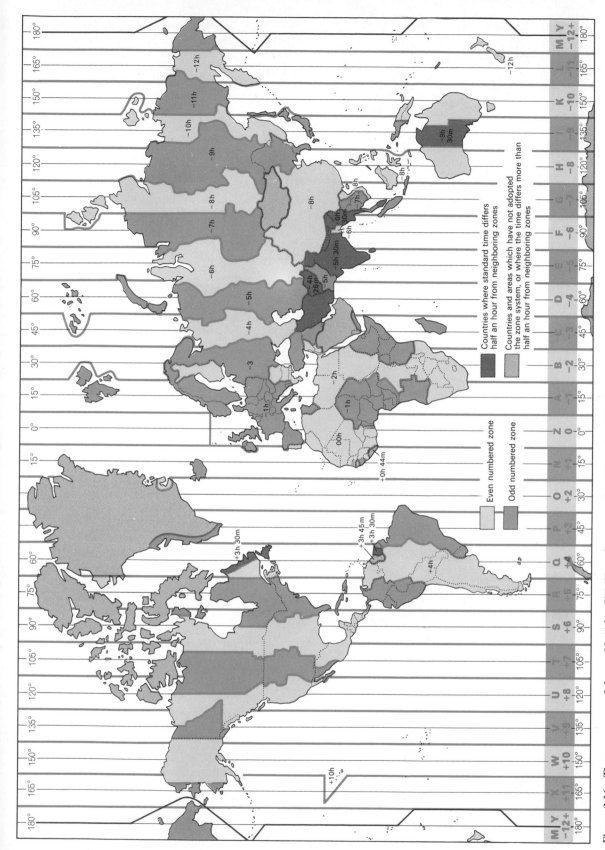

Figure 2.16   Time-zone map of the world. (After U.S. Navy Oceanographic Office, No. 5192.)

Countries where standard time differs
half an hour from neighboring zones

Countries and areas which have not adopted
the zone system, or where the time differs more than
half an hour from neighboring zones

Even numbered zone

Odd numbered zone

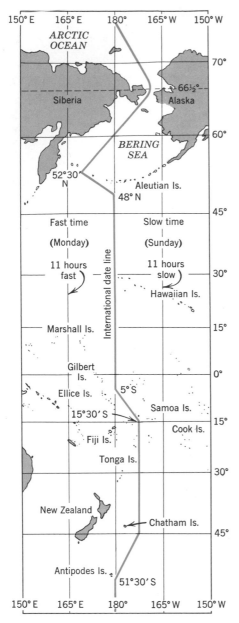

Figure 2.17   The International Date Line.

on the east (American) side. For example, if it is Monday on the Asiatic side of the 180th meridian, it is Sunday on the American side. Confusion can be avoided and the correct dates consistently obtained if it is remembered that by counting hours from the Greenwich meridian eastward around the globe by way of Asia, the 180th meridian has the time 12 hours fast, hence that the Asiatic side of the meridian must be a day ahead of the other side.

In the days of very slow trans-Pacific travel by sailing vessel and low-powered steamship, an entire calendar day was merely omitted on a westbound voyage and an entire day repeated on an eastbound voyage. The change was made at any convenient place and time in midocean and usually planned so as to have neither two Sundays in one week nor a week without a Sunday.[5]

It was due to their failure to advance the calendar by a whole day that the crew of Magellan's only surviving ship, reaching Seville after circumnavigating the globe in a westward direction, found that in Spain it was September 8, 1522, whereas by their own reckoning it was only September 7 of that year.

When traveling across the Pacific in today's high-speed jet aircraft, the 24-hour correction is made at the time the 180th meridian is crossed. Suppose that the plane is traveling eastward toward North America and crosses the meridian at 4:00 P.M., (standard time 12 hours fast) on a Tuesday. At the instant of crossing, the time becomes 4:00 P.M. Monday. When traveling westward to the Orient the time moves ahead by a full day. For example, if the meridian is crossed at 9:30 A.M. Wednesday, the time becomes 9:30 A.M. Thursday.

Because of these peculiar properties, the 180th meridian was designated the *International Date Line* by the International Meridian Conference held in Washington, D.C., in 1884 (Figure 2.17). It is one of the fortuitous occurrences of modern civilization that, after the Greenwich meridian had come into widespread use in English-speaking countries as the international basis for the reckoning of longitude, the 180th meridian should have been found to fall in an almost ideal location—squarely in the middle of the world's largest expanse of ocean. Nevertheless, the International Date Line has had to deviate both eastward and westward to permit certain land areas and groups of islands to have the same calendar day (Figure 2.17). By an eastward bulge passing through Bering Strait, the easternmost part of Siberia is included in the Asiatic side, and a westward deflection of the line allows the Aleutian Islands to be included

[5] Willis E. Johnson, 1907, *Mathematical geography*, American Book Co., New York. See pp. 96–103.

meridian, we find that the 180th meridian is again number 12, but that the time is 12 hours slow. Both results are, of course, correct; and the explanation becomes obvious when we note that the difference in time between 12 hours fast and 12 hours slow is 24 hours, or a full day. At the precise instant when the noon meridian coincides with the Greenwich meridian, the 180th meridian coincides with the midnight hour meridian. At this instant, and only at this instant, the same calendar day exists on both sides of the meridian. At all other times the calendar day on the west (Asiatic) side of the 180th meridian is one day ahead of that with the Alaskan peninsula. A few degrees south of the equator the date line is shifted eastward $7\frac{1}{2}°$ and thus avoids cutting through the Ellice, Wallis, Fiji, and Tonga island groups, which have the same day as New Zealand.

Duration of days on the globe

One of the most curious aspects of global time is the manner in which calendar days appear and disappear on the globe. If a day lasts 24 hours for a specific place on the earth, and the same series of hours reaches places lying farther west at a later time, it follows that a calendar day exists more than 24 hours for the earth as a whole. To develop this idea fully use the following visual aids:

Holding a small globe in your hands so that the Pacific Ocean is before you, imagine that the international date line is a narrow slit extending from pole to pole on the 180th meridian. (For the time being let it be assumed that the line has no deviations from the meridian.) Imagine further that calendar days—Monday, Tuesday, etc.—issue from the long slit and spread westward over the globe like a thin film. The leading edge of this film extends from pole to pole and corresponds to the midnight time meridian. The film, which is now issuing from the slit, can be designated Monday, and all global areas which it covers have the calendar day Monday. Traveling at the rate of 15° of longitude per hour, the midnight meridian, or leading edge of the film, will require twelve hours to arrive at the Greenwich meridian. At this precise moment it is midnight in England and the calendar day Monday covers the eastern hemisphere. The western hemisphere still has the calendar day Sunday, which may be pictured as retreating ahead of Monday. Twelve hours later the imaginary film which we are calling Monday has spread over the entire globe and the edge of the film has reached the slit at the 180th meridian.

At this precise instant Monday envelops the entire earth and is the only calendar day present anywhere.

Because no calendar day can ever cross the date line, we will have to picture Monday as disappearing into the slit whence it had begun to issue 24 hours earlier. As it does so, the next calendar day, Tuesday, is beginning to issue from the slit and to spread across the Pacific Ocean toward Asia, just as Monday had done earlier. Now we are prepared to answer the intriguing question: what is the total number of hours that the calendar day Monday, whose progress we are observing, will exist on the globe? The answer is that Monday will exist 48 hours. It has required 24 hours for the film representing Monday to spread around and completely cover the earth, and it requires an additional 24 hours for it to disappear back into the slit. Thus Monday is present on the earth's surface for a continuous period of 48 hours, although for any specified place on the globe it can last only 24 hours.

REFERENCES FOR FURTHER STUDY

Hosmer, G. L. and J. M. Robbins (1948), *Practical astronomy*, fourth edition, John Wiley and Sons, New York, 355 pp. See Chapters 1 and 5.

Mehlin, T. G. (1959), *Astronomy*, John Wiley and Sons, New York, 391 pp.

Strahler, A. N. (1969), *Physical geography*, third edition, John Wiley and Sons, New York, 733 pp. See Chapters 4, 5.

Royal Astronomical Society of Canada, *The observer's handbook*, issued annually, Univ. of Toronto Press.

U.S. Naval Observatory, *The air almanac*, issued annually, U.S. Govt. Printing Office, Washington, D.C.

World Journal Tribune, *The world almanac and book of facts*, issued annually, Newspaper Enterprise Association, New York.

REVIEW QUESTIONS

1. Why are earth-sun relationships important in the study of geography?

2. What is meant by rotation of the earth? What is the period of rotation with reference to the sun?

3. What is the direction of rotation of the earth? Does the earth's rotation have any observable effects on objects at or near the earth's surface aside from the astronomical effects?

4. Describe and explain the Foucault pendulum experiment. What does it prove? Could you determine your latitude by means of a Foucault pendulum? How?

5. What is meant by revolution of the earth? In what two ways can the length and starting point of a year be reckoned? Which one is used in our calendar system? How long is this type of year? How must it be periodically corrected to fit our calendar? What is the direction of the earth's revolution?

6. What form has the earth's orbit? What is perihelion? What is aphelion? On what dates do they occur? What distances separate earth and sun at perihelion and at aphelion? What effect does this have on the seasons?

7. Describe the way in which the earth's axis is tilted with respect to the plane of the earth's orbit. What is the angle of tilt? Does this angle change throughout the year?

8. Describe each of the two solstice and two equinox positions of the earth with respect to the sun, giving the dates of each in correct sequence.

9. What is the circle of illumination? Where is the circle of illumination located on the date of equinox? On December solstice? On June solstice?

10. Describe the conditions of global illumination, length of day and night, and path of sun in the sky at June solstice. Give details for the equator, Tropics of Cancer and Capricorn, Arctic and Antarctic Circles, and poles.

11. Describe the conditions of global illumination, length of day and night, and path of sun in the sky at equinox. Give details for the equator, Tropics of Cancer and Capricorn, Arctic, and Antarctic Circles, and poles. How is the sun's noon altitude related to latitude on the date of equinox?

12. Describe the path of the sun in the sky throughout the year at the north pole. How do conditions differ at the south pole?

13. Describe the path of the sun in the sky at the Arctic Circle at equinox and at each solstice.

14. Describe in a general way how the compass direction of sunrise and sunset varies with season of year and with latitude.

15. Explain how longitude is related to time. How many degrees of longitude are the equivalent of one hour of time? How many hour meridians are there on the globe? In which direction do they travel?

16. Do places located east of you have a time that is earlier or later than your time?

17. What is local time? On what is it based? Can places that differ in longitude have the same local time? Can places that differ in latitude have the same local time?

18. What is standard time? Explain the system of standard time used in the United States. Which meridians are used?

19. How are boundaries between standard time zones determined? Give examples of various kinds of boundaries used in the United States.

20. What is daylight saving time? Why is it used? When is it used?

21. Explain the system of world time zones now in general use. On what prime meridian is it based? What is fast time? Slow time? What advantages are there in conforming to the world time zone system?

22. What is the International Date Line? Describe its location and form. Explain the difference in calendar days on the two sides of this line. Which side has the earlier days? Is it possible for the same day to exist simultaneously on both sides of the line?

23. How long does a given calendar day exist on the globe? Describe the life history of one calendar day as it envelopes the globe and then disappears.

CHAPTER 3
The Earth's Atmosphere

MAN lives at the bottom of an ocean of air; he is an air-breather dependent upon favorable conditions of pressure, temperature, and chemical composition of the atmosphere that surrounds him. He also lives on the solid outer surface of the earth, upon which he is dependent for food, clothing, shelter, and means of movement from place to place. But the air and the land are not two entirely separate realms; they constitute an interface across which there is a continual flux of matter and energy. Man's surface environment is a shallow but highly complex zone in which atmospheric conditions exert control upon the land surface, while at the same time the surface of the land exerts an influence upon the properties of the immediately adjacent atmosphere.

Essentially the same statements apply with respect to the surface of the oceans and the atmospheric layer above it. Man utilizes the surface of the sea as a source of food and a means of transportation. There is a continual flux of energy and matter between the sea surface and the lower layer of the atmosphere. Here, again, we find an interface of vital concern to Man. The sea influences the atmosphere above it while the atmosphere influences the sea beneath it.

Our object in this book is to examine the atmosphere and oceans with particular reference to air-land and air-sea interfaces which are so vital to Man. To the geographer, the distributions of physical properties of the ocean and atmosphere are matters of special interest, concerned as he is with spatial relationships on a global scale. The physical geographer seeks to describe and explain the manner in which the environmental elements of weather and climate change with latitude and season, and with geographical position in relation to oceans and continents. The geographer seeks out the broad patterns of similar regions and attempts to define their boundaries and organize them into systems of classification.

States of matter

Study of the atmosphere, ocean, and land requires continual application of the principles relating to three basic states of matter: *gaseous state*, *liquid state*, and *solid state*. A *gas* is a substance that expands easily to fill any small empty container, is readily compressible, and usually much less dense than liquids and solids of the same chemical composition. While the atmosphere is largely in the gaseous state, it also contains varying amounts of substances in the liquid and solid states.

A *liquid* is a substance that flows freely in response to unequal stresses, but characteristically maintains a free upper surface. Liquids are compressed only slightly with strong stresses. Liquids have densities closely comparable with solids of the same composition.

Although the world ocean is largely composed of water in the liquid state, it also contains substances in the gaseous and solid states. Both gases and liquids belong to the class of *fluids*. Layers of fluids tend to assume positions of equilibrium at rest in which a less dense fluid overlies a more dense fluid.

Solids are substances that resist changes of shape and volume and are typically capable of withstanding large unequal stresses without yielding. When yielding occurs, it is usually by sudden breakage. Although the earth's crust is largely in the solid state, it also contains substances in both gaseous and liquid states.

A further observation concerning states of matter is that a *change of state* is possible and occurs frequently in the world of nature. Most important and widespread is the change of state of water from water vapor (a gas) to liquid water and *vice versa*, and from liquid water to ice (solid state) and *vice versa* (Chapter 6). These changes of state require either an input of heat energy or the disposal of heat energy, depending upon the direction of the change.

The broad generalizations and principles that have been set down here concerning states of matter are further refined, explained, qualified, and applied in the ensuing chapters of this book.

Composition of the atmosphere

The earth's atmosphere consists of a mixture of various gases surrounding the earth to a height of many miles. Held to the earth by gravitational attraction, this envelope of air is densest at sea level and thins rapidly upward. Although almost all of the atmosphere (97 percent) lies within 18 mi (29 km) of the earth's surface, the upper limit of the atmosphere can be drawn approximately at a height of 6000 mi (10,000 km), a distance approaching the diameter of the earth itself. The science of *meteorology* deals with the physics of this atmosphere.

From the earth's surface upward to an altitude of about 50 mi (80 km) the chemical composition of the atmosphere is highly uniform throughout in terms of the proportions of its component gases. The name *homosphere* has been applied to this lower, uniform layer, in contrast to the overlying *heterosphere*, which is nonuniform in an arrangement of spherical shells.

Pure, dry air of the homosphere consists largely of *nitrogen* (78.084 percent by volume) and *oxygen* (20.946 percent). Nitrogen does not easily enter into chemical union with other substances, and can be thought of as primarily a neutral filler substance. In contrast, oxygen is highly active chemically and combines readily with other elements in the process of *oxidation*. Combustion of fuels represents a rapid form of oxidation, whereas certain forms of rock decay (weathering) represent very slow forms of oxidation.

The remaining 0.970 percent of the air is mostly *argon* (0.934 percent). *Carbon dioxide*, although constituting only about 0.033 percent, is a gas of great importance in atmospheric processes because of its ability to absorb heat and thus to allow the lower atmosphere to be warmed by heat radiation coming from the sun and from the earth's surface.

Green plants, in the process of *photosynthesis*, utilize carbon dioxide from the atmosphere, converting it with water into solid hydrocarbons. A pronounced rise in the carbon dioxide content of the atmosphere has been noted since 1900 and is a possible result of Man's combustion of vast quantities of wood, coal, petroleum, and natural gas. In this change, we may find an example of Man's impact upon his environment, for an increase in carbon dioxide, as well as in atmospheric dusts, can result in appreciable increases in average atmospheric temperatures.

The remaining gases of the homosphere are *neon, helium, krypton, xenon, hydrogen, methane,* and *nitrous oxide*, listed in decreasing order of percentage by volume. Altogether, these constituents total slightly less than 0.003 percent by volume. All of the component gases of the homosphere are perfectly diffused among one another, so as to give the pure, dry air a definite set of physical properties, just as if it were a single gas.

The heterosphere

The heterosphere, encountered at about 55 mi (90 km) above the earth's surface, consists of four gaseous layers, each of distinctive composition (Figure 3.1). Lowermost is the *molecular nitrogen layer* consisting dominantly of molecules of nitrogen (N_2) and extending upward to about 125 mi (200 km). Above this height lies the *atomic oxygen layer*, consisting dominantly of oxygen atoms (O). Between about 700 mi (1100 km) and 2200 mi (3500 km) lies the *helium layer*, composed dominantly of helium atoms (He). Above this region lies the *atomic hydrogen layer*, consisting of hydrogen atoms (H). No definite outer limit can be set to the hydrogen layer. A height of 6000 mi (10,000 km) may perhaps be taken as an arbitrary limit, for here the density of the hydrogen atoms is approximately the same as that found throughout interplanetary space. However, hydrogen atoms rotating with the earth, and hence belonging to the earth's atmosphere, may exist as far out as 22,000 mi (35,000 km).

It should be noted that the four layers described above have transitional boundary zones, rather than sharply defined surfaces of separation. The arrangement of gases is in order of their weights: molecular nitrogen, the heaviest, is lowest; atomic hydrogen, the lightest, is outermost. It should further be kept in mind that, at the extremely high altitudes of the heterosphere, the density of the

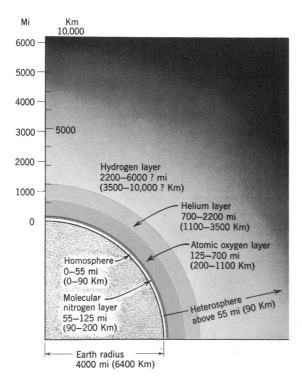

Figure 3.1 Homosphere and heterosphere. (Based on data of R. Jastrow, N.A.S.A., and M. Nicolet.)

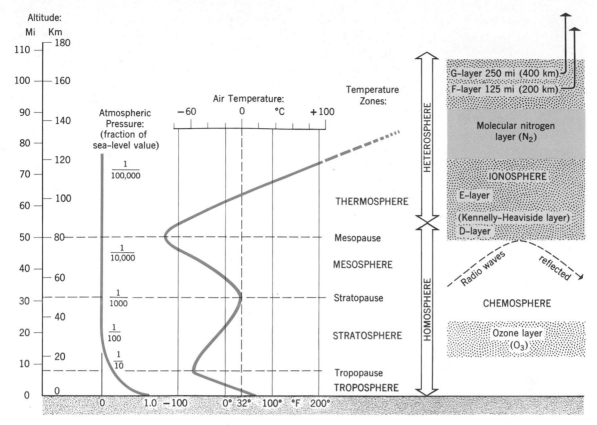

Figure 3.2 Structure of the atmosphere. (After A. N. Strahler, *The Earth Sciences*, Harper and Row, New York.)

gases molecules and atoms is extremely low. For example, at 60 mi (96 km) close to the base of the heterosphere, the atmosphere has a density of only about one millionth that at sea level. Atoms and molecules of the heterosphere are neutral in charge and turn with the earth's rotation.

Subdivisions of the homosphere

The atmosphere has been subdivided into layers according to temperatures and zones of temperature change. Three temperature zones lie within the homosphere; a fourth is assigned to the lower heterosphere. Figure 3.2 shows how temperature is related to altitude. Starting at the earth's surface, temperature falls steadily with increasing altitude at the fairly uniform average rate of $3\frac{1}{2}$ F° per 1000 ft (6.4 C° per km). This rate of temperature drop is known as the *normal temperature lapse rate*. Departures from this rate will be observed, depending upon geographical location and season of year. The layer in which the normal lapse rate applies is known as the *troposphere*, the properties of which are discussed in detail below. Figure 3.3 shows details of a typical atmospheric sounding in middle latitudes.

The normal lapse rate gives way rather abruptly

at a height of 8 to 9 mi (12.5 to 15 km) to a layer, known as the *stratosphere*, in which temperature holds essentially constant with increasing height (see Figure 3.3). The level at which the troposphere gives way to the stratosphere is termed the *tropopause*. Figure 3.4 shows that the elevation of the tropopause is least at the poles, 5 to 6 mi (8 to 10 km), whereas at the equator, the tropopause is encountered at 10 mi (17 km). If the troposphere is thought of as a complete surface in three dimensions, it resembles an oblate ellipsoid with a polar flattening and an equatorial bulge.

Seasonal changes in the elevation of the tropopause are marked in middle and high latitudes. For example, at 45° latitude the average altitude in January is 8 mi (12.5 km), but rises to 9 mi (15 km) in July. Temperatures at the tropopause are markedly lower at the equator than at the poles, as shown in Figure 3.4. At first glance, this relationship may seem strange, accustomed as we are to considering the equatorial region to be hot and the poles cold. However, with a constant temperature lapse rate assumed, the higher the tropopause, the colder will be the air.

Upward through the stratosphere there is a

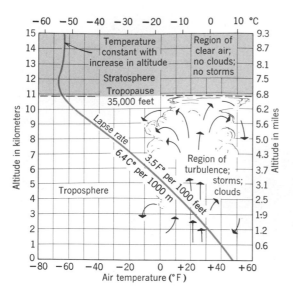

Figure 3.3 A typical lapse rate curve.

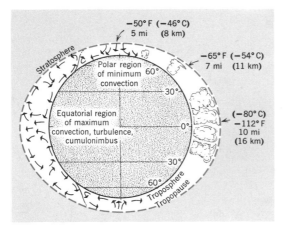

Figure 3.4 Schematic cross section of the troposphere. Figures give height and temperature of the tropopause.

slow rise in temperature until a value of about 32°F (0°C) is reached at about 30 mi (50 km). Here, at the *stratopause*, a reversal to falling temperature sets in. Temperature decreases through the overlying *mesosphere*, a layer extending upward to about 50 mi (80 km), where a low point of −120°F (−83°C) is reached. This level of temperature minimum and reversal is termed the *mesopause*. With further increasing elevation, a steep climb in temperature is observed within the *thermosphere*. As previously noted, the thermosphere lies within the heterosphere. Hence, the mesopause may be regarded as coincident with the upper limit of the homosphere. Within the thermosphere, temperatures reach 2000 to 3000°F (1100 to 1650°C), but such figures have little meaning when we consider that the density of the

air is so slight as to approach a vacuum. Very little heat can be held or conducted by air of such low density.

The troposphere and Man

It is the lowermost atmospheric layer, the troposphere, that is of most direct importance to Man in his environment at the bottom of the atmosphere. Virtually all phenomena of weather and climate that materially affect Man take place within the troposphere.

In addition to pure dry air, the troposphere contains *water vapor*, a colorless, odorless gaseous form of water which mixes perfectly with the other gases of the air. The degree to which water vapor is present is designated as the *humidity* and is of primary importance in weather phenomena. Water vapor can condense into clouds and fog. If condensation is excessive, rain, snow, hail, or sleet, collectively termed *precipitation*, may result. Where water vapor is present only in small proportions, extreme dryness of air typical of the hot deserts results. There is, in addition, a most important function performed by water vapor. Like carbon dioxide, it is capable of absorbing heat, which penetrates the atmosphere in the form of radiant energy from the sun and earth. Water vapor gives to the troposphere the qualities of an insulating blanket, which prevents the rapid escape of heat from the earth's surface.

The troposphere contains myriads of tiny dust particles, so small and light that the slightest movements of the air keep them aloft. They have been swept into the air from dry desert plains, lake beds and beaches, or explosive volcanoes. Strong winds blowing over the ocean lift droplets of spray into the air. These may dry out, leaving as residues extremely minute crystals of salt which are carried high into the air. Forest and brush fires are yet another important source of atmospheric dust particles. Countless meteors, vaporizing from the heat of friction as they enter the upper layers of air, have contributed dust particles.

Dust in the troposphere contributes to the occurrence of twilight and the red colors of sunrise and sunset, but the most important function of dust particles is not observable and is rarely appreciated. Certain types of dust particles serve as *nuclei*, or centers, around which water vapor condenses to produce cloud particles. This is illustrated in the air over industrial cities which discharge much chemically active dust into the air. So effective are these dusts in causing moisture to collect around them that almost perpetual dense haze, or *smog*, hangs over the city.

The stratosphere and higher layers are virtually free of water vapor and dust. Clouds are rare and storms are absent in the stratosphere, although winds of high speed are observed.

Atmospheric pressure

Although we are not constantly aware of it, air is a tangible, material substance. At sea level, the atmosphere exerts a pressure of about 15 lb per square inch (about 1 kg per square centimeter) on every solid or liquid surface exposed to it. Because this pressure is exactly counterbalanced by the pressure of air within liquids, hollow objects, or porous substances, its ever-present weight creates no special concern. The pressure on one square inch of surface can be thought of as the actual weight of a column of air one inch in cross section extending upward to the outer limits of the atmosphere. Air is readily compressible. That which lies lowest is most greatly compressed and is, therefore, densest. In an upward direction, both density and pressure of the air fall off rapidly.

The meteorologist uses another method of stating the pressure of the atmosphere, based on a classic experiment of physics first performed by Torricelli in the year 1643. A glass tube about 3 ft (1 m) long, sealed at one end, is completely filled with mercury. The open end is temporarily held closed. Then the tube is inverted and the end is immersed into a dish of mercury. When the opening is uncovered, the mercury in the tube falls a few inches, but then remains fixed at a level about 30 in (76 cm) above the surface of the mercury in the dish (Figure 3.5). Atmospheric pressure now balances the weight of the mercury column.

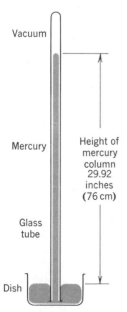

Figure 3.5   Principle of the mercurial barometer.

Figure 3.6   Top and base of standard mercurial barometer. (U.S. Weather Bureau.)

Should the air pressure increase or decrease, the mercury level will rise or fall correspondingly. Here, then, is an instrument for measuring air pressure and its variations.

Any instrument that measures atmospheric pressure is a *barometer*. The type devised by Toricelli is known as the *mercurial barometer*. With various refinements over the original simple device it has become the standard instrument (Figure 3.6). Pressure may be read in inches or centimeters of mercury, the true measure of the height of the mercury column. Standard sea-level pressure is 29.92 in. on this scale. In metric units this is 76 cm (760 mm).

Another unit has been introduced by meteorologists. This is the *millibar* (mb). One inch of mercury is equivalent to about 33.9 mb. Standard sea-level pressure is 1013.2 mb, and each $\frac{1}{10}$ inch of mercury is equal to about 3 mb (0.1 in. $=$ 3.39 mb). In this book, both systems of stating air pressure will be used.

Another type of barometer is the *aneroid barometer* (Figure 3.7). It consists of a hollow metal chamber partly emptied of air and sealed. The walls of the chamber are flexible, so that the chamber expands and contracts as the outside air pressure varies. These movements operate a hand

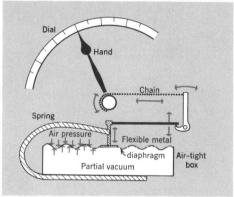

Figure 3.7 Aneroid barometer (above) and schematic diagram of workings (below). (Photograph by U.S. Weather Bureau. Diagram after Wold, *College Physics*.)

which is read against a calibrated circular dial. The aneroid is compact and easily carried in a plane or on the person.

Vertical distribution of pressure

Figure 3.8 shows how pressure falls with increasing altitude. For every 900 ft (275 m) of rise in elevation, the mercury column falls $\frac{1}{30}$ of its height. As the graph shows (by a steepening of the curve), the rate of drop of the mercury becomes less and less with increasing altitude until, beyond a height of 30 mi (50 km), decrease is extremely slight.

The effects of decreased air pressure upon human physiology, upon the boiling point of water, and upon the gain and loss of atmospheric heat are treated in Chapter 12.

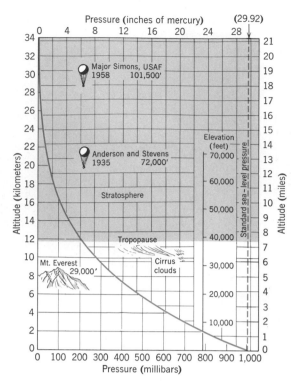

Figure 3.8 Decrease of air pressure with elevation. (Data from Humphreys, *Physics of the Air*.)

Phenomena of the outer atmosphere

Man continues to extend his activities into more distant layers of the atmosphere. Of particular interest to students of geography are advances in use of communications satellites and remote sensing instruments which scan the earth's surface. Weather satellites continually orbit the earth, providing photographs that materially assist in weather forecasting and in the early detection of tropical storms. Certain physical phenomena of the outer atmospheric regions are thus of importance in the broad framework of physical geography.

Of particular interest in the development of radio communication on a global scale is a layer known as the *ionosphere*, located in the altitude range of 50 to 250 mi (80 to 400 km). Notice that this position coincides with the bottom of the heterosphere in which the molecular nitrogen layer and the atomic oxygen layer are found (see Figure 3.2). Furthermore, the ionosphere is essentially identical in position with the lower thermosphere. The ionosphere consists of a number of layers in which the process of *ionization* takes place. Here highly energetic gamma rays and X-rays from the solar radiation spectrum are absorbed by molecules and atoms of nitrogen and oxygen. In the absorption process, each molecule or atom gives up an electron, becoming a positively charged *ion*. The electrons thus released form an electric current that flows freely on a global scale within the ionosphere. Of particular interest in the geography of radio communication is the ability of the layers of ions to reflect radio waves and thus to turn them back toward the earth. Most of the important reflection of long-wave radio waves takes place in the lower part of the ionosphere, which bears the name of *Kennelly-Heaviside layer*. Without such reflection, long-distance radio communication would not be possible. Because the process of ionization requires direct solar radiation, the ionospheric layers, of which there are five, are developed on the sunlight side of the earth (Figure 3.9). On the dark side, under nighttime conditions, the layers tend to weaken and disappear.

Yet another phenomenon, one of vital concern to Man and all other life forms on earth, is the presence of an *ozone layer* largely occurring in the region from 12 to 21 mi (20 to 35 km) elevation, but also extending upward to an elevation of 30 to 35 mi (50 to 55 km) (see Figure 3.2). The ozone layer thus extends from the upper stratosphere into the mesosphere. The ozone layer is a region of concentration of the form of oxygen molecule known as *ozone*, (O_3), in which three oxygen atoms are combined instead of the usual two atoms (O_2). Ozone is produced by the action of ultraviolet rays upon ordinary oxygen atoms. The ozone layer thus serves as a shield, protecting the troposphere and earth's surface from most of the ultraviolet radiation found in the sun's radiation spectrum. If these ultraviolet rays were to reach the earth's surface in full intensity, all exposed bacteria would be destroyed and animal tissues severely burned. Thus the presence of the ozone layer is an essential element in Man's environment. It is also interesting to note that the high temperatures of the mesosphere are produced by the absorption of the ultraviolet rays in the upper part of the ozone layer. The level of greatest ozone concentration is at its highest altitude in low latitudes (28 mi; 48 km), but descends to the lowest altitude in arctic latitudes (22 mi; 35 km). There are also marked seasonal variations in altitude in middle latitudes.

The magnetosphere

The earth can be thought of as a simple bar magnet, the axis of which approximately coincides with the earth's geographic axis (Figure 3.10). Magnetism is generated within the earth's metallic core, a central spherical body about half of the earth's diameter. The earth's magnetic axis is inclined several degrees with respect to the geographic axis. Hence the north and south *magnetic poles* do not coincide with the geographic north and south poles, nor does the *magnetic equator* coincide with the earth's geographic equator.

Lines of force of the earth's magnetic field,

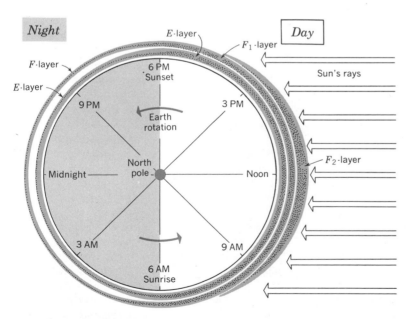

Figure 3.9 This diagram of the ionospheric layers is a cross section through the earth's equator. The observer looks down upon the cross section from a point over the north pole. (Based on a figure by B. F. Howell, Jr., 1959, *Introduction to Geophysics*, McGraw-Hill, N.Y. From A. N. Strahler, *The Earth Sciences*, Harper and Row, N.Y.)

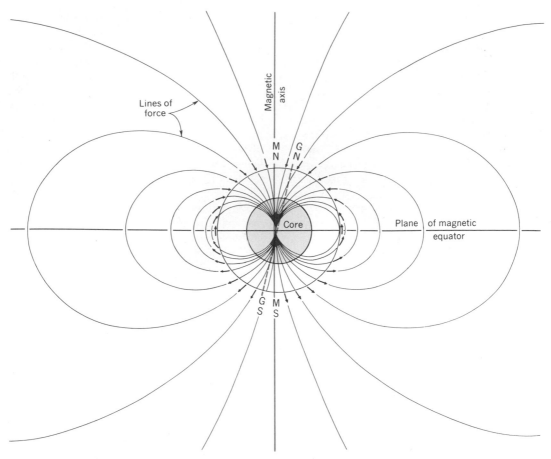

Figure 3.10 Lines of force of the earth's magnetic field are shown diagrammatically in a cross section drawn through the magnetic and geographic poles. The small arrows show the inclination of lines of force at surface points over the globe. (From A. N. Strahler, *The Earth Sciences*, Harper and Row, N.Y.)

shown in Figure 3.10, pass outward through the earth's surface and into surrounding space. A magnetic compass needle, which is nothing more than a delicately balanced bar magnet, orients itself in a position of rest parallel with the lines of force. The lines of force, which extend out into space, comprise the earth's *external magnetic field.* If we assume, for purposes of comparison, that the earth's atmosphere extends outward to a distance equal to twice its own radius, or 8000 mi (13,000 km) it becomes evident that the magnetic field extends far beyond the outermost limits of the atmosphere. The effective limit of the external magnetic field lies perhaps 40,000 to 80,000 mi (64,000 to 130,000 km) from the earth. All of the region within this limit is described as the *magnetosphere;* its outer boundary, the *magnetopause.*

The simplest geometrical model for the shape of the magnetosphere would be a doughnut-shaped ring surrounding the earth. The plane of the ring would lie in the plane of the magnetic equator, while the earth would occupy the opening in the center of the doughnut. Actually, this ideal shape does not exist because of the action of the *solar wind,* a more or less continual flow of electrons and protons emitted by the sun. Pressure of the solar wind acts to press the magnetopause close to the earth on the side nearest the sun (Figure 3.11). Here the distance to the magnetopause is on the order of 10 earthradii (about 40,000 mi, or 64,000 km). Lines of force in this region are crowded together and the magnetic field is intensified. On the opposite side of the earth, in a line pointing away from the sun, the magnetopause is drawn far out from the earth and the force lines are greatly attenuated. The extent of this "tail" is not known, but the entire shape of the magnetosphere has been described as resembling a comet. Length of the magnetic tail has been estimated to be at least 4 million miles (6,400,000 km) and is possibly vastly longer.

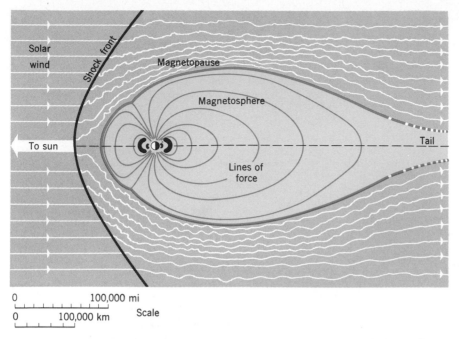

Figure 3.11 Magnetosphere and magnetopause. The Van Allen radiation belts are shown as black areas on either side of earth. (After C. O. Hines, *Science*, 1963, and B. J. O'Brien, *Science*, 1965.)

Radiation belts

In 1958, satellites *Explorer I* and *III*, carrying Geiger counters, sent to earth information concerning the existence of a region of intense radioactivity within the magnetosphere. It was soon discovered that two ring-shaped belts of radiation existed, one lying within the other (Figure 3.11). These rings were named the *Van Allen radiation belts*, after the physicist who first described them. An inner belt was found to lie about 2300 mi (2600 km) from the earth's surface; an outer and much more intense belt at about 8000 to 12,000 mi (13,000 to 19,000 km) distance.

The Van Allen radiation belts represent concentrations of charged particles—protons and electrons—trapped within lines of force of the earth's external magnetic field. These highly energetic particles are derived from the sun and are trapped upon entering the magnetopause. Intensity of trapped radiation fluctuates over a wide range. Solar flares from the sun's surface, occurring at irregular intervals, send bursts of ion clouds toward the earth. At such times, the intensity of trapped particle radiation is greatly increased. One manifestation of such events is the *aurora*, which is most intense over arctic and antarctic latitudes. On earth, severe disturbances to the magnetic field, known as *magnetic storms*, accompany the arrival of ion clouds from solar flares and seriously disrupt radio communication.

REFERENCES FOR FURTHER STUDY

Nelson, J. H., L. Hurwitz, and D. G. Knapp (1962), *Magnetism of the earth*, Coast and Geodetic Survey, Publ. 40-1, U.S. Govt. Printing Office, Washington, 79 pp.

Jacobs, J. A. (1963), *The earth's core and geomagnetism*, The Macmillan Company, New York, 137 pp. See Chapters 5 and 6.

Bates, D. R. (1964), *The planet earth*, second edition, Pergamon Press, Oxford, 370 pp.

Riehl, H. (1965), *Introduction to the atmosphere*, McGraw-Hill Book Co., New York, 365 pp.

Craig, R. A. (1968), *The edge of space, exploring the upper atmosphere*, Doubleday and Co., New York, 150 pp.

1. Describe Man's environment on earth in terms of the realms of land, sea, and air. In what way is physical geography concerned with sciences that treat the atmosphere and oceans?

2. Describe the gaseous, liquid, and solid states of matter. Do these states of matter comprise three complete separate and distinct realms of planet Earth? How are energy changes related to changes of state?

3. Compose definitions of the terms *atmosphere* and *meteorology*. What is the basis for subdividing the atmosphere into the homosphere and the heterosphere?

4. List the nonvarying component gases of the homosphere in order of decreasing proportion by volume, giving approximate percentages for each. Evaluate each gas in terms of its importance to plant and animal life on earth.

5. Describe the layers of the heterosphere in order from lowest to highest. Are the layers sharply divided from one another? Explain the development of layers. How far from earth does the atmosphere extend?

6. Describe the subdivisions of the homosphere. What conditions of temperature prevail in each zone? At what altitude is the tropopause encountered over the equator? over the poles? over middle latitudes? What changes in altitude of the tropopause take place throughout the year?

7. Why is the troposphere of special interest to geographers? What are the constituents of the troposphere, in addition to gases of the pure, dry air? What importance has each constituent?

8. What is the cause of atmospheric pressure? How is pressure measured? Describe Torricelli's experiment and explain the principle involved. Describe a mercurial barometer.

9. What value is given to standard sea-level pressure in inches and centimeters of mercury? What is the equivalent value in millibars? Give a good reason for choosing to use millibars instead of inches or centimeters of mercury.

10. Explain the principle of the aneroid barometer. What advantages has this type over the mercurial barometer? Explain how the aneroid barometer can be used as an altimeter for determining elevation in aircraft.

11. At what rate does air pressure diminish with increasing altitude? Is the rate constant? Does the pressure-altitude curve reveal the position of the tropopause? Explain how air density is related to air pressure.

12. Explain the formation of the ionosphere. At what level does the ionosphere occur? What relation exists between intensity of development of ionospheric layers and the exposure of the atmosphere to sunlight? How does the ionosphere promote long-distance radio communication? What is the Kennelly-Heaviside layer?

13. What is the ozone layer? At what altitude is it found? How is ozone produced? How does the ozone layer influence the environment of plant and animal life at the earth's surface?

14. Describe the earth's magnetic field. Define the magnetic poles and magnetic equator.

15. Describe the magnetosphere. What is the magnetopause? How is the shape of the magnetosphere influenced by the solar wind?

16. What are the Van Allen radiation belts? How were they first discovered? What is the source of radioactivity in these belts? In what way does the activity of the charged particles of the radiation belts differ from that of the gas atoms and molecules of the heterosphere? In what way are aurorae and magnetic storms related to the magnetosphere and to the Van Allen radiation belts?

CHAPTER 4
Heating and Cooling of the Earth's Surface

THE physical geographer seeks to describe and explain the conditions of Man's natural environment. So far as the atmospheric influences are concerned, it is the elements of weather and climate that directly concern Man. These subjects are descriptive and explanatory phases of the sciences of meteorology and climatology.

Weather may be defined as the state of the atmosphere at a given point in time and may be described for a single observing station or for any specified area of the earth's surface. In contrast, *climate* is the characteristic condition of the atmosphere deduced from long periods of repeated observations. Climate includes not only an analysis of average values, but also the departures from those averages and the probabilities of recurrence of particular sets of observations. It follows then that information about climate is derived from weather information; that the former represents a generalization, whereas the latter deals with the specific event.

A statement of weather is a summation of a number of descriptive parameters, often referred to as *weather elements*, falling into several groups, as follows: (1) *air temperature*; (2) *air pressure*; (3) *winds*—the direction and speed of horizontal air movement; and (4) *atmospheric moisture*, including (*a*) *humidity* (measure of water vapor content), (*b*) *clouds* and *fog*, when present, and (*c*) *precipitation*, the falling of liquid or solid water particles. Determination of the weather elements is made by instrumental and direct visual observations at a fixed station on the ground and can be extended upward by means of instrument packages raised by balloon. Ideally, a complete picture of the weather must be three dimensional, extending vertically through the troposphere as well as horizontally over a network of stations. The area between observing stations is treated by means of *interpolation*, and consists of the drawing of lines of equal value (*isopleths*) to show the estimated values of the various weather elements over an entire region and the directions and rates of change of those values from one place to another. Preparation of *weather maps*, which show the simultaneous weather conditions by means of station data and isopleths, constitutes a branch of weather science known as *synoptic meteorology*. Weather maps can be constructed not only for the ground surface but also for various altitude levels.

A statement of climate is based upon the same groups of elements that comprise weather. However, certain categories of information receive major emphasis. In many systems of climate, air temperatures and precipitation amounts are the sole basis of defining climate types and differentiating one from another. The nature of climate data is treated in detail in Chapter 8.

Concept of a global heat budget

This chapter deals primarily with the weather element of air temperature and with related temperatures of land and water surfaces. Everyone is familiar with the measurement of air temperature by thermometer. Most of us routinely receive and interpret air temperature readings as a guide to planning our dress and outdoor activities. Yet we must remember that the air temperature is simply an indicator of the quantity of heat energy within the air. This heat, which may be called *sensible heat*, exists by reason of the *kinetic energy* (energy of motion) of gas molecules. As temperature rises, the energy of molecular motion increases. Therefore attention must be focused upon heat energy and the manner in which it is received and transmitted, exchanged with liquid and solid surfaces, and stored in latent form.

We are all familiar with the cyclic nature of temperature changes. There is a daily rhythm of rise and fall of temperature as well as a seasonal rhythm. There are also systematic average changes in air temperatures from equatorial to polar latitudes as well as from oceanic to continental surfaces. Correspondingly, the lower atmosphere and the surfaces of the lands and oceans must be receiving and giving up heat energy in daily and seasonal cycles. There must also be great differences in the quantities of heat received and given up as one travels from low latitudes to high latitudes.

Despite the existence of thermal cycles and latitudinal contrasts in temperature, human history as well as the geologic record indicate an overall uniformity of the global thermal environment through time. It is apparent that the earth as a planet maintains a *heat balance,* for if this were not the case, the gradual drift toward either increasing heat or decreasing heat would ultimately render the earth's surface too hot or too cold to support life.

For all practical calculations, the sole source of heat energy of the earth's surface is the sun. Minute quantities of heat that flow toward the earth's surface from internal radioactive and volcanic sources can be disregarded as inconsequential. Solar heat is intercepted by our spherical planet and the level of heat energy tends to be raised. At the same time, our planet radiates heat into outer space, a process that tends to diminish the level of heat energy. Incoming and outgoing processes are simultaneously in action. In one place and time more heat is being gained than lost; in another place and time more heat is being lost than gained.

These considerations lead to the concept of a global *heat budget,* analogous in many respects to the money budget of a complex institution such as a large corporation or a government. A first premise of the heat-budget concept is that, over a long period of time, the average level of heat energy for the system as a whole remains absolutely constant. All departures from the average, whether in cycles of short period or long period are balanced out by equal and opposite departures. Furthermore, because equatorial regions receive much more heat than is lost directly to space and polar regions lose much more heat than is received, there must be included in the system mechanisms of heat transfer adequate to export excess heat from one region and to import heat into a region of deficiency. On our planet, motions of the atmosphere and oceans act as heat-transfer mechanisms. Thus a study of the earth's heat budget will not be complete until the patterns of global air and water circulation are described and explained in Chapter 5.

Storage of heat energy in the latent form is an important part of the earth's heat budget. It was noted, in the introduction to Chapter 3, that changes of state between gaseous, liquid, and solid states are accompanied by the taking up of heat energy or release of heat energy. Water in its three states—as water vapor in the atmosphere and as liquid and solid water in the oceans and over the land surfaces—absorbs and liberates heat as it changes from one state to another. Consequently, a study of the earth's heat budget will not be complete until the processes of change of state of water in the atmosphere are examined in Chapters 6, 7, and 9.

Upon reflection, it becomes apparent that the movement of water through atmosphere, oceans, and upon the lands comprises a system of equal importance to the flow of heat and that the activities of these two systems are closely intermeshed. The concept of a *water budget* with a *water balance* can be developed and takes its place beside the heat budget (Chapters 9 and 20). The heat budget can be thought of as dealing with energy and the water budget with matter. Together, these two great systems of energy and matter form a single, grand planetary system and permit us to relate and explain many of the environmental phenomena of our earth within a single unified framework.

A systematic approach to the earth's heat budget begins with an examination of the input, or source, of energy from solar radiation. This radiation is traced as it penetrates the earth's atmosphere and is absorbed or transformed. We then turn to the mechanism of output of heat energy by the earth as a secondary radiator.

Solar radiation

Our sun, a star of about medium mass and temperature, as compared with the overall range of stars, has a surface temperature of about $11,000°$F $(6000°$C). The highly heated, incandescent gas that comprises the sun's surface emits a form of energy known as *electromagnetic radiation.* This form of energy transfer can be thought of as a collection, or *spectrum,* of waves of a wide range of lengths traveling at the uniform velocity of 186,000 mi (300,000 km) per second. The energy travels in straight lines radially outward from the sun, and requires about $9\frac{1}{3}$ minutes to travel the 93 million miles (150 million kilometers) from sun to earth. Although the solar radiation travels through space without energy loss, the intensity of radiation within a beam of given cross section (such as one square inch) decreases inversely as the square of the distance from the sun. The earth thus intercepts only about one two-billionth of the sun's total energy output.

The solar radiation spectrum consists of (*a*) X-rays, gamma rays, and ultraviolet rays, carrying about 9 percent of the total energy, (*b*) visible light rays, 41 percent, and (*c*) invisible infrared and heat rays, 50 percent. Table 4.1 gives the wavelengths in microns, of the various parts of the spectrum. A micron is equivalent to one ten-thousandth of a centimeter. Hereafter, the term *shortwave* radiation is applied to the visible and ultraviolet portion of the spectrum (wavelengths less than 0.7 microns) as distinct from the *infrared* portion (wavelengths longer than 0.7 microns). We see that the total energy of the radiation spectrum is about

TABLE 4.1

	Wavelength	Total Energy (%)
(Shortest)		
X-rays and gamma rays	1/2000 to 1/100 micron	9
Ultraviolet rays	0.2 to 0.4 micron	
Visible light rays	0.4 to 0.7 micron	41
(Longest)		
Infrared rays	0.7 to 3000 microns	50

equally divided between shortwave and infrared portions.

The source of solar energy is in the sun's interior where, under enormous confining pressure and high temperature, hydrogen is converted to helium. In this fusion process, a vast quantity of heat is generated and finds its way by convection and conduction to the sun's surface. Because the rate of production of energy is constant, the output of solar radiation is also unvarying, to the best of scientific knowledge. Hence, at the average distance from the sun, the amount of solar energy received upon a unit area of surface held at right angles to the sun's rays is also unvarying. Known as the *solar constant*, this radiation rate has a value of 2 gram calories per square centimeter per minute. It is assumed, of course, that the radiation is measured beyond the limits of the earth's atmosphere so that none has been lost. One gram calorie per square centimeter constitutes a unit measure of heat energy termed the *langley*. Therefore, we can say that the solar constant is equal to 2 langleys per minute. In English heat units, the solar constant is equivalent to 430 Btu per square foot per hour. Orbiting space satellites equipped with suitable instruments for measuring electromagnetic radiation intensity have provided precise data on the solar constant.

Insolation over the globe

Because the earth is a sphere (disregarding oblateness), only one point on earth—that upon which the sun's noon rays are perpendicular—presents a surface at right angles to the sun's rays. In all directions away from this point, the earth's curved surface becomes turned at a decreasing angle with respect to the rays until the circle of illumination is reached. Along that great circle the rays are parallel with the surface. (See Chapter 2 for a discussion of the relation between the earth and the sun's rays throughout the year.)

Let us now assume that the earth is a perfectly uniform sphere with no atmosphere. Only at the subsolar point will solar energy be intercepted at the maximum rate of 2 langleys per minute. (Hereafter the term *insolation* will be used to mean the interception of solar energy by an exposed surface.) At any particular place on the earth, the quantity of insolation received in one day will then depend upon two factors: (1) the angle at which the sun's rays strike the earth, and (2) the length of time of exposure to the rays. These factors are varied by latitude and by the seasonal changes in the path of the sun in the sky.

Figure 4.1 shows that intensity of insolation is greatest where the sun's rays strike vertically, as they do at noon at the latitude equal to the sun's declination ranging between the tropics of Cancer and Capricorn. With diminishing angle, the same amount of solar energy spreads over a greater area of ground surface. Hence, on the average, the polar regions receive the least heat per unit area. This fact helps to explain the general distribution of average air temperatures over the globe, from a maximum at low latitudes to a minimum near either pole.

We have seen in Chapter 2 that, because of the inclination of the earth's axis, the angle of the sun's noon rays shifts through a total range 47° from one solstice to the next. This cycle does not make the yearly total of insolation for the entire globe different from an ideal situation in which the earth's axis would not be inclined, but it does cause a great difference in both latitudinal and seasonal distribution of the insolation.

Consider first that, if the earth's axis were perpendicular to the orbital plane, the poles would not receive any insolation, regardless of time of year, whereas the equator would receive an unvarying maximum. In other words, equinox conditions (shown in Figures 2.9 and 2.12) would apply every

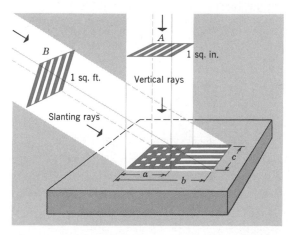

Figure 4.1 The angle of the sun's rays determines the intensity of insolation upon the ground.

day of the year. The earth's inclination, by exposing the poles alternately to the sun, redistributes the yearly total insolation toward higher latitudes, but deducts somewhat from the equatorial zone.

Second, the earth's inclination produces seasonal differences in insolation at any given latitude, and these differences increase toward the poles, where the ultimate in opposites (six months of day; six of night) is reached. Along with the variation in angle of the sun's rays operates another factor, the duration of daylight. At the season when the sun's path is highest in the sky, the length of time it is above the horizon is correspondingly greater. The two factors thus work hand in hand to intensify the contrast between amounts of insolation at opposite solstices.

A three-dimensional diagram (Figure 4.2) shows how insolation varies with latitude and with season of year. Figure 4.3 shows graphs of insolation at various selected latitudes from equator to north pole. These diagrams show insolation at the outer limits of the atmosphere and thus would apply at the ground surface only for an earth imagined to have no atmosphere to absorb or reflect radiation. Notice that the equator receives two maximum periods (corresponding with the equinoxes, when the sun is overhead at the equator) and two minimum periods (corresponding to the solstices, when the sun's declination is farthest north and south of the equator). At the Arctic Circle, $66\frac{1}{2}°$ N, insolation is reduced to nothing on the day of the winter solstice, and with increasing latitude poleward this period of no insolation becomes longer. All latitudes between the Tropics of Cancer and Capricorn have two maxima and two minima, but one maximum becomes dominant as the tropic is approached. From $23\frac{1}{2}°$ to $66\frac{1}{2}°$ there is a single continuous insolation cycle with maximum at one solstice, minimum at the other.

Figure 4.4 presents these insolation data in yet another way. Lines of equal total daily solar radia-

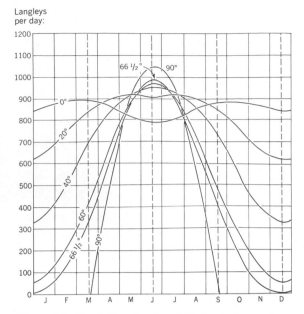

Figure 4.3   Insolation at various latitudes in the northern hemisphere. (From A. N. Strahler, *The Earth Sciences*, Harper and Row, New York.)

tion are drawn upon a graph in which latitude and the yearly calendar form the coordinates. This graph may be thought of as the base upon which the three-dimensional diagram of Figure 4.2 is constructed, while the curves in Figure 4.3 may be thought of as cross sections taken across Figure 4.4 at specified latitudes.

World latitude zones

The sun's path in the sky, which determines the flow of solar energy reaching the earth's surface—and hence governs the thermal environment of Man—provides a basis for dividing the globe into latitude zones (Figure 4.5). It is not intended that the specified zone limits be taken as absolute and binding, but rather that the system be considered as a convenient terminology for referring to world areas throughout this text.

The *equatorial zone* lies astride the equator and extends to 10° latitude north and south. Within this zone, the sun throughout the year provides intense insolation, while day and night are of roughly equal duration. Astride the Tropics of Cancer and Capricorn are the *north tropical zone* and *south tropical zone* respectively, spanning the latitude belts 10° to 25° north and south. In this zone, the sun takes a path close to the zenith at one solstice and is appreciably lower at the opposite solstice. Thus a marked seasonal cycle exists, but is combined with a potentially large total annual insolation. The geography student should note that literary usage and some geographical

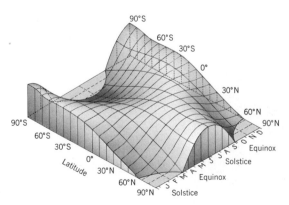

Figure 4.2   Insolation at various latitudes and seasons. (After W. M. Davis.)

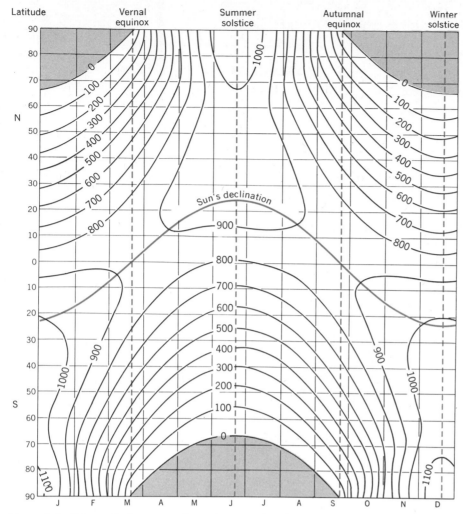

Figure 4.4 Solar radiation received on a horizontal surface outside the earth's atmosphere. (Data after S. Fritz, 1951, *Compendium of Meteorology*. From A. N. Strahler, *The Earth Sciences*, Harper and Row, New York.)

works differ from what is described here, for the word *tropics* has been widely used to denote the entire belt of 47 degrees of latitude between the tropics of Cancer and Capricorn. Such, indeed, is the definition of "tropics" found in most dictionaries, but that usage, if classically correct, ill fits the needs of physical geography, as will become apparent in the study of climate, soil, and natural vegetation.

Immediately poleward of the tropical zones are transitional regions which have become widely accepted among geographers as the *subtropical zones*. For convenience, these zones are here assigned the latitude belts 25° to 35° north and south, but it is understood that the adjective "subtropical" as applied to geographical regions may extend a few degrees farther poleward or equatorward of these parallels.

The *middle-latitude zones*, lying between 35°

and 55° north and south latitude represent regions in which the sun's path shifts through a relatively large range of noon altitudes, so that seasonal contrasts in incoming solar energy are strong. Strong seasonal differences in lengths of day and night exist as compared with the tropical zones.

Bordering the middle-latitude zones on the poleward side are the *subarctic zones*, 55° to 60° north and south latitudes, transitional between middle-latitude and arctic zones.

Astride the Arctic and Antarctic circles, 66½° north and south latitudes, lie the *arctic zones*, which may be further differentiated, if desired, into an *arctic zone* and an *antarctic zone*. The latitudinal extent of the arctic zones is here specified as 60° to 75° north and south, but these limits should not be imposed severely. The arctic zones have an extremely large yearly variation in lengths of day and night, yielding enormous contrasts in

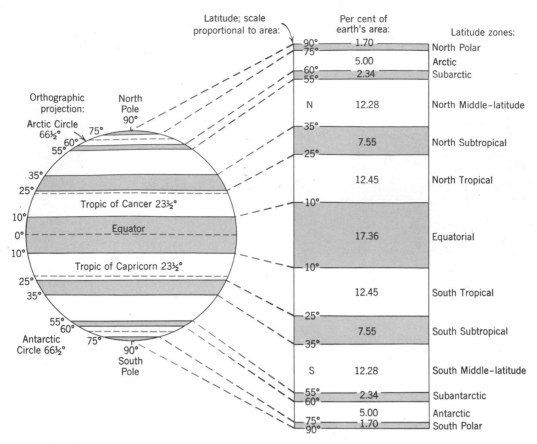

Figure 4.5 A geographical system of latitude zones.

incoming solar energy from solstice to solstice. Notice that classical usage, as found in standard dictionaries, considers the "arctic" or "arctic region" as the entire area from arctic circle to north pole, and "antarctic" in a corresponding sense for the southern hemisphere. As with "tropics," "arctic" and "antarctic" in the literary sense are not well suited to the needs of physical geography.

The *polar zones*, north and south, are circular areas between 75° latitude and the poles. Here the polar solar regime of a six-month day and six-month night is predominant and yields the ultimate in seasonal contrasts of incoming solar energy.

Insolation losses in the atmosphere

As the sun's radiation penetrates the earth's atmosphere, a series of selective depletions and diversions of energy take place. At an altitude of 95 mi (150 km), the radiation spectrum possesses almost 100 percent of its original energy, but in penetration to an altitude of 55 mi (88 km) absorption of X-rays is almost complete and some of the ultraviolet radiation has been absorbed as well. As noted in Chapter 3, the ionosphere is

developed in this region of the atmosphere by the effect of the highly energetic X-rays, gamma rays, and ultraviolet rays upon molecules and atoms of nitrogen and oxygen.

As solar radiation penetrates into deeper and denser atmospheric layers, gas molecules cause the visible light rays to be turned aside in all possible directions, a process known as *Rayleigh scattering*. Where dust particles are encountered in the troposphere, further scattering occurs. The total process may be described as *diffuse reflection*. That the clear sky is blue in color is explained by Rayleigh scattering of the shorter visible wavelengths. These predominantly blue light waves reach our eyes indirectly from all parts of the sky. The red wavelengths and infrared rays are less subject to scatter and largely continue in a straight-line path toward earth. The setting sun appears red because a part of the red rays escape deflection from the direct line of sight.

As a result of all forms of shortwave scattering, some solar energy is returned to space and forever lost, while at the same time some scattered shortwave energy also is directed earthward. The latter is referred to as *diffuse sky radiation*, or *down scatter.* An additional but minor cause of energy loss is that which occurs in the ozone layer (see Chapter 3), as oxygen molecules are broken into atoms and reformed into ozone molecules.

Another form of energy loss takes place as the sun's rays penetrate the atmosphere. Both carbon dioxide and water vapor are capable of directly absorbing infrared radiation. Absorption results in a rise of sensible temperature of the air. Thus some direct heating of the lower atmosphere takes place during incoming solar radiation. Although carbon dioxide is a constant quantity in the air (0.033 percent by volume) the water vapor content varies greatly from place to place, being as low as 0.02 percent under desert conditions to as high as 1.8 percent in humid equatorial regions. Absorption correspondingly varies from one global environment to another.

All forms of direct absorption listed above, namely X-ray, gamma ray, and ultraviolet absorption in the ionosphere and ozone layer, combined with direct longwave absorption by carbon dioxide, water vapor, and other gas molecules and dust particles, is estimated to average as little as 10 percent for conditions of clear, dry air, to as high as 30 percent when a cloud cover exists.

Figure 4.6 shows in a highly diagrammatic way the range of values of the various forms of reflection and absorption that may occur. When skies are clear, reflection and absorption combined may total about 20 percent, leaving as much as 80 percent to reach the ground.

Yet another form of energy loss must be brought into the picture. The upper surfaces of clouds are extremely good reflectors of shortwave radiation. Air travelers are well aware of how painfully brilliant the sunlit upper surface of a cloud deck can be when seen from above. Cloud reflection can account for a direct turning back into space of from 30 to 60 percent of total incoming radiation (Figure 4.6). Thus we see that, under conditions of a heavy cloud layer, the combined reflection and absorption from clouds alone can account for a loss of from 35 to 80 percent of the incoming radiation and allow from 45 to 0 percent to reach the ground.

The surfaces of the land and ocean reflect some shortwave radiation directly back into the atmosphere. This quantity, which is very small, may be combined with cloud reflection in evaluating total reflective losses.

The percentage of radiant energy reflected back by a surface is termed the *albedo.* This is an important property of the earth's surface because it determines the relative rate of heating of the surface when exposed to insolation. Albedo of a water surface is very low (2 percent) for nearly vertical rays, but high for low-angle rays. It is also extremely high for snow or ice (45 to 85 percent). For fields, forests, and bare ground the albedos

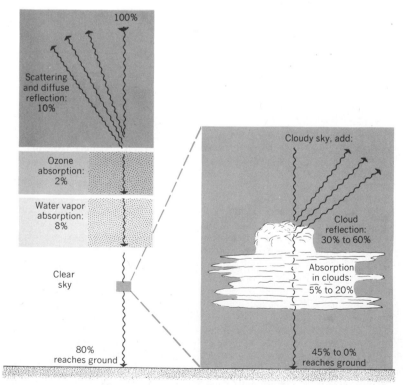

Figure 4.6 Losses of incoming solar energy by scattering, reflection and absorption. (From A. N. Strahler, *The Earth Sciences*, Harper and Row, New York.)

TABLE 4.2[a]

	Percent	Langleys per Minute
Total incoming energy lost or absorbed in atmosphere	100	0.485
LOST OR ABSORBED IN ATMOSPHERE		
Absorption by atmosphere, including clouds	19	0.092
Diffuse reflection loss to space	9	0.044
Cloud and ground reflection loss	25	0.121
Total losses	53	0.257
RECEIVED AT GROUND		
From direct penetration	41	0.199
From diffuse sky radiation	6	0.029
Total received at ground	47	0.228
RETURNED TO ATMOSPHERE		
By long-wave radiation (net value)	14	0.069
By latent heat transfer (evaporation and condensation)	23	0.111
By conduction (net value)	10	0.048
Total returned by ground	47	0.228

[a] Data from H. G. Houghton, 1954, "On the Annual Heat Balance of the Northern Hemisphere," *Journal of Meteorology*, Vol. 11, pp. 1–9.

are of intermediate value, ranging from as low as 3 percent to as high as 25 percent.

Attempts have been made to calculate average annual values for percentage of losses of incoming solar radiation in the northern hemisphere. One such estimate is given in the upper part of Table 4.2.

Ground radiation and atmospheric heating

Any substance that possesses heat sends out energy of electromagnetic radiation from its surface. The amount of radiant energy thus sent out is directly proportional to the fourth power of the absolute temperature of the substance. Also, the lower the temperature of the radiating material, the longer are the wavelengths of the rays emitted.

The ground or ocean surface, possessing heat derived originally from absorption of the sun's rays, continually radiates this energy back into the atmosphere, a process known as *ground radiation* or *terrestrial radiation*. This infrared radiation consists of wavelengths longer than 3 or 4 microns and is referred to here as *longwave radiation*. The atmosphere also radiates energy both toward the

earth and outward into space where it is lost. Note that longwave radiation is quite different from reflection, in which the rays are turned back directly without being absorbed. Longwave radiation from both ground and atmosphere continues during the night, when no solar radiation is being received.

Energy radiated from the ground is easily absorbed by the atmosphere because it consists largely of very long wavelengths (4 to 30 microns), in contrast to the visible light rays (0.4 to 0.7 microns) and shorter infrared rays (0.7 to 3.0 microns) which make up almost all of the entering solar radiation. Absorption of longwave radiation by water vapor and carbon dioxide takes place largely in wavelengths from 4 to 8 microns and 12 to 20 microns. However, radiation in the range of wavelengths between 8 and 12 microns passes freely through the earth's atmosphere and into outer space. From one-quarter to one-third of the longwave radiation that is directed outward leaves the atmosphere in this manner. Thus the atmosphere receives heat by an indirect process in which the radiant energy in shortwave form is permitted to pass through, but that in longwave form is not all permitted to escape. For this reason, the lower atmosphere with its water vapor and carbon dioxide acts as a blanket which returns heat to the earth and helps to keep surface temperatures from dropping excessively during the night or in winter at middle and high latitudes. Somewhat the same principle is employed in greenhouses and in homes using the solar-heating method (Figure 4.7). Here the glass permits entry of shortwave energy. Accumulated heat cannot escape by mixing with cooler air outside. The expression *greenhouse effect* has been used by meteorologists to describe the atmospheric heating principle.

Returning to consideration of the earth's total heat budget, it is a requirement of a heat balance that just as much energy, on the average, over long periods of time is sent out from the entire planet Earth into space as is received from the sun. It can also be inferred that the earth's total surface (including both land and water surfaces) on the yearly average must return to the atmosphere just as much energy as it receives, otherwise the surface temperature would rise to intense heat or sink to extreme cold.

The earth's surface gives back heat energy to the atmosphere not only by longwave ground radiation, described above, but also by two other heat-transfer mechanisms. The first of these is by *latent heat* associated with evaporation and subsequent condensation of water. As water evaporates from water surfaces and moist soil, heat energy enters into a latent (stored) form in water vapor. This water vapor diffuses and mixes with the lower atmosphere, carrying the latent heat with it. Condensation in clouds and precipitation in the form of rain and snow releases the latent heat into sensible heat form, raising the temperature of the atmosphere. (This process is treated in further detail in Chapters 6 and 9.)

The second additional mechanism for transfer of heat energy from ground to atmosphere is by direct conduction. Heat is transferred directly from land or sea surfaces to the air in contact with it. Turbulent air motions accompanying winds mix the heated air with higher layers. Of course, when the ground is colder than the air above it, conduction acts in reverse and the ground receives heat from the air.

The estimate in Table 4.2 gives 47 percent as the proportion of incoming solar energy received at the earth's surface as an annual average for the northern hemisphere. Table 4.2 also gives corresponding estimates of the proportions whereby this 47 percent is disposed of. We see that the net loss upward from longwave radiation is 14 percent. By net loss is meant here the difference between upward and downward longwave radiation. From latent heat transfer the loss upward is 23 percent; the net loss from direct conduction of sensible heat is 10 percent.

Although the budget has been stated in percentage units of the original 100 percent of incoming energy, it is easily possible to use absolute units of energy flow, namely, langleys per minute. Our 100 units of incoming solar energy are equivalent

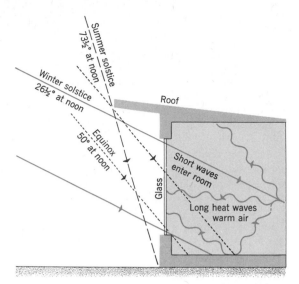

Figure 4.7   Principle of solar heating of a house.

to an average annual value of 0.485 langleys per minute for the northern hemisphere. The figure of 47 percent is equivalent to 0.228 langleys per minute. The remaining equivalents are listed in Table 4.2.

Effect of latitude

Thus far, in examining the total heat budget of the earth, including its atmosphere, consideration has not been given to the effect of latitude. Obviously, an annual average for one entire hemi-

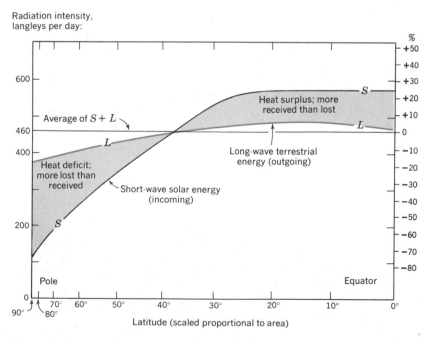

Figure 4.8   The earth's heat budget. (Data after H. G. Houghton, 1954. From A. N. Strahler, *The Earth Sciences*, Harper and Row, New York.)

sphere does not reveal the fact that low latitudes have a large annual radiation surplus, while high latitudes have a large annual deficit. Figure 4.8 is an attempt to evaluate the imbalances of incoming and outgoing energy according to latitude for the earth's surface and the entire atmosphere. On this graph, latitude is scaled on the horizontal axis in proportion to the earth's surface area between any two given parallels of latitude. The vertical axis is on a percentage basis. A plus value represents a radiation surplus relative to the average value, which is the zero line. A negative value represents a deficit. One curve is for incoming shortwave energy; another is for outgoing longwave radiation. The areas lying between these two lines represent the amounts of surplus and deficit; the two areas are equal. Equatorward of about 38° latitude, an annual surplus of radiation exists; poleward of that parallel is an equal deficit. The radiation deficit increases to a maximum at the pole.

It is obvious that moderate temperatures cannot be maintained over the entire globe unless heat is transferred from low to high latitudes. Such heat transfer, in fact, does occur through the action of wind and ocean current systems. These vast movements are discussed in Chapter 5.

Seasonal imbalances in the heat budget

Just as the annual seasonal cycle of incoming radiation shows increasing summer-to-winter contrasts as it is followed from equatorial to polar regions (Figure 4.3), so the radiation budget incurrs increasing summer surpluses and winter deficits from low to high latitudes. Figure 4.9 is a graph in which the difference between incoming and outgoing radiation is plotted against latitude. One curve represents July, the other January. Close to the equator, a heat surplus is shown for both seasons. Poleward of about 20° latitude, the winter season yields a negative value, or deficit, which increases to either pole. Summer surpluses, behaving somewhat differently with latitude, increase to

maxima at 30 to 35° latitude, then fall off with high latitude. The curves actually show negative values in summer, poleward of about 65° latitude. When we look for that latitude at which the summer surplus just equals the winter deficit, it will be found to lie between 35° and 40° north and south. Referring back to Figure 4.8, the latitude at which incoming and outgoing curves cross is 38° N, which agrees well with Figure 4.9. We must keep in mind that the data of Figures 4.8 and 4.9 are generalizations and estimates based upon highly incomplete global data.

Land and water differences

As a final consideration in analyzing the earth's heat budget, it is necessary to introduce the fact that land and water surfaces have quite different properties in absorption and radiation of heat. The general law may be stated as follows. Land surfaces are rapidly and intensely heated under the sun's rays, whereas water surfaces are only slowly and moderately heated. On the other hand, land surfaces cool off more rapidly and reach much lower temperatures than water surfaces when solar radiation is cut off. Temperature contrasts are therefore great over land areas, but only moderate over water areas. It is further true that the larger the mass of land, the greater are seasonal temperature contrasts. Because the heating of ground and water surfaces controls heating of the atmosphere above, the same observations apply to air temperature as to surface temperature.

An explanation of the law of land and water contrasts may be found in the application of certain simple principles of physics (Figure 4.10). Water is transparent and permits heat rays to penetrate many feet, thus distributing the heat through a thickness of several feet of water. The ground surface, being opaque, absorbs heat only at the surface, which thus attains a higher temperature than the water surface. Ocean waters are mixed by vertical rising and sinking motions in the surface layer, allowing heat to be distributed and stored through a great mass of water, but no such movement can occur in the ground. Water surfaces permit continual evaporation, which is a cooling process and serves to alleviate the surface heat. Ground surfaces, which are commonly moist and covered by vegetation, also permit cooling by evaporation, but to a lesser degree than ocean surfaces. As a further cause of contrast, water must absorb almost five times as much heat energy in order to rise in temperature the same amount as dry soil or rock. If heat is being applied equally to both substances, the ground will attain a high temperature long before the water will; *specific heat* of the water is said to be great, that of rock or soil to be small.

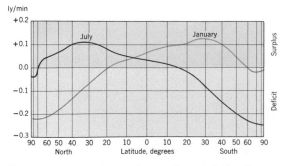

Figure 4.9   Net radiation from pole to pole for January and July. (Data by G. C. Simpson, 1929. After B. Haurwitz and J. M. Austin, 1944, *Climatology*, McGraw-Hill Book Company, N.Y.)

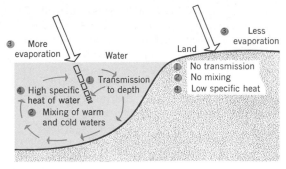

Figure 4.10   Contrasts in heating of land and water.

Measurement of air temperature

The remainder of this chapter deals with the surface thermal environment, specifically, with sensible heat expressed in air temperatures. At every recording weather station, the temperature of the air is read at regular intervals from thermometers mounted inside a boxlike shelter built several feet off the ground (Figure 4.11). The instruments are protected from direct sunlight, but air is allowed to circulate freely through the shelter. Standard equipment consists of a pair of *maximum-minimum* thermometers, one of which shows the maximum temperature, the other the minimum, that have occurred in the period since last reset. In addition, an automatic recording thermometer, called a *thermograph*, may be used to draw a continuous temperature record on a piece of graph paper (Figure 4.12).

Throughout this book, the *Fahrenheit* temperature scale is used—the common everyday scale among the general public in the United States. Freezing temperature is 32°; the boiling point is 212° on this scale (Figure 4.13). The U.S. Weather Bureau uses this scale, as do many descriptive elementary textbooks of weather and climate published in this country. One advantage of this scale is that the general reader can relate Fahrenheit temperatures to degrees of bodily comfort or discomfort. The *Centigrade*, or *Celsius*, temperature scale, in which 0° is freezing point and 100° the boiling point, is favored in England and on

Figure 4.11   Standard U.S. Weather Bureau shelter. Rain gauge at left. (U.S. Weather Bureau.)

the European continent and is normally used in physics, chemistry, and advanced meteorology. Fahrenheit and Centigrade (Celsius) measurements are distinguished by the letter "F" or "C" following the figure.

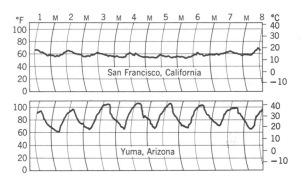

Figure 4.12   Thermograph trace sheets show temperatures for one week. (After Kincer, U.S. Dept. of Agriculture.)

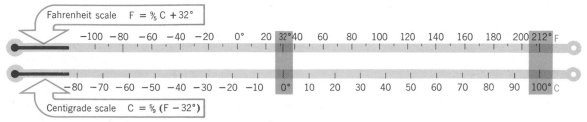

Figure 4.13   Fahrenheit and Centigrade (Celsius) temperature scales.

Annual cycle of air temperature

In order to build statistical information about temperatures for longer periods of time than a single day, a unit known as the *mean daily temperature* is used. The U.S. Weather Bureau follows a very simple method of obtaining the mean daily temperature, using readings made once a day from the maximum-minimum thermometers. The maximum and minimum temperatures for one day are added together and divided by 2. If the mean daily temperatures are collected for many years and averaged for each calendar day or month, then plotted on a graph, a smooth curve of annual temperature is obtained. Figure 4.14 shows such curves for two places at about 40° latitude. Concordia, Kansas, has a midcontinent location; San Luis Obispo, California, at about the same latitude, is near a great ocean body.

Although insolation reaches a maximum at summer solstice, the hottest part of the year on land is about a month later, owing to the lag of peak output of heat energy from the ground beyond the time of peak insolation. One contributing cause of this lag lies in the absorption of heat early in the summer by the soil layer. Air temperature maximum, closely coinciding with maximum ground output of heat, is correspondingly delayed. (Bear in mind that this cycle applies to middle and high latitudes, but not to the region between Tropics of Cancer and Capricorn.) Simi-

larly, the coldest time of year for land areas is January, about a month after winter solstice, because the ground surface continues to lose heat even after insolation begins to rise.

Over the oceans there are two differences. (1) Maximum and minimum temperatures are reached about a month later than on land—in August and February, respectively—because water bodies heat or cool much more slowly than land areas. (2) The yearly range is less than over land, following the law of temperature differences between land and water surfaces. Coastal regions are usually influenced by the oceans to the extent that maximum and minimum temperatures occur later than in the interior. This principle shows nicely for the summer maximum at San Luis Obispo (Figure 4.14).

Several other monthly temperature figures are required for a complete statement of the annual cycle of temperature characteristics of a place. These are named and illustrated in Figure 4.15.

Air temperature maps

The distribution of air temperatures over large areas can best be shown by a map composed of *isotherms*. Similar to topographic contours, whose meaning and construction are discussed at length in Appendix II, isotherms are drawn to connect all points having the same temperature. Figure 4.16 shows a weather map upon which the observed air temperatures have been recorded in the

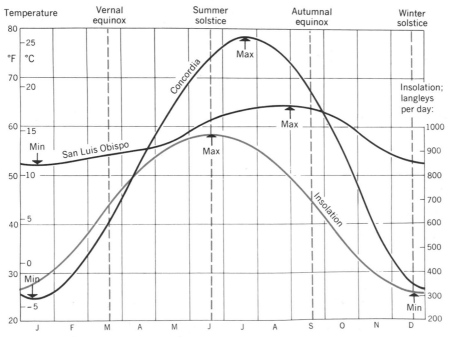

Figure 4.14 The annual cycle of temperature in middle latitudes. (Data of U.S. Dept. of Agriculture. From A. N. Strahler, *The Earth Sciences*, Harper and Row, New York.)

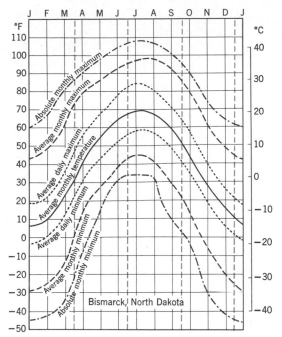

Figure 4.15 Seven annual temperature curves. (After J. B. Kincer, U.S. Dept. of Agriculture.)

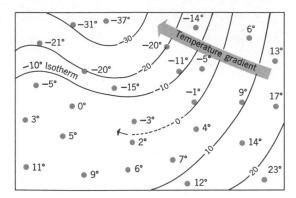

Figure 4.16 Construction of isotherms.

correct places. These may represent single readings taken at the same time everywhere, or they may represent the averages of many years of records for a particular day or month of a year, depending upon the purposes of the map. Usually, isotherms representing 5° or 10° differences are chosen, but they can be drawn for any selected temperatures. The isotherms pass through the observing stations only when the station readings coincide with the value selected for an isotherm. Otherwise it is necessary to draw the isotherms by estimating their proper position between stations. The value of isothermal maps is that they make clearly visible the important characteristics of the prevailing temperatures. Centers of high or low temperature are clearly outlined. Zones of gradation are readily seen. From a mere mass of figures on a map, these features are not easily grasped.

World temperature distribution

Isothermal maps of the world for January and July are shown in Figures 4.17 and 4.18. Annual temperature range is shown in Figure 4.19. Isotherms have a general east-west trend around the earth because of the general decrease of insolation from equator to polar regions. The east-west trend and parallelism of isotherms are best developed in the southern hemisphere, south of the 25th parallel, where land areas are small. In the northern hemisphere, isotherms show wide northward and southward deflections where they pass from

a land area to an ocean area, particularly in January, when land and ocean surface temperatures are brought most strongly into contrast.

The land-water effect is represented diagrammatically in Figure 4.20, for the northern hemisphere. The January isotherm is deflected southward over the land, northward over the water. Temperatures along a single parallel are low on land but high on water. In July the reverse is true, with the isotherm pushed far north over the continent.

Throughout the year, isotherms shift through several degrees of latitude, following the declination of the sun but lagging behind a month or so in time. Over large water areas, such as the south Pacific, the annual shift amounts to only about five degrees of latitude, whereas over landmasses, such as Africa, this shift is as much as 20° latitude. (Examine the change in position of the 70°F (21°C) isotherms over Africa.) This difference in amount of latitude shift is also explained by the rapidity and intensity with which lands are heated and cooled as compared with ocean areas.

Certain definite centers of high and low temperature occur and are shown by isotherms which are completely closed to form oval or irregular-shaped enclosures. Notice that all of them are over landmasses. In July, high-temperature centers occur over the southwestern United States, North Africa, and southwestern Asia. In January, a continental center of low temperature occurs over Siberia and is strongly developed with the average January temperatures lower than −50°F (−46°C). A corresponding region of low temperature, marked off by the closed isotherm of −30°F (−34°C), occurs in northernmost North America. It is not so well developed as that of Asia because of the presence of considerable areas of Arctic Ocean among the islands of the northern fringe of the landmass and the smaller size of the North American landmass.

Permanent centers of low temperature exist over

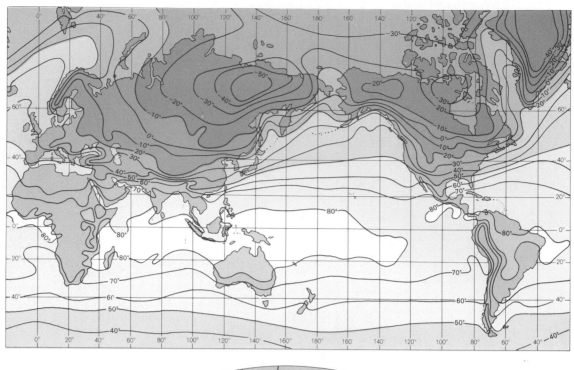

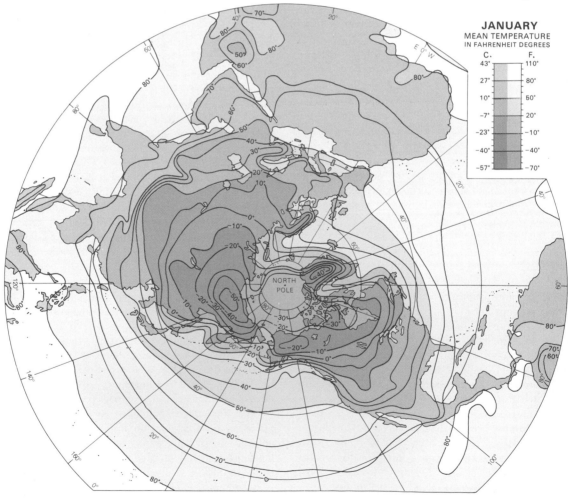

JANUARY
MEAN TEMPERATURE
IN FAHRENHEIT DEGREES

C.		F.
43°		110°
27°		80°
10°		50°
−7°		20°
−23°		−10°
−40°		−40°
−57°		−70°

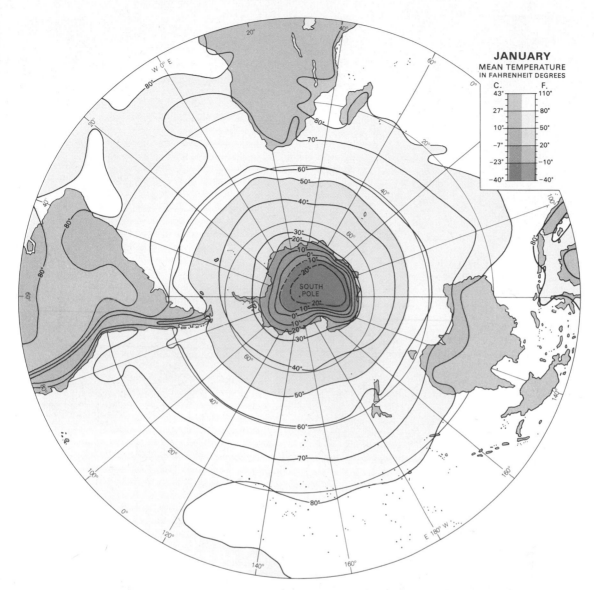

Figure 4.17 (*above and left*)   January temperatures in degrees Fahrenheit. (Compiled by John E. Oliver from data by World Climatology Branch, Meteorological Office, *Tables of Temperature,* 1958, Her Majesty's Stationery Office, London; U.S. Navy, 1955, *Marine Climatic Atlas,* Washington, D.C.; and P. C. Dalrymple, 1966, American Geophysical Union. Cartography by John Tremblay.)

Greenland and Antarctica, the two regions of massive icecaps. Temperatures over Greenland do not, however, reach the extreme low of northern Siberia in January, although the annual average temperature of the icecap is much lower. Analyzed from the polar projection in Figure 4.17, the January region of extreme cold is a distorted ellipse with its long axis extending across the Arctic Ocean from eastern Siberia to northern Canada, but with a sharp, deep reentrant over Greenland. By contrast, the Antarctic region yields concentric isotherms neatly centered on the south pole.

A comparison of winter temperatures at the two poles is instructive, since one lies in a region of deep ocean and the other in the heart of a continent at high elevation. Because of heat conduction through the floating sea ice, not more than 15 ft (5 m) thick, north-polar January mean temperature is probably about −30°F (−35°C). By contrast, the July average at the south pole is about −75°F (−60°C), partly because heat which radiates rapidly from the elevated plateau is not replaced from below. The true cold pole of the antarctic continent is centered 7° to 8° distant from the geographic south pole (Figure 4.18). Here the July mean is below −90°F (−68°C).

The annual range of monthly mean temperatures

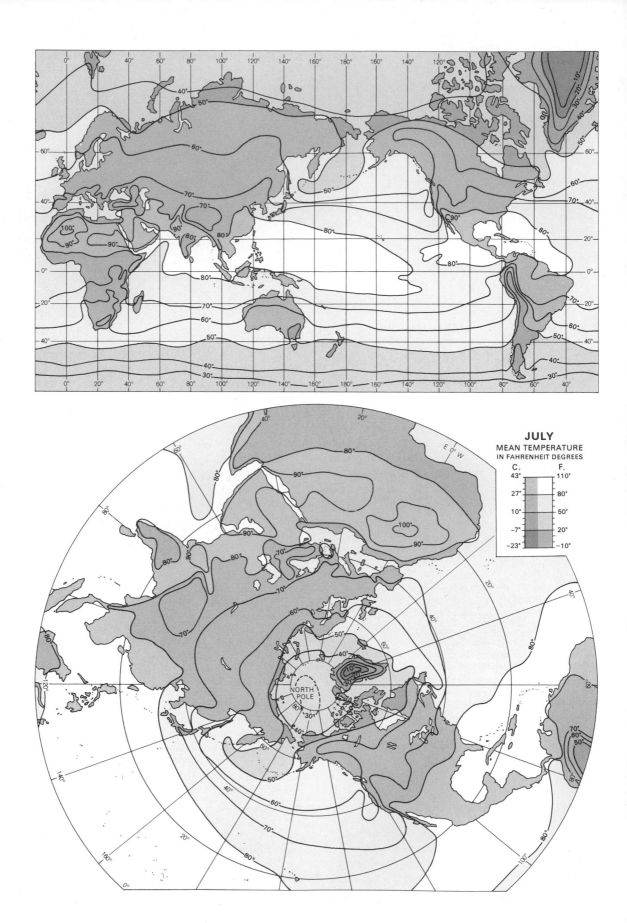

JULY
MEAN TEMPERATURE
IN FAHRENHEIT DEGREES

C.	F.
43°	110°
27°	80°
10°	50°
-7°	20°
-23°	-10°

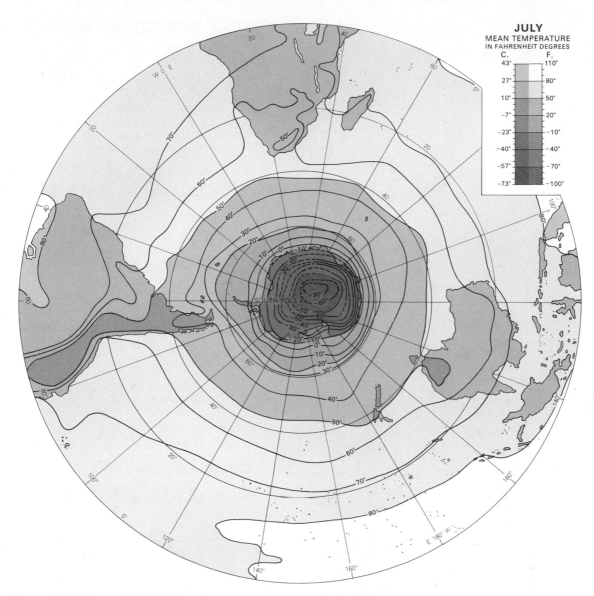

Figure 4.18 (*above and left*) July temperatures in degrees Fahrenheit. (Data sources same as Figure 4.17.)

at any desired location may be roughly computed from the January and July maps but is more conveniently analyzed on a world map of annual temperature ranges (Figure 4.19). The lines on this map are drawn through points of equal range and may be called *corange lines*. In northern Siberia, the range is about 110 F° (61 C°), greatest of any place on earth. Next are north-central Canada, just west of Hudson Bay, and Greenland with ranges of 70 F° (40 C°). The Arctic Ocean and surrounding continental fringes have a comparable range. Then follow Africa, South America, and Australia, with maximum ranges of about 30 F° (17 C°). An equatorial belt, about 35° of latitude in width over the oceans and about 10° wide across Africa and South America, has an annual range of 5 F°

(2.8 C°) or less.

Ocean currents locally exert a noticeable modification upon the isotherms. The North Atlantic current, which runs northeastward close to the British Isles and the Norwegian coast, causes a sharp northeastward bend of isotherms in winter, when temperature contrasts are generally most marked. An opposite, or equatorward, deflection of isotherms occurs along the western coasts of South America, North America (July), and Africa, where cold currents cause the air temperatures to be lower than usual for tropical latitudes.

Daily cycle of air temperature

If a thermometer is read every hour or half hour throughout 24 hours and the readings are plotted

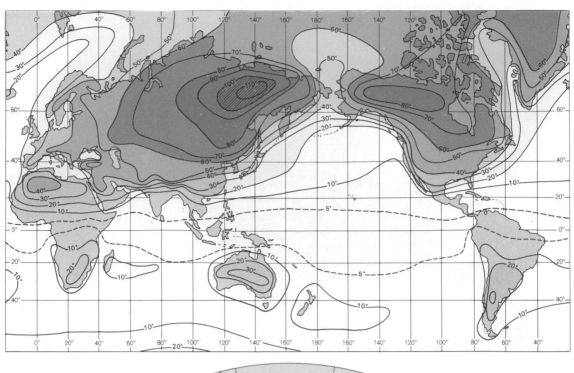

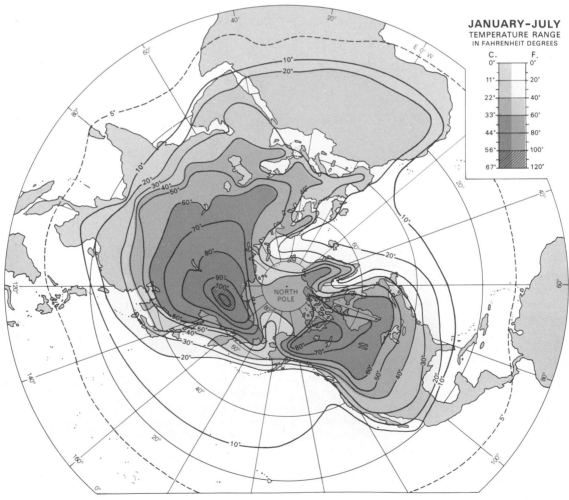

JANUARY–JULY
TEMPERATURE RANGE
IN FAHRENHEIT DEGREES

C.		F.
0°		0°
11°		20°
22°		40°
33°		60°
44°		80°
56°		100°
67°		120°

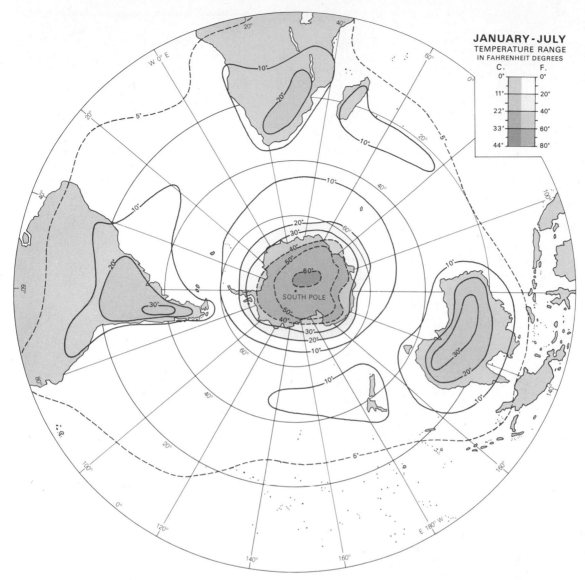

Figure 4.19 (*above and left*)   Annual range of air temperature. (Same data source as in Figures 4.17 and 4.18.)

on a graph, the curve for a clear day typically shows one low point near sunrise and one high point in midafternoon with a fairly smooth curve throughout. This rhythmic rise and fall of air temperature is termed the *daily*, or *diurnal*, *air temperature cycle*.

Figure 4.21 relates the typical diurnal air temperature curve (bottom) with cycles of both incoming and outgoing heat energy. The diagrams are schematic and apply to a typical middle-latitude location (40° to 45° latitude). We assume equinox conditions (March 21 or September 23), with the time of sunrise and sunset as 6:00 A.M., and 6:00 P.M., respectively. The uppermost graph shows incoming shortwave radiation throughout a clear day. Both direct and indirect shortwave energy are

included. Disregarding small amounts of indirect shortwave energy before sunrise and after sunset, the curve begins at 6:00 A.M., reaches a maximum at noon, and ends at 6:00 P.M.

The middle graph shows the net radiation of energy to and from the ground surface. All wavelengths are included. We see here only the difference between incoming and outgoing radiant energy. Where the curve is above the zero line, excess energy is passing upward from ground to atmosphere; where it is below the zero line, excess energy is passing from atmosphere to ground. These positive and negative values are labeled *surplus* and *deficit*, respectively. When a surplus is present, the temperature of the air layer above the ground tends to rise; when a deficit exists, the air tends

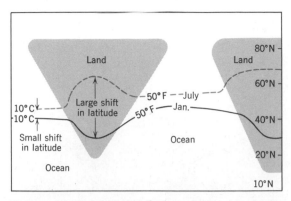

Figure 4.20 Seasonal shift of an isotherm.

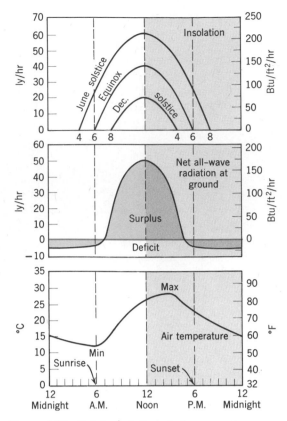

Figure 4.21 Relation of diurnal air temperature curve to insolation and net ground radiation, middle latitudes at equinox.

to be cooled. The net all-wave radiation curve tends to be symmetrical with respect to noon (the maximum point) and to be nearly flat during hours of darkness. As the graph shows, a surplus normally sets in about one hour after sunrise and thereafter increases rapidly. A deficit sets in about an hour or so before sunset.

The typical diurnal air temperature curve, shown in the lowermost graph of Figure 4.21, is not symmetrical. The minimum point occurs at about sunrise. Temperature rises steeply as the radiation surplus increases rapidly. Air temperature continues to rise past noon, for the radiation surplus, although beginning to diminish, is still large. The air temperature maximum typically occurs between 2:00 and 4:00 P.M. Thereafter, the air temperature begins to fall, even though a radiation surplus exists, as the middle curve shows. If air temperature depended only upon ground radiation, the maximum temperature would occur later in the day, perhaps at about 5:00 P.M. under the equinox conditions illustrated. However, another factor has entered the picture. During the early afternoon, mixing of the lower air in turbulent motions increases in intensity, carrying the heated air upward and replacing it with cooler air. The effect of this mixing is to cause the air temperature curve to begin to drop long before the radiation surplus has ended.

Under conditions of June solstice, incoming solar radiation is greatly increased (upper graph in Figure 4.21); the insolation commences much earlier and ends much later. The surplus part of the net ground radiation curve (not shown) is similarly broadened and raised in height. The hour of minimum temperature is set back correspondingly to perhaps 4:00 A.M. However, the hour of maximum temperature remains essentially the same. At December solstice, corresponding reductions of incoming energy and a narrowing and lowering of the part of the surplus radiation curve occur. The time of minimum temperature is advanced correspondingly.

Figure 4.22 shows the average diurnal cycle of air temperature at two places, one of interior continental location in a dry climate, the other very close to the ocean water on a windward coast. Notice that the hour of minimum temperature changes with solstice and equinox, but that the hour of maximum temperature remains fairly constant. These two graphs show the contrasting effects of water and land in controlling the daily and annual temperature ranges.

Temperature inversion and frost

Although air temperature typically falls with increasing elevation, weather conditions at night in the lower air on land are typically such that, instead of falling, the temperature first rises with increasing height above ground before beginning to drop off into the normal lapse rate (Figure 4.23). This condition is termed *temperature inversion* and signifies that warmer air overlies colder air.

Low-level temperature inversion, or *ground inversion*, commonly results at night from rapid heat loss by radiation from the ground surface and basal air layer upward into space. Rapid radiation

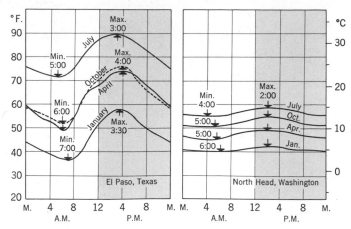

Figure 4.22 The average daily march of air temperatures for two stations. (After J. B. Kincer, U.S. Dept of Agriculture.)

loss is favored by calm air and clear skies. Over snow-covered surfaces on clear winter nights, inversion is particularly marked. If the heat loss is great during a night in spring or early fall, the air temperature close to the ground may drop below freezing, resulting in a killing frost which damages sensitive crops. There are other causes and varieties of both temperature inversions and killing frosts, but these involve movements of masses of air and are not phenomena of heat radiation.

Killing frost may be prevented in orchards and citrus groves by causing a circulation of air that mixes the warmer air above with the cold-air layer near the ground. Where the cold-air layer is thin this effect may be accomplished by the use of powerful motor-driven propellors that circulate air much as does a fan in a room.

REFERENCES FOR FURTHER STUDY

Kincer, J. B. (1928), Temperature, sunshine and wind, *Atlas of American Agriculture*, U.S. Govt. Printing Office, Washington, D.C., 40 pp.

Johnson, J. C. (1954), *Physical meteorology*, Technology Press of M.I.T. and John Wiley and Sons, New York, 393 pp.

Gates, D. M. (1962), *Energy exchange in the biosphere*, Harper and Row, New York, 151 pp.

Geiger, R. (1965), *The climate near the ground*, fourth edition, Harvard University Press, Cambridge, Mass., 611 pp.

Sellers, W. D. (1965), *Physical climatology*, University of Chicago Press, 272 pp.

Petterssen, S. (1969), *Introduction to meteorology*, edition, McGraw-Hill Book Co., New York, 333 pp.

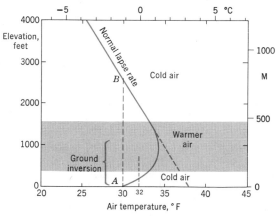

Figure 4.23 A low-level temperature inversion.

1. Define weather and climate. What is the difference between the two?

2. List the weather elements. How does synoptic meteorology treat the weather elements?

3. Explain the concept of the earth's heat budget. How can the earth maintain a heat balance?

4. From what source does the earth's atmosphere receive its heat energy? Describe the solar radiation spectrum. What is the solar constant?

5. What factors determine the amount of insolation received each day at a given place on the earth? On the average, throughout the year, how is insolation related to latitude?

6. How does the occurrence of the seasons (changing declination of the sun) affect insolation in equatorial latitudes? in middle latitudes (40° to 50°)? in near-polar regions?

7. Name the world latitude zones in order from equator to poles. Give the approximate latitude span and insolation conditions for each zone.

8. Describe the losses in solar radiation in the troposphere. Why should insolation received at the ground surface vary from day to day?

9. How is the earth's atmosphere heated? What has ground radiation to do with atmospheric heating? Is direct reflection from the earth's land or water surfaces important in atmospheric heating?

10. What is the principle of solar heating in a home? What is the greenhouse effect?

11. What effect has latitude upon the annual totals of heat surplus and heat deficit?

12. Describe the seasonal imbalances in heat budget components in middle and high latitudes.

13. Explain the basic differences of land and water surfaces as regards their properties for absorbing and transmitting insolation. How may these differences be expected to influence air temperatures over continental areas as contrasted with ocean areas?

14. Under what conditions is air temperature measured at standard U.S. Weather Bureau stations? What instrument makes a continuous automatic temperature record?

15. How do Fahrenheit and Centigrade temperature scales compare? Explain how temperature can be converted from one scale to the other. Describe the Kelvin scale.

16. What is the mean daily temperature? How is it computed by the U.S. Weather Bureau?

17. How does the annual temperature curve for a middle-latitude place with an inland location differ from that of a place located on a seacoast? Is there a difference in the times at which maximum and minimum temperatures normally occur in these two locations? Why?

18. What are isotherms? How are they drawn? In what way are isothermal maps more useful in the study of weather and climate than maps showing only the temperature figures?

19. What is the general trend of isotherms on the globe? What influence has the changing sun's declination upon the isotherms of monthly mean temperatures? Is the latitudinal shift of isotherms greater over land or water areas? Explain your answer.

20. What effect have the land masses of North America and Asia upon the isotherms for January and July? Where is the earth's greatest annual temperature range experienced? Where would you expect to find a minimum annual range of temperature?

21. Describe the normal daily temperature curve. When are maximum and minimum air temperatures normally reached? How is the air temperature cycle related to the cycle of insolation and to net all-wave ground radiation? Describe the effect of changing seasons in middle latitudes upon the time of minimum daily temperature.

22. What is temperature inversion? Explain how a ground inversion occurs. How are killing frosts produced by radiation of heat?

CHAPTER 5

Winds and the Global Circulation

UP to this point in our study of the atmosphere, little or nothing has been said of air in motion. In Chapter 3 the structure of the atmosphere was described as if motion were absent. In Chapter 4 the daily and seasonal rhythms of heating and cooling were examined, but without explanation of the air motions that such thermal imbalances produce. We therefore turn next to the subject of winds and the global circulation.

The basic principles involved in motion of the earth's atmosphere are these: Nonuniform heating and cooling of the earth's atmosphere sets up nonuniform horizontal distributions of barometric pressure, which in turn tend to set the air in motion. When air motion occurs, heat is transferred from one place to another. Heat exchange by such atmospheric mixing is of great significance in the world patterns of weather and climate, and in the earth's heat budget. Air in motion also transports water vapor from place to place, and this process is likewise of great significance in weather and climate. The role of water with its changes of state in the atmosphere is treated in Chapter 6.

Horizontal distribution of air pressure

Although the vertical decrease of barometric pressure described in Chapter 3 is of interest and importance in many phases of weather science and aviation, it is not the vital element in understanding weather conditions at ground level or world climate types and their distribution. Weather stations that are appreciably above sea level, as most are, have their barometer readings corrected in such a way as to reduce them to sea-level equivalents. Thus, surface weather maps and most climate charts show pressure conditions as if the earth's surface were all at sea level. This adjustment is necessary in order to reveal pressure differences due to factors other than elevation. The recording barograph makes a continuous record of pressure and is an important instrument in weather observing stations (Figure 5.1).

Comparison of barometric pressures (reduced to sea level) recorded simultaneously at several different observing stations, or a comparison of a sequence of pressure readings taken several hours apart at the same station will show small differences that are of great importance in analyzing weather conditions. If 1013 mb (29.92 in., 76 cm) is taken as standard sea-level pressure, readings higher than this will frequently be observed in middle latitudes, occasionally up to 1040 mb (30.7 in., 78 cm) or higher. These pressures are designated as *high*. Pressures ranging down to 982 mb 29 in., 74 cm) or below are *low*. As explained later, low-pressure centers normally are commonly associated with unsettled, cloudy, or stormy weather conditions, whereas high-pressure centers generally have fair, dry weather. This explains why mariners have always regarded a falling barometer as a sign of approaching bad weather.

During winter in middle latitudes, a series of observations taken at one locality will reveal that spells of cold dry weather are commonly accompanied by high pressures, warm or mild spells with cloudiness and precipitation, by low pressures.

Isobaric maps

Pressure conditions can be shown on a map by means of *isobars*, which are lines connecting all places having the same barometric pressure. On the daily weather map, which shows conditions

Figure 5.1   Recording barograph. (U.S. Weather Bureau.)

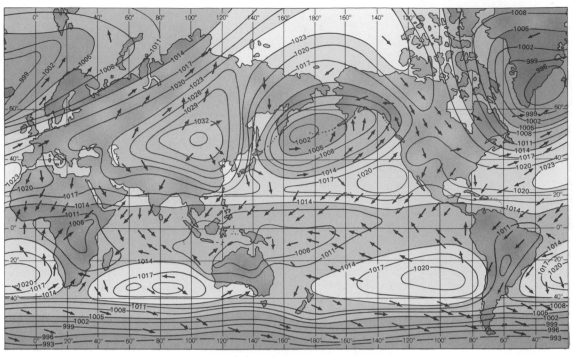

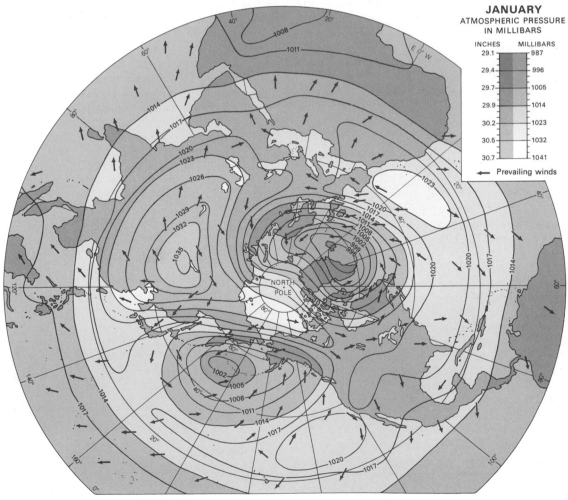

JANUARY
ATMOSPHERIC PRESSURE
IN MILLIBARS

INCHES	MILLIBARS
29.1	987
29.4	996
29.7	1005
29.9	1014
30.2	1023
30.5	1032
30.7	1041

⟵ Prevailing winds

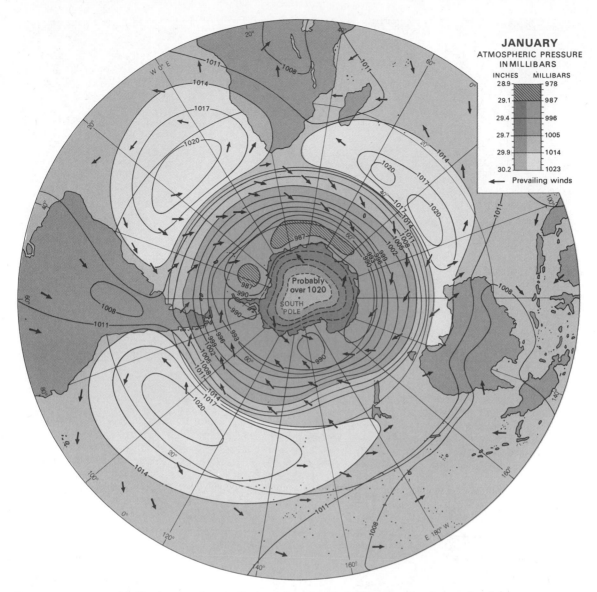

Figure 5.2 (*above and left*)   Average January barometric pressures (millibars, reduced to sea level) and surface winds. Wind arrows in polar regions largely inferred from isobars. (Compiled by John E. Oliver from data by Y. Mintz, G. Dean, R. Geiger, and J. Blüthagen. Cartography by John Tremblay.)

for a specified time only, the isobars are essential in showing the location of moving centers of high or low pressure. On climate maps the isobars show average pressures, computed from the accumulated data of many years. It is to the average world conditions that attention is now directed. Figures 5.2 and 5.3 show conditions in January and July, the months in which temperature extremes are reached.

Figure 5.4 will be useful in converting millibars to inches.

World pressure belts

In the equatorial zone is a belt of somewhat lower than normal pressure, between 1011 and 1008 mb (29.9 and 29.8 in., 76 and 75.7 cm), which is known as the *equatorial trough*. Lower pressure is made conspicuous by contrast with belts of higher pressure lying to the north and south and centered on about latitudes 30° N and S. These are the *subtropical belts of high pressure*. In the southern hemisphere this belt is clearly defined but contains centers of high pressure, termed *pressure cells*. In the northern hemisphere in summer the high-pressure belt is dominated by two oceanic cells, one over the eastern Pacific, the other over the eastern North Atlantic. Average pressures exceed 1026 mb (30.3 in., 77.0 cm) in the centers of the cells.

Poleward of the subtropical high-pressure belts

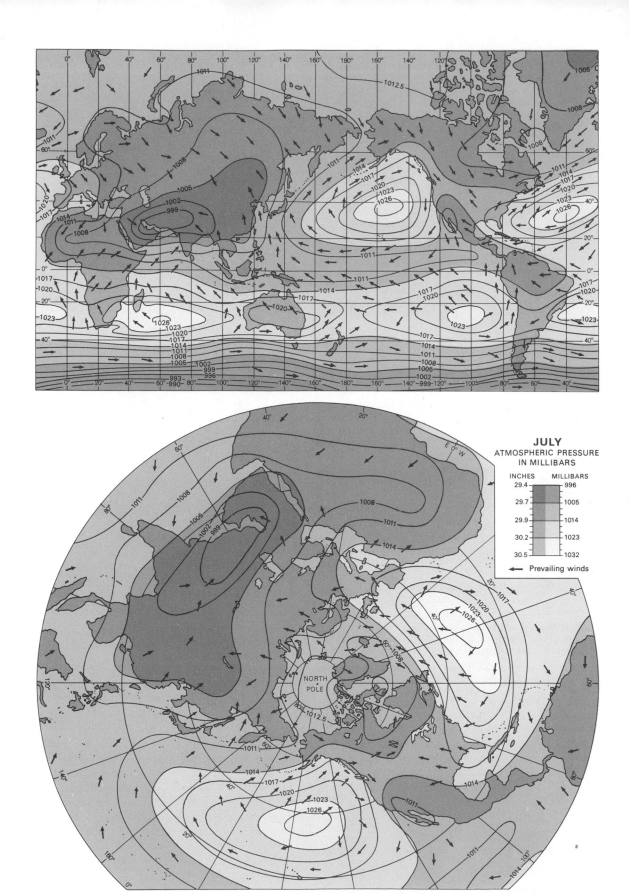

JULY
ATMOSPHERIC PRESSURE
IN MILLIBARS

INCHES	MILLIBARS
29.4	996
29.7	1005
29.9	1014
30.2	1023
30.5	1032

⟵ Prevailing winds

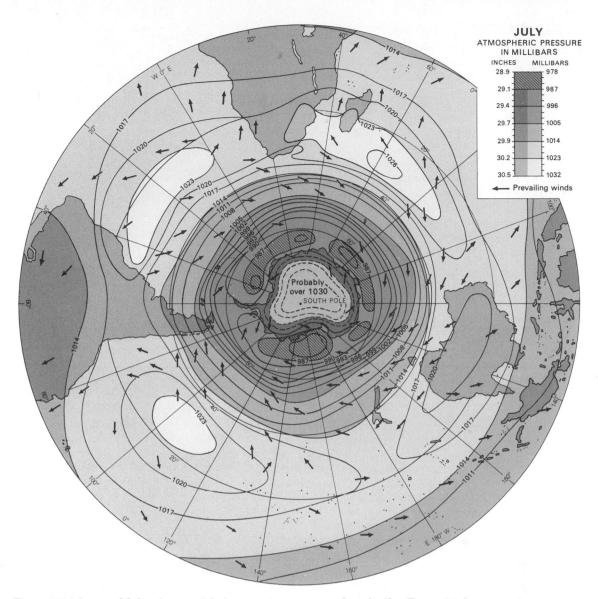

JULY
ATMOSPHERIC PRESSURE
IN MILLIBARS

INCHES		MILLIBARS
28.9		978
29.1		987
29.4		996
29.7		1005
29.9		1014
30.2		1023
30.5		1032

◄— Prevailing winds

Probably over 1030
SOUTH POLE

Figure 5.3 (*above and left*)   Average July barometric pressures and winds. (See Figure 5.2 for sources.)

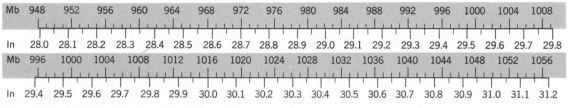

Mb	948	952	956	960	964	968	972	976	980	984	988	992	996	1000	1004	1008

In	28.0	28.1	28.2	28.3	28.4	28.5	28.6	28.7	28.8	28.9	29.0	29.1	29.2	29.3	29.4	29.5	29.6	29.7	29.8

Mb	996	1000	1004	1008	1012	1016	1020	1024	1028	1032	1036	1040	1044	1048	1052	1056

In	29.4	29.5	29.6	29.7	29.8	29.9	30.0	30.1	30.2	30.3	30.4	30.5	30.6	30.7	30.8	30.9	31.0	31.1	31.2

Figure 5.4   Scale for conversion of millibars to inches.

are broad belts of low pressure, extending roughly from the middle-latitude zone to the arctic zone but centered and intensified in the subarctic zone at about the 60th parallels of latitude. In the southern hemisphere, over the continuous expanse of southern ocean the *subantarctic low-pressure belt* is especially defined with average pressures as low as 984 mb (29.1 in., 73.9 cm). The polar zones have permanent centers of high pressure known as the *polar highs*, better illustrated by the south polar zone where the high contrasts strongly with the encircling subantarctic low.

The pressure belts shift seasonally through several degrees of latitude, just as do the isotherm belts that accompany them. More attention will be given this shift in the explanation of climates.

Northern hemisphere pressure centers

The vast landmasses of North America and Asia, separated by the North Atlantic and North Pacific oceans, exert such a powerful control over pressure conditions in the northern hemisphere that the belted arrangement typical of the southern hemisphere is absent.

Land areas develop high-pressure centers at the same time that winter temperatures fall far below those of adjacent oceans. In summer, land areas develop low-pressure centers, at which season land-surface temperatures rise sharply above temperatures over the adjoining oceans. Ocean areas show centers of pressure opposite to those on the lands, as seen in the January and July isobaric maps. In winter, pressure contrasts are greater, just as temperature contrasts are greater. Over north central Asia is developed the *Siberian high*, with pressure average exceeding 1035 mb (30.6 in., 77.7 cm). Over central North America is a clearly defined, but much less intense, ridge of high pressure, called the *Canadian high*. Over the oceans are the *Aleutian low* and the *Icelandic low*, named after the localities over which they are centered. These two low-pressure areas have much cloudy, stormy weather in winter, whereas the continental highs characteristically have a large proportion of clear, dry days.

Figure 5.5 shows diagrammatically the pressure centers as they appear grouped around the north pole. Highs and lows occupy opposite quadrants.

In summer, pressure conditions are exactly the opposite of winter conditions. Asia and North America develop lows, but the low in Asia is more intense. It is centered in southern Asia where it is fused with the equatorial low-pressure belt. Over the Atlantic and Pacific oceans are two well-developed cells of the subtropical belt of high pressure, shifted northward of their winter position and considerably expanded. These are termed the

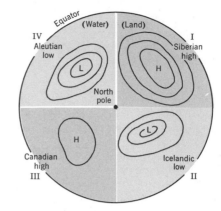

Figure 5.5 Northern hemisphere pressure centers in January.

Azores (or *Bermuda*) *high* and the *Hawaiian high* respectively.

Relation of winds to barometric pressure

Figure 5.6 is an isobaric map, showing centers of high and low pressure. From the center of the high there is progressively lower pressure, or a *barometric slope*, outward in any direction from the center. This slope is also termed the *pressure gradient*. Direction of pressure gradient, indicated by broad arrows, is always at right angles to the isobars. Whether there are pressure centers or belts, a pressure gradient always exists, running from higher to lower pressure. Closely spaced isobars indicate that the gradient is strong and that pressure changes occur rapidly within a short horizontal distance. Widely spaced isobars indicate a weak gradient.

The widespread and persistent winds of the earth are those movements of air set up in response to pressure differences. A *pressure gradient force*, acting in the direction of the pressure gradient, tends to start the flow of air from higher to lower pressure. Strong pressure gradients, indicated by

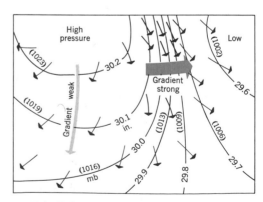

Figure 5.6 Relation of winds to isobars.

closely crowded isobars on the map, cause strong winds, whereas widely spaced isobars can be expected to yield weak winds. Calms exist in the centers of high pressure.

The Coriolis force and its effect on winds

If the earth did not rotate upon its axis, winds would follow the direction of pressure gradient. Instead, earth rotation produces another force, the *Coriolis force*, which tends to turn the flow of air. The direction of action of the Coriolis force is stated in Ferrel's law: Any object or fluid moving horizontally in the northern hemisphere tends to be deflected to the right of its path of motion, regardless of the compass direction of the path. In the southern hemisphere a similar deflection is toward the left of the path of motion. The Coriolis force is absent at the equator but increases progressively poleward.

On Figure 5.7 the various arrows show how an initial straight line of motion is modified by the deflective force. Note especially that the compass direction is not of any consequence. If we face down the direction of motion, turning will always be toward the right hand in the northern hemisphere. Because the deflective force is very weak, it is normally apparent only in freely moving fluids such as air or water. Ocean current patterns are greatly affected by it, and streams occasionally will show a tendency to undercut their right-hand banks in the northern hemisphere. Driftwood floating in rivers at high northerly latitudes con-

centrates along the right-hand edge of the stream. Rifle bullets are slightly deflected over long ranges.

Applying these principles to the relation of winds to pressure (Figure 5.8), the gradient force (acting in the direction of the pressure gradient) and the Coriolis force (acting to the right of the path of flow) quickly reach a balance, or equilibrium, when the wind has been turned to the point that it flows in a direction at right angles to the pressure gradient, that is, parallel with the isobars. The ideal wind in this state of balance with respect to the two forces is termed the *geostrophic wind* for cases in which the isobars are straight. Where isobars are curved, centrifugal force must also be taken into account, but, in general, air flow at high altitudes parallels the isobars (Figure 5.8). The rule for the relation of wind to pressure in the northern hemisphere is known as *Ballot's law;* it states: Stand with your back to the wind and the low pressure will be on your left, the high on your right.

Near the earth's surface, at levels from the ground upward to about 2000 or 3000 ft (600 or 900 m), yet another force modifies the wind direction. This is the force of friction of the air with the ground. It acts in such a way as to counteract in part the Coriolis force and to prevent the wind from being deflected until parallel with the isobars. Instead, the wind blows obliquely across the iso-

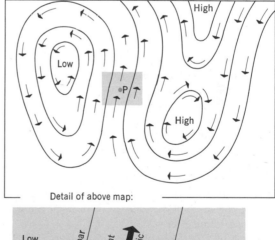

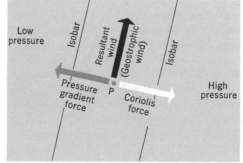

Detail of above map:

Figure 5.8 Wind follows isobars at high levels.

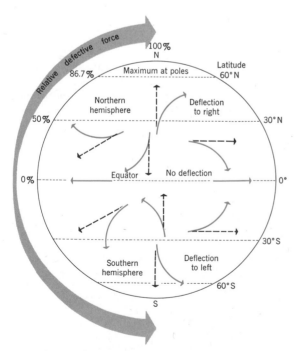

Figure 5.7 Deflective force of the earth's rotation.

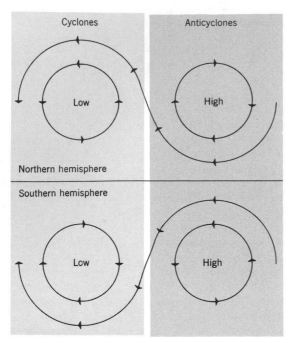

Figure 5.9 Winds at high elevation around cyclones and anticyclones.

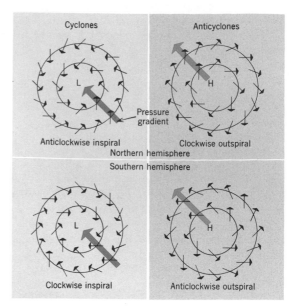

Figure 5.10 Surface winds within cyclones and anticyclones.

bars, the angle being from 20° to 45°. Figure 4.9 illustrates surface winds and is typical of conditions found on the surface weather map. The angle is large for rugged terrain; small for smooth surfaces, such as water or a flat plain. Wind speed is reduced in proportion to ground friction.

Cyclones and anticyclones

In the language of meteorology and climatology, a center of low pressure is designated a *cyclone*; a center of high pressure is an *anticyclone*. Cyclones and anticyclones may be of the stationary, or semipermanent type, such as the Aleutian low and Siberian high described under the topic of global pressure centers, or they may be rapidly moving pressure centers such as create the weather disturbances described in Chapter 6. It will be of great help to the student of weather and climate to have a clear picture of the pattern of winds in cyclones and anticyclones of both northern and southern hemispheres.

As shown in Figure 5.8 air moving at high elevations, where surface friction does not act, parallels the isobars. We can thus give the rule that in the northern hemisphere winds move anticlockwise (counterclockwise) about a cyclone; clockwise about an anticyclone. In the southern hemisphere these patterns are exactly reverse: clockwise about a cyclone; anticlockwise about an anticyclone. The simple upper-air map in Figure 5.9 shows how these rules apply.

For surface winds, which move obliquely across the isobars, the systems for cyclones and anticyclones in both hemispheres are shown in Figure 5.10. To construct the wind arrows, first draw a long arrow to show the direction of pressure gradient. Next, draw short, straight arrows through the point where the gradient arrow crosses each isobar. Be sure that the wind arrows are deflected to the right in the northern hemisphere and to the left in the southern. After carrying out this operation on four sides of a pressure center the pattern will emerge and can easily be completed. To avoid confusion, keep the wind arrows short and straight, as they are on weather maps. When the pattern is complete, it will be seen that the winds in a cyclone in the northern hemisphere show an anticlockwise inspiral; in an anticyclone, a clockwise outspiral. Note the reversal between the labels "anticlockwise" and "clockwise" in the southern hemisphere. In both hemispheres the surface winds spiral inward upon the center of the cyclone, hence the air is *converging* upon the center and must also rise to be disposed of. For the anticyclone, by contrast, surface winds spiral out from the center, which represents a *diverging* of air flow and must be accompanied by a sinking (subsidence) of air in the center of the anticyclone to replace the outmoving air.

Measurement of winds

A description of winds requires measurement of two quantities, direction and speed. Direction is easily determined by a *wind vane*, known to all because it is frequently mounted on rooftops and is one of the commonest of the amateur weather instruments. Wind direction is stated in terms of the direction from which the wind is coming. Thus an east wind comes from the east, but the direction of air movement is toward the west. The direction of movement of low clouds is an excellent indicator of wind direction and can be observed without the aid of instruments.

Speed of wind is measured by an *anemometer*. There are several types. The commonest one seen at weather stations is the *cup anemometer*. It consists of three or four hemispherical cups mounted as if at the ends of spokes of a horizontal wheel (Figure 5.11). The cups travel with a speed proportional to that of the wind. Other types of anemometers depend on measuring the force exerted by the wind upon an exposed surface.

For wind velocities at higher levels a small hydrogen-filled balloon, whose rate of ascent is known, is released into the air and observed through a telescope. Knowing the balloon's vertical position by measuring the elapsed time, an observer can calculate the horizontal drift of the balloon downwind. For modern, upper-air measurements of wind velocity and direction the balloon carries a target that reflects radar waves and can thus be followed when the sky is overcast.

Conventional symbols have been established to show wind direction and speed on weather and climate maps. For single observations, such as those on a daily weather map, an arrow of the type shown in Figure 5.12 is used. The circle takes the place of the arrow point and is centered on the observing station. Attached to the shaft of the

Figure 5.11   A cup-type anemometer. (Courtesy of U.S. Weather Bureau.)

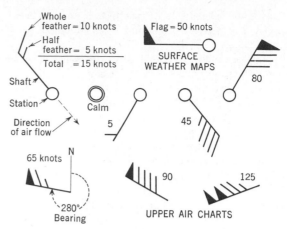

Figure 5.12   Standard weather map symbols.

arrow are short lines, or feathers, and pennants, whose numbers show wind speed in knots (nautical miles per hour) according to a predetermined code.

For wind observations collected over a long period of time a device known as a *wind rose* is constructed. One is illustrated in Figure 5.13. Wind directions are reduced to eight compass sectors shown by lines radiating from the central point. The percentage of total length of time during which wind blows from these sectors of the compass is indicated by length of the lines. On charts showing wind roses there is printed on the margin a scale from which the percentages may be measured. Small feather lines attached to the ends of the radial lines are used to show the average wind speed during the period for which the wind rose is constructed. The Beaufort scale of wind force is used.

Earth's surface wind systems

Prevailing surface winds during the months of January and July are suggested by arrows on the pressure maps, Figures 5.2 and 5.3. A highly diagrammatic representation of the wind systems

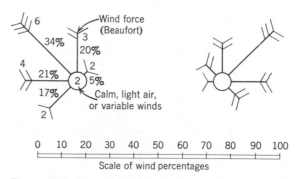

Figure 5.13   The wind rose.

in Figure 5.14 shows the earth as if no land areas existed to modify the belted arrangement of pressure zones.

In the equatorial trough of low pressure, intense solar heating causes the moist air to break into great convection columns, so that there is a general rise of air. This region, lying roughly between 5° S and 5° N lat., was long called the *equatorial belt of variable winds and calms,* or the *doldrums.* There are no prevailing surface winds here, but a fair distribution of directions around the compass (Figure 5.15). Calms prevail as much as a third of the time. Violent thunderstorms with strong squall winds are common. Centrally located on a belt of low pressure, this zone has no strong pressure gradients to induce a persistent flow of wind.

North and south of the doldrums are the *trade wind* belts, covering roughly the zones lying between 5° and 30° N and S. The trades are a result of a pressure gradient from the subtropical belt of high pressure to the equatorial trough of low pressure. In the northern hemisphere, air moving equatorward is deflected by the earth's rotation to turn westward. Thus, the prevailing wind is from the northeast and the winds are termed the *northeast trades.* In the southern hemisphere, deflection of the moving air to the left causes the *southeast trades.* Trade winds are noted for their steadiness

and directional persistence. Figure 5.15 shows wind roses within the trade wind belts. Note that most winds come from one quarter of the compass.

The system of doldrums and trades shifts seasonally north and south, through several degrees of latitude, as do the pressure belts that cause them. Because of the large land areas of the northern hemisphere, there is a tendency for these belts to be shifted farther north in the northern hemisphere in summer (July) than they are shifted south in winter (January). The trades are best developed over the Pacific and Atlantic oceans, but are upset in the Indian Ocean region by the proximity of the great Asiatic landmass.

The trade winds provided a splendid avenue for westward travel in the days of sailing vessels. Steadiness of wind and generally clear weather made this a favorite zone of mariners. Crossing of the doldrums was hazardous because of the possibility of being becalmed for long periods and because of the uncertainty of wind direction. The trade wind belts are not altogether favorable for navigation and flying, however, because over certain oceanic portions, at certain seasons of the year, terrible tropical storms known as hurricanes or typhoons occur (Chapter 7).

Between latitudes 30° and 40° are what have long been called the *subtropical belts of variable*

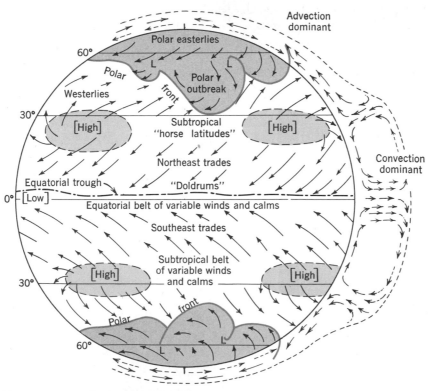

Figure 5.14 The general scheme of atmospheric circulation.

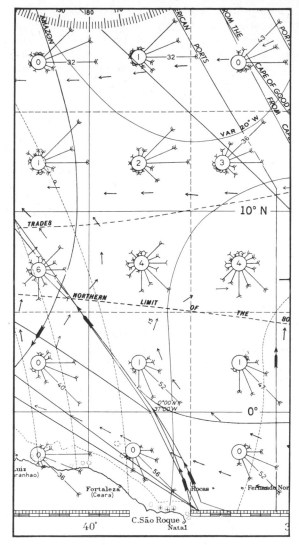

Figure 5.15 This portion of an Oceanographic Office pilot chart of the North Atlantic for July shows parts of both the northeast and southeast trade wind belts, with the narrow doldrum belt lying between. The area shown here extends from 5°S lat. to 20° N lat. Wind roses occupy 5° squares. (U.S. Navy Oceanographic Office, Chart 1400.)

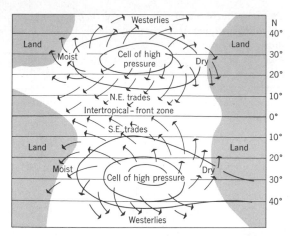

Figure 5.16 Semi-permanent centers of high pressure and surface winds.

winds and calms, or *horse latitudes,* coinciding with the subtropical high-pressure belt. Instead of being continuous even belts, however, the high-pressure areas are concentrated into distinct anticyclones or cells, located over the oceans. Figure 5.16 shows anticyclones in the northern and southern hemisphere and the resultant surface winds. The apparent outward spiraling movement of air is directed equatorward into the easterly trade wind system; poleward into the westerly wind system. The cells of high pressure are most strongly developed in the summer (January in the southern hemisphere, July in the northern). There is also

a latitudinal shifting following the sun's declination. This amounts to less than 5° in the southern hemisphere, but it is about 8° for the strong Hawaiian high located in the northeastern Pacific.

Wind roses for the horse latitudes are shown in Figure 5.17. Winds are distributed around a considerable range of compass directions. Calms prevail as much as a quarter of the time. The cells of high pressure have generally fair, clear weather, with a strong tendency to dryness. Most of the world's great deserts lie in this zone and in the adjacent trade wind belt. An explanation of the dry, clear weather lies in the fact that the anticyclonic cells are centers of descending air, settling from higher levels of the atmosphere and spreading out near the earth's surface. Descending air, as explained more fully in the next chapter, becomes increasingly dry.

Between latitudes 35° and 60°, both N and S, is the belt of the *westerlies,* or *prevailing westerly winds.* Moving from the subtropical anticyclones toward the subarctic lows, these surface winds are shown on Figure 5.14 to blow from a southwesterly quarter in the northern hemisphere, from a northwesterly quarter in the southern hemisphere. This generalization is somewhat misleading, however, because winds from polar directions are frequent and strong. It is more accurate to say that within the westerly wind belt, winds blow from any direction of the compass, but that the westerly components are definitely predominant. Storm winds are common in this belt, as are frequent cloudy days with continued precipitation. Weather is highly changeable.

In the northern hemisphere, landmasses cause considerable disruption of the westerly wind belt, but in the southern hemisphere, between the latitudes 40° and 60° S, there is an almost unbroken belt of ocean. Here the westerlies gain great strength and persistence, giving rise to the mariner's expressions, "the roaring forties," "the furious fifties," and "the screaming sixties." This belt was extensively used for sailing vessels traveling eastward from the South Atlantic Ocean to Australia, Tasmania, New Zealand, and the southern

Pacific islands. From these places it was then easier to continue eastward around the world to return to European ports. Rounding Cape Horn was relatively easy on an eastward voyage, but in the opposite direction, in the face of prevailing stormy westerly winds, was fraught with great danger.

Although the westerly wind belts no longer exert a strong influence over the routes of modern ocean vessels, they are important in long-distance flying. Transoceanic and transcontinental flights in the easterly direction require less fuel and a shorter time. On westward flights, strong head winds may eat dangerously into the fuel supply on the plane and in any event necessitate reduced pay loads.

A wind system often termed the *polar easterlies* has been described as characteristic of the arctic and polar zones (Figure 5.14). The concept is greatly oversimplified, if not actually erroneous, for winds in these regions take a variety of directions, as dictated by local weather disturbances. Perhaps in Antarctica, where an icecapped landmass rests squarely upon the pole and is surrounded by a vast oceanic expanse, the outward spiraling flow of polar easterlies is a valid concept. Deflected to the left in the southern hemisphere, the radial winds would spiral counterclockwise, producing a system of southeasterly winds.

Monsoon winds of Asia and North America

Frequent reference has been made to the powerful control which Asia and North America exert upon conditions of temperature and pressure in the northern hemisphere. Because pressure conditions control winds, it is obvious that these areas must develop wind systems relatively independent of the belted system of earth winds so well illustrated in the southern hemisphere.

In summer, southern Asia develops a cyclone into which there is a considerable flow of air. This may be a *heat low*, or *thermal low*, limited to the lower levels of the atmosphere (Figure 5.18). From the Indian Ocean and the southwestern Pacific warm, humid air moves northward and northwestward into Asia, passing over India, Indochina, and China (See Figure 5.3.) This air flow constitutes the *summer monsoon*, which is accompanied by heavy rainfall in southeastern Asia.

In winter, Asia is dominated by a strong center of high pressure, from which there is an outward flow of air reversing that of the summer monsoon (see Figures 5.2, 5.18). Blowing southward and southeastward toward the equatorial oceans, this *winter monsoon* brings dry, clear weather for a period of several months.

North America, being smaller in extent, does not have the remarkable extremes of monsoon winds experienced by southeastern Asia, but there

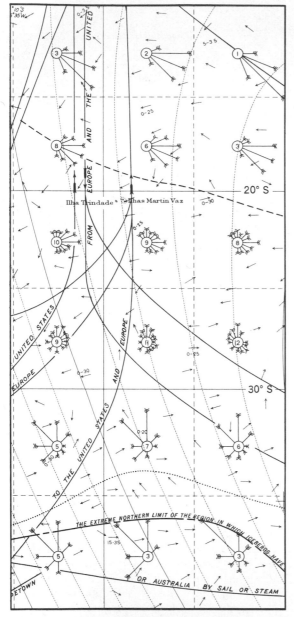

Figure 5.17 The subtropical belt of variable winds and calms (horse latitudes) lies in the center of this portion of an Oceanographic Office pilot chart of the South Atlantic for June–July–August. To the north are the steady southeast trades; to the south, the variable northwesterlies. The area shown here extends from 10° to 40° S lat. Wind roses occupy 5° squares. (U.S. Navy Oceanographic Office, Chart 2600.)

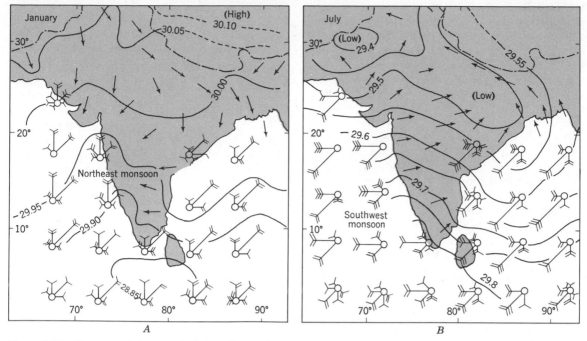

Figure 5.18   Monsoon winds over the Indian Peninsula. (Data from U.S. Navy Oceanographic Office and Blanford.)

is nevertheless an alternation of temperature and pressure conditions between winter and summer. Analysis of wind records shows that in summer there is a prevailing tendency for air originating in the Gulf of Mexico to move northward across the central and eastern part of the United States, whereas in winter there is a prevailing tendency for air to move southward from sources in Canada (Figures 5.2 and 5.3). Australia, too, shows a monsoon effect, but being south of the equator it reverses the conditions of Asia.

Global circulation systems

The surface wind systems thus far described represent only a shallow basal air layer a few thousands of feet thick, whereas the troposphere is from 5 to 12 mi (8 to 20 km) thick. What is the nature of air flow at these higher levels? Since the Second World War, a vast network of observing stations has been taking upper-air observations extending up to 80,000 ft (25,000 m) or higher by means of the *radiosonde*, a compact set of weather instruments that automatically sends back information on pressure, temperature, and humidity by code from a self-contained radio transmitter. Carried aloft by a gas-filled balloon, the radiosonde transmitter can be tracked by means of a radio direction finder. From observations taken simultaneously at many places, a weather map of upper air conditions can be drawn. It has been found that there are slowly moving high- and

low-pressure systems aloft, but that these are generally simple in pattern with smoothly curved isobars. Winds, which may be extremely strong and follow the isobars closely, move counterclockwise around the lows (northern hemisphere), but clockwise around the highs, as shown in Figure 5.8.

The general or average pattern of upper air flow is sketched in Figure 5.19. Two systems dominate.

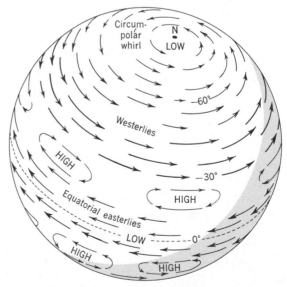

Figure 5.19   Schematic representation of circulation in the upper part of the troposphere, 20,000 to 40,000 ft (6 to 12 km).

One is the system of *westerlies* blowing in a complete circuit about the earth from about latitude 20° almost to the poles. At high latitudes these westerlies constitute a *circumpolar whirl*, coinciding with a great polar low-pressure center. Toward low latitudes the pressure rises steadily at a given altitude, to form two high-pressure ridges at latitudes 15° to 20° N and S. These are the high-altitude parts of the subtropical highs, but are shifted somewhat equatorward. In the high-pressure zones, wind velocities are low, just as in the horse latitudes at sea level. Between the high-pressure ridges is a trough of weak low pressure, in which the winds are easterly, comprising the second major circulation system of the globe, termed the *equatorial easterlies*. At lower elevation their influence spreads into somewhat higher latitudes as the trade winds.

The jet stream

The upper-air westerlies tend to form somewhat serpentine or meandering paths, giving rise to slowly moving *upper air waves*, in which the winds are turned first equatorward, then poleward (Figure 5.20). Associated with the development of such upper air waves at altitudes of 30,000 to 40,000 ft (10,000 to 12,000 m) are narrow zones in which wind streams attain velocities up to 200 to 250 knots (350–450 km per hr). This phenomenon, named the *jet stream*, consists of pulselike movements of air following a broadly curving track (Figure 5.20). In cross section the jet may be likened to a stream of water moving through a hose, the center line of highest velocity being surrounded by concentric zones of less rapidly moving fluid, as pictured in Figure 5.21.

It is by means of the upper air waves that warm air of the tropics is carried far north at the same time that cold air of polar regions is brought equatorward. In this way horizontal mixing, or *advection*, develops on a gigantic scale and serves to provide heat exchange between regions of high and low insolation.

Figure 5.21 The jet stream. (From U.S. Weather Bureau.)

Local winds

In certain favorable localities, *local winds* are generated by immediate influences of the surrounding terrain, rather than by the large-scale pressure systems that produce global winds and large traveling storms. Local winds are important to the geographer interested in climates, because the characteristic repetition of occurrence of a local wind is itself an important aspect of the local climate.

One simple system of local winds, widely occurring along coasts, is the *land and sea breeze* (Figure 5.22), in which the wind blows from sea to land during the day, but from land to sea at night. The daytime wind, or sea breeze, is caused by a landward-sloping barometric pressure gradient developed as a result of daytime heating of the lower air layer over the land. Because the adjacent sea surface is not so intensely heated it remains relatively cooler. In response to the local pressure gradient, air flows landward at low level. The sea breeze is comparatively cool and brings pleasant relief to a narrow coastal belt on hot summer afternoons. At night the temperature differences are reversed because of more rapid cooling of the land surface. The pressure gradient is now from

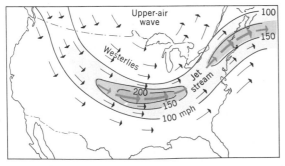

Figure 5.20 The jet stream, shown by lines of equal wind speed. (After U.S. Weather Bureau.)

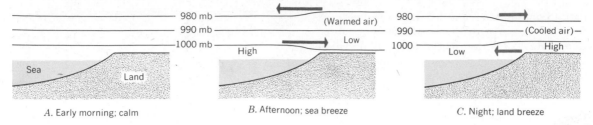

980 mb		980	
990 mb	(Warmed air)	990	(Cooled air)
1000 mb	Low	1000	High
	High		Low

Sea · Land

A. Early morning; calm · B. Afternoon; sea breeze · C. Night; land breeze

Figure 5.22   Sea breeze and land breeze. (After S. Petterssen: from A. N. Strahler, *The Earth Sciences*, Harper and Row, New York.)

land to sea, causing a seaward flow of air—the land breeze.

Mountain and *valley winds* are local winds following a daily alternation of direction in a manner similar to the land and sea breezes. The air moves from valleys, upward over rising mountain slopes, toward the summits during the day, when slopes are intensely heated by the sun. The air then moves valleyward, down the ground slopes, when the same slopes have been cooled at night by radiation of heat from ground to air. These winds are therefore responding to local pressure gradients set up by heating or cooling of the lower air.

Still another group of local winds are known as *drainage winds*, or *katabatic winds*, in which cold air flows under the influence of gravity from higher to lower regions. Such cold, dense air may accumulate in winter over a high plateau or high interior valley. When general weather conditions are favorable some of this cold air spills over low divides or through passes to flow out upon adjacent lowlands as a strong, cold wind. Drainage winds occur in many mountainous regions of the world and go by various local names. The *bora* of the northern Adriatic coast and the *mistral* of southern France are well-known examples. In southern California there issues on occasion from the Santa Ana Valley a strong, dry east wind, the *Santa Ana*, which blows across the coastal lowland. This air is of desert origin and may carry much dust and silt in suspension. On the icecaps of Greenland and Antarctica, powerful katabatic winds move down the gradient of the ice surface and are funneled through coastal valleys to produce powerful blizzards lasting for days at a time.

Still other types of local winds, bearing such names as *foehn* and *chinook*, result when strong regional winds passing over a mountain range are forced to descend on the lee side with the result that the air is heated and dried. These winds are explained in Chapter 6.

Ocean currents

The circulation of the oceans is a matter of great interest to the geographer seeking an understand-ing of climates. Warm water currents bring a moderating influence to coasts in arctic latitudes; cool currents greatly alleviate the heat of tropical deserts along narrow coastal belts.

The direction of motion of surface ocean currents is stated in terms of compass direction toward which the water is moving. Much information on the direction and velocity of currents has been gained from *drift bottles*, sealed bottles containing return-addressed cards, set adrift in the hope that the bottle will be found at a distant point and the card returned. To measure currents, anchored vessels may put down *current meters* which tell the direction and speed of water flow. These devices look much like the current meter used in stream gauging (Figure 20.8). Much of our knowledge of currents has been built up from navigational observations. The direction and speed of a current can be computed when a ship's actual position is determined astronomically after a day's run. The true position is compared with the supposed position reckoned from the known rate and direction of course held through the preceding day. The difference in the two positions gives a measure of oblique drifting as a result of traveling at an angle to the current direction. (Figure 5.23).

Virtually all of the important surface currents of the oceans are set in motion by prevailing surface winds. Energy is transferred from wind to water by the frictional drag of the air blowing over the water surface. Because of the Coriolis force, the water drift is impelled toward the right of its

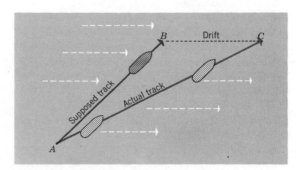

Figure 5.23   Calculation of drift.

path of motion (northern hemisphere), and therefore the current at the water surface is in a direction about 45° to the right of the wind direction. Under the influence of winds, currents may tend to bank up the water close to the coast of a continent, in which case the force of gravity, tending to equalize the water level, will cause other currents to be set up.

Density differences may also cause flow of ocean water. Such differences arise from greater heating by insolation, or greater cooling by radiation, in one place than another. Thus, surface water chilled in the arctic and polar seas will sink to the ocean floor, spreading equatorward and displacing upward the less dense, warmer water.

Still another controlling influence upon water movements is the configuration of the ocean basins and coasts. Currents initially caused by winds impinge upon a coast and are locally deflected to a different path or confined in straits or gulfs.

The combined action of wind and density differences sets up an oceanic circulation system including not only horizontal motions, but upwelling and down-sinking motions besides. In this introductory study, emphasis is on the currents of a shallow surface water zone, inasmuch as these motions directly affect marine navigation and have a strong climatic influence upon the overlying layer of atmosphere.

Generalized scheme of ocean currents

To illustrate surface water circulation an idealized ocean extending across the equator to latitudes of 60° or 70° on either side may be taken for illustration (Figure 5.24). Perhaps the most outstanding features are the circular movements, or *gyres*, around the subtropical highs, centered about 25° to 30° N and S. An *equatorial current* marks the belt of the trades. Whereas the trades blow to the southwest and northwest, obliquely across the parallels of latitude, the water movement follows the parallels. Thus, ocean currents trend at an angle of about 45° with the prevailing surface winds, because of the deflective force of the earth's rotation.

A slow eastward movement of water over the zone of the westerly winds is named the *west-wind drift*. It covers a broad belt between 35° and 45° in the northern hemisphere, and between 30° or 35° and 70° in the southern where open ocean exists in the higher latitudes.

The equatorial currents are separated by an *equatorial countercurrent*. This is well developed in the Pacific, Atlantic, and Indian oceans (Figure 5.25).

Along the west sides of the oceans in low latitudes the equatorial current turns poleward, form-

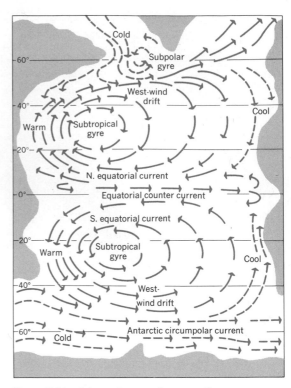

Figure 5.24   Schematic map of system of ocean currents.

ing a warm current paralleling the coast. Examples are the *Gulf Stream* (*Florida* or *Caribbean* stream), the *Japan current* (*Kuroshio*) and the *Brazil current*, which bring higher than average temperatures along these coasts. This is illustrated in Figure 4.18 by the southward bulge of the 70°F (21°C) July isotherm along the west side of the South Atlantic.

The west-wind drift, upon approaching the east side of the ocean, is deflected both south and north along the coast. The equatorward flow is a cool current, produced by upwelling of colder water from greater depths. It is well illustrated by the *Humboldt current* (Peru current) off the coast of Chile and Peru; by the *Benguela current* off the southwest African coast; by the *California current* off the west coast of the United States; and by the *Canaries current*, off the Spanish and North African coast. Note that this cold upwelling causes a marked equatorward deflection of the isotherms, illustrated in Figure 4.17 by the northward bend in the 70°F (21°C) January isotherm along the east side of the South Pacific and South Atlantic oceans.

In the northern eastern Atlantic Ocean, the westwind drift is deflected poleward as a relatively warm current. This is the *North Atlantic current*, which spreads around the British Isles, into the North Sea, and along the Norwegian coast. The

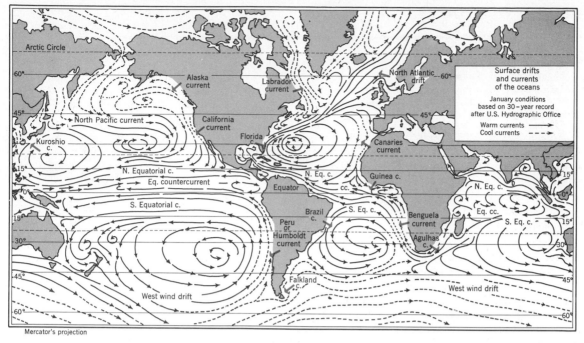

Arctic Circle

60°

45°

North Pacific current

Kuroshio c.

15°

N. Equatorial c.

Eq. countercurrent

S. Equatorial c.

15°

30°

45°

West wind drift

60°

Mercator's projection

Alaska current

Labrador current

California current

Florida c.

Equator

cc.

Brazil c.

Peru or Humboldt current

Falkland c.

North Atlantic drift   60°

Canaries current

N. Eq. c.

Guinea c.

S. Eq. c.

Benguela current

Agulhas c.

West wind drift

45°

N. Eq. c.

Eq. cc.

S. Eq. c.

15°

30°

Surface drifts and currents of the oceans

January conditions based on 30-year record after U.S. Hydrographic Office

Warm currents ⟶
Cool currents ⇢

Figure 5.25   Surface drifts and currents of the oceans (January). (After U.S. Navy Oceanographic Office.)

ice-free port of Murmansk, on the Arctic Circle, has year-round navigability by way of this coast. Note that in Figure 4.17 isotherms are deflected northward where they cross this current. In winter this effect is much more pronounced than in summer.

In the northern hemisphere, where the polar sea is largely landlocked, cold water flows equatorward along the west side of the large straits connecting the Arctic Ocean with the Atlantic basin. Three principal cold currents are the *Kamchatka current*, flowing southward along the Kamchatka Peninsula and Kurile Islands; the *Greenland current*, flowing south along the east Greenland coast through the Denmark Strait; and the *Labrador current*, moving south from the Baffin Bay area through Davis Strait to reach the coasts of Newfoundland, Nova Scotia, and New England.

In both the north Atlantic and Pacific oceans the Icelandic and Aleutian lows in a very rough way coincide with two centers of counterclockwise circulation involving the cold arctic currents and the west-wind drifts.

The antarctic region has a relatively simple current scheme consisting of a single *antarctic*

circumpolar current moving clockwise around the antarctic continent in latitudes 50° to 65° S, where a continuous expanse of open ocean occurs.

Oceanographers today recognize that oceanic circulation involves the complex motions of water masses of different temperature and salinity characteristics. The very simple outline of surface currents given here does not take into account the movements of these water masses at different depths.

REFERENCES FOR FURTHER STUDY

Cotter, C. H. (1965), *The physical geography of the oceans*, American Elsevier Publ. Co., New York, 317 pp. See Chapters 10, 12, and 13.

Donn, W. L. (1965) *Meteorology*, third edition, McGraw-Hill Book Co., New York, 484 pp.

Riehl, H. (1965), *Introduction to the atmosphere*, McGraw-Hill Book Co., New York, 365 pp.

Reiter, E. R. (1967), *Jet streams*, Doubleday and Co., New York, 189 pp.

Petterssen, S. (1969), *Introduction to meteorology*, third edition, McGraw-Hill Book Co., New York, 333 pp.

REVIEW QUESTIONS

1. What range of sea-level air pressures is normally found to occur from day to day and season to season at a given place? What is considered high pressure? What is low pressure?

2. What are isobars? Explain how an isobaric map is prepared.

3. Describe the principal pressure belts of the globe, giving the latitude and approximate pressures for each.

4. Why do the pressure belts shift in latitude throughout the year? Is this shift over as great a latitude as the sun's declination?

5. Why do the landmasses of North America and Asia disrupt the belted pressure pattern of the globe? What pressures occur on these land areas in the winter solstice season? In the summer solstice season?

6. What pressure centers dominate the North Atlantic and North Pacific oceans during the summer months? During the winter months? Give the names of the lows and highs.

7. Explain the relation of pressure gradient to isobars and to the production of winds. What is the Coriolis force? State Ferrel's law. How are wind directions modified near the earth's surface?

8. What is a cyclone? An anticyclone? Describe the pattern of surface winds in cyclones and anticyclones in both northern and southern hemispheres.

9. With what instruments are winds measured? What information is needed about winds? How is the direction of a wind designated? How can wind direction and speed in the upper air levels be determined?

10. Explain how a wind rose is constructed. What symbols are used on weather maps to show surface winds? To show winds at high levels?

11. What conditions of winds and calms prevail in the equatorial trough? Why are steady, prevailing winds absent here?

12. Describe the trade winds. Which way do they blow in the northern hemisphere? In the southern hemisphere? Explain the direction of the trade winds. How were the trade winds used by mariners in the days of sailing vessels?

13. What are the subtropical belts of variable winds and calms? Describe the cells of high pressure of which this belt is composed. How do they shift in latitude seasonally?

14. Describe the westerly wind belts. How do the westerlies compare with the trades for constancy of direction and strength? How does the westerly wind belt influence transoceanic sailing and flying?

15. Describe the monsoon wind systems of southeastern Asia. What general type of weather condition is associated with the summer monsoon? With the winter monsoon? Why is a monsoon system not so strongly developed on other continents, such as Africa?

16. How is weather information obtained from the upper levels of the troposphere? How are winds related to isobars at high levels? How does air circulate around a center of low pressure? Around a high?

17. Describe the general global circulation in the middle and upper troposphere. What are the westerlies? The equatorial easterlies? What is the circumpolar whirl? With what barometric pressure is it associated?

18. Describe the jet stream and its relation to upper air waves of the westerlies.

19. What are local winds? Explain the land and sea breeze; the mountain and valley breeze. Where are katabatic (drainage) winds found? Give examples.

20. How are the direction and velocity of ocean currents measured? Discuss the causes of surface ocean currents. Show how the Coriolis force modifies the direction of flow.

21. Sketch a hypothetical ocean modeled after the Pacific or Atlantic and indicate the general scheme of surface currents. Label fully.

22. List three warm currents and three cool currents, giving the location and direction of flow of each. What effect have these currents upon isotherms of air temperatures?

23. What current systems conduct very cold water into the northern Atlantic Ocean? What current prevails in the southern ocean?

CHAPTER 6

Moisture, Clouds, and Precipitation

THE physical nature and *importance* of *water vapor* in the atmosphere has already been touched upon in an earlier chapter, in which the capacity of water vapor to absorb radiant heat energy was described. We will consider here the proportions in which water vapor is held in the air and the manner in which vapor may condense to form clouds, fog, rain, and snow.

The *amount* of *water vapor* that may be present in the *air* at a given time varies widely from place to place. It ranges from virtually nothing in the cold, dry air of arctic regions in winter to as much as 4 or 5 percent of a given volume of the atmosphere in humid, hot tropical areas.

Water vapor enters the atmosphere by *evaporation* from exposed water surfaces such as oceans, lakes, rivers, or moist ground. Some is supplied by plants which. *transpire* water as a physiological function (Chapter 15). With large expanses of ocean and densely forested humid lands over the globe, there is no lack of evaporating surfaces.

Water states and heat

Water occurs in three states, (1) frozen as ice, a crystalline *solid*, (2) *liquid* as water, and (3) *gaseous* as water vapor (Figure 6.1). From the gaseous vapor state, molecules may pass into the liquid state by *condensation*, or, if temperatures are below the freezing point, they can pass by *sublimation* directly into the solid state to form ice crystals. By *evaporation*, molecules can leave a water surface to become gas molecules in water vapor. The analogous change from ice directly into water vapor is also designated sublimation. Then, of course, water may pass from liquid to solid state by *freezing*, and from solid state to liquid state by *melting*. All of this can be represented by a triangle in which the three states of water form the corners. Arrows show the six possible changes of state.

Of great importance in weather science are the exchanges of heat energy accompanying changes of state. For example, when water evaporates, *sensible heat*, which we can feel and measure by thermometer, passes into a hidden form held by the water vapor and known as the *latent heat of vaporization*. This results in a drop in temperature of the remaining liquid. The cooling effect produced by

evaporation of perspiration from the skin is perhaps the most obvious example. For every gram of water that is evaporated, about 600 calories change into the latent form. In the reverse process of condensation, an equal amount of energy is released to become sensible heat and the temperature rises correspondingly. Similarly, the freezing process releases heat energy in the amount of about 80 calories per gram of water, whereas melting absorbs an equal quantity of heat. This is referred to as the *latent heat of fusion*. When sublimation occurs, the heat absorbed by vaporization, or released by crystallization, is still greater for each gram of water, for the latent heats of vaporization and fusion are added together.

Humidity

The term *humidity* simply refers to the degree to which water vapor is present in the air. For any specified temperature there is a definite limit to the quantity of moisture that can be held by the air. This limit is known as the *saturation point*. The proportion of water vapor present relative to the maximum quantity is the *relative humidity*, expressed as a percentage. At the saturation point, relative humidity is 100 percent; when half of the total possible quantity of vapor is present, relative humidity is 50 percent, and so on.

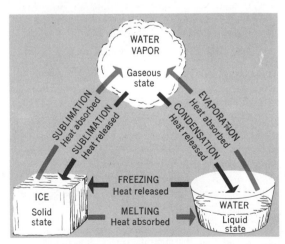

Figure 6.1 Three states of water. (After C.A.A., U.S. Dept. Commerce.)

A change in relative humidity of the atmosphere can be caused in one of two ways. If an exposed water surface is present, the humidity can be increased by evaporation. This is a slow process, requiring that the water vapor diffuse upward through the air. The other way is through a change of temperature. Even though no water vapor is added, a lowering of temperature results in a rise of relative humidity. This is automatic and is a logical consequence of the fact that the capacity of the air to hold water vapor has been lowered by cooling; thus the existing amount of vapor represents a higher percentage of the total capacity of the air. Similarly, a rise of *air temperature* results in decreased relative humidity, even though no water vapor has been taken away. The principle of *relative humidity change caused by temperature change* is illustrated by a graph of these two properties throughout the day (Figure 6.2). As air temperature rises, relative humidity falls, and vice versa.

A simple example may be given to illustrate these principles. At a certain place the temperature of the air is 60°F (16°C), the relative humidity 50 percent. Should the air become warmed by the radiant energy from the sun and ground surface to 90°F (32°C), the relative humidity automatically drops to 20 percent, which is very dry air. Should the air become chilled during the night and its temperature fall to 40°F (5°C), the relative humidity will automatically rise to 100 percent, the saturation point. Any further cooling will cause condensation of the excess vapor into liquid form. As the air temperature continues to fall, the humidity remains at 100 percent, but condensation continues. This may take the form of minute droplets of dew or fog. If the temperature falls below freezing, condensation occurs as frost upon exposed surfaces.

The term *dew point* is applied to the critical temperature at which the air is fully saturated, and below which *condensation* normally occurs. An excellent illustration of condensation due to cooling is seen in summertime when beads of moisture form on the outside surface of a pitcher or glass filled with ice water. Air immediately adjacent to the cold glass or metal surface is sufficiently chilled to fall below the dew point temperature, causing moisture to condense on the surface of the glass.

In understanding the relationships between atmospheric temperature and relative humidity, a very homely analogy may prove helpful. An ordinary sponge, if left to soak up water, will take up moisture to its full capacity. This is analogous to the manner in which air will gradually increase its humidity to the saturation point, if allowed to stand over a water surface and to maintain a constant temperature. If the sponge is now lifted out of the water and held carefully in the hand it will continue to hold the absorbed water. Suppose now the sponge is slowly squeezed. Water is expelled. In a like manner, the lowering of temperature below dew point of the saturated air expels moisture by condensation of the excess water vapor. After most of the water has been squeezed from the sponge it may be released, but not permitted to touch the water. This is analogous to a rise of air temperature back to the previous starting point, but without being permitted to take up water. This condition would prevail over an interior desert region. The air, like the sponge, is now holding only a small fraction of its total possible moisture content. If it is again cooled no condensation will result from the air until a temperature below the previous minimum is reached, just as no water will be released from the sponge until it is squeezed even harder than previously.

Absolute humidity

Although relative humidity is an important indicator of the state of water vapor in the air, it is a statement only of the relative quantity with respect to a saturation quantity. The actual quantity of moisture present is denoted by *absolute humidity*, defined as the weight of water vapor contained in a given volume of air. Weight is stated in grams, volume in cubic meters. For any specified air temperature, there is a maximum weight of water vapor that a cubic meter of air can hold (the saturation quantity). Figure 6.3 is a graph showing this maximum moisture content of air for a wide range of temperatures.

In a sense, the absolute humidity is a geographer's yardstick of a basic natural resource—water—to be applied from equatorial to polar regions. It is a measure of the quantity of water that can be extracted from the atmosphere as precipitation. Cold air can supply only a small quantity of rain or snow; warm air is capable of supplying huge quantities.

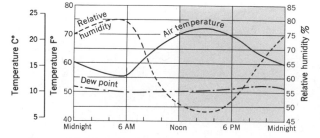

Figure 6.2 Relative humidity, temperature, and dew point for May at Washington, D.C. (After U.S. Weather Bureau; from A. N. Strahler, *The Earth Sciences*, Harper and Row, New York.)

Vapor pressure

In Chapter 3, it is explained that the weight of a column of atmosphere counterbalances the mercury column in a barometer and that variations in mercury height measure changes in air pressure. When water vapor is added to otherwise pure, dry air, the water molecules diffuse perfectly among the other gas molecules. That part of the total barometric pressure that is due to the water vapor alone is termed the *vapor pressure*. For cold, dry air, the vapor pressure may be as low as 0.05 in (0.013 cm); for very warm, moist air of the equatorial regions, it may be as high as 0.80 in (2 cm).

Figure 6.3 shows the maximum possible vapor pressure for air of a range of temperatures from very cold to very warm. Both vapor pressure and absolute humidity tell the quantity of water vapor present in the air, but in somewhat different ways.

One disadvantage of using absolute humidity in the study of atmospheric moisture is that when air rises or sinks in elevation, it undergoes corresponding volume changes of expansion or compression. Thus the absolute humidity cannot remain a constant figure for the same body of air. Modern meteorology therefore makes use of another measure of moisture content, *specific humidity*, which is the ratio of weight of water vapor to weight of moist air (including the water vapor). This ratio

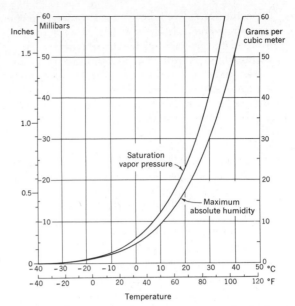

Figure 6.3 Maximum vapor pressure and absolute humidity. (From A. N. Strahler, *The Earth Sciences*, Harper and Row, New York.)

is stated in units of grams of water vapor per kilogram of moist air. When a given parcel of air is lifted to higher elevations without gain or loss of moisture the specific humidity remains constant, despite volume increase.

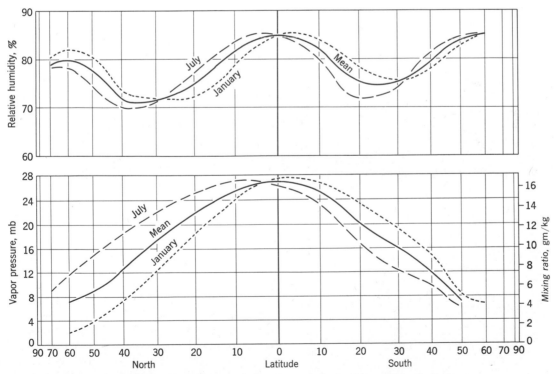

Figure 6.4   Relation of relative humidity to latitude (above) and of mixing ratio to latitude (below). Values of mixing ratio are numerically similar to values of specific humidity. (After Haurwitz and Austin, 1944, *Climatology*.)

Specific humidity is often used to describe the moisture characteristics of a large mass of air. For example, extremely cold, dry air over arctic regions in winter may have a specific humidity of as low as 0.2 grams per kilogram, whereas extremely warm moist air of tropical regions may hold as much as 18 grams per kilogram. The total natural range on a world-wide basis is such that the largest values of specific humidity are from 100 to 200 times as great as the least.

Figure 6.4 is a graph of average specific *humidity and latitude*, showing that the warm air of equatorial regions normally holds vastly more water vapor than cold air of arctic and polar regions.

How humidity is measured

Humidity of the air can be measured in two ways. An instrument known as a *hygrometer* indicates relative humidity on a calibrated dial. One simple type uses a strand of human hair which lengthens and shortens according to the relative humidity and thereby activates the dial (Figure 6.5). A continuous record of humidity can be obtained by means of a recording *hygrograph*. Using the same basic mechanism as the hygrometer, a continuous, automatic record is drawn by a pen on a sheet of paper attached to a rotating drum.

A different principle is applied in the *sling psychrometer*. This instrument is simply a pair of *thermometers* mounted side by side (Figure 6.6). One is of the ordinary type; the other has a piece

Figure 6.6 Sling psychrometer. (U.S. Weather Bureau.)

of wet cloth around the bulb. If the air is fully saturated (relative humidity 100 percent), there will be no evaporation from the wet cloth and both thermometers will read the same. If, however, the air is not fully saturated, evaporation will occur, cooling the cloth-covered thermometer below the temperature shown on the ordinary thermometer. Because the rate of evaporation depends on *dryness* of the air, the difference in temperature shown by the two thermometers will increase as relative humidity decreases.

Standard tables are available to show the relative humidity for a given combination of *wet- and dry-bulb temperatures*. In order to be sure that maximum possible evaporation is taking place, the two thermometers are attached by a swivel joint to a handle by which the thermometers can be swung around in a circle by hand. Other types have a fan to blow air past the wet thermometer bulb.

How *condensation* occurs

Falling rain, snow, sleet, or hail, referred to collectively as *precipitation*, can result only where large masses of air are experiencing a steady drop in temperature below the dew point. This condition cannot be brought about by the simple process of chilling of the air through loss of heat by long-wave radiation during the night. Instead, it is necessary that the large mass of air be rising to higher elevations. This statement requires that a new principle of weather science be explained.

One of the most important laws of meteorology is that rising air experiences a drop in temperature, even though no heat energy is lost to the outside (Figure 6.7). The drop of temperature is a result of the decrease in air pressure at higher elevations, permitting the rising air to expand. Individual molecules of the gas are more widely diffused and do not strike one another so frequently, hence imparting a lower sensible temperature to the gas. When no condensation is occurring, the rate of drop of temperature, termed the *dry adiabatic rate*, is about $5\frac{1}{2}$ F° per 1000 feet of vertical rise of air. In metric units the rate is 1 C° per 100 meters. The dew point also declines with rise of air; the rate is 1 F° per 1000 ft (0.2 C° per 100 m).

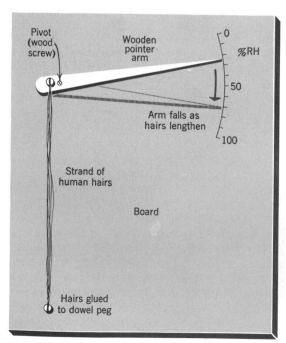

Pivot (wood screw)

Wooden pointer arm

0

%RH

50

Arm falls as hairs lengthen

100

Strand of human hairs

Board

Hairs glued to dowel peg

Figure 6.5 A simple hygrometer.

If water vapor in the air is condensing, the adiabatic rate is less, about 3.2 F° per 1000 ft (0.6 C° per 100 m), owing to the partial counteraction of temperature loss through the liberation of latent heat during the condensation process. This modified rate is referred to as the *wet adiabatic*, or *saturation adiabatic* rate (Figure 6.7). Adiabatic cooling rate should not be confused with the normal lapse rate, explained in Chapter 3. The normal lapse rate applies only to still air whose temperature is measured at successively higher levels.

Where *condensation* is occurring directly in the form of snow (ice crystals), the *adiabatic rate* is intermediate in value between dry and saturated rates (Figure 6.7).

The various ways in which large masses of air are induced to rise to higher elevations are treated more fully in a later paragraph. Because the actual fall of rain, snow, or other forms of precipitation is preceded by the formation of clouds, it is desirable first to consider the various types of clouds and their significance.

Clouds

Clouds consist of extremely tiny *droplets* of water, 0.0008 to 0.0024 in (0.02 to 0.06 mm) in diameter, or minute crystals of ice. These are sustained by the slightest upward movements of air. In order for cloud droplets to form, it is necessary that microscopic dust particles serve as centers, or *nuclei*, of *condensation*. *Dusts* with a high affinity for water are abundant throughout the atmosphere.

Where the air temperature is well below freezing, clouds may form of tiny crystals. Water in such minute quantities can remain liquid far below normal freezing temperatures. Thus, water droplets exist at temperatures down to 10°F (−12°C); a mixture of water droplets and ice crystals from 10° to −20°F (−12 to −30°C) or even lower; and predominantly *ice crystals* below −20°F (−30°C). Below −40°F (−40°C) all of the cloud is ice. *Clouds* appear *white* when thin or when the sun is shining upon the outer surface. When dense and thick, clouds appear gray or black underneath simply because this is the shaded side.

Cloud types may be classified on the basis of two characteristics: general form and altitude. On the basis of form there are two major groups: *stratiform* or layered types, and *cumuliform* or massive, globular types (Figure 6.8).

The stratiform clouds are blanketlike, often covering vast areas, but are fairly thin in comparison to horizontal dimensions. Stratiform clouds are subdivided according to the level of elevation at which they lie. The highest type is the *cirrus* cloud and its related forms, *cirrostratus* and *cirrocumulus* (Figures 6.8 and 6.9). These are roughly within the altitude range of 20,000 to 40,000 ft (6000

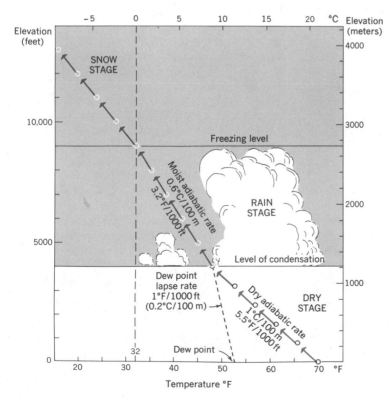

Figure 6.7 Adiabatic changes of temperature in a rising air mass. (From A. N. Strahler, *The Earth Sciences*, Harper and Row, New York.)

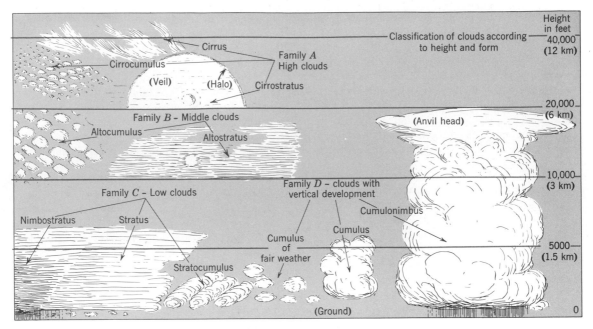

Figure 6.8 Cloud types are grouped into *families* according to height range and form.

to 12,000 m) and are composed of *ice crystals*. *Cirrus* is a delicate, wispy cloud, often forming streaks or stringers across the sky. It does not interfere with the passage of sunlight or moonlight and appears to the ground observer to be moving very slowly, if at all. *Cirrostratus* is a more com-

plete layer of cloud, producing a *halo* about the sun or moon. Where the layer consists of closely packed globular pieces of cloud, arranged in groups or lines, the name cirrocumulus is given. This is the *mackerel sky* of popular description.

At *intermediate height range*, from 6500 to

A. Cirrus in parallel trails and small patches. (Photograph by F. Ellerman. Courtesy of U.S. Weather Bureau.)

B. Cirrocumulus (*above*) with tufted cirrus (*below*). (Photograph by F. Ellerman. Courtesy of U.S. Weather Bureau.)

C. Thin altostratus with fractostratus patches below. (Photograph by G. A. Clarke. Courtesy of U.S. Weather Bureau.)

D. Altocumulus; active form. (Photograph by C. F. Brooks. Courtesy of U.S. Weather Bureau.)

E. Stratus, a uniform layer extending below the level of the hilltop, with shreds of fractostratus visible against the hillside. (Photograph by G. A. Clarke. Courtesy of U.S. Weather Bureau.)

F. Stratocumulus in irregular horizontal rolls. (Photograph by W. J. Humphreys. Courtesy of U.S. Weather Bureau.)

G. Cumulus of fair weather. (Photograph by H. T. Floreen. Courtesy of U.S. Weather Bureau.)

Figure 6.9 (*above and left*) Cloud types.

H. Cumulonimbus, an isolated thunderstorm showing rain falling from base. (Photograph by Air Service, U.S. Navy. Courtesy of U.S. Weather Bureau.)

20,000 ft (2000 to 6000 m), are the *altostratus* and *altocumulus* clouds. Altostratus is a blanket layer, often smoothly distributed over the entire sky. It is grayish in appearance, usually has a smooth underside, and will often show the sun as a bright spot in the cloud. Altocumulus is a layer of individual cloud masses, fitted closely together in geometric pattern. The masses appear white, or somewhat gray on the shaded sides, and blue sky is seen between individual patches or rows. Altostratus is commonly associated with the development of bad weather, whereas altocumulus is usually characteristic of generally fair conditions.

In the low cloud group, from ground level to 6500 ft (2000 m), are *stratus, nimbostratus,* and *stratocumulus* clouds. Stratus is a dense, low-lying dark-gray layer (Figure 6.9E). If rain or snow is falling from this cloud, it is termed nimbostratus, the prefix *nimbo* merely meaning that precipitation is coming from the cloud. Stratocumulus is a low-lying cloud layer consisting of distinct grayish masses of cloud between which is open sky. The individual masses often take on the form of long rolls of cloud, oriented at right angles to the direction of wind and cloud motion. Stratocumulus is generally associated with fair or clearing weather, but sometimes rain or snow flurries issue from individual cloud masses.

Fog is simply a form of stratus cloud lying very close to the ground. One type, known as a *radiation fog*, is formed at night when temperature of the basal air falls below the dew point. Another type, *advection fog*, results from the movement of warm, moist air over a cold or snow-covered ground surface. Losing heat to the ground, the air layer undergoes a drop of temperature below the dew point, and condensation sets in. A similar type of advection fog is formed over *oceans* where air from over a warm *current* blows across the cold surface of an adjacent cold current. Fogs of the Grand Banks off Newfoundland are largely of this origin because here the cold *Labrador current* comes in contact with warm waters of *Gulf Stream* origin.

The cumuliform clouds tend to display a height as great as, or greater than, their horizontal dimensions. *Cumulus* is a white, woolpack cloud mass, often showing a flat base and a bumpy upper surface somewhat resembling a head of cauliflower (Figure 6.9G). These clouds look pure white on the side illuminated by the sun, but may be gray or black on the shaded or underneath side. Small cumulus clouds are associated with fair weather. Under different conditions, discussed below, individual masses grow into *cumulonimbus*, the *thunderstorm cloud* mass of enormous size which brings heavy rainfall, thunder and lightning, and

gusty winds (Figure 6.9H). A large cumulonimbus cloud may extend from a height of 1000 to 2000 ft (300 to 600 m) at the base up to 30,000 or 40,000 ft (9000 to 12,000 m). When seen from a great distance, the top of the cumulonimbus cloud is pure white, but to observers beneath, the sky may be darkened to almost nighttime blackness. More will be said of this cloud in connection with thunderstorms.

Forms of precipitation

Precipitation results when condensation is occurring rapidly within a cloud. *Rain* is formed when cloud droplets in large numbers are caused to coalesce into drops too large to remain suspended in the air. The drops may then grow by colliding with other drops and joining with them to become as large as 0.25 in (7 mm) in diameter; but above this size they are unstable and break into smaller drops. Falling droplets less than 0.02 in (0.5 mm) in diameter make a *drizzle*.

Sleet, as the term is used in the United States, consists of pellets of ice produced from freezing of rain. The raindrops form in an upper, warmer layer, but fall into an underlying cold air layer. (Elsewhere in the English-speaking world sleet means a mixture of rain and snow.)

Snow consists of masses of crystals of ice, grown directly from the water vapor of the air, where air temperature is below freezing. Individual *snow crystals*, which can be carefully caught upon a black surface and examined with a strong magnifying glass, develop in six-sided, flat crystals, or as prisms. They display infinite variations in their beautiful symmetrical patterns (Figure 6.10).

Hail consists of rounded lumps of ice, having an internal structure of concentric layers, much like an onion. Ordinarily the ice is not clear but has a frosted appearance. *Hailstones* range from 0.2 to 2 in (0.5 to 5 cm) in diameter and may be extremely destructive to crops and light buildings (Figure 6.11). Hail occurs only from the cumulonimbus cloud type, inside of which are extremely strong updrafts of air. Raindrops are carried up to high altitudes, are frozen into ice pellets, then fall again through the cloud. Suspended in powerful updrafts, the hail stone grows by the attachment and freezing of droplets, much as ice accumulates on the leading edge of an airplane wing. Eventually the hailstone escapes from the updrafts and falls to earth.

When rain falls upon a ground surface that is covered by an air layer of below-freezing temperature, the water freezes into clear ice after striking the ground or other surfaces such as trees, houses, or wires (Figure 6.12). The coating of ice that results is called a *glaze*, and an *icing storm* is said

Figure 6.10 *Snowflakes.* (Photograph by Ewing Galloway.)

to have occurred. Actually no ice falls, so that ice glaze is not a form of precipitation. Icing storms cause great damage, especially to telephone, telegraph, and power wires and to tree limbs. Roads and sidewalks are made extremely hazardous.

How *precipitation is measured*

Precipitation is generally stated in units of inches or centimeters that fall per unit of time. One inch of rainfall, for example, is a quantity sufficient to cover the ground to a *depth* of one

Figure 6.11 These hailstones, larger than hens' eggs (arrow), fell at Girard, Illinois, on August 13, 1929. (U.S. Weather Bureau.)

inch, provided that none is lost by runoff, evaporation, or sinking into the ground. A simple form of *rain gauge* can be operated merely by setting out a straight-sided, flat-bottomed pan and measuring the depth to which water accumulates during a particular period. Unless this period is short, however, evaporation seriously upsets the results. Furthermore, very small amounts of rainfall, such as 0.1 in (0.25 cm), would make too thin a layer to be accurately measured. To avoid this difficulty, as well as to reduce evaporation loss, good rain gauges are made in the form of a cylinder whose base is a funnel leading into a narrow pipe (Figure 6.13). A small amount of rainfall will fill the narrow pipe to a considerable height, thus making it easy to read accurately, once a simple scale has been provided for the pipe. This gauge requires frequent emptying unless it is equipped with automatic devices for this purpose.

Snowfall is *measured* by melting a sample column of snow and reducing it to an equivalent in water. Thus, rainfall and snowfall records may be combined for purposes of comparison. Ordinarily, a ten-inch layer of snow is assumed to be equivalent to one inch of rainfall, but this ratio may range from 30 to 1 in very loose snow to 2 to 1 in old, partly melted snow.

Precipitation-producing conditions

Thus far it has been stated that precipitation results when air rises and is adiabatically cooled

Figure 6.12 Heavily coated wires and branches caused heavy damage in eastern New York State in January 1943, as a result of this icing storm. (Courtesy of U.S. Weather Bureau, and New York Power and Light Co., Albany, N.Y.)

Figure 6.13   A tipping-bucket rain gauge. (U.S. Weather Bureau.)

As the air rises, it is cooled adiabatically so that eventually it will reach the same temperature as the surrounding air and come to rest. Before this happens, however, it may be cooled below the dew point. At once condensation begins, and the rising air column appears as a cumulus cloud whose flat base shows the critical level above which *condensation* is occurring (Figure 6.14). The bulging "cauliflower" top of the cloud represents the top of the rising warm air column, pushing into higher levels of the atmosphere. Should this convection column continue to develop, the cloud may grow to a *cumulonimbus* mass, or *thunderstorm*, from which heavy rain will issue.

The picture outlined above contains a serious defect. Any alert person will wonder why convection continues vigorously after the cloud has grown so large as to shade the ground, or after the cloud has drifted downwind away from the originally heated spot on the ground. Actually, the unequal heating of the ground served only as a trigger effect to release a spontaneous updraft, fed by *latent heat* energy liberated from the *condensing* water vapor. Recall that for every gram of water formed by condensation 600 calories of *heat* are released.

The graph of Figure 6.15*A* is a plot of altitude against air temperature. The small circles represent a small parcel of air being forced to rise steadily higher, following the same *dry adiabatic rate of* cooling shown in Figure 6.7. To the right of this line is a solid line showing the temperature of the undisturbed surrounding air; it is the normal lapse rate, such as is shown in Figure 3.3. Suppose that the air parcel is lifted from a point near the ground, where its temperature is 90°F (32°C). After the air parcel has been carried up 2000 ft (600 m), its temperature has fallen about 11 F° (6 C°) and is now 79°F (26°C); whereas the surrounding air is cooler by only about 7 F° (4 C°), and has a temperature of 83°F (28°C). The air parcel would thus be cooler than the surrounding air at 2000 ft (600 m), and if no longer forcibly carried upward, would tend to sink back to the ground. These conditions represent *stable* air, not likely to produce convection cells, because the air would resist lifting.

below the dew point so rapidly that not only do clouds form, but rain, snow, or hail is produced as well. Consider, then, how large masses of air are actually induced to rise to higher elevations. The three possible ways are (1) *convectional*, (2) *orographic*, and (3) *cyclonic* or *frontal*.

Convectional precipitation results from a *convection cell*, which is simply an updraft of warmer air, seeking higher altitude because it is lighter than surrounding air (Figure 6.14). The cell is completed by a downdraft of cooler, denser air. Suppose that on a clear, warm summer morning the sun is shining upon a landscape consisting of patches of open fields and woodlands. Certain of these types of surfaces, such as the bare ground, heat more rapidly and transmit radiant heat to the overlying air. Air over a warmer patch is thus warmed more than adjacent air and begins to rise in a tall column, much as hot air and smoke rise in a chimney. Vertical currents of this type are often called *thermals* by sailplane pilots who use them to obtain lift.

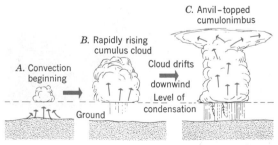

Figure 6.14   Convectional rainfall.

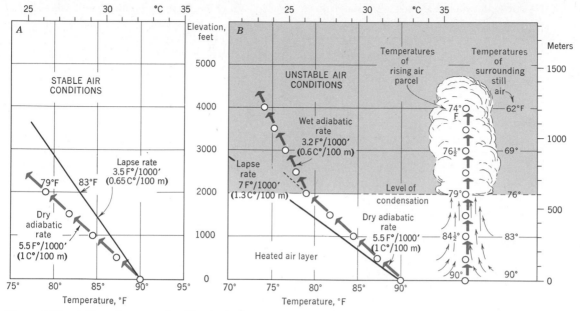

Figure 6.15 *A,* Forced ascent of stable air. *B,* Spontaneous rise of unstable air.

When the air layer near the ground is excessively heated by the sun, the lapse rate is increased (solid line reclines to lower slope, Figure 6.15*B*). The air parcel near the ground begins to rise spontaneously because it is lighter than air over adjacent, less intensely heated ground areas. Although cooled adiabatically while rising, the air parcel at 1000 ft (300 m) has a temperature of 85°F (29°C), but this is well above the temperature of the surrounding still air. The air parcel, therefore, is lighter than the surrounding air and continues its rise. At 2000 ft (600 m), the dew point is reached and condensation sets in. Now the rising air parcel is cooled at the reduced *wet adiabatic rate* of 3.2 F° per 1000 feet (0.6 C° per 100 m), because the *latent heat* liberated in condensation offsets the rate of drop due to expansion. At 3000 ft (900 m), the rising air parcel is still several degrees warmer than the surrounding air, and therefore continues its spontaneous rise.

The air described here as spontaneously rising during condensation is *unstable* in properties. In such air the updraft tends to increase in intensity as time goes on, much as a bonfire blazes with increasing ferocity as the updraft draws in greater supplies of oxygen. Of course, at very high altitudes, the bulk of the water vapor having condensed and fallen as precipitation, the energy source is gone; the convection cell then weakens and air rise finally ceases.

Unstable air, given to spontaneous convection in the form of heavy *showers* and *thunderstorms,* is most likely to be found in warm, humid areas such as the equatorial and tropical oceans and their bordering lands throughout the year, and the middle-latitude regions during the summer season.

The second precipitation-producing mechanism is described as *orographic,* which means "related to mountains." Prevailing winds or other moving masses of air may be forced to flow over mountain ranges (Figure 6.16). As the air rises on the windward side of the range, it is cooled at the adiabatic rate. If cooling is sufficient, precipitation will result. After passing over the mountain summit, the air will begin to descend the lee side of the

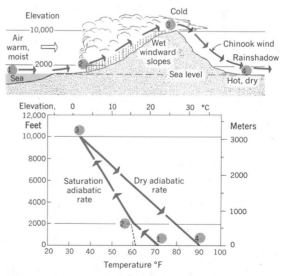

Figure 6.16 Forced ascent of oceanic air masses, producing precipitation and a rainshadow desert. (From A. N. Strahler, *The Earth Sciences,* Harper and Row, New York.)

range. Now it will undergo a warming through the same adiabatic process and, having no source from which to draw up moisture, will become very dry. A belt of dry climate, often called a *rainshadow*, may exist on the lee side of the range. Several of the important dry deserts of the earth are of this type.

Dry, warm *foehn winds* (Europe) and *chinook winds* (northwestern United States), which occur on the lee side of a mountain range, may cause extremely rapid evaporation of snow or soil moisture. These winds result from turbulent mixing of lower and upper air in the lee of the range. The upper air, which has little moisture to begin with, is greatly dried and heated when swept down to low levels.

An excellent illustration of orographic precipitation and rainshadow occurs in the far west of the United States. Prevailing westerly winds bring moist air from the Pacific Ocean over the coast ranges of central and northern California and the great Sierra Nevada Range, whose summits rise to 14,000 ft (4000 m) above sea level. (See Figure 12.15.) Heavy rainfall is experienced on the windward slopes of these ranges, nourishing great forests. Passing down the steep eastern face of the Sierras, air must descend nearly to sea level, even below sea level in Death Valley. The adiabatic heating thus caused, and a consequent drop in humidity, produces part of America's great desert zone, covering a strip of eastern California and all of Nevada.

Much orographic rainfall is actually of the convectional type, in that it takes the form of heavy convectional showers and thunderstorms. The storms are induced, however, by the forced ascent of unstable air as it passes over the mountain barrier.

A third type of precipitation is *cyclonic*. This topic cannot be fully understood until the entire subject of cyclonic storms and fronts has been developed. It will suffice here to note that in middle and high latitudes much of the precipitation occurs in cyclonic storms or eastward-moving centers of low pressure into which air is converging and being forced to rise.

Thunderstorms

A *thunderstorm* is an intense local storm associated with a large, dense cumulonimbus cloud in which there are very strong updrafts of air (Figures 6.9*H*, 6.14). *Thunder* and *lightning* normally accompany the storm, and rainfall is heavy, often of cloudburst intensity, for a short period. Violent surface winds may occur at the onset of the storm. A thunderstorm is not the cyclonic type of storm described in Chapter 7 because there is no inspiraling wind system. Instead, it may be described as a *convective storm* because it is essentially a powerful updraft of air seeking a higher elevation. The cause of precipitation and spontaneous rise of air has been explained in the discussion of unstable air and convectional precipitation.

Thunderstorm development has been studied by coordinated use of aircraft, ground weather stations, radiosonde stations, and long-range radar. Most thunderstorms consist of several *cells*, each cell being an updraft, usually with an associated downdraft. Individual cells pass through a life cycle as shown in Figure 6.17, and a storm may contain cells in various stages of development. First of three stages is the *cumulus stage*, in which the cell is a simple updraft with a steadily rising cloud top. The updraft tends to draw in air from the sides as it rises. In the *mature stage*, falling rain has created a downdraft by frictional drag on the air. Reaching the ground and spreading out, this air is cool. Upon first arriving at the ground, the

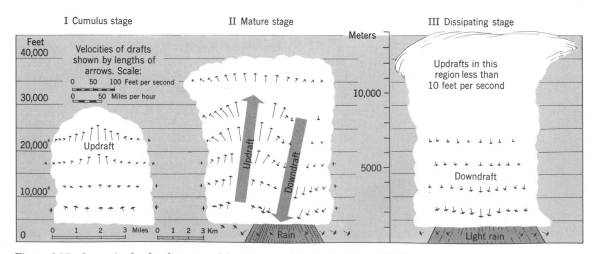

Figure 6.17   Stages in the development of thunderstorm cell. (After Horace R. Byers, 1949.)

downdraft produces strong gusts which are normally felt before or during the start of the rain.

The *dissipating stage* of the thunderstorm cell is reached after the *downdraft* has spread across the lower levels of the cell and updraft has ceased. The upper part of the *cloud* has now taken on a mushroom or *anvil form* and continues to spread out, forming an altostratus or cirrostratus layer. In a large, complex thunderstorm new cells are forming while others are dissipating, thus maintaining the storm in a condition of constant activity as it drifts along with the regional air movement.

The principal *updraft*, or chimney, of a thunderstorm cell may have amazingly rapid flow of air. The fact that 3 in (8-cm) *hailstones* require updrafts of 120 mi (200 km) per hour in order to be kept aloft while forming indicates something of the intensity. Updrafts of 70 mi (112 km) per hour are not unusual in the mature stage of the storm. In general, warm air is required for thunderstorms because only warm air can hold large quantities of water vapor. Thus thunderstorms are commonest at low latitudes and are virtually absent in polar regions. In middle latitudes thunderstorms are commonest during the summer.

Thunderstorms may be classified into several types, based on the mechanism or cause of initial lift of the air column that sets off the spontaneous growth of the storm. One common type is the *thermal*, or *air mass thunderstorm*, which is set off by thermal convection caused by solar heating of the ground and lower air layer. Figure 6.14 illustrates the process. Storms of this type are often widely scattered over a large region covered by warm, moist air. The time of occurrence is typically in the late afternoon because air temperatures near the ground reach their highest levels at this time.

In the *orographic thunderstorm*, air is forced to rise over a mountain barrier, as shown in Figure 6.16. If the air is warm and moist, with unstable properties, it readily breaks into heavy showers and thunderstorms. The torrential *monsoon rains* of the Asiatic and East Indian mountain ranges are largely of this type. For example, Cherrapunji, a hill station facing the summer monsoon air drift in northeast India, averages 426 in (1082 cm) of rainfall annually. In the arid southwestern United States, isolated mountain ranges and high plateaus receive abundant summer rain in the form of thunderstorms orographically induced. Here rich forests grow while the surrounding lower areas are barren or sparsely vegetated deserts.

Still other types of thunderstorms are set off by the forced rise of a layer of warm air over a layer of cold air. Such storms, which may be described as *frontal thunderstorms*, are explained in the following chapter.

World distribution of rainfall

Average annual rainfall for the earth is shown in Plate 1 (folded map at end of book). Lines termed *isohyets* have been drawn through places having the same annual average rainfall. This map shows how the principles of precipitation are applied to world regions.

Rainfall is very heavy, more than 80 in (200 cm), in the *equatorial zone*, where high temperatures and great oceanic expanses provide large amounts of water vapor and generally unstable atmospheric conditions. Virtually all this rainfall is convectional in origin, although the presence of mountainous belts serves locally to add the orographic effect.

Rainfall is very light over the subtropical high-pressure belts, or cells, owing to the general pattern of subsiding air warmed adiabatically and thus made extremely dry. The *deserts* of North Africa, Arabia, and Iran lie along this zone as well as those of Australia, South Africa, and the west coast of South America. This zone of aridity is extended equatorward into the trade wind belt.

The *trade winds* should not be thought of as necessarily dry winds. Wherever they blow from the ocean across a coastal belt which is hilly or mountainous, fairly heavy orographic precipitation is caused. One good illustration is Central America, another is Madagascar. In both places the eastern, or windward, coasts receive rainfall of more than 80 in (200 cm) annually.

On the east side of the oceanic subtropical highs are dry coastal belts, such as the Peruvian coast of South America and the Namib Desert of southwest Africa. Here air subsides and is adiabatically heated.

The *monsoon* winds of Asia largely control *rainfall* of the southeastern part of that continent. In summer the flow of moist tropical air from the Indian Ocean and western Pacific encounters various mountain chains to give very heavy orographic precipitation. This is vividly illustrated by the narrow strips having precipitation over 80 in (200 cm), coinciding with the Himalayas in northern India and their southeastward extensions into Burma and Malaya, and with the high western edge of the Deccan Plateau along the west coast of peninsular India. The rugged mountain chains of the East Indies receive *orographic rainfall* associated with both sets of monsoon winds which blow between Australia and Asia.

Precipitation in the middle latitudes shows clearly the effect of the prevailing westerly winds. Western continental coasts within the latitude range 35° to 60° appear as narrow strips of heavy precipitation. Most striking are the British Columbia-southeastern Alaska coast in the northern hemisphere, and the Chilean coast of southernmost South America. Here, copious orographic precipitation is induced by coastal ranges which force a rise of moist air moving in from the Pacific Ocean. The effect is less striking on the Atlantic coast of Europe and the British Isles where relief is less, but it is nevertheless clearly evident. Note, too, that New Zealand, south of 40° S lat., has a marked zoning of precipitation from west to east.

Rainshadow deserts are developed in the westerly wind belt, in the lee of coastal ranges. The best illustration is the desert Great Basin of Nevada, already cited as an example of the principle of rainshadow development. A similar relation exists in southernmost South America but is not so well developed. Even the Iberian Peninsula shows semiaridity in its eastern half, as compared with a moist western coast. The vast continent of Eurasia exhibits a generally increasing aridity from west to east. Not only is moist air from the Atlantic dried as it travels eastward, but moist tropical air from the Indian Ocean is blocked by mountain barriers on the south.

Both the eastern United States and the eastern part of Asia, including much of China, Japan, Korea, and Manchuria, are moist, even though located on the eastern, or lee, side of the continents. This is explained by the prevalence of humid masses of air which move northward from the subtropical oceans during the summer months as a part of the monsoon circulation.

In arctic regions, precipitation is very small in terms of annual total, as the map clearly shows. The atmosphere here is at prevailingly low temperatures and thus does not hold large quantities of water vapor from which precipitation can be produced. At the same time, low temperatures reduce evaporation to such an extent that there is generally abundant soil moisture and surface water in summer and snow and ice in winter.

REFERENCES FOR FURTHER STUDY

Byers, H. R. and R. R. Braham, Jr. (1949), *The thunderstorm*, U.S. Dept. of Commerce, Weather Bureau, U.S. Govt. Printing Office, Washington, D.C., 287 pp.

World Meteorological Organization (1956), *International cloud atlas*, 2 volumes, English language edition, Geneva, Switzerland.

Byers, H. R. (1959), *General meteorology*, McGraw-Hill Book Co., New York, 540 pp. See Chapters 8, 18, 20.

Battan, L. J. (1962), *Cloud physics and cloud seeding*, Doubleday and Co., New York, 144 pp.

U.S. Dept. of Commerce and Federal Aviation Agency (1965), *Aviation Weather*, U.S. Govt. Printing Office, Washington, D.C., 299 pp.

Petterssen, S. (1969), *Introduction to meteorology*, third edition, McGraw-Hill Book Co., New York, 333 pp. See Chapters 4–8.

1. In what ways is the presence of water vapor in the air important in weather and climate? What are the sources of atmospheric moisture?

2. Describe the three states in which water may exist. How, and in what amounts, is heat liberated or absorbed as water passes from one state to another?

3. Define relative humidity. What effect has a change of air temperature upon the relative humidity of the air? What is the dew point temperature? Explain the phenomenon of condensation of water vapor.

4. What is absolute humidity? Vapor pressure? Specific humidity? What use have these measures of water vapor? What ranges of specific humidity may be expected from equatorial to polar regions?

5. How is humidity of the air measured? Explain the principle of the hygrometer and psychrometer.

6. Explain the process whereby condensation and precipitation occur as a result of the lifting of a mass of air. What is the adiabatic rate? How does this differ from the normal lapse rate?

7. Of what are clouds composed? How are clouds classified? Name the common cloud types, give their height ranges, and describe their forms. What are stratiform and cumuliform clouds?

8. Name and describe the various forms of precipitation. Is an icing storm a form of precipitation? How does hail form?

9. In what units is rainfall measured? Describe a rain gauge. How is snowfall measured? How does 1 foot of snow compare with 1 foot of rainfall for relative amounts of water?

10. What are the three ways in which precipitation can occur on a large scale? Describe how a convection column of air operates to produce rainfall. What keeps the updraft in operation during a convectional storm? What is an unstable air mass?

11. Explain how orographic precipitation occurs. Why is there often a dry area, or rainshadow, on the lee side of a mountain chain? Is the air which has reached a rainshadow area normally warmer or cooler than it was at the same elevation over the coastal zone from which it came? Explain.

12. Describe the internal meteorological conditions of a thunderstorm. What size and height do these storms commonly attain? How is hail formed in a thunderstorm? What keeps a thunderstorm active? What is the source of energy?

13. Name three common types of thunderstorms. What can be said about time of day and geographical regions of occurrence of thunderstorms?

14. How can rainfall distribution be represented on a map? Briefly summarize the distribution of annual rainfall over the globe, using the pressure and wind belts as a basis for your discussion. What regions are most favorably located for heavy rainfall? How do the arctic regions compare in total annual precipitation with the equatorial and tropical regions?

CHAPTER 7
Cyclonic Storms, Air Masses, Weather Fronts

MUCH of the unsettled, cloudy weather experienced in middle and high latitudes is associated with traveling cyclones. The convergence of masses of air toward these centers is accompanied by lift of air and adiabatic cooling, which, in turn, produces cloudiness and precipitation. By contrast, much fair, sunny weather is associated with traveling anticyclones in which the air tends to subside and spread outward, causing adiabatic warming, a process that is unfavorable to the development of clouds and precipitation.

Cyclones may be very mild in intensity, passing with little more than a period of cloud cover and light rain or snow. On the other hand, if the pressure gradient is strong, winds ranging in strength from moderate to gale force may accompany the cyclone. In such a case, the disturbance may be called a *cyclonic storm*.

Moving cyclones fall into three general classes. (1) The *middle-latitude cyclone* (or *extratropical cyclone*) is typical of middle and high latitudes. It ranges in severity from a weak disturbance to a powerful storm. (2) The *tropical cyclone* is found in low latitudes over ocean areas. It ranges from a mild disturbance to the terribly destructive *hurricane*, or *typhoon*. (3) The *tornado*, although a very small storm, is an intense cyclonic vortex of enormously powerful winds. It is on a very much smaller scale of magnitude than other types of cyclones and must be treated separately.

The thunderstorm, described in Chapter 6, is a localized disturbance connected with a large cumulonimbus cloud in which there is a rapid convectional rise of air. It lacks the cyclonic spiral flow of winds. Thunderstorms often occur in large numbers within a single cyclonic storm, and occasionally tornadoes develop within the thunderstorms at the same time.

Middle-latitude cyclones

Before the advent of modern weather theory it was well known that within the middle latitudes, 35° to 65° N and S, weather changes are associated with moving centers of low and high barometric pressure, which can be drawn on the weather map and followed from day to day in the course of their eastward travel.

Lows, or cyclones, consist of oval-shaped concentric isobars, usually elongate on a northeast to southwest axis. The lows travel 25 to 30 mi (40 to 50 km) per hour, on the average, and have diameters of 500 to 1000 mi (800 to 1600 km). Cloudiness and precipitation generally are located in the eastern or southeastern half, and the northwestern part is usually a zone of clearing weather.

Between successive lows occur anticyclones, or simply *highs*, bringing cool, fair weather with distinctly drier air than in the lows. In the United States, the highs characteristically move in from the north, whereas the lows more often develop in southerly areas and move northeast, although there are many exceptions.

With the aid of periodic weather maps showing the positions of highs and lows, successful weather forecasting was carried on by experienced men who had come to know the habits of lows as a result of many years of observing their behavior. It was known that lows follow certain common paths, such as moving along the St. Lawrence Valley.

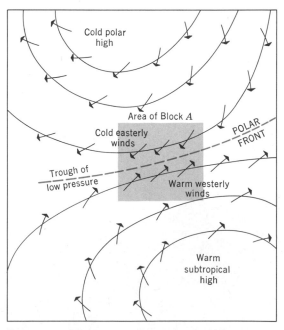

Figure 7.1 The trough between two high-pressure regions is a likely zone for development of a wave cyclone.

Wave theory of cyclones

The simple analysis of weather outlined above leaves many puzzling things unexplained. During the First World War a Norwegian meteorologist, J. Bjerknes, brought forward a new theory to explain moving cyclones and anticyclones. In middle latitudes, there exists a line of contact between cold air of arctic or polar origin and warm air of subtropical origin. This fluctuating line of contact was named the *polar front* (Figure 7.1). To any large body of the lower atmosphere having fairly uniform conditions of temperature and moisture, the term *air mass* was applied. We can say that a front separates two air masses of unlike properties.

The term *front*, used by Bjerknes, was particularly apt because of the resemblance of this feature to the fighting fronts in western Europe, then active. Just as vast armies met along a sharply defined front which moved back and forth, so

masses of cold polar air meet in conflict with warm, moist air from the subtropical regions. Instead of mixing freely, these unlike air masses remain clearly defined, but interact along the polar front in great whorls whose structure is not unlike the form of an ocean wave seen in cross section.

A series of individual blocks, Figure 7.2 shows the various stages in the life history of an extratropical (middle-latitude) cyclone. At the start of the cycle the polar front is simply a smooth boundary along which air of unlike qualities is moving in opposite directions, as shown in Figure 7.1. In Block *A* of Figure 7.2, the polar front shows a bulge, or *wave*, beginning to form. Cold air is turned in a southerly direction, warm air in a northerly direction, as if each would penetrate the domain of the other. The situation is very much as if two individuals enter a revolving door from opposite sides. Pivoting on a center pin, the door permits each to pass into the area vacated

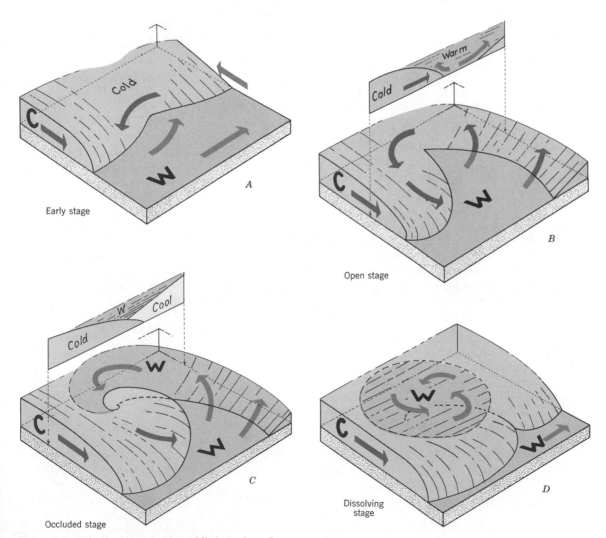

Figure 7.2   The development of a middle-latitude cyclone.

by the other. It is now necessary to digress from the series of diagrams of Figure 7.2 to consider what occurs when cold air moves into an area of warm air, or vice versa.

Cold and warm fronts

The structure of a frontal contact zone in which cold air is invading the warm-air zone is shown in Figure 7.3. A front of this type is termed a *cold front*. The colder air mass, being heavier, remains in contact with the ground and forces the warmer air mass to rise over it. The slope of the cold front surface is greatly exaggerated in the figure, being actually of the order of slope of 1 in 40 to 1 in 80 (meaning that the slope rises 1 foot vertically for every 80 feet of horizontal distance). Cold fronts are associated with strong atmospheric disturbance; the warm air thus lifted often breaks out in violent thunderstorms. In some instances these violent thunderstorms occur along a line well in advance of the cold front, a *squall line*. Such lines of thunderstorms can be seen on the radar screen (Figure 7.4).

Figure 7.5 illustrates a *warm front* in which warm air is moving into a region of colder air. Here, again, the cold air mass remains in contact with the ground, and the warm air mass is forced to rise as if ascending a long ramp. Warm fronts have lower slopes than cold fronts, being of the order of 1 in 80 to as low as 1 in 200. Moreover, warm fronts are commonly attended by stable atmospheric conditions and lack the turbulent air motions of the cold front. Of course, if the warm air is unstable, it will develop convection cells and there will be heavy showers and thunderstorms. To the list of thunderstorm types described in Chapter 6 can now be added the thunderstorms of both cold front and warm front origin.

Cold fronts normally move along the ground at a faster rate than warm fronts. Hence, when both types are in the same neighborhood, as they are

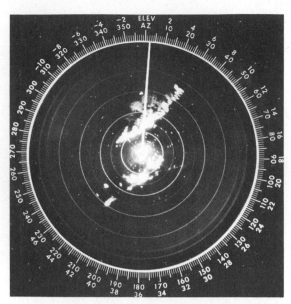

Figure 7.4 Photograph of a radar screen on which lines of thunderstorms show as bright patches. The heavy circles are spaced 50 nautical miles (80 km) apart. (Courtesy of U.S. Weather Bureau.)

in the cyclonic storm, the cold front may overtake the warm front. A curious combination known as an *occluded front* then results (Figure 7.6). The colder air of the fast-moving cold front remains next to the ground, forcing both the warm air and the less cold air to rise over it. The warm air mass is lifted completely free of the ground. The relations between warm, cold, and occluded fronts will now be introduced into the life history of the cyclone.

In Figure 7.2*B* the wavelike disturbance along the polar front has deepened and intensified. Cold air is now actively pushing southward along a cold front; warm air is actively moving northeastward along a warm front. Each front is convex in the direction of motion. The zone of precipitation is

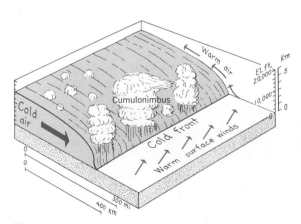

Figure 7.3 A cold front.

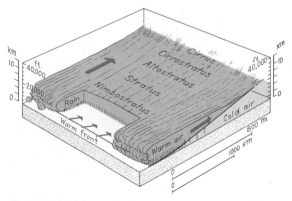

Figure 7.5 A warm front.

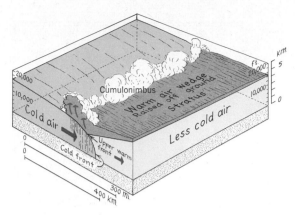

Figure 7.6 An occluded warm front.

now considerable, but wider along the warm front than along the cold front. In a still later stage the more rapidly moving cold front has reduced the zone of warm air to a narrow sector. In Block *C*, the cold front has overtaken the warm front, producing an occluded front and forcing the warm air mass off the ground, isolating it from the parent region of warm air to the south. The source of moisture and energy thus cut off, the cyclonic storm gradually dies out and the polar front is reestablished as originally (Block *D*).

Cyclones on the daily weather map

Further details and characteristics of a wave cyclone are illustrated in Figure 7.7, in which two weather maps are shown. These have been redrawn with only slight changes from the U.S. Weather Bureau daily maps for April 3 and 4, 1963. Map *A* shows a cyclone in a stage approximately equivalent to Block *B* of Figure 7.2. The storm is centered over western Minnesota and is moving northeastward. Note the following points: (*a*) Isobars of the low are closed to form an oval-shaped pattern. (*b*) Isobars make a sharp V where crossing cold and warm fronts. (*c*) Wind directions, indicated by arrows, are at an angle to the trend of the isobars and form a pattern of counterclockwise inspiraling. (*d*) In the warm-air sector there is northward flow of warm, moist tropical air toward the direction of the warm front. (*e*) There is a sudden shift of wind direction accompanying the passage of the cold front, as indicated by the widely different wind directions at stations close to the cold front, but on opposite sides. (*f*) There is a severe drop in temperature accompanying the passage of the cold front, as shown by differences in temperature readings at stations on either side of the cold front. (*g*) Precipitation, shown by shading, is occurring over a broad zone near the warm front and in the central area of the cyclone, but extends as a thin band down the length of

the cold front. (*h*) Cloudiness, shown by degree of blackness of station circles, is great over the entire cyclone. (*i*) The low is followed on the west by a high (anticyclone) in which low temperatures and clear skies prevail. (*j*) The 32° F (0° C) isotherm crosses the cyclone diagonally from northeast to southwest, showing that the southeastern part is warmer than the northwestern part.

A cross section through Map *A* along the line AA′ shows how the fronts and clouds are related. Along the warm front is a broad area of stratiform clouds. These take the form of a wedge with a thin leading edge of cirrus. Westward this thickens to altostratus, then to stratus, and finally to nimbostratus with steady rain. Within the warm air mass sector the sky may partially clear with scattered cumulus. Along the cold front are violent thunderstorms with heavy rains, but this is along a narrow belt that passes quickly.

The second weather map, Map *B*, shows conditions 24 hours later. The cyclone has moved rapidly northeastward into Canada, its path shown by the line labeled *storm track*. The center has moved about 800 mi (1300 km) in 24 hours, a speed of just over 40 mi (65 km) per hour. The cyclone has occluded. An occluded front replaces the separate warm and cold fronts in the central part of the disturbance. The high-pressure area, or tongue of cold polar air, has moved in to the west and south of the cyclone, and the cold front is passing over the eastern and Gulf Coast states. Within closed isobars around the anticyclone the skies are clear and winds are weak. In another day the entire storm will have passed out to sea, leaving the eastern United States in the grip of cold but clear weather. A cross section below the map shows conditions along the line BB′, cutting through the occluded part of the storm. Note that the warm air mass is being lifted higher off the ground and is giving heavy precipitation.

Long observation on the movements of cyclones and anticyclones has revealed that certain tracks are most commonly followed. Figure 7.8 is a map of the United States showing these common paths. Notice that whereas some cyclonic storms travel across the entire United States from places of origin in the North Pacific, such as the Aleutian low, others originate in the Rocky Mountain region, the central states, or the Gulf Coast. Most tracks converge toward the northeastern United States and pass out into the North Atlantic, where they tend to concentrate in the region of the Icelandic low. General world distribution of paths of cyclonic storms is shown in Figure 7.9. Notice the heavy concentration in the neighborhood of the Aleutian and Icelandic lows. Extratropical cyclones commonly form in succession to travel in a chain across the north Atlantic and north

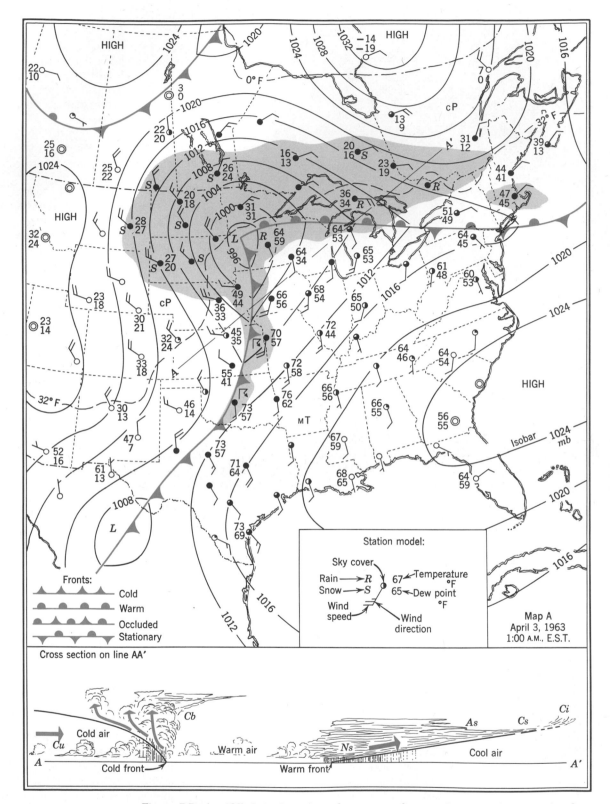

Figure 7.7 A middle-latitude cyclone shown on surface weather maps for successive days. Pressures are given in millibars; temperatures in degrees Fahrenheit. Areas experiencing

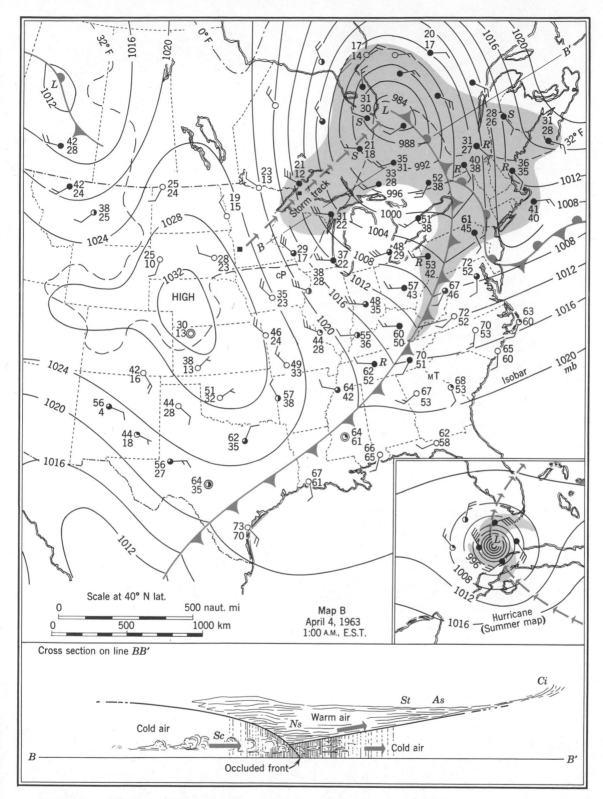

Scale at 40° N lat.

| 0 | 500 naut. mi |
| 0 | 500 | 1000 km |

Map B
April 4, 1963
1:00 A.M., E.S.T.

Hurricane
(Summer map)

Cross section on line *BB'*

precipitation are shaded. See Figure 5.12 for explanation of wind symbols. (Modified and simplified from Daily Weather Map of the U.S. Weather Bureau.)

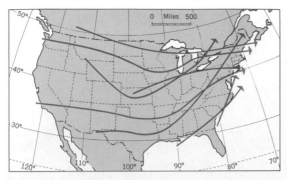

Figure 7.8 Common tracks taken by middle-latitude cyclones passing across the United States. (After Bowie and Weightman.)

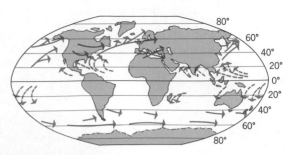

Figure 7.9 Principal tracks of middle-latitude cyclones are shown by solid lines; those of tropical cyclones by dashed lines. (After Petterssen, *Introduction to Meteorology*.)

Pacific oceans. Figure 7.10, a world weather map, shows several such *cyclone families*.

In the southern hemisphere, storm tracks are more nearly along a single lane, following the parallels of latitude. This appears to be the result of uniform ocean surface throughout the middle latitudes, only the southern tip of South America breaking the monotonous oceanic expanse. Furthermore, the polar-centered icecap of Antarctica provides a centralized source of polar air.

Air masses and source regions

The wave theory of extratropical cyclones places much emphasis upon the interaction of air masses, which differ in physical properties but which are essentially of uniform nature internally. The general subdivision of air masses of middle latitudes into two major groups, *polar* and *tropical*, has already been emphasized in explaining the development of the cyclonic storms. The distinctive prop-

erties of an air mass are developed when the air layer lies stagnant over a body of land or ocean, the *source region*. As the air mass leaves the source region, to travel across other kinds of land or ocean surfaces, the air is cooled or heated, and gains or loses moisture, as the case may be. Thus, an air mass evolves or changes as it follows a characteristic path, or *trajectory*, which may take it thousands of miles from the source region.

The polar air masses are divided into *maritime* and *continental* types. The continental polar air masses of North America (symbol, cP) originate over north-central Canada and are characterized by low temperature and low moisture content (Figure 7.11). These air masses form tongues of cold air, which periodically extend south and east from the source region to produce anticyclones accompanied, in winter, by low temperatures and clear skies. Over the North Pacific and Bering Straits originate the maritime polar air masses

Figure 7.10 A daily weather map of the world for a given day during July or August might look like this map, which is a composite of typical weather conditions. (After M. A. Garbell.)

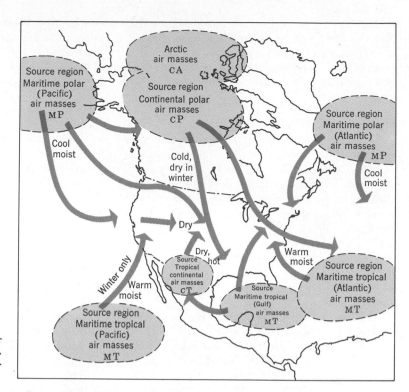

Figure 7.11 North American air mass source regions and trajectories. (After Haynes, U.S. Dept. Commerce.)

(symbol, MP). With ample opportunity to absorb moisture both over the source region and throughout their travel southeastward to the west coast of North America, these air masses are characteristically cool and moist, with a tendency in winter to become unstable, giving heavy precipitation over coastal ranges. Note that the polar air masses (MP, cP) originate in the subarctic latitude zone (Figure 4.5), not in the polar latitude zone. The meteorological definition of the word "polar" for air masses has long been in use and has international acceptance, hence cannot be changed to conform with the latitude zones defined in Chapter 4.

Another maritime polar air mass of the North American region originates over the North Atlantic Ocean (Figure 7.11). It, too, is cool and moist. Because of its location, east of the North American continent, it only occasionally reaches the United States when drawn in over New England by a favorably placed cyclone lying south of New England. This situation brings on a *northeaster*, a storm characterized by northeast winds, penetrating cold, and misting rain. The northeaster is restricted to the northeastern states.

Of the tropical air masses, there are also maritime and continental types. The commonest visitor to the central and eastern states is the maritime tropical air mass (symbol, MT) from the Gulf of Mexico (Figure 7.11). It moves northward, bringing warm, moist, unstable air over the eastern part of the country. In the summer, particularly, this air mass brings hot, sultry weather to the east. It gives frequent thunderstorms. Closely related is a maritime tropical air mass from the Atlantic Ocean east of Florida, over the Bahamas. It has characteristics similar to those of the tropical Gulf air mass and brings similar weather conditions.

During the summer there originates over northern Mexico, western Texas, New Mexico, and Arizona a tropical continental air mass (symbol, cT) which is hot and dry. This air mass does not travel widely, but governs weather conditions over the source region.

Over the Pacific Ocean in the cell of high pressure located to the southwest of Lower California is a source region of another maritime tropical air mass. It visits the United States only in winter and affects only the southern coast of California.

Besides the air masses of middle latitudes already described, there are additional groups typical of polar and equatorial locations. Over the Arctic Ocean and its bordering lands of the arctic zone there develops the arctic air mass (symbol, A) which is extremely cold and stable. When this air mass invades the United States, it produces a severe cold wave. The Antarctic continent provides a source region for extremely cold air masses, designated as antarctic air masses (symbol, AA).

Over equatorial oceans, in the trough of low pressure toward which the trade winds converge, air becomes very moist, warm, and unstable. It is then classified as an equatorial air mass (symbol, мE). Invasion of southeastern Asia by equatorial air masses occurs during the wet monsoon season and constitutes the source of moisture for torrential rains.

Tropical and equatorial weather

As the study of weather conditions has been extended along modern lines into the low latitudes of the globe, largely as a result of enormously increased aircraft operation during the Second World War, a much clearer picture has emerged of the relations of air masses and storms in these regions.

Figure 7.10 shows a typical set of weather conditions in the subtropical and equatorial parts of the world, in addition to showing the general scheme of traveling cyclonic waves in middle and high latitudes.

Fundamentally the arrangement consists of two rows of high-pressure cells, one or two cells to each land or ocean body. The northern row lies approximately along the Tropic of Cancer; the southern row, along the Tropic of Capricorn. Between the subtropical highs lies the equatorial trough of low pressure. Toward this trough converge the north-

east and southeast trade winds. For this reason the trough is called the *Intertropical convergence zone*. At higher levels in the troposphere, the air flow is almost directly from east to west in the form of persistent *tropical easterlies*, described in Chapter 4.

One of the simplest forms of weather disturbance is an *easterly wave*, a slowly moving trough of low pressure within the belt of tropical easterlies. These waves occur in a zone 5° to 30° latitude over oceans both north and south of the equator, but not over the equator itself. Figure 7.12 is a simplified weather map of an easterly wave, showing isobars, winds, and the zone of rain. The wave is simply a series of indentations in the isobars to form a shallow pressure trough. Note that it travels westward, perhaps 200 to 300 mi (325 to 500 km) per day. Air flow tends to converge on the eastern, or rear, side of the wave axis. This causes the moist air to be lifted and to break out into scattered showers and thunderstorms. The rainy period may last for a day or two.

Another related disturbance is an *equatorial wave*, or *weak equatorial low*, which forms in the center of the equatorial trough (Figure 7.13). Although the normal air flow is from east to west, there develops an eddy in the tropical easterlies.

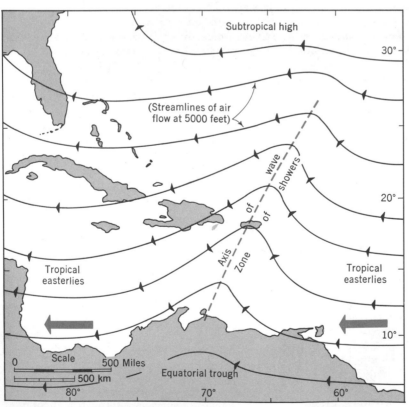

Figure 7.12   An easterly wave passing over the West Indies. (After Riehl.)

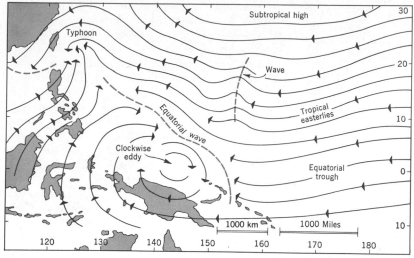

Figure 7.13   An equatorial wave. (After Riehl.)

Here air flow is locally reversed and tends to run opposite to the main stream. The result is a weak low-pressure center, the equatorial wave, into which moist equatorial air masses converge, causing rainfall from many individual convectional storms within the low. Several such weak lows are shown on the world weather map (Figure 7.10), lying along the equatorial trough. Because the map is for a day in July or August, the trough is shifted well north of the equator.

Another distinctive feature of tropical weather is the occasional penetration of powerful tongues of cold polar air from the middle latitudes into very low latitudes. These are known as *polar outbreaks* and bring unusually cool, clear weather with strong, steady winds moving behind a cold front with squalls. The polar outbreak is best developed in the Americas. Outbreaks which move southward from the United States over the Caribbean Sea and Central America are called *northers* or *nortes*; those which move north from Patagonia into tropical South America are called *pamperos* or *friagems*. One such outbreak is shown over South America on the world weather map (Figure 7.10).

Tropical cyclones

One of the most powerful and destructive types of cyclonic storms is the *tropical cyclone*, otherwise known as the *hurricane* or *typhoon*. The storm develops over oceans in latitudes 8° to 15° N and S, but not close to the equator, where the Coriolis force is extremely weak. In many cases an easterly wave simply deepens and intensifies, growing into a deep, circular low. High sea-surface temperatures which are over 80°F (27°C) in these latitudes, are of basic importance in the environ-ment of storm origin. Warming of air at low level creates instability and predisposes toward storm formation. Once formed, the storm moves westward through the trade wind belt. It may then curve northwest and north, finally penetrating well into the belt of westerly winds. The tropical cyclone is an almost circular storm center of extremely low pressure into which winds are spiraling with great speed, accompanied by very heavy rainfall (Figure 7.14). The diameter of the storm may be 100 to 300 mi (150 to 500 km); the wind velocities range from 75 to 125 mi (120 to 200 km) per hour, sometimes much more; and the barometric pressure in the center commonly falls to 965 mb (28.5 in, 72.4 cm) or lower (Figure 7.15).

A brief description of the passage of a tropical cyclone at sea might be as follows. During the day

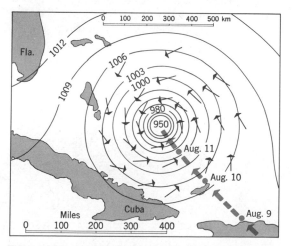

Figure 7.14   A hurricane (tropical cyclone) of the West Indies.

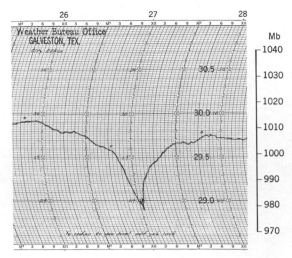

Figure 7.15 Trace sheet from a barograph at Galveston, Texas, during the hurricane of July 27, 1943. (U.S. Weather Bureau.)

preceding the storm the air is generally calm, the pressure somewhat above normal, and the sky shows cirrus clouds in long streamers, seeming to originate from a distant point on the horizon. The cirrus may be veil-like, giving a halo to the sun or moon and producing a red sunset. A long swell is felt on the sea, this being the train of dying storm waves that have outrun the slowly moving storm center. As the storm approaches, the barometer begins to fall. Wind springs up. A great dark wall of cloud approaches. When this envelopes the ship, torrential rainfall begins. The wind rises quickly to terrifying intensity. Great waves break over the vessel; the spray blows in continuous sheets which reduce visibility almost to zero.

This terrible storm continues for several hours and is abruptly followed by total calm and clearing skies, and sometimes by a sharp rise in temperature. The barometer has now reached its lowest point and the vessel is in the calm *central eye* of the storm. This is merely a hollow vortex produced by the rapid spiraling of air in the storm, comparable to the funnel-like air hole in the center of a vortex of water passing down a drain. Although the air is clear and calm, seas are mountainous and rise in great peaklike masses which are of gravest peril to the vessel. The period of calm may last a half hour. Then a great dark wall of cloud strikes the vessel and winds of high velocity again set in, but this time, in reverse direction to those of the first half of the storm. For several hours more the full fury of the storm rages, then gradually the winds abate, the clouds break, and fair weather returns.

World distribution of tropical cyclones is limited to six regions, all of them over tropical and subtropical oceans (Figure 7.16): (1) West Indies, Gulf of Mexico, and Caribbean Sea; (2) western North Pacific, including the Philippine Islands, China Sea, and Japanese Islands; (3) Arabian Sea and Bay of Bengal; (4) eastern Pacific coastal region off Mexico and Central America; (5) south Indian Ocean, off Madagascar; and (6) western South Pacific, in the region of Samoa and Fiji Islands and the east coast of Australia. Curiously enough, these storms are unknown in the South Atlantic. Tropical cyclones never originate over land, although they often penetrate the margins of continents.

Paths, or tracks, of tropical cyclones of the North Atlantic (Figure 7.17) show that most of the storms originate at 10° to 20° latitude, travel westward and northwestward through the trades, then turn northeast at about 30° to 35° latitude into the zone of the westerlies. Here the intensity lessens and the storms change into typical middle-latitude cyclones. In the trade wind belt the cyclones travel some 6 to 12 mi (10 to 20 km) per hour, in the westerlies, from 20 to 40 mi (30 to 60 km) per hour.

The occurrence of tropical cyclones is restricted to certain seasons of year, depending on the global location of the storm region. Those of the West

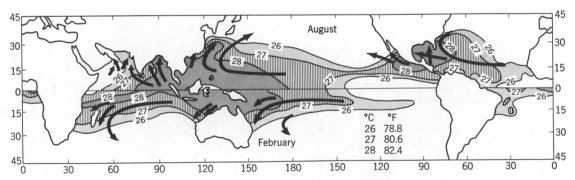

Figure 7.16 Typical paths of tropical cyclones in relation to sea surface temperatures (°C) in summer of the respective hemisphere. (After Palmen, 1948.)

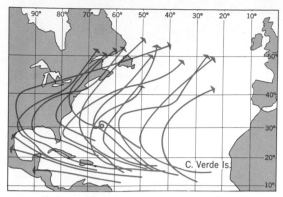

Figure 7.17   Tracks of some typical hurricanes occurring during August. (After U.S. Navy Oceanographic Office.)

Figure 7.18   This devastation along the south coast of Haiti was caused by Hurricane Flora on October 3, 1963. (Miami News photo by Chief Photographer Jay Spencer.)

Indies, and off the western coast of Mexico, occur largely from May through November, with maximum frequency in late summer or early autumn. Those of the western North Pacific, Bay of Bengal, and Arabian Sea are spread widely through the year but are dominant from May through November. Those of the South Pacific and south Indian oceans occur from October through April. Thus, they are restricted to the warm season in each hemisphere.

The geographical importance of tropical cyclones lies in their tremendously destructive effect upon island and coastal habitations (Figure 7.18). Wholesale destruction of cities and their inhabitants has been reported on several occasions. A terrible hurricane which struck Barbados in the West Indies in 1780 is reported to have torn stone buildings from their foundations, destroyed forts, and carried cannon more than a hundred feet from their locations. Trees were torn up and stripped of their bark. More than 6000 persons perished there.

Coastal destruction by storm waves and greatly raised sea level is perhaps the most serious effect of tropical cyclones. Where water level is raised by strong wind pressure, great breaking storm waves attack ground ordinarily far inland of the limits of wave action. A sudden rise of water level, known as a *storm surge*, may take place as the hurricane moves over a coastline. Ships are lifted bodily and carried inland to become stranded. If high tide accompanies the storm the limits reached by inundation are even higher. The terrible hurricane disaster at Galveston, Texas, in 1900 was wrought largely by a sudden storm surge inundating the low coastal city and drowning about 6000 persons. At the mouth of the Hooghly River on the Bay of Bengal, 300,000 persons died as a result of inundation by a 40-foot (12-meter) storm surge which accompanied a severe tropical cyclone in 1737. Low-lying coral atolls of the western Pacific may be entirely swept over by wind-driven sea water, washing away palm trees and houses and drowning the inhabitants.

Of importance, too, is the large quantity of rainfall produced by tropical cyclones. A considerable part of the summer rainfall of certain coastal regions can often be traced to a few such storms.

Tornadoes

The smallest but most violent of all known storms is the *tornado*. It seems to be a typically

Figure 7.19   This funnel cloud was photographed by William L. Males at Amarillo, Oklahoma, on May 4, 1961. The tornado was less than one mile from the observer when the picture was taken.

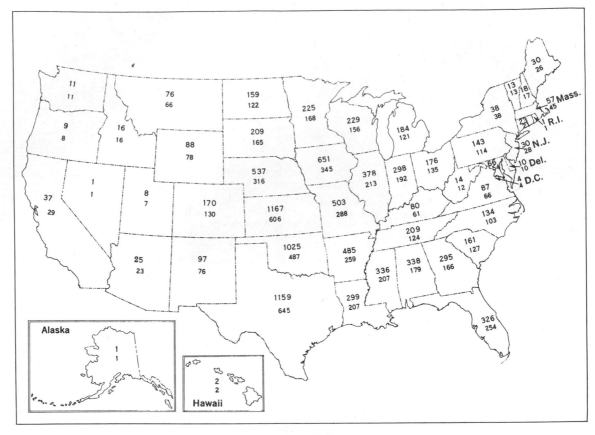

Figure 7.20 Tornado frequency by states for the period 1916–1960. The upper figure shows the total number of tornadoes reported in each state during the 45-year period; the lower figure shows the total number of days in the same period on which one or more tornadoes were reported in some part of the state. (From U.S. Weather Bureau.)

American storm, being most frequent and violent in the United States, although occurring in Australia in substantial numbers and reported occasionally in other places in middle latitudes. Tornadoes are also known throughout tropical and subtropical regions of the globe.

The tornado is a small, intense cyclone in which the air is spiraling at tremendous velocity. It appears as a dark *funnel cloud* (Figure 7.19), hanging from a large cumulonimbus cloud. At its lower end the funnel may be from 300 to 1500 ft (90 to 460 m) in diameter. The funnel appears dark because of the density of condensing moisture, dust, and debris swept up by the wind.

Wind speeds in a tornado exceed anything known in other storms. Estimates of wind speed run to as high as 500 mi (800 km) per hour. There is, in addition, a violent updraft in the funnel. As the tornado moves across the country the funnel writhes and twists. The end of the funnel cloud may alternately sweep the ground, causing complete destruction of anything in its path, and rise in the air to leave the ground below unharmed. Tornado destruction occurs both from the great

wind stress and from the sudden reduction of air pressure in the vortex of the cyclonic spiral. Closed houses literally explode. It is even reported that the corks will pop out of empty bottles, so great is the difference in air pressure.

Tornadoes occur as parts of powerful cumulonimbus clouds in the squall line that travels in advance of a cold front. They seem to originate where turbulence is greatest. They are commonest in the spring and summer but occur in any month. Where maritime polar (mP) air lifts warm, moist tropical (mT) air on a cold front, conditions may become favorable for tornadoes. They occur in greatest numbers in the Mississippi Valley region and are rare over mountainous and forested regions. They are almost unknown from the Rocky Mountains westward and are relatively few on the eastern seaboard (Figures 7.20 and 7.21).

Devastation from a tornado is complete within the narrow limits of its path (Figure 7.22), but fortunately the storms are uncommon and the total danger small. Even in states having the most tornadoes, deaths from automobile and other accidents are a very much more serious danger.

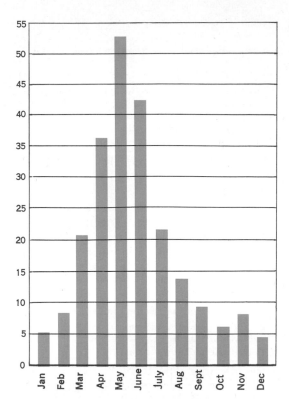

Figure 7.21 Average number of tornadoes reported in each month in the United States for the period 1916–1960. (From U.S. Weather Bureau.)

Figure 7.22 Tornado devastation at Ionia, Iowa, on April 23, 1948, included this store in which two persons were killed. (Courtesy of U.S. Weather Bureau and *Des Moines Register.*)

Storm cellars built completely below ground provide satisfactory protection if they can be reached in time. Although a tornado can often be seen or heard approaching, a cold front passing during the hours of darkness, as it often does, may present no warning. The U.S. Weather Bureau maintains a tornado forecasting and warning system. Whenever weather conditions conspire to favor tornado development, the danger area is alerted and systems for observing and reporting a tornado are set in readiness. Communities in the paths of tornadoes may thus be warned in time for inhabitants to take shelter.

Waterspouts are similar in structure to tornadoes but form at sea under cumulonimbus clouds. They are smaller and less powerful than tornadoes. Sea water may be lifted 10 ft (3 m) above the sea surface, and the spray is carried higher. Water-

spouts are commonly found in subtropical waters of the Gulf of Mexico and off the southeastern coast of the United States and seem to result from air turbulence occurring when continental air masses spread out over these oceans.

REFERENCES FOR FURTHER STUDY

Flora, S. D. (1954), *Tornadoes of the United States*, Univ. of Oklahoma Press, Norman, 221 pp.

Flora, S. D. (1956), *Hailstorms of the United States*, Univ. of Oklahoma Press, Norman, 201 pp.

Battan, L. J. (1961), *The nature of violent storms*, Anchor Science Study Series, Doubleday and Co., New York, 158 pp.

Dunn, G. E. and B. I. Miller (1964), *Atlantic hurricanes*, Louisiana State Univ. Press, Baton Rouge, 377 pp.

Riehl, H. (1965), *Introduction to the atmosphere*, McGraw-Hill Book Co., New York, 365 pp.

Helm, T. (1967), *Hurricanes: weather at its worst*, Dodd, Mead and Co., New York, 234 pp.

REVIEW QUESTIONS

1. What varieties of cyclonic storms exist? Is the thunderstorm a cyclonic storm?

2. Describe the weather conditions associated with a middle-latitude cyclone, or low, in the central and eastern United States. Include discussion of the pressure pattern, winds, cloudiness and precipitation areas, temperatures, size, and rate of travel.

3. Describe the weather conditions associated with an anticyclone, or high, such as might normally follow the cyclone mentioned in the previous question.

4. Explain the wave theory of middle-latitude weather which was originated by J. Bjerknes. What is the polar front? What is an air mass?

5. How does a cyclone develop and change along the polar front? Explain a cold front and the air mass relationships involved in it. In a similar manner, describe and explain a warm front and an occluded front.

6. Describe the changes in wind direction and strength, cloudiness and precipitation, and temperature that an observer experiences when a middle-latitude cyclone passes with its center north of the observer. Do the same for a cyclone in which the center passes south of the observer.

7. What is the prevailing direction of movement of cyclonic storms in the United States? Do storm tracks show any tendency to concentrate at particular places in the northern hemisphere? If so, where? Do you think this may explain the average pressure conditions in these areas as shown on the world pressure maps?

8. How are air masses classified and designated? Discuss the air masses of North America, giving their source regions, paths of movement, and generally associated weather characteristics. What air masses are found over polar regions? over the equatorial belt?

9. What is an easterly wave? What weather does it bring? What is an equatorial wave? What weather does it bring?

10. Describe polar outbreaks of tropical regions. What names do these outbreaks have locally in the Central American area? in South America?

11. Describe the weather elements in a tropical cyclone, or hurricane. Where do these storms originate? What paths do they normally follow? Explain the calm central eye of a tropical cyclone.

12. Name the regions of the world where tropical cyclones occur. Tell the seasons of occurrence of storms in each locality. How is season of storms related to position of the equatorial trough?

13. Describe some of the destructive effects of tropical cyclones along low-lying coasts. How is ocean level affected by these storms?

14. What is a tornado? Where and how does it occur? What wind velocities are thought to exist in a tornado? Why are tornadoes in the United States commonest in the spring? What is a waterspout?

CHAPTER 8
Climate Classification and Climatic Regimes

THE importance of climate as a geographical factor is so marked, and reaches into so many aspects of human life, that overemphasis would be difficult.

Climate determines to a large extent the type of soil and natural vegetation in a given region and hence influences the utilization of the land, whether for crop cultivation, forest, or grazing. Together with terrain factors of relief and steepness of slope, then, climate in part determines the ability of land to support a population. To be sure, the introduction of engineering works has provided irrigation and power where it might not otherwise be available, but the distribution of world population still reflects strongly the advantages of favorable climate and terrain.

Climate exerts an important influence on physiology. Cold weather, particularly that involving alternations of cold and mild periods with varying degrees of cloudiness, precipitation, and windiness, is believed to stimulate human mental and physical activity.

A contributing factor to economic development in middle latitudes is that freezing temperatures reduce the activity of many organisms responsible for disease. In warm, humid tropical and equatorial zones a great variety of parasitic organisms thrives, together with numerous host animals necessary in their life cycles. Malaria and yellow fever, borne by the mosquito, and sleeping sickness, carried by the tsetse fly, are well-known examples. Fungi in excessively humid, warm areas hasten the corrosion and decay of clothing, furniture, and machines.

Classification of climate

As in dealing with all other natural phenomena, the scientist attempts to devise schemes of classification that will include all variations in climate, yet permit them to be placed in several clearly defined and easily distinguished groups. Success is only partially achieved. Nature resists simple categorical schemes.

Maps showing the extent of climate zones are often misleading in that a solid, thin line separates two zones whose shading or color pattern is uniform to the line of contact. Actually, most of these boundaries are transitional, so that intergradation of patterns or colors might be more suitable. Only where the abrupt face of a mountain range creates a sharp climate boundary are the conventional map symbols appropriate. A second misleading quality of most climate maps and classifications is the restriction of the climate zones to land areas. Oceans, though having smaller contrasts in climate, nevertheless can be fitted into the general scheme.

If the weather elements—temperature, pressure, winds, and precipitation—are taken as bases of classification, some useful systems may be devised. The distribution of natural vegetation and soils might suggest still other types of classifications.

Temperature as a basis of climate classification

The general parallelism of isotherms with parallels of latitude has already been pointed out in an earlier chapter and was perhaps the first basis of climate classification. In Chapter 4 a series of latitude zones was defined: equatorial, tropical, subtropical, middle-latitude, subarctic, arctic, and polar (Figure 4.5). Unfortunately, because of the contrasting thermal properties of land and water bodies, we cannot correlate temperature regimes directly with latitude zones.

With intelligent application temperature has been made a fundamental factor in most climate classifications. Three major climate groups can be defined: (1) winterless climates of low latitudes, (2) climates of middle latitudes with both a summer and a winter season, and (3) summerless climates of high latitudes. A winterless climate is commonly defined as one in which no month of the year has a monthly mean temperature lower than 64.4°F (18°C). The approximate position of the appropriate isotherm is shown on a world map (Figure 8.1). Note that the isotherm shows a considerable latitude range, bending equatorward over the cool west-coast ocean currents and over the land in north Africa and Australia.

A summerless climate is commonly defined as one in which no month has a monthly mean temperature higher than 50°F (10°C). The appropriate isotherm, shown in Figure 8.1, has a wide latitude range in the northern hemisphere, being

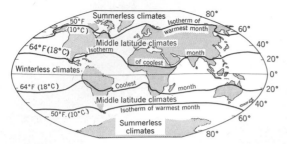

Figure 8.1 Climate groups based on temperature.

farthest poleward over the landmasses of North America and Eurasia, but dipping to middle latitudes over the intervening oceans. The 50°F (10°C) isotherm of the warmest month closely coincides with the northernmost limit of tree growth, hence it separates the regions of boreal forests from the treeless arctic tundra. Here then, is an example of a thermal climatic boundary selected to coincide with a natural vegetation boundary.

Climates having both a summer and a winter season lie in the regions between the two boundary isotherms described above and constitute a middle-latitude group.

Temperature alone as a basis of climate classification is unsatisfactory because humid and desert regions receive no distinction. An obvious remedy is a further subdivision according to precipitation.

Precipitation as a basis for climate classification

The profound effect of precipitation on nature of vegetation, drainage systems, soil moisture, and ground water, makes it desirable to consider the amount and seasonal distribution of rainfall and snowfall as a basis of climate classification.

Five precipitation ranges are as follows:

Climatic Type	Rainfall Type	Annual Rainfall Inches	Cm
Arid	Scanty	0–10	0–25
Semi-arid	Light	10–20	25–50
Subhumid	Moderate	20–40	50–100
Humid	Heavy	40–80	100–200
Very wet	Very heavy	More than 80	More than 200

A climate map on this basis would be the same as the mean annual rainfall map (Plate 1). An important refinement of this classification would be the subdivision of classes according to distribution of precipitation throughout the year, whether uniform or seasonal.

As a useful climate classification this one fails because it groups cold arctic climates together with the hot deserts of low latitudes. Evaporation, which determines what proportion of the rainfall

will remain in the ground, is controlled by air temperature. In general, cold climates are effectively humid with the same meager precipitation that produces very dry deserts in hot subtropical and tropical latitudes. It would seem, then, that a useful climate system must combine temperature and precipitation classes. Although such combining will increase the number of climate types recognized, each type will be an effective description of a distinctive environment.

An example of intelligent combination of monthly mean temperature and precipitation data to derive climatic zones is illustrated in Table 8.1. The temperature units are fixed in equal increments of 10 C° (18 F°), whereas the precipitation range is varied to suit temperature range. Consequently, "humid" denotes 4 to 12 in (10 to 30 cm) of precipitation in a month when monthly mean temperature is above 86°F (30°C), but denotes only 1 to 3 in (2.5 to 7.5 cm) when temperature mean is between 14° and 32°F (−10° and 0°C). Climate is described according to such combinations of forms as "warm-dry," "cool-humid," or "cold-dry." Because this system is applied to a given month of the year, a world climate map based upon it shows conditions only for the specified month (Figure 8.2). Twelve such maps would be needed to give an analysis of world climate throughout the year. The system was used by the U.S. Air Force for a set of northern hemisphere monthly climatic maps published in 1947.

TABLE 8.1 CLIMATIC ZONES BASED ON AVERAGES OF A GIVEN MONTH[a]

	Dry	Humid	Wet	
Above 30°C				Hot (Above 86°F)
	Under 4 in (10 cm)	4–12 in (10–30.5 cm)	Over 12 in (30.5 cm)	
30°C				86°F
	Under 3 in (7.5 cm)	3–12 in (7.5–30.5 cm)	Over 12 in (30.5 cm)	Warm
20°C				68°F
	Under 2 in (5 cm)	2–8 in (5–20 cm)	Over 8 in (20 cm)	Mild
10°C				50°F
	Under 1 in (2.5 cm)	1–5 in (2.5–12.7 cm)	Over 5 in (12.7 cm)	Cool
0°C				32°F
	Under 1 in (2.5 cm)	1–3 in (2.5–7.5 cm)	Over 3 in (7.5 cm)	Cold
−10°C				14°F
		Very cold (−4° to 14°F)		
−20°C		−4°F		
		Extreme cold (−40° to −4°F)		
−40°C		−40°F		
		Ultra cold (Below −40°F)		

[a]After U.S. Air Force, Aeronautical Chart Service, 1947.

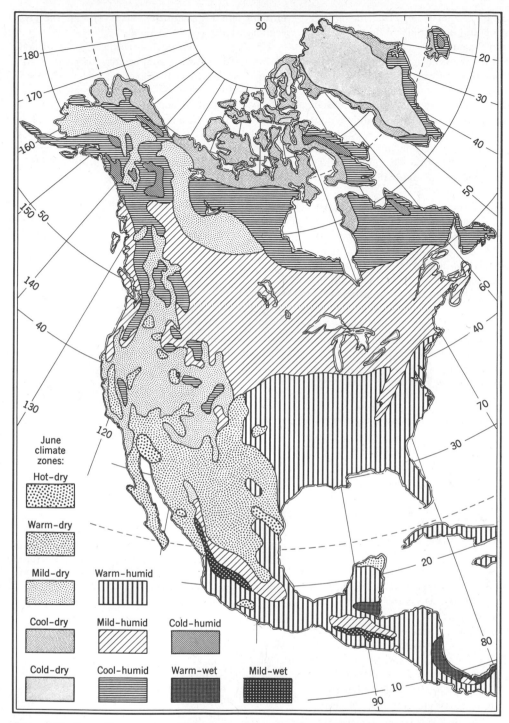

June climate zones:

Hot–dry

Warm–dry

Mild–dry   Warm–humid

Cool–dry   Mild–humid   Cold–humid

Cold–dry   Cool–humid   Warm–wet   Mild–wet

Figure 8.2   June climate zones of North America, using the combinations of precipitation and temperature given in Table 8.1. (After Aeronautical Chart Service, U.S. Air Force, 1947.)

Vegetation and soils as a basis of climate classification

Botanists and geographers have long been aware that plants are highly responsive to differences in climate. With each plant species is associated a particular combination of climatic elements most favorable to its growth, as well as certain extremes of heat, cold, or drought beyond which it cannot survive. Plants tend to adapt their physical forms to meet the stresses of climate, hence we find that there is a wide range of forms taken by assemblages of the dominant plant species, and that the overall form patterns or habits closely reflect climate. These subjects are treated in Chapters 15 and 16.

There is much to recommend a climate classification based on the different forms taken by groups of plants. On the other hand, such plant forms represent responses to climate, rather than causes of global climate variation. A fundamental principle of scientific classification is that the setting up of classes of things is better done according to the causes of the class differences than according to the effects that such differences produce.

Since the late nineteenth century, soil scientists have recognized that the several fundamental classes of mature soils, seen on a world-wide scope, are more strongly controlled by climatic elements than by any other single factor. (See Chapters 13 and 14.) But plants also contribute to determining the properties of the very soils upon which they depend. Thus, as in the case of vegetation, soils reflect differences in climate but do not cause those differences. A climate classification based upon soils would be highly meaningful, but not scientifically as desirable as a climate classification based upon causes.

The Köppen climate classification system

The foregoing concepts of devising climate classes that combine temperature and precipitation characteristics, but of setting limits and boundaries fitted into known vegetation and soil distributions, were actually carried out in 1918 by Dr. Wladimir Köppen of the University of Graz in Austria. The classification was subsequently revised and extended by Köppen and his students to become the most widely used of climatic classifications for geographical purposes.

The Köppen system is strictly empirical. This is to say that each climate is defined according to fixed values of temperature and precipitation, computed according to the averages of the year or of individual months. In such a classification, no concern whatever need be given to the causes of the climate in terms of pressure and wind belts, air masses, fronts, or storms. It is possible to assign a given place to a particular climate subgroup solely on the basis of the records of temperature and precipitation of that place, provided, of course, that the period of record is long enough to yield meaningful averages. Air temperature and precipitation are the most easily obtainable surface weather data, requiring only simple equipment and a very elementary observer education. A climate system based on these data has a great advantage, in that the areas covered by each subtype of climate can be computed or estimated for large regions of the world.

The Köppen system features a shorthand code of letters designating major climate groups, subgroups within the major groups, and further subdivisions to distinguish particular seasonal characteristics of temperature and precipitation.

Five major climate groups are designated by capital letters as follows (Figure 8.3). Groups *A*, *C*, and *D* have sufficient heat and precipitation for growth of high-trunk trees (e.g., forest and woodland).

A *Tropical climates.* Average temperature of every month is above 64.4°F (18°C). These climates have no winter season. Annual rainfall is large and exceeds annual evaporation.

B *Dry climates.* Potential evaporation exceeds precipitation on the average throughout the year. No water surplus; hence no permanent streams originate in *B* climate zones.

C *Warm temperate (mesothermal)* [3] *climates.* Coldest month has an average temperature under 64.4°F (18°C), but above 26.6°F (−3°C); at least one month has an average temperature above 50°F (10°C). The *C* climates thus have both a summer and a winter season.

D *Snow (microthermal)* [4] *climates.* Coldest month average temperature under 26.6°F (−3°C). Average temperature of warmest month above 50°F (10°C), that isotherm coinciding approximately with poleward limit of forest growth.

E *Ice climates.* Average temperature of warmest month below 50°F (10°C). These climates have no true summer.

Note that four of these five groups (*A*, *C*, *D*, and *E*) are defined by temperature averages, whereas one (*B*) is defined by precipitation-to-evaporation ratios. This may seem to be a fundamental inconsistency.

Subgroups within the five major groups are designated by a second letter according to the following code:

S *Steppe climate:* A semi-arid climate with about 15 to 30 in (38 to 76 cm) of rainfall annually at low latitudes. Exact rainfall boundary determined by formula taking temperature into account.

W *Desert climate:* Arid climate. Most regions included have less than 10 in (25 cm) of rainfall annually.

---

[3] The middle latitudes are anything but "temperate" in seasonal and day-to-day weather. *Mesothermal* is widely substituted for *temperate* to imply intermediate temperatures as compared with the extreme heat of dry deserts and the extreme cold of polar and arctic climates.

[4] The term *microthermal*, meaning "small heat," is widely substituted in the United States, along with *mesothermal*, in the titles of Köppen groups *D* and *C*, respectively.

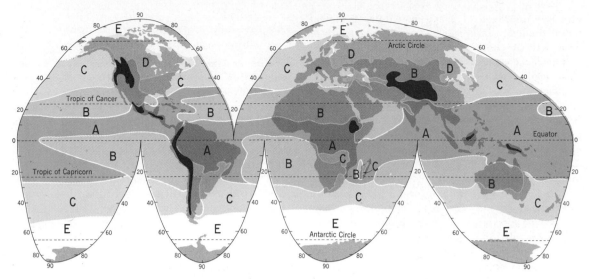

Figure 8.3   Highly generalized world map of major climatic regions according to the Köppen classification. Highland areas in black. (Based on Goode Base Map. Copyright by the University of Chicago. Used by permission of the University of Chicago Press.)

Exact boundary with steppe climate determined by formula.

(The letters S and W are applied only to the dry B climates, yielding two combinations, BS and BW.)

f Moist. Adequate precipitation in all months. No dry season. This modifier is applied to A, C, and D groups.

w Dry season in winter of the respective hemisphere (low-sun season).

s Dry season in summer of the respective hemisphere (high-sun season).

m Rainforest climate despite short, dry season in monsoon type of precipitation cycle. Applies only to A climates.

From combinations of the two letter groups, eleven distinct climates emerge as follows:

Af Tropical rainforest (Also Am, a variant of Af).
Aw Tropical savanna.
BS Steppe climate.
BW Desert climate.
Cw Temperate rainy (humid mesothermal) climate with dry winter.
Cf Temperate rainy (humid mesothermal) climate moist all seasons.
Cs Temperate rainy (humid mesothermal) climate with dry summer.
Df Cold snowy forest (humid microthermal) climate moist in all seasons.
Dw Cold snowy forest (humid microthermal) climate with dry winter.
ET Tundra climate.
EF Climates of perpetual frost (icecaps).

To differentiate still more variations in temperature or other weather elements. Köppen added a third letter to the code group. Meanings are as follows:

a With hot summer; warmest month over 71.6°F (22°C) (C and D climates).

b With warm summer; warmest month below 71.6°F (22°C) (C and D climates).

c With cool, short summer; less than four months over 50°F (10°C) (C and D climates).

d With very cold winter; coldest month below −36.4°F (−38°C) (D climates only).

h Dry-hot; mean annual temperature over 64.4°F (18°C) (B climates only).

k Dry-cold; mean annual temperature under 64.4°F (18°C) (B climates only).

As an example of a complete Köppen climate code, BWk would refer to *cool desert climate*, Dfc would refer to *cold, snowy forest climate with cool, short summer*. The complete list of climate code designations is included in the summary table of climates to follow. A world map of climates following a later revision of the Köppen system (the Köppen-Geiger system) appears as Plate 2.

An explanatory-descriptive climate system

Earlier chapters on weather elements, global circulation, air mass properties and source regions, and cyclonic storms have armed the reader with many principles concerning the causes of weather patterns and their seasonal variations. It would be a pity if this knowledge were left unused with no application in the systematic study of world climates, in favor of a purely empirical system based on temperature and precipitation averages expressed by code letters.

The most satisfying systems of natural science classification are those described as *genetic*, meaning that the genesis, or origin, of the phenomena is put first in designing the classification units. Because a genetic system gives an explanation of the things classified, it may be considered as *explanatory*. If the explanation is carried out largely in verbal statements (as distinct from numerical or mathematical statements) it may be labeled as *descriptive*. This general approach to classification can therefore be designated as *explanatory-descriptive* in contrast to the *empirical-quantitative* approach followed in the Köppen system.

In the chapters to follow, world climates are treated by an explanatory-descriptive system based on causes and effects. It will not be difficult to introduce the Köppen code symbols into such a system by way of interrelating the two. In fact, it will soon become apparent that the explanatory-descriptive system is simply providing a reasonable scientific explanation for the existence of Köppen's climate groups and subgroups.

Air mass source regions and frontal zones as a basis of classification

The system of climate analysis used in the succeeding chapters is based on the location of air mass source regions and the nature and movement of air masses, fronts, and cyclonic storms. A schematic global diagram, Figure 8.4, shows the principles of the classification. A world map, Figure 8.5, shows the general locations of each of fourteen fundamental types.

Three major climate groups are shown in Figure 8.4. Group I includes the tropical air mass source regions and the equatorial trough or convergence zone between. *Equatorial* is here defined as pertaining to the latitude zone within a few degrees north and south of the equator; *tropical* is defined as referring to the two latitude zones included within a few degrees north and south of the tropics of Cancer and Capricorn (Figure 4.5).

Climates of Group I are controlled by the dynamic subtropical high-pressure cells, or anticyclones, which are regions of air subsidence and are basically dry, and by the great equatorial trough of convergence that lies between them.

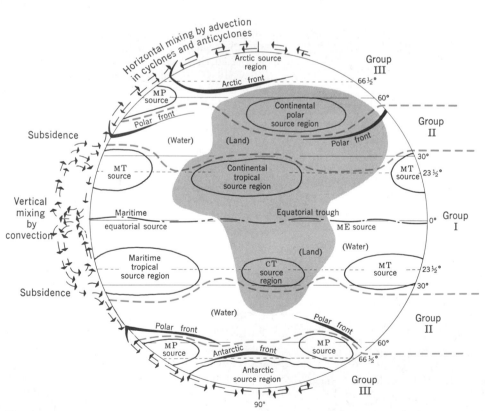

Figure 8.4 Global diagram of major climate groups. (After Petterssen and others.)

Though it is true that air of polar origin occasionally invades the tropical and equatorial latitudes, it may be said of the climates of Group I that they are almost wholly dominated by tropical and equatorial air masses.

Climates of Group II are in a zone of intense interaction between unlike air masses: the *polar front zone*. Tropical air masses moving poleward and polar air masses moving equatorward are in conflict in this zone, which contains a procession of eastmoving cyclonic storms. Locally and seasonally either tropical or polar air masses may dominate in these regions, but neither has exclusive control.

Climates of Group III are dominated by polar and arctic (including antarctic) air masses. The two polar continental air mass source regions of northern Canada and Siberia fall into this group,

but there is no southern hemisphere counterpart to these continental centers. In the belt of the 60th to 70th parallels, air masses of arctic origin meet polar continental air masses along an *arctic front zone*, creating a series of east-moving cyclones.

Fourteen major climate types are listed below. Where a compound climate type is formed by the alternation of two simple types, it follows them on the list. Simple names are applied to these climate types following common usage among geographers. Numbers correspond with those in Figure 8.5. Köppen symbols are added to the table to show close correspondence of these fourteen types with his eleven climates and their subtypes. Note, however, that the Köppen symbols and climate types can be completely disregarded without affecting the explanatory-descriptive treatment in terms of air masses and frontal zones.

GROUP I: LOW-LATITUDE CLIMATES (CONTROLLED BY EQUATORIAL AND TROPICAL AIR MASSES)

Climate Name		Köppen Symbol	Air Mass Source Regions and Frontal Zones, General Climate Characteristics
1. Wet Equatorial Climate 10° N–10° S lat. (Asia 10°—20° N)	Af Am	Tropical rainforest climate, and Tropical rainforest climate, monsoon type	Equatorial trough (convergence zone) climates are dominated by warm, moist tropical maritime (mT) and equatorial (mE) air masses yielding heavy rainfall through convectional storms. Remarkably uniform temperatures prevail throughout the year.
2. Trade Wind Littoral Climate 10°—25° N and S lat.	Af-Am	Included in climates	Tropical easterlies (trades) bring maritime tropical (mT) air masses from moist western sides of oceanic subtropical high-pressure cells to give narrow east-coast zones of heavy rainfall and uniformly high temperatures. Rainfall shows strong seasonal variation.
3. Tropical Desert and Steppe Climates 15°—35° N and S lat.	BWh BSh	Desert climate, hot, and Steppe climate, hot	Source regions of continental-tropical (cT_s) air masses in high-pressure cells at high level over lands astride the Tropics of Cancer and Capricorn give arid to semi-arid climate with very high maximum temperatures and moderate annual range.
4. West Coast Desert Climate 15°—30° N and S lat.	BWk BWh	Desert climate, cool, and Desert climate, hot (BWn in earlier versions, n meaning frequent fog)	On west coasts bordering the oceanic subtropical high-pressure cells, subsiding maritime tropical (mT_s) air masses are stable and dry. Extremely dry, but relatively cool, foggy desert climates prevail in narrow coastal belts. Annual temperature range is small.
5. Tropical Wet-Dry Climate 5°–25° N and S lat.	Aw Cwa	Tropical rainy climate, savanna; also Temperate rainy (Humid mesothermal) climate, dry winter, hot summer	Seasonal alternation of moist mT or mE air masses with dry cT air masses gives climate with wet season at time of high sun, dry season at time of low sun.

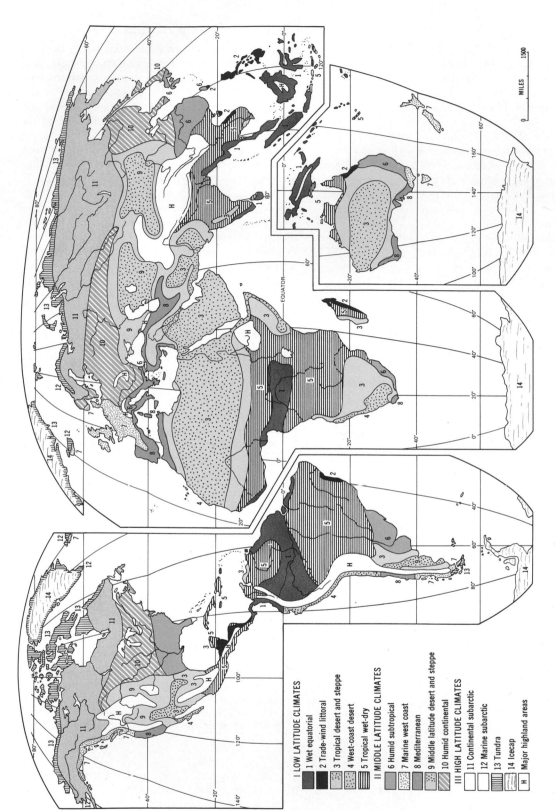

Figure 8.5 Generalized and simplified world map showing distribution of fourteen climates. In many respects, these climate regions correspond with regions as defined by G. T. Trewartha. (Aitoff's equal-area projection adapted by V. C. Finch.)

I LOW LATITUDE CLIMATES
1 Wet equatorial
2 Trade-wind littoral
3 Tropical desert and steppe
4 West-coast desert
5 Tropical wet-dry
II MIDDLE LATITUDE CLIMATES
6 Humid subtropical
7 Marine west coast
8 Mediterranean
9 Middle latitude desert and steppe
10 Humid continental
III HIGH LATITUDE CLIMATES
11 Continental subarctic
12 Marine subarctic
13 Tundra
14 Icecap
H Major highland areas

0   MILES   1500

GROUP II: MIDDLE-LATITUDE CLIMATES (CONTROLLED BY BOTH TROPICAL AND POLAR AIR MASSES)

6. Humid Subtropical Climate 20°–35° N and S lat.	Cfa	Temperate rainy (Humid mesothermal) climate, hot summers	Subtropical, eastern continental margins dominated by moist maritime (mT) air masses flowing from the western sides of oceanic high-pressure cells. In high-sun season, rainfall is copious and temperatures warm. Winters are cool with frequent continental polar (cP) air mass invasions. Frequent cyclonic storms.
7. Marine West Coast Climate 40°–60° N and S lat.	Cfb Cfc	Temperate rainy (Humid mesothermal) climate, warm summers, and same but cool, short summers	Windward, middle-latitude west coasts receive frequent cyclonic storms with cool, moist maritime polar (mP) air masses. These bring much cloudiness and well-distributed precipitation, but the winter maximum. Annual temperature range is small for middle latitudes.
8. Mediterranean Climate 30°–45° N and S lat.	Csa Csb	Temperate rainy (Humid mesothermal) climate, dry, hot summer, and same, but dry, warm summer	This wet-winter, dry-summer climate results from seasonal alternation of conditions causing climates 4 and 7; mP air masses dominate in winter with cyclonic storms and ample rainfall, mT_s air masses dominate in summer with extreme drought. Moderate annual temperature range.
9. Middle-latitude Desert and Steppe Climates 35°–50° N and S lat.	BWk BWk' BSk BSk'	Desert climate, cool same, but cold; and Steppe climate, cool same, but cold	Interior, middle-latitude deserts and steppes of regions shut off by mountains from invasions of maritime air masses (mT or mP), but dominated by continental tropical (cT) air masses in summer and continental polar (cP) air masses in winter. Great annual temperature range; hot summers, cold winters.
10. Humid Continental Climate 35°–60° N lat.	Dfa Dfb Dwa Dwb	Cold, snowy forest (Humid microthermal) climate, moist all year, hot summers, and same, but warm summers; also Cold, snowy forest (Humid microthermal) climate, dry winters, hot summers, and same, but warm summers	Located in central and eastern parts of continents of middle latitudes, these climates are in the polar front zone, the battle ground of polar and tropical air masses. Seasonal contrasts are strong and weather highly variable. Ample precipitation throughout the year is increased in summer by invading maritime tropical (mT) air masses. Cold winters are dominated by continental polar (cP) air masses invading frequently from northern source regions.

GROUP III: HIGH-LATITUDE CLIMATES (CONTROLLED BY POLAR AND ARCTIC AIR MASSES)

11. Continental Sub-arctic Climate 50°–70° N lat.	*Dfc*	Cold, snowy forest (Humid microthermal) climate, moist all year, cool summers, and	This climate lies in source region of continental polar (cP) air masses, which in winter are stable and very cold. Summers are short and cool. Annual temperature range is enormous. Cyclonic storms, into which maritime polar (mP) air is drawn, supply light precipitation, but evaporation is small and the climate is therefore effectively moist.
	Dfd	Same, but very cold winters, also	
	Dwc	Cold, snowy forest (Humid microthermal) climate, dry winter, cool summer, and	
	Dwd	same, but very cold winter	
12. Marine Subarctic Climate 50°–60° N and 45°–60° S lat.	*ET*	Polar, tundra climate	Located in the arctic frontal zones of the winter season, these windward coasts and islands of subarctic latitudes are dominated by cool mP air masses. Precipitation is relatively large and annual temperature range small for so high a latitude.
13. Tundra Climate North of 55° N South of 50°S	*ET*	Polar, tundra climate	The arctic coastal fringes lie along a frontal zone, in which polar (mP, cP) air masses interact with arctic (A) air masses in cyclonic storms. Climate is humid and severely cold with no warm season or summer. Moderating influence of ocean water prevents extreme winter severity as in Climate 11.
14. Icecap Climate (Greenland, Antarctica)	*EF*	Polar climate, perpetual frost	Source regions of arctic (A) and antarctic (AA) air masses situated upon the great continental icecaps have climate with annual temperature average far below all other climates and no above-freezing monthly average. High altitudes of ice plateaus intensify air mass cold.
Highland Climates			Cool to cold moist climates, occupying high-altitude zones of the world's mountain ranges, are localized in extent and not included in classification system.

Climatic regimes

Whatever system of climate classification a geographer may choose to adopt, a study of the combined monthly temperature and precipitation characteristics will reveal that a climate expresses one or more of seven basic *regimes*. These regimes are of particular importance in an understanding of soils and plant geography because each regime induces a characteristic response from soils and vegetation.

The climatic regimes are particularly well displayed by means of a *thermohyet diagram* (see Figure 8.6). Similar diagrams have been previously referred to as *climographs*. The thermohyet diagram shows the monthly means of air temperature and precipitation plotted in combination as points representing each month of the year. The months are connected by lines to show the cycle of monthly averages throughout the year. The thermohyet diagram used throughout this chapter has temperature and precipitation scales of special design to expand the low-precipitation range and compress the low-temperature range in order to give more space on the graph where it is most needed. Data used on the thermohyet diagram are for the most part the averages of many years of record and thus provide an expression of the characteristic yearly regime, or climatic cycle.

1. Middle-latitude equable regime

One concept of climate description, expressed in the words *equability* and *inequability*, relates to the degree of uniformity and lack of uniformity, respectively, in the temperature and precipitation characteristics of a station. An *equable climate* exhibits a small annual range in monthly means of temperatures or of precipitation, or of both. On the thermohyet diagram, equability is expressed by a diagram covering a small surface area. Several examples are shown in Figure 8.6.

Because temperature and precipitation are largely independent constituents of climate, equability might conceivably be present (1) in both monthly mean temperature and monthly mean precipitation, (2) in temperature, but not in precipitation; and (3) in precipitation, but not in temperature. The first case, which can be described as one of *total equability*, is rare in nature, but can be found in middle latitudes in certain insular or windward coast locations. The five examples shown in Figure 8.6*A–E* (Auckland, Punta Arenas, Hasselbough Bay, South Orkneys, and Los Evangelistas) are all in regions of prevailing westerlies and are exposed constantly to moist maritime air masses. They represent the *middle-latitude equable regime* of climate.

The equability of Hasselbough Bay, Macquarie Islands, is truly remarkable, with its annual range of only 10 F° (6 C°); but the record of Los Evangelistas, Chile, is almost unbelievable. This station lies close to sea level on a small island off the west coast of South America at about $52\frac{1}{2}$° S. latitude. Its annual temperature range is only $7\frac{1}{2}$ F° (4 C°), while the total annual precipitation of 119 in (300

cm) is so evenly distributed that no month averages less than 8 in (20 cm) or more than 12 in (30 cm)! Below freezing temperatures are rare despite an annual mean temperature of 43° F (6°C). No month of the year averages fewer than 20 days with more than 0.04 in (0.1 cm) of precipitation; which is to say that it rains about two days out of three on the average throughout the year.

These examples of total equability are all the more remarkable when we consider that in middle latitudes the annual cycle of insolation shows a wide range (Figure 4.3), tending to cause a strong annual temperature range. In all five cases shown in Figure 8.6*A–E*, the moderating influence of the vast Southern Ocean effectively suppresses the insolation cycle.

2. Equatorial regime

Great uniformity of monthly mean temperatures, but rarely of monthly precipitation means, is characteristic of the equatorial belt, which may be said to possess a high degree of thermal equability. The climatic regime associated with these very low latitudes is the *equatorial regime*. Two examples, Figure 8.6 *F, G*, are selected for exceptionally low annual ranges in monthly mean rainfall. Figure 8.7 shows the more commonly observed conditions, consisting of exceptional thermal equability superimposed upon a wide range in monthly rainfall averages. The result is a thermohyet diagram which is almost a horizontal

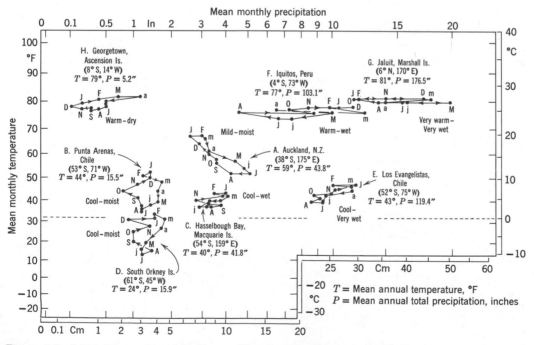

Figure 8.6 Selected examples of highly equable climatic regimes shown by thermohyet diagrams. (Data from Meterological Office of Great Britain and G. T. Trewartha.)

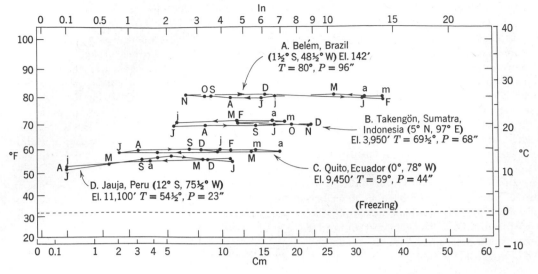

Figure 8.7 Examples of stations exhibiting the wet equatorial climatic regime. (Data from Meteorological Office of Great Britain.)

straight line on the graph. Note the wide ranges in monthly rainfalls, combined with large annual totals.

With increasing altitude the entire cycle is pushed down and to the left on the graph, but otherwise little changed, as shown in succession by Takengön, Sumatra; Quito, Ecuador; and Jauja, Peru (Figure 8.7, *B, C, D*).

Equability in temperature but not in precipitation is found only in the very dry deserts and is treated below under the desert regime. Georgetown, Ascension Island (Figure 8.6*H*), is an exceptional case of total equality in equatorial latitudes, but with a very dry climate.

3. Tropical wet-dry regime

A *tropical wet-dry regime* is distinctive because a very rainy season at time of high sun alternates with a very dry season at time of low sun, usually to the accompaniment of a moderate seasonal range in temperatures. Thermohyets of tropical wet-dry stations are shown in Figure 8.8. Paraná, Brazil, tends toward thermal equability, but this equability is offset by the great contrast in drought and rains. Kano, Nigeria, is in many respects the most extreme of low-latitude examples, for it has a rainless period when the sun's declination is most southerly. In striking contrast is the extremely wet period following summer solstice. Obviously, a regime of such great precipitation contrasts will place a severe stress upon plants and animals in the dry season.

A noteworthy feature of the tropical wet-dry regime is a thermal peak about during April and May (northern hemisphere). The temperatures moderate somewhat after May, because of arrival

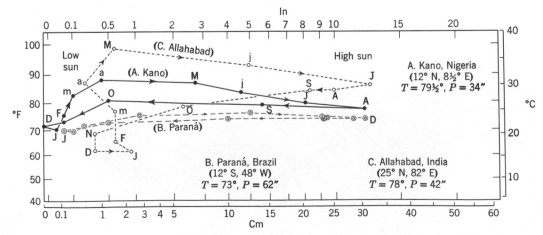

Figure 8.8 Examples of stations exhibiting the tropical wet-dry climatic regime. (Data from Meteorological Office of Great Britain.)

of cloud cover and the cooling effect of falling rain and its evaporation. Thus we find a three-season cycle in certain tropical regions: cool-dry, hot-dry, warm-wet. In north India, particularly, the thermohyet diagram forms a figure-eight loop of distinctive shape, as shown in the data for Allahabad, India (Figure 8.8*C*). Here the thermal range is exaggerated by the higher latitude and is superimposed on a strong Asiatic monsoon.

4. Mediterranean regime

The *Mediterranean regime* (*dry-summer subtropical regime*) is characterized by a very dry, warm summer alternating with a wet, mild winter. The precipitation extremes of dry and wet season are not greatly different from those of the tropical wet-dry regime, but are in exactly opposite phase in terms of the sun's declination. Moreover, the thermal cycle of the Mediterranean regime tends to be rather equable because of a west coast location in the zone of prevailing westerlies.

Thermohyet diagrams (Figure 8.9) show two varieties of the Mediterranean regime. Santiago, Chile, illustrates the cooler, drier variety found along the west coasts of continents. Siracuse, Sicily, is a warmer, wetter variety with greater temperature range and warmer summers. A distinctive feature of the thermohyet diagram is that the loop takes the form of an ellipse whose long axis is inclined downward to the right. The same tendency is found in thermohyet diagrams of any west coast station in the belt of westerly winds, whether a dry season exists or not. (See Vancouver, B.C., Figure 8.9*C* and Auckland, N.Z., Figure 8.6*A*.)

As in the case of the tropical wet-dry regime, the Mediterranean regime places a severe stress upon plants and animals during the almost rainless summer, but in neither regime is cold an important stress.

5. Continental regime

The *continental regime* is dominated by a large annual temperature range, often accompanied by a severely cold winter. The thermohyet diagrams of Figure 8.10 show a distinctive elongation and are diagonally oriented, running from upper right to lower left. The significance of the diagonal trend of the diagram is that the warm season has appreciably greater precipitation than the cold season. The continental regime thus usually exhibits seasonal contrasts both in temperature and precipitation. There is, however, a considerable difference in overall aridity or humidity of climate from one position or another within a continent.

As Figure 8.10 shows, the centers of the continents tend to be dry (Yakutsk, U.S.S.R.) whereas the more easterly locations have increasingly more precipitation throughout the year (Omaha, Nebraska; Shanghai, China). As we would expect, increased latitude brings a marked overall lowering of the temperatures and at the same time increases the annual range. In one respect the continental regime resembles the tropical wet-dry regime: the greatest precipitation falls in the high-sun period. One might thus expect a complete gradation of the tropical wet-dry regime poleward into the continental regime, where the climate is traced along the central and eastern parts of a continent from tropical into middle-latitude zones.

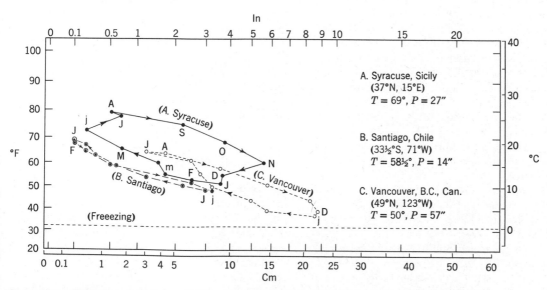

Figure 8.9 Examples of stations exhibiting the Mediterranean climatic regime. (Data from Meteorological Office of Great Britain.)

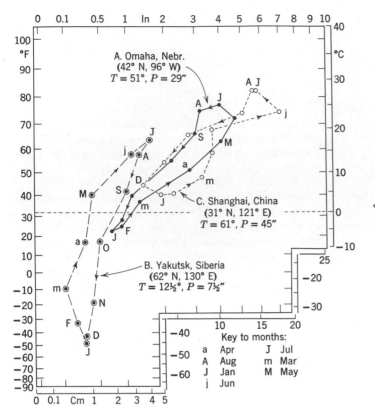

Figure 8.10 Examples of stations exhibiting the continental climatic regime. (Data from G. T. Trewartha and Meteorological Office of Great Britain.)

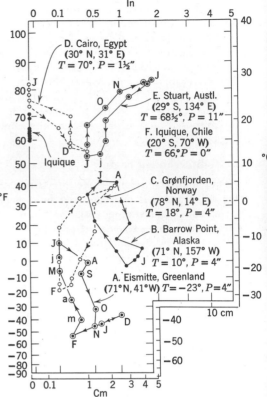

Figure 8.11 Examples of stations exhibiting desert a polar climatic regimes. (Data from G. T. Trewartha a Meteorological Office of Great Britain.)

6. Polar regime

A *polar regime* exists poleward of the Arctic and Antarctic circles in both hemispheres. This regime is, of course, one of year-round low temperatures. Only three or four months may have means above freezing. The thermohyet diagrams of representative stations from the Arctic coastal fringes are thus situated far down on the left of the chart (Figure 8.11A, B, C). Precipitation is measurable but small in all months. The extreme range in intensity of insolation throughout the year is responsible for a large annual temperature range of 40 to 50 F° (22 to 28 C°), but this range is greatly exceeded in the continental regime at lower latitudes in Canada and Siberia (see Yakutsk, Figure 8.10B).

By far the most intense polar regime is found at high altitudes in the interiors of the great icecaps of Greenland and Antarctica. The thermohyet of one year's record at Eismitte, over 9000 ft (2700 m) in elevation in central Greenland, shows that only two monthly means were above 0°F (−18°C), whereas four months averaged −40°F (−40°C) or below! Because no plant life can survive under icecap conditions, this icecap

form of the polar regime constitutes a total cold desert.

7. Desert regime

The *desert regime* combines extremely low precipitation with temperatures ranging from warm to very hot. As a result, a typical thermohyet diagram lies to the far left and generally rather high up (Figure 8.11D, E, F). The annual cycles of many desert stations have no month with more than 0.2 in (0.5 cm) average rainfall. Rainfall is sporadic, so that an average of 0.1 in (0.25 cm) for a given month may represent the precipitation of a single storm preceded and followed by many years in which that month was rainless. With regard to temperatures, the desert regime shows the influence of one of the other three regimes: Effect of the continental regime is felt in the interiors of large landmasses, as seen in the thermohyet diagram for Stuart, Australia; here the annual range is 30 F° (17 C°).

Effect of the Mediterranean regime is seen on the diagram for Cairo, Egypt, which has a rainless summer but a marked rainfall maximum in winter. Thermal equability of low latitudes is seen in the thermohyet diagram of Iquique, Chile, with only 11 F° (7 C°) range. The desert regime is one of great stress upon plants and animals, primarily because of the unavailability of water throughout the year, and also, in many locations because of the extremely great heat of air and soil at time of high sun.

Combinations of climatic regimes

The seven climatic regimes have specific areas of optimum development and might be said to form a world climate classification of seven basic types—each one readily understood in terms of the annual cycle of insolation, the properties of land and water bodies, the prevailing winds, and the interplay of air masses and their conflicts in fronts and storms. On the other hand, large land areas lie in intermediate regimes, comprised of two or more of the seven basic regimes. It is for this reason that we have chosen a climate classification with fourteen classes, rather than seven.

Where a region possesses a climatic regime combining two or more of the basic regimes, we can use modifying phrases derived from the names of the seven regimes. For example, Cairo, Egypt (Figure 8.11D), has "a desert regime showing a Mediterranean tendency"; Allahabad, India (Figure 8.8C), has "a tropical wet-dry regime with a marked continental influence"; Yakutsk, Siberia (Figure 8.10B) has "a continental regime with a polar influence."

As explained in Chapter 16, structure and distribution of the several basic forms of natural vegetation strongly reflect the influences of the seven basic climatic regimes. Soil development is similarly influenced. This close association of natural vegetation and soils with climate is ample justification for an analysis of world climates in terms of the fundamental regimes.

REFERENCES FOR FURTHER STUDY

Kendrew, W. G. (1953), *The climates of the continents*, Oxford University Press, London, 607 pp.

Köppen-Geiger (1954), Klima der Erde (map), Justus Perthes, Darmstadt, Germany; American distributor, A. J. Nystrom and Co., Chicago.

Dansereau, Pierre (1957), *Biogeography, an ecological perspective*, The Ronald Press Co., New York, 394 pp. See Chapter 2.

Koeppe, E., and G. C. DeLong (1958), *Weather and climate*, McGraw-Hill Book Co., New York, 341 pp.

Trewartha, G. T. (1968), *An introduction to climate*, fourth edition McGraw-Hill Book Co., New York, 408 pp.

Petterssen, S. (1969), *Introduction to meteorology*, third edition McGraw-Hill Book Co., New York, 333 pp. See Chapter 17.

REVIEW QUESTIONS

1. Discuss the importance of climate as a force influencing soils, natural vegetation, agriculture, physiology, disease, and other factors that affect human life.

2. What difficulties stand in the way of devising satisfactory climate classifications? In what way can climate maps be misleading?

3. Explain how temperature can be used as a basis for a climate classification. Why is this element alone unsatisfactory as a basis for a climate system?

4. How can annual precipitation be used as a basis for a climate classification? Why does the factor of evaporation make this classification of limited value?

5. Why is it that the distribution of natural vegetation types forms a workable basis for distinguishing climate zones?

6. Explain the basic principles underlying the Köppen climate system. What are the five major climate groups? What code system is used to designate eleven climates? Name these and give symbols for each.

7. What do each of the following letters mean in the Köppen system: *a, b, c, d, h, k?*

8. Explain how a climate classification system can be based on the air mass source regions, patterns of air mass flow, and frontal zones. What advantage has a natural science classification that is based on origin of the things classified as against one that is based on numerical values?

9. How may the climates of the globe be classified into three major groups on the basis of air mass types?

10. What five climates comprise the group controlled by equatorial and tropical air masses? How are they related to the system of pressure and wind belts? What Köppen climates are equivalent?

11. What five climates are controlled by both tropical and polar air masses? Explain briefly how each type is determined. What Köppen climates are equivalent?

12. What four climates are controlled by polar and arctic air masses? Explain how the large northern hemisphere landmasses and icecaps control these climates. What is the arctic front zone? Is there a comparable front zone in the southern hemisphere? What are the equivalent Köppen climates?

13. Explain what is meant by "climatic regimes." Why is the study of climatic regimes a matter of importance in geography?

14. Name the six climatic regimes. For each regime state the essential characteristics and global location.

15. What is a "thermohyet diagram"? What climate information is given on the horizontal and vertical scales of this diagram? Why are the units not uniformly spaced on these scales?

16. Can the climate of a given place show the influence of more than one basic regime? Give examples.

CHAPTER 9
Soil Water and the Water Balance

A FUNDAMENTAL concept of climatology now coming increasingly under consideration by geographers is that the availability of water to plants and animals is a more important factor in classifying climates than precipitation itself, which is only the amount of water received from the sky. Much of the water received as precipitation is lost in a variety of ways and is not usable for plants and animals. Just as in a fiscal budget, when the monthly or yearly loss of moisture exceeds the precipitation a budgetary deficit results; when precipitation exceeds the losses, a budgetary surplus results. An analysis of the water balance, or water budget, of a given area of the earth's surface is actually arrived at in much the same way as the understanding of a fiscal budget and requires only additions and subtractions of amounts within fixed periods of time, such as the calendar month or year. On the other hand, to understand why such gains and losses occur involves study of the physical processes affecting water in its vapor, liquid, and solid states not only in the atmosphere but also in the soil and rock and in exposed water of streams, lakes, and glaciers. The science of *hydrology* treats such relationships of water as a complex but unified system on the earth. In this chapter, our primary concern is with the hydrology of the soil zone.

Surface and subsurface water

We may classify water according to whether it is *surface water*, flowing exposed or ponded upon the land, or *subsurface water*, occupying openings in the soil, overburden, or bedrock. That which is held in the soil within a few feet of the surface is termed *soil water*, and is the particular concern of the botanist, soil scientist, and agricultural engineer. That which is held in the openings of the bedrock or deep within thick layers of transported overburden is usually referred to as *ground water*, studied by the geologist, who is concerned with the storage and flow of this water in various kinds of rocks and structures. Surface water in streams, lakes, and glaciers, and the geologic aspects of ground water and wells are considered in later chapters.

The hydrologic cycle

Water of oceans, atmosphere, and lands moves in a great series of continuous interchanges of both geographic position and physical state, known as the *hydrologic cycle* (Figure 9.1). A particular molecule of water might, if we could trace it continuously, travel through any one of a number of possible circuits involving alternately the water vapor state and the liquid or solid state. A good place to start the description of the hydrologic cycle would be the oceans, which cover nearly three-quarters of the globe. It has been estimated that some 80,000 cu mi (335,000 cu km) of water are evaporated each year from the oceans, and another 15,000 cu mi (65,000 cu km) from the lakes and moist land surfaces of the continents. The total yearly evaporation figure 95,000 cu mi (400,000 cu km) must exactly balance, on the average, the total quantity of water restored to the earth's surface by condensation from the atmosphere. About 24,000 cu mi (100,000 cu km) of water falls as precipitation upon the earth's land surfaces each year—enough to cover an area the

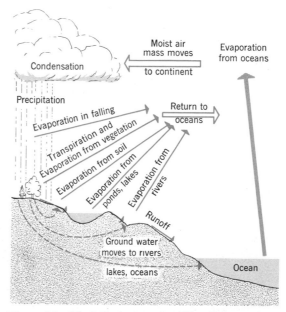

Figure 9.1   The hydrologic cycle. (After Holtzman.)

size of Texas to a depth of 475 ft (145 m). We see that considerably more water falls upon the lands than is returned to the atmosphere by evaporation from the lands. Therefore, much of what falls returns to the sea in liquid form.

Those parts of the hydrologic cycle dealing with water vapor, its movement toward the lands in maritime air masses, and eventual precipitation from clouds, has been treated in Chapters 6 and 7 as a phase of meteorology. Some of the falling precipitation evaporates directly before reaching the ground. Part of that which reaches the ground may be quickly returned to the atmosphere in water vapor form by evaporation from vegetative and soil surfaces, but if rainfall is heavy and long continued, much will soak in to the soil to become a part of the soil water. If close to the ground surface, soil water may be returned to the atmosphere by plants which take up the water through their roots and release it into the air through their leaves. Some soil water will evaporate directly into the soil air, which permeates soil openings.

Should rain continue to fall, water will percolate through the soil under gravitational pull and will reach the bedrock, or will penetrate deep deposits of transported overburden. This ground water moves very slowly but at length emerges in streams, lakes, or even from the ocean floor in the form of seepages and springs.

If the ability of the soil to receive and transmit heavy and long-continued rainfall is exceeded, surface flow occurs, conducting the water directly down slopes into streams and lakes. From these exposed surfaces the water may evaporate directly into the atmosphere or may reach the oceans by stream flow. We have now traced water through its full hydrologic cycle.

Evaporation and the earth's heat budget

The annual heat budget of the earth is outlined in Chapter 4. In Table 4.2, under the heading of energy returned to the atmosphere, the amount returned through the process of latent heat transfer (evaporation from ground and subsequent condensation in atmosphere) is given as 23 percent of total incoming energy, equivalent to 0.111 langleys per minute for the northern hemisphere. The importance of this heat-exchange process becomes evident when we note from Table 4.2 that the quantity of energy involved is almost twice that of longwave ground radiation (14 percent) and more than twice that transferred by conduction (10 percent).

Attempts have been made to estimate energy loss by latent heat transfer over the entire global surface. Such maps are not complete, but reveal striking differences from place to place. Table 9.1 lists some representative values for heat loss from evaporation in various global environments. Data have been reduced to equivalents in langleys per minute, to correspond with Table 4.2.

Knowledge of world distribution of isotherms, belts of pressure and winds, and air mass regions (Chapters 4 through 7) can be put to good use in interpreting the information in Table 9.1. Highest values of latent heat transfer by evaporation occur in the equatorward parts of the subtropical high-pressure cells. Here dry air in combination with intense insolation and a warm ocean surface gives maximum yearly evaporation rates. Evaporation over equatorial oceans, although higher than average, is less than over tropical oceans because the equatorial air holds a large amount of water vapor and there is greater cloud cover. In contrast,

TABLE 9.1[a]

Region	Latitude Belt	Energy Flux (Langleys Per Minute)
Equatorial oceans	5° S to 5° N	0.11 to 0.15
Tropical South Indian and South Pacific oceans	10 to 20° S	0.23 to 0.27
Tropical North Atlantic and North Pacific oceans	15 to 25° N	0.15 to 0.19
Tropical North Africa and Arabian peninsula	20 to 30° N	Under 0.02
Middle-latitude oceans	40 to 60° N, S	0.04 to 0.11
Middle-latitude interior of North America and Asia	40 to 60° N	0.02 to 0.04
Arctic and polar regions	Above 60° N, S	Under 0.02

[a] Data from M. I. Budyko, 1958.

the tropical deserts give very low values because soil is dry and there is almost no surface water to be evaporated. Continental interiors of middle latitudes are also low in annual evaporation, partly because of long cold winters. Arctic and polar regions have very low values because of prevailingly low air and ground temperatures.

Infiltration and runoff

Most soil surfaces in their undisturbed, natural states are capable of absorbing the water from light or moderate rains, a process known as *infiltration*. Such soils have natural passageways between poorly fitting soil particles, as well as larger openings, such as earth cracks resulting from soil drying, borings of worms and animals, cavities left from decay of plant roots, or openings made by heaving and collapse of soil as frost crystals alternately grow and melt. A mat of decaying leaves and stems breaks the force of falling drops and helps to keep these openings clear. If rain falls too rapidly to be passed downward through these soil openings the excess amount flows as a surface water film or sheet down the direction of ground slope, a runoff process termed *overland flow* (Figure 9.2).

As stated in Chapter 6, rainfall is measured in units of inches or centimeters per hour. This is the depth to which water will accumulate in each hour if rain is caught in a flat-bottomed, straight-sided container, assuming none to be lost by evaporation or splashing out. Similarly, infiltration is stated in inches or centimeters per hour and might be thought of as the rate at which the water level in the same container might drop if the water were leaking through a porous base. Runoff, also stated in inches (cm) per hour, may be thought of as the amount of overflow of the container per hour when rain falls too fast to be disposed of by leaking through the base.

Now, it is an important fact about soils that their *infiltration capacity*, or ability to infiltrate rainfall, is usually great at the start of a rain which has been preceded by a dry spell, but drops rapidly as the rain continues to fall and to soak into the soil. After several hours the soil's infiltration capacity becomes almost constant. The reason for the high starting value and its rapid drop is, of course, that the soil openings rapidly become clogged by particles brought from above, or tend to close up as the colloidal clays take up water and swell. From this we can easily reason that a sandy soil with little or no clay will not suffer so great a drop in infiltration capacity, but will continue to let the water through indefinitely at a generous rate. In contrast, the clayrich soil is quickly sealed to the point that it allows only a very slow rate of infiltration. This principle is

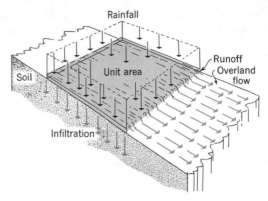

Figure 9.2 Rainfall, infiltration, and overland flow. (From A. N. Strahler, *The Earth Sciences*, Harper and Row, New York.)

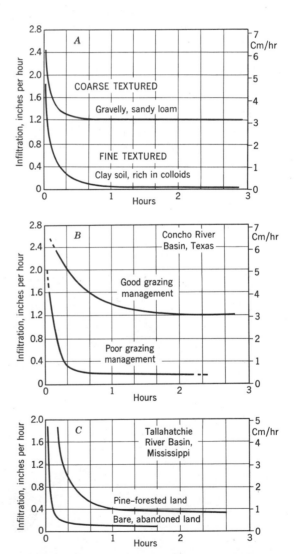

Figure 9.3 Infiltration rates vary greatly according to soil texture and land use. (Data from Sherman and Musgrave; Foster.)

illustrated by the graph in Figure 9.3*A* showing the infiltration curves of two soils, one sandy, one rich in clay.

It also follows that a sandy soil may be able to infiltrate even a heavy, long-continued rain without any surface runoff occurring, whereas the clay soil must divert much of the rain into overland flow, a process that may lead to erosion by gullies. Many forms of artificial disturbance of natural soils tend to decrease the infiltration capacity and to increase the amount of surface runoff (Figure 9.3*B*, *C*). Cultivation tends to leave the soil exposed so that rain beat quickly seals the soil pores. Fires, by destroying the protective vegetation and surface litter, also expose the soil to rain beat. Trampling by livestock will tamp the porous soil into a dense, hard layer. It is little wonder, then, that man has, through his farming and grazing practices, radically changed the original proportions of infiltration to runoff in such a way as to result in severe erosion damage and at the same time to decrease the reserves of soil moisture which might otherwise sustain plant growth and stream flow in droughts.

Evaporation and transpiration

Between periods of rain, water held in the soil is gradually given up by a twofold drying process.

First, direct evaporation into the open air occurs at the soil surface and progresses downward. Air also enters the soil freely and may actually be forced alternately in and out of the soil by atmospheric pressure changes. Even if the soil did not "breathe" in this way, there would be a slow diffusion of water vapor surfaceward through the open soil pores. Ordinarily only the first foot (30 cm) of soil is dried by evaporation in a single dry season, but in the prolonged drought of deserts, drying will extend to depths of many feet. Second, plants draw the soil water into their systems through vast networks of tiny rootlets. This water, after being carried upward through the trunk and branches into the leaves, is discharged in the form of water vapor, through leaf pores into the atmosphere, a process termed *transpiration*. Further details of plant transpiration are given in Chapter 15.

In studies of climatology and hydrology it is convenient to use the term *evapotranspiration* to cover the combined moisture loss from direct evaporation and the transpiration of plants. The rate of evapotranspiration slows down as soil moisture supply becomes depleted during a dry summer period because plants employ various devices to reduce transpiration. In general, the less moisture remaining, the slower is the loss through evapotranspiration. Consequently, it is necessary to

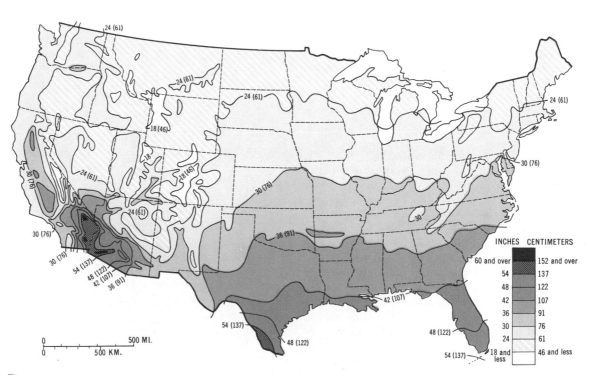

Figure 9.4   Average annual potential evapotranspiration in the United States. (Courtesy of *The Geographical Review,* **vol. 38, 1948. Copyrighted by the American Geographical Society of New York.)**

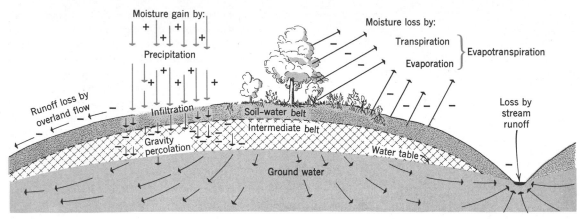

Figure 9.5 The soil-water belt occupies an important position in the hydrologic cycle.

define two forms of evapotranspiration: (1) *Potential evapotranspiration* is the maximum loss of water possible under the given conditions of plant cover and climatic factors, assuming that we can continue to supply the soil by irrigation with all of the water which the plants can use and the soil pores can hold. (2) *Actual evapotranspiration* is the true or observed quantity of evapotranspiration, decreasing in rate as the soil moisture is depleted. Figure 9.4 is a map of the United States showing the average annual potential evapotranspiration. This map is based on air temperatures and latitude; it does not take into account wind or varying humidities. The quantities shown have, nevertheless, been found to agree rather well with direct measurements of evapotranspiration made in irrigated areas, and with estimates derived from data on stream flow.

Figure 9.5 shows diagrammatically the various terms explained up to this point and serves to give a more detailed picture of that part of the hydrologic cycle involving the soil. As the plus signs show, the soil belt gains water through precipitation and infiltration. As the minus signs show, the soil loses water through transpiration, evaporation, and overland flow, and by gravity percolation downward through the soil to the ground water zone below.

Moisture in the soil

When infiltration occurs during heavy and prolonged rains (or when a snow cover is melting) the water is drawn downward by gravity through the soil pores, wetting successively lower layers. Soon the soil openings are filled with water moving downward, except for some air entrapped in the form of bubbles. Then the percolation continues downward into the bedrock. Suppose now that the rain stops and a period of several days of dry weather follows. The excess soil water continues to drain downward, but some water clings to the

soil particles and completely resists the pull of gravity through the force of *capillary tension*. We are all familiar with the way in which a water droplet seems to be enclosed in a "skin" of surface molecules, drawing the droplet together into a rounded outline, so that it clings to the side of a glass indefinitely without flowing down. Similarly, tiny films of water adhere to the soil grains, particularly at the points of grain contacts, and will stay until disposed of by evaporation or by absorption into plant rootlets.

When a soil has first been saturated by water, then allowed to drain under gravity until no more water moves downward, the soil is said to be holding its *field capacity* of water. This takes no more than two or three days for most soils. Most is drained out within one day. Field capacity is measured in units of depth, usually inches or centimeters, just as with precipitation. This means that for a given cube of soil, say 12 in on a side (1 cu ft), if we were to extract all of the field moisture, it might form a layer of water 3-in deep in a pan 1-ft square. This would be equivalent to complete absorption of a 3-in rainfall by a completely dry 12-in layer of soil. (Metric units of volume and depth may be substituted.)

Field capacity of a given soil depends largely on its texture. Sandy soil has a very low field capacity; clay soil has high field capacity. This effect is shown in Figure 9.6, a graph in which field capacity is plotted against soil texture, from coarse to fine. It should also be noted that sandy soils reach their field capacity very quickly, both because of the ease with which the water penetrates and the low quantity required. Clay soils take long rain periods to reach field capacity because the infiltration is slow and the total quantity required to be absorbed is great.

Agricultural scientists also use a measure of soil moisture termed the *wilting point*. This is the quantity of soil water below which plants will be

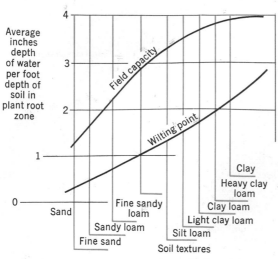

Figure 9.6 Field capacity and wilting point. (After Smith and Ruhe, *Yearbook of Agriculture*, 1955.)

unable to extract further moisture from the soil and the foliage will wilt. As Figure 9.6 shows, the wilting point also depends upon particle size.

The soil water cycle

Equipped with the foregoing explanations of processes and terms relating to water gains and losses in the soil, we can turn next to consider the annual water budget of the soil, involving principles of great concern not only in plant geography and agriculture, but in the further study of ground water, surface runoff, stream flow, and therefore of the sculpturing of land slopes.

Figure 9.7 shows the annual cycle of soil water for the year 1944 at an agricultural experiment station in Coshocton, Ohio. If we follow the

changes in this example, the cycle it shows can be considered generally representative of conditions in humid, middle-latitude climates where there is a strong temperature contrast between winter and summer. Let us start with the early spring (March). At this time the evaporation rate is low, because of low temperatures. The abundance of melting snows and rains has restored the soil moisture to a surplus quantity. For two months the quantity of water percolating through the soil and entering the ground water keeps the soil pores nearly filled with water. This is the time of year when one encounters soft, muddy ground conditions, whether driving on dirt roads or walking across country. This, too, is the season when runoff is heavy and major floods may be expected on larger streams and rivers. In terms of the soil water budget, a *water surplus* exists.

By May, the rising air temperatures, increasing evaporation, and the full growth of plant foliage, bringing on heavy transpiration, have reduced the soil moisture to a quantity below field capacity, although it may be restored temporarily by unusually heavy rains in some years. By midsummer, a state of heavy *water deficiency* exists in the water budget. Even the occasional heavy thunderstorm rains of summer cannot restore the water lost by steady and heavy evapotranspiration. Small springs and streams dry up, the soil becomes firm and dry. By November (and sometimes in September), however, the soil moisture again begins to increase. This is because the plants go into a dormant state, sharply reducing transpiration losses, while, at the same time, falling air temperatures reduce evaporation. By late winter, usually in February at this location, the field capacity of the soil is again restored.

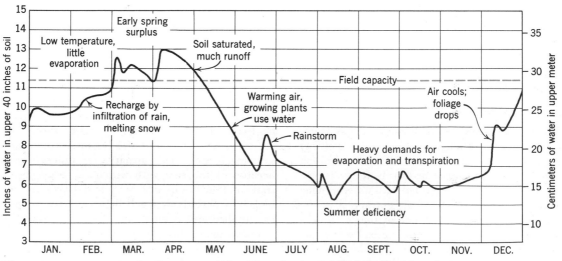

Figure 9.7 Annual cycle of soil moisture. (After Thornthwaite and Mather, *The Water Balance,* 1955.)

The water budget

With the above example of soil moisture changes throughout a year's time, we turn to a more generalized concept of the water budget in terms of the two essential parts of the budget: (1) precipitation, representing the income, and (2) potential evapotranspiration, representing the expenditure. These two quantities will be considered on a month-to-month basis throughout the year, but average monthly values for a record of many years' duration will be used instead of the data of but a single year.

Figure 9.8 shows the water budget for Seattle, Washington, a typical station of the marine west-coast climate. Two graphs are superimposed: (*A*) average monthly precipitation (solid line), and (*B*) average monthly potential evapotranspiration (hollow line). Each graph takes a series of steps up and down across the twelve-month period. (Essentially, these are bar graphs whose height can be read against the scale of inches or centimeters on the side of the diagram.) Note first that the precipitation is highest in the months of November through February but declines to low values in summer. In almost completely opposite phase, the evapotranspiration rises from low values in winter to maximum values in summer. This form is just what would be expected in middle latitudes because potential evapotranspiration increases with rising air temperatures and with increased number of hours that the sun is in the sky.

Starting in January, the graph shows that precipitation is much greater in that month than potential evapotranspiration. There exists a large *water surplus*, and this escapes by percolating down through the soil or flows over the ground to the nearest stream. As spring progresses, potential evapotranspiration rises rapidly, while precipitation declines. In May potential evapotranspiration is the larger value, requiring that the supply of stored soil moisture be drawn upon. In a sense, soil moisture is like a savings bank account, to be drawn upon when monthly expenses exceed income. Through June the soil moisture is sufficient to make up the lack of precipitation. But soil moisture is limited to 4 in (10 cm) of water and is entirely used up by early July.[1] Here, on the graph, a vertical line has been drawn and marks the beginning of a *water deficiency*. In terms of one's personal money budget, the water deficiency corresponds to going in debt, but with the reasonable assurance that a surplus later in the year will enable the loan to be paid back. Altogether, a

[1] It is assumed here for simplicity that the soil moisture is used at a steady rate equal to the potential evapotranspiration rate until it is all gone, whereas actually the rate of use declines as the remaining quantity declines.

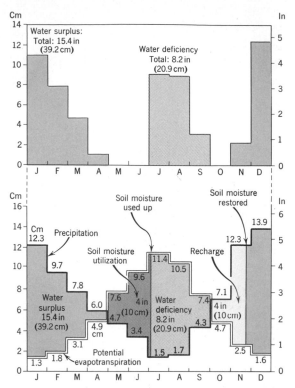

Figure 9.8 The water budget for Seattle, Washington. (After C. W. Thornthwaite, *Geographical Review*, 1948.)

water deficiency of about 8 in (21 cm) is incurred by the end of September. In October the monthly precipitation is again greater than potential evapotranspiration. Now the excess is used to restore, or *recharge*, the soil moisture, but it is not until late in November that this is accomplished (vertical line on graph) and a water surplus is resumed. Returning to an analogy with one's personal budget, the savings account has been restored before beginning repayment on the loan.

Summarizing the annual water budget at Seattle, it is evident that the climatic regime it represents is one of both a surplus and a deficiency, occurring in opposite seasons. On a separate graph in the upper part of Figure 9.8 are shown the monthly surpluses and deficiencies together with the totals of each. The total surplus was 15.4 in (39.2 cm); the total water deficiency was 8.2 in (20.9 cm). Obviously, considerable irrigation is needed in July, August, and September, if plant growth is to be sustained at the maximum possible rate. The analysis which we have made of the water budget at Seattle is, therefore, of great practical value to gardeners and farmers in that area.

These water budget principles with concepts of moisture deficiencies and surpluses were developed by Dr. C. Warren Thornthwaite, a climatologist who applied his knowledge to practical problems of agriculture as well as to a world-wide system

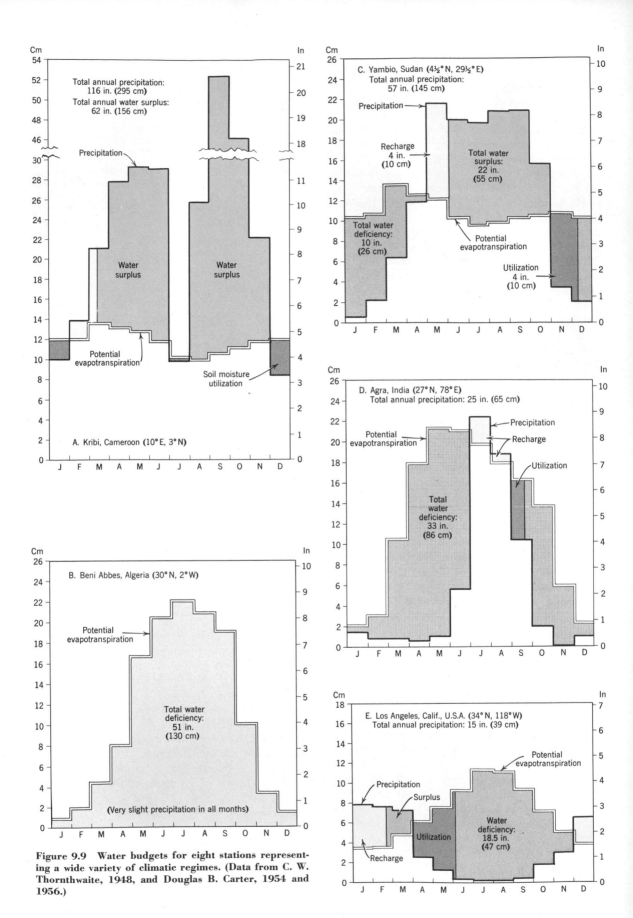

Figure 9.9 Water budgets for eight stations representing a wide variety of climatic regimes. (Data from C. W. Thornthwaite, 1948, and Douglas B. Carter, 1954 and 1956.)

of climate classification based on the relations between precipitation and evapotranspiration. Some further refinements have been made in estimating the amount of soil moisture remaining as the deficiency sets in, but these are beyond the scope of the present discussion.

Climatic regimes and the water budget

Water budgets for some other representative climatic regimes are shown in Figure 9.9. These give some idea of the great contrasts that are possi-

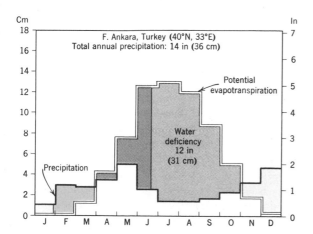

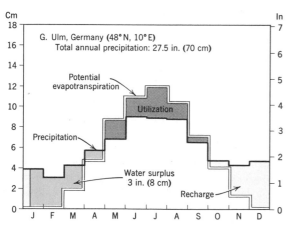

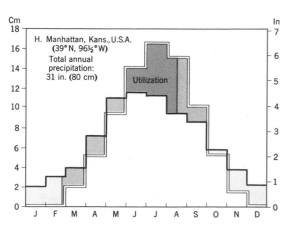

ble in water surpluses and deficiencies. The vast water surplus of a wet equatorial climate is illustrated by data for Kribi, Cameroon, an African west coast station at 3° N lat. (Figure 9.9*A*). Rainfall totals about 116 in (295 cm) for the year and has two strong maxima, one in the period April through June, the other in September and October. During these periods large water surpluses occur, the annual total surplus being 64 in (162 cm). The graph of potential evapotranspiration shows only a small annual variation because of uniformly high air temperatures.

Although soil moisture is drawn upon in December and January, no water deficiency occurs. Obviously in such a climatic regime soil moisture will be plentiful for plant growth year-around; runoff to streams will be copious.

A tropical desert regime is seen in the data for Beni Abbes, Algeria, at 30° N lat. (Figure 9.9*B*). Here the total annual precipitation averages about 1 in (2.5 cm). Throughout the entire year potential evapotranspiration greatly exceeds rainfall; a large water deficiency prevails—over 50 in (127 cm).

The tropical wet-dry climatic regime of low latitudes is illustrated by Yambio, Sudan, at 4½° N lat. (Figure 9.9*C*). Potential evapotranspiration here follows a cycle much like that of Kribi, but the dry season of low sun brings on a large water deficiency. Rains of the high-sun period make a substantial water surplus. Agra, India, at 27° N lat. illustrates the wet-dry regime with Asiatic monsoon influence (Figure 9.9*D*). The potential evapotranspiration cycle closely resembles that of Beni Abbes, which lies about at the same latitude. The summer monsoon at Agra brings rainfall of 8 to 9 in (20 to 30 cm) per month in July and August, sufficient to exceed potential evapotranspiration and bring a short period of soil moisture recharge. This soil moisture is used up in September. No water surplus can develop, whereas the water deficiency, totaling 33 in (86 cm), prevails for more than nine months of the year.

The Mediterranean (dry-summer subtropical) regime is illustrated by Los Angeles, California (Figure 9.9*E*). Here the large summer deficiency of 18.5 in (47 cm) results from the fact that the high summer rate of potential evapotranspiration coincides with the summer precipitation minimum. The small surplus of 1.4 in (3.6 cm) endures little over a month's time. The overall plan of the Los Angeles water budget resembles that of Seattle (Figure 9.8) so far as the phase of the cycles is concerned, but with a major contrast in water surpluses.

Ankara, Turkey, illustrates a water budget having the same phasing and roughly the same values of surplus and deficiency as seen in the Mediterranean regime of Los Angeles, but with important differences in precipitation distribution. The cold winter months at Ankara have little or no potential evapotranspiration.

Continentality of climatic regime is seen in the water budgets of Ulm, Germany, and Manhattan, Kansas (Figure 9.9*G, H*). Both localities have a summer maximum of precipitation coinciding with a summer maximum of potential evapotranspiration. The result is that there is no water deficiency at Ulm and only a small deficiency at Manhattan. Both places show a small water surplus in late winter and early spring.

Each type of climate has its own characteristic annual water budget. To attempt to present the full spectrum of budgets would go far beyond the scope of this brief survey. From the examples shown, however, the importance of Thornthwaite's system of analysis of soil water conditions should be evident. The method is quantitative, giving definite estimates of the amounts of water gained and lost, and thereby making possible realistic estimates of the water available or needed for use in agriculture, as well as of the water surpluses which may be drawn upon for irrigation and hydroelectric systems. Thornthwaite's system of water-balance evaluation is particularly effective in estimating the stress (or lack of it) which climate exerts upon vegetation, as will be shown in Chapter 16.

Ground water

Water that is drawn downward by gravity through the soil zone to lower levels becomes part of the ground water body, the relations of which are shown in Figure 9.10. Strictly speaking, *ground water* is that part of subsurface water which fully saturates the pore spaces of the rock or its overburden and which behaves in response to gravitational force. The ground water occupies the *zone of saturation*. Above it is the *zone of aeration*, in which water does not fully saturate the pores. We have seen that the soil water belt is the uppermost layer of the zone of aeration and that moisture is held in this belt by capillary force in tiny films adhering to the soil particles. A similar condition prevails through the underlying *intermediate belt*. The sole basis for distinguishing these two belts is that the soil belt represents a shallow zone of moisture usable by plants, whereas the intermediate belt is too deep for capillary water to be returned to the atmosphere by either direct evaporation or transpiration. The depth of the zone of aeration may be very shallow or missing (when the ground water is close to the surface in low, flat regions), or up to several hundred feet thick in hilly or mountainous regions with low ground water level.

At the base of the zone of aeration is the *capillary fringe*, a thin layer in which the water has been drawn upward from the ground water body through capillary force. The action is much like the rise of kerosene in a lamp wick, or of water in a blotter whose edge is immersed. Water in the capillary fringe largely fills the soil pores, hence, is continuous with the ground water body. Thickness of the capillary fringe depends on the soil texture, because capillary rise is higher when the openings are smaller. Thus, in a silty material the capillary fringe may be 2 ft (0.6 m) thick, but only a fraction of an inch (1 cm) thick in coarse sand or fine gravel with large pore spaces.

Ground water in the zone of saturation moves under the force of gravity and therefore its upper surface, the *water table*, tends to become a horizontal surface, just as with a free water body such as a lake. But because water moves very slowly through the rock, the water table actually maintains a sloping surface, highest under the hill tops and divides, lowest in valleys. Ground water is discussed further in Chapter 20.

Effluent and influent streams

An important contrast between regions of arid and humid climates lies in the nature of stream flow. The manner in which the channels of streams and their associated landforms differ in the contrasting climates is discussed in later chapters on geomorphology. From the standpoint of soils and the physical environments of natural vegetation, an important consideration is the manner in which the water enters and leaves a stream channel.

In a humid region with a high water table sloping toward the stream channels, ground water

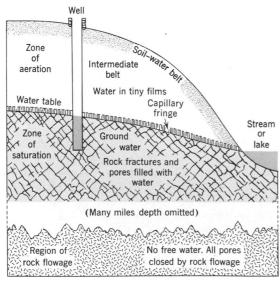

Figure 9.10 Ground-water and soil-water belts and zones. (After Ackerman, Colman and Ogrosky.)

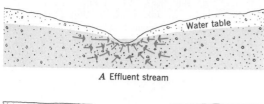

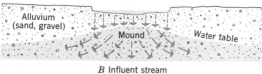

A Effluent stream

B Influent stream

Figure 9.11   Effluent and influent streams. (From. A. N. Strahler, *The Earth Sciences*, Harper and Row, New York.)

moves steadily toward the channels, into which it seeps, producing permanent, or *perennial* streams. Such streams are described as *effluent* (Figure 9.11*A*). Surplus water in a humid region— both that which flows over the ground in wet weather, and that which seeps into the channel from the ground water table—escapes by through-flowing channels to regions of lower elevation and eventually to the sea. In the course of such drainage, dissolved substances (ions and colloids) that have been removed from the soil are taken out of the region, permanently removing them from the soil (Chapter 13).

In arid regions, where streams flow across plains of gravel and sand, water is lost from the channels by seepage to a water table that lies below the level of the streams. Such streams are designated as *influent* (Figure 9.11*B*). This water gradually moves away from the zone of the channel and eventually returns to the surface in flat-floored basins known as *playas* (Chapter 22), where evaporation takes place and salts are left behind to form the saline soils and salt crusts.

Water-table lakes, marshes, and bogs

In humid climates with a large annual surplus in the water budget and with generally high water tables, the various events of geologic history, such as erosion and deposition by wind and by ice sheets of the most recent glacial stage, have created natural depressions with floors at or below the water table elevation. Seepage of ground water, as well as direct runoff of precipitation, maintains these free water surfaces permanently throughout the year. Examples of fresh water-table ponds are found widely distributed in North America and Europe where plains of glacial sand and gravel contain natural pits and hollows left by the melting of stagnant ice masses. (See Chapter 23.) Figure 9.12 is a portion of a topographic map of a number of small fresh-water ponds on Cape Cod, within a mile or two (2–3 km) of the Atlantic Ocean. The surface elevation of these ponds is 8 ft (2.5 m) and is maintained by the supply of fresh ground water which excludes salt water by causing the ground water to flow gradually from land to ocean.

Many former fresh-water water-table ponds have, since they were formed by glacial ice, become partially or entirely filled by the organic matter from growth and decay of water-loving plants. The result is a *bog* whose surface lies close to water table (Chapter 15).

Marshes and swamps, where water stands at or

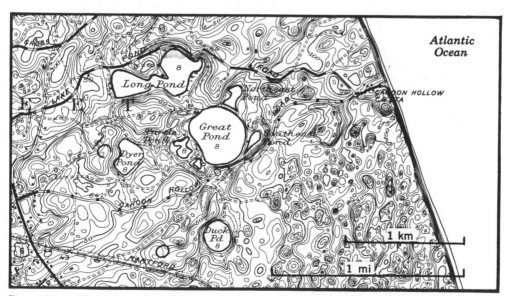

Figure 9.12   Portion of a topographic map of Cape Cod, Mass., showing freshwater ponds close to the ocean shoreline. The figure in each pond gives elevation in feet. (U.S. Geological Survey.)

close to the ground surface over a broad area, represent the appearance of the water table at the surface. Such areas of poor surface drainage have a variety of origins. For example, the broad, shallow fresh-water marshes of the Atlantic and Gulf coastal plain represent regions only recently emerged from the sea. Others are created by the shifting of river channels on floodplains (Chapter 21).

REFERENCES FOR FURTHER STUDY

Thornthwaite, C. W. (1948), An approach toward a rational classification of climate, *Geographical Review*, Vol. 38, pp. 55–94.

Thornthwaite, C. E., and J. R. Mather (1955), *The water balance*, Drexel Institute of Technology, Laboratory of Climatology, Publ. in *Climatology*, vol. 8, No. 1, Centerton, N.J., 86 pp.

U.S. Dept. Agriculture (1955), *Water; Yearbook of Agriculture, 1955*, U.S. Govt. Printing Office, Washington, D.C., 751 pp.

Todd, D. K. (1959), *Ground water hydrology*, John Wiley and Sons, New York, 336 pp.

Chang, Jen-Hu (1968), *Climate and agriculture; an ecological survey*, Aldine Publ. Co., Chicago, 304 pp. See Chapters 12, 13, and 19.

REVIEW QUESTIONS

1. Why is an understanding of soil water important to the student of soils? Of plant geography? Of economic geography?

2. Distinguish between surface water and subsurface water; between soil water and ground water.

3. Give a complete general account of the hydrologic cycle. What quantities of water are evaporated annually from the oceans? From the lands? How is this balanced by precipitation and runoff?

4. What is infiltration? What is overland flow? In what units is each stated? How does infiltration capacity change when rain falls? Explain. How do soils differ in this respect? In what ways does man alter infiltration capacity?

5. To what depth does soil moisture evaporate? How do atmospheric pressure changes assist in this process?

6. Explain the process of transpiration by plants. What quantities of soil water are thus lost? To what depth does this loss extend? Define evapotranspiration. Distinguish between potential and actual evapotranspiration.

7. How is soil water held in place? What is the source of soil moisture? What is field capacity? How does it vary with soil texture? What is the wilting point?

8. Describe the annual cycle of soil water in humid, middle-latitude climates. When and why do water surpluses and deficits occur?

9. Sketch a typical annual water budget cycle by months, showing precipitation and potential evapotranspiration. Label water surpluses and deficiencies. When is irrigation needed?

10. How does the annual water budget differ between regions of tropical desert climate and those of a tropical wet-dry climate? Describe other climate contrasts as shown in their respective water budgets.

11. Define ground water. How does one distinguish between zones of saturation and aeration? Between soil water and intermediate water belts? What is the capillary fringe? How does its thickness vary with soil texture? What is the water table?

12. Describe effluent and influent streams and their relation to climate. How are soil-forming processes tied in with such streams?

13. What are water-table lakes? Do they show great seasonal changes of water level? What is a bog? A marsh?

CHAPTER 10

Equatorial and Tropical Climates

CLIMATES of group I are those of low latitudes. They are controlled largely by equatorial and tropical air masses.

1. Wet equatorial climate (Af, Am)

Bringing together the information on global pattern of weather elements contained in previous chapters and focusing attention on the equatorial belt of the earth, it will be seen that within 5° of latitude north and south of the equator the following conditions prevail:

(*a*) Temperatures average close to 80°F (27°C) for every month (Figures 4.17 and 4.18), so that the annual range is extremely small.

(*b*) Air pressures average between 1009 and 1012 mb, or slightly less than normal sea-level pressure (Figures 5.2 and 5.3).

(*c*) The general flow of air is from east to west in the tropical easterlies at high altitude, but somewhat more equatorward in the surface trade winds, which originate in the subtropical high-pressure cells. Here, then, is a region of converging warm, unstable air masses meeting along the equatorial trough (Chapter 7).

(*d*) Examination of the world rainfall map (Plate 1) shows this to be a belt of heavy rainfall, exceeding 80 in (200 cm) annually for the most part. This belt has been explained in Chapter 7 as convectional rainfall from great cumulonimbus masses in weak, slowly moving easterly waves and equatorial lows.

From the above information we can compile a description of the wet equatorial climate. Average annual temperatures are close to the 80°F (27°C) mark; seasonal range of temperature is so slight as to be imperceptible because the sun is high throughout the year. Rainfall is heavy during the entire year, but with considerable differences in monthly averages because of the seasonal shifting of the equatorial convergence zone and a consequent variation in air mass characteristics.

Figure 10.1 is a monthly average temperature-rainfall graph for Iquitos, Peru, a typical wet equatorial station located at about 3° S lat. in the broad basin of the Amazon River. Note that the annual range in temperature is only 4 F° (2.2 C°)

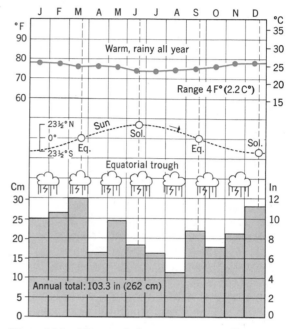

Figure 10.1 Climograph for Iquitos, Peru, 3½° S. lat., upper Amazon River basin. (Data from Blair.)

and that the annual rainfall total is more than 100 in (250 cm). In all but one month the monthly rainfall averages more than 6 in (15 cm). According to Köppen's definition of this *Af* climate, no month averages less than 2.4 in (6 cm) of rainfall.

Some idea of the extreme monotony of the daily temperature cycle in this climate can be had from Figure 10.2, which shows minimum and maximum daily temperatures for two months at Panama, 9° N lat. An especially noteworthy feature is that the daily range is normally from 15 to 20 F° (8 to 11 C°), a vastly greater range than the annual range of monthly mean temperatures. In other words, daily variations far exceed seasonal variations in the wet equatorial climate. The thermohyet diagram for Belém, Brazil (Figure 8.7*A*) illustrates well the extreme equability of temperature to be expected.

An illustration of month-to-month differences in average monthly rainfall in the wet equatorial climate is given in Figure 10.3. All the stations shown here lie within a few degrees of the equator

149

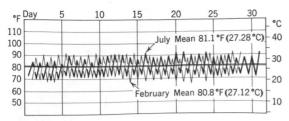

Figure 10.2 July and February temperatures at Panama, 9° N lat. (After Mark Jefferson, *Geographical Review*.)

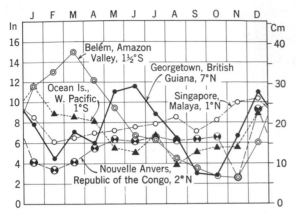

Figure 10.3 Monthly rainfall averages for stations in wet equatorial climates. (Data from Trewartha.)

and are basically similar in climate. The fact that some stations show one maximum and one minimum in the rainfall graph, whereas others have two maxima and two minima is not easily explained, but is probably attributable to a rather complex series of seasonal changes in air mass movements and positions of the equatorial trough.

Because of the abundant rainfall and prevailingly warm temperatures, the equatorial region is characterized by growth of *rainforest*, or *selva*, described in detail in Chapter 16, a vegetation type unexcelled for luxuriance of tree growth and numbers of species (Figures 16.2, 16.3, 16.4). Broadleaf trees rise to heights of 100 to 150 ft (30–45 m), forming a dense leaf canopy through which little sunlight can reach the ground. Giant *lianas* (woody vines) hang from the trees. The forest is evergreen, although individual species have a rhythm of leaf-shedding (Figure 10.4).

Rainforest is the home of the small forest animals, of which the monkeys are perhaps the best representatives, taking advantage of the continuous forest canopy for living and traveling. Birds, too, are numerous in species and spectacularly plumaged.

With copious rainfall and prevailingly high temperatures, chemical processes are continuously active on the rocks and soils in the wet equatorial regions. Leaching out of all soluble constituents

of the deeply decayed rock results in a distinctive type of soil, termed a *latosol* (color plate). Reddish or yellowish, and often containing irregular nodules of reddish iron hydroxides, this soil is especially rich in hydroxides of iron, manganese, and aluminum. These have been left behind in the soil after the soluble minerals (including silica) have been carried down through the soil and into the streams and rivers. Large concentrations of the iron, manganese, or aluminum minerals occurring as lenses or layers in the soil are termed *laterite*. These minerals may be extracted as ores of commercial value. *Bauxite* is the foremost ore of aluminum in use today, being extracted, for example, from the Guiana coast of South America (Figure 10.5). Manganese ores are likewise of great value, but lateritic iron ore is as yet not so widely exploited because of iron ore abundance in other forms.

Vegetation and climate work hand in hand to make lateritic soils and ore bodies. The soil-forming regime of laterization is discussed in Chapter 13. At the prevailingly warm temperatures, bacteria in the upper soil layer are unusually vigorous and consume virtually all dead vegetation. Thus *humus*, the black, partially decomposed plant matter present in most soils of middle-latitude and arctic climates, is almost entirely absent on well-drained sites. Rotting and decay of bedrock may be very deep in wet equatorial regions of low topographic relief. It is reported that even to depths as great as 300 ft (90 m) the rock has been found to be soft and crumbly from chemical action.

Stream flow tends to be fairly constant and extremely copious because a large water surplus exists throughout much of the year and provides ample runoff (Figure 9.9*A*). River channels are lined along the banks with dense vegetation. Sand bars or sand banks are not so conspicuous as in drier regions. Floodplains have meanders and many swampy sloughs where the river channels have shifted their courses (Figure 10.4). Although water is abundant, river systems such as the Amazon carry relatively little material in chemical solution. This has been explained as resulting from the thorough leaching of soils, which have little soluble mineral matter left to contribute.

Human transportation is by streams whose sluggish courses can be utilized by dugout canoes or shallow-draft river steamers. The use of light aircraft has, however, greatly simplified long-distance transportation. Villages are characteristically built close to the river banks.

Not all equatorial rainforest areas are of low topographic relief. Hilly or mountainous belts have very steep slopes, on which flows, slides, and

Figure 10.4 This air view over the rainforest of the Amazon Basin shows the Rio Negro, a tributary of the Amazon River. The view is southward from a point about 0°23′ S, 64°05′ W. (Photograph by Colonel Richmond, courtesy of the American Geographical Society.)

Figure 10.5 Bauxite mining in British Guiana. These thick layers of bauxite ore accumulated as sediments derived from the erosion of lateritic soils of interior uplands. (Photograph by courtesy of Aluminium Limited, Canada.)

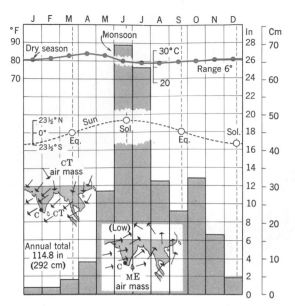

Figure 10.6 Climograph for Cochin, India, 10° N. lat. (Data from Clayton, Smithsonian Institution.)

within moist maritime equatorial air masses moving northward from the Indian Ocean.

2. Trade-wind littoral climates (var. of Af and Am)

A study of the world rainfall map, Plate 1, shows that along the east coasts of Central and South America, Madagascar, Indochina, the Philippines, and northeastern Australia, in latitudes 10° to 25°, there are narrow belts of heavy rainfall. These are coasts exposed to moist maritime tropical air masses brought by the tropical easterlies, or trade winds, from the oceanic subtropical highs. These air masses are abundantly supplied with moisture, as is typical of maritime tropical air on the western sides of the high-pressure cells. When these air masses encounter the hill and mountain slopes of the coasts, a heavy orographic precipitation results. Easterly waves cause periods of rainy weather. In addition, the high-sun solstice period sees the onset of tropical cyclones, to which these coasts are vulnerable.

Because of its coastal position with respect to the easterlies, this climate may be called a *trade-wind littoral climate.* Under the Köppen system it is included with the rainforest climates *Af* and *Am.* It is treated as a separate climate here because the precipitation characteristics differ from the wet equatorial climate, and the temperatures show a somewhat more pronounced annual range.

The temperature and rainfall graph for a representative station of this climate is shown in Figure 10.7. Belize, British Honduras, is located in Cen-

avalanches of soil and rock frequently occur, stripping away all forest and soil down to the bedrock.

Of economic importance are several forest products. Hardwood lumber, such as mahogany, ebony, or rosewood, is a valuable tropical product. Quinine, cocaine, and other drugs come from the bark and leaves of tropical plants; cocoa comes from the seed kernel of the cacao plant. Rubber, made from the sap of the rubber tree, is now largely an economic product of Malaya, Sumatra, and Ceylon, although the tree comes from South America, where it was first exploited.

Although most of the wet equatorial climate lies in a belt 10° of latitude on either side of the equator, the Malabar coast of India and the coasts of Burma and Thailand, located between 10° and 25° N lat., have a warm, wet climate with large annual total of rainfall, which also supports rainforest. This Asiatic climate may be considered as a special wet monsoon type. Under the Köppen system it is designated by the symbol *Am,* with the stipulation that the average rainfall of the driest month shall be less than 2.4 in (6 cm). (See Figure 10.21.) Figure 10.6 shows the temperature-precipitation graph of a wet monsoon climate. A short dry season occurs in the low-sun season, when the winter monsoon is in effect and continental air masses move southward from interior Asia. This dry period is too brief to deplete soil moisture and ground water reserves. Rainforest therefore thrives. Note that the summer (rainy) monsoon brings enormous quantities of rainfall in June and July. This rain is derived from convectional storms

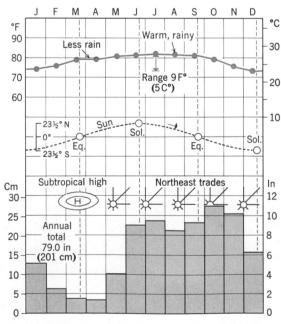

Figure 10.7 Climograph for Belize, British Honduras, 17° N. lat. (Data from Trewartha.)

tral America at 17° N lat. The total rainfall is great, almost 80 in (200 cm) annually, and rain is abundant in most months. A tendency toward dryness, typical of the tropical wet-dry climates (which generally elsewhere lie at this same latitude), is seen in the low rainfall of February, March, and April. The temperature cycle has a range of only 9 F° (5 C°) because of the moderating influence of the nearby ocean but this range is conspicuously greater than that of the wet equatorial climate.

The warm, wet climates of trade-wind coasts support a tropical rainforest vegetation, somewhat similar to the equatorial rainforest. Differences are explained in Chapter 16.

3. Tropical desert and steppe climates (BWh, BSh)

In strong contrast to the wet equatorial climate are the very dry climates controlled by subsiding, outwardly moving air of the continental high-pressure cells which dominate much of the earth's land areas in the latitude belts 15° to 35°, roughly centered on the Tropics of Cancer and Capricorn. Here lie the source regions of the continental tropical (*cTs*) air masses. The vast deserts of north Africa, Arabia, Iran, and West Pakistan exemplify this climate type, as do also the Sonoran Desert of the southwestern United States and northern Mexico, the interior Kalahari Desert of South Africa, and the Australian Desert.

Within this region of general aridity we can recognize the truly arid zones, or *deserts (BWh)*, in which annual rainfall is less than 10 in (4 cm), and the semi-arid zones, or *steppes (BSh)*, in which rainfall is from 10 to 30 in (4 to 12 cm) annually.

In the continental interiors in these tropical regions, far removed from oceanic sources of moisture, extreme aridity prevails. Average annual rainfall is often less than 5 in (2 cm); at some localities a period of several years may pass without measurable rainfall. As an example, at In Salah, Algeria, the average annual rainfall for a 15-year record was 0.6 in (1.5 cm). Other Algerian stations had even less. Because these are regions of subsiding air in the general global scheme of circulation, adiabatic heating reduces the relative humidity of the air to low levels much of the time. At the drier stations in the Sahara Desert, relative humidity at 1:00 P.M. averages 25 to 30 percent for the year, but from 15 to 20 percent in the hottest months. Although in the summer low pressure due to ground-surface heating develops over the tropical landmasses (as seen in Figures 5.2 and 5.3), this is only a low-level condition, with the permanent highs persisting at higher levels of the atmosphere.

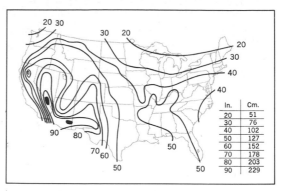

Figure 10.8 Annual evaporation (inches) from a free water surface in the United States. (After Mead.)

The capacity of tropical desert air for evaporation of exposed water surfaces is enormous (Figure 10.8). In the Sonoran Desert, annual evaporation from a free water surface exceeds 90 to 100 in (230 or 250 cm) annually, or about twenty times as much as falls in rain. It is obvious that the full amount of evaporation that is possible does not take place in the tropical deserts. Once the stream channels and soil have become dry after a rain, further evaporation is limited to a small amount of moisture slowly brought to the surface locally by capillary movement from moist soil or rock at depth. Figure 9.9*B* shows the water balance for Beni Abbes, Algeria, and illustrates well the prevailing water deficiency of the tropical desert climate.

Although dryness is the dominant characteristic of the continental tropical air mass source regions, sporadic heavy rainfall does occur from violent convectional storms. Penetration of maritime tropical or equatorial air may be responsible for such storms. During a single cloudburst confined to an area of a few square miles, the major portion of rainfall of one or more year's total may fall, producing debris floods in the stream channels.

The concept of variability in rainfall is an important one in climate study. By variability we mean the degree to which the rainfall of the individual years differs from the average value computed over a long period of years. Figure 10.9 illustrates this principle. The yearly amounts of rainfall are shown by bars for three stations for a period of many years. At the top the record for Padang, Sumatra, gives data for a wet equatorial climate; the bottom graph gives rainfall for Abbassia, Egypt (near Cairo), a tropical desert station. The middle graph is of a combined climate type (tropical wet-dry climate). We say that the rainfall variability of the wet equatorial climate is small because the individual yearly amounts were, in this case, never more than twice the average or less

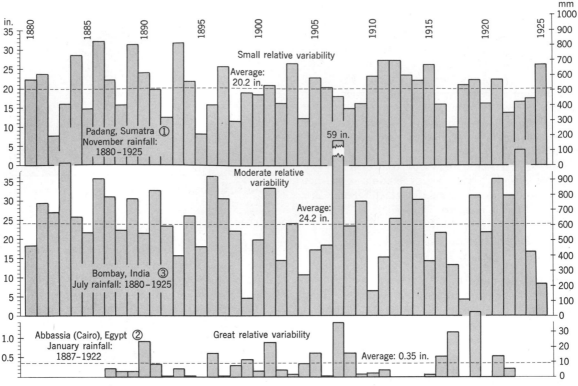

Figure 10.9 These three graphs illustrate the concept of variability of rainfall by showing the actual amount of rain received each year during the month which, on the average, is the rainiest month at that place. (Data from H. H. Clayton, Smithsonian Institution.)

than a third of the average. On the other hand, Abbassia shows great variability because, although several years had no rainfall at all, five of the years had more than twice the average amount and two of the years had four or more times the average.

Rainfall variability has been estimated for the world in terms of percentage departure from the normal value (Figure 10.10). As would be expected, the tropical desert areas lying near the tropics of Cancer and Capricorn have the highest variability; the equatorial belt of heavy rainfall, the least variability of the low-latitude part of the globe.

Consider next the temperature conditions of the dry continental tropical air mass source regions. Figure 10.11 shows the yearly march of temperatures at Yuma, Arizona, a representative North American station of the tropical desert (*BWh*) climate type. Two points are noteworthy: (*a*) temperatures are very high during the period of high sun; (*b*) annual range is moderately strong. Normally the annual range of temperature (hottest month average minus coolest month average) in these climates is 30 to 40 F° (17 to 22 C°) and is directly related to the height of the sun in the sky.

For those who consider southern Arizona to be a hot desert, the data of Bou-Bernous, Algeria, may

be enlightening (Figure 10.12). Here the mean daily temperature averaged over 100°F (38°C) in July—a full 10 F° (5.5 C°) higher than Yuma, Arizona.

More interesting, perhaps, than annual range is the normal daily range of temperatures in the tropical deserts. Figure 10.13 shows maximum and minimum daily readings for the months of January and July at Phoenix, Arizona. Note that the daily range is often as much as 35 F° (22 C°), the average daily range about 30 F° (17 C°). In no other climate does so great a daily range occur. This is explained by the rapid nightly heat loss from the ground and lower air layers because of the low water-vapor content of the air. On the other hand, insolation during the day is extremely intense and air temperatures soar to great heights. An all-time record range was recorded in Bir Milrha in the Sahara Desert, south of Tripoli, where 31°F and 99°F (−0.6°C and 37.2°C) were recorded on the same day, a range of 68 F° (37.8 C°).[1]

The world's highest temperatures are recorded in certain parts of the continental tropical deserts.

[1]G. T. Trewartha, *An Introduction to Weather and Climate*, McGraw-Hill Book Co., New York, 1943, p. 367.

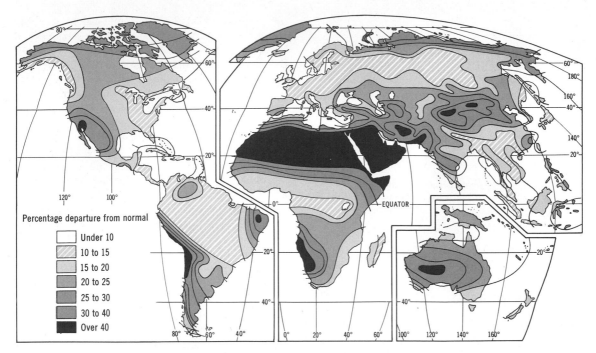

Figure 10.10   Precipitation variability map of the world. (After William Van Royen, 1954, *Atlas of the World's Agricultural Resources*, Prentice-Hall, Englewood Cliffs, N.J.)

Percentage departure from normal

	Under 10
	10 to 15
	15 to 20
	20 to 25
	25 to 30
	30 to 40
	Over 40

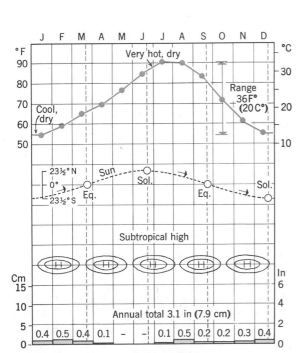

Figure 10.11   Climograph for Yuma, Arizona, 33° N. lat. (Data from Trewartha.)

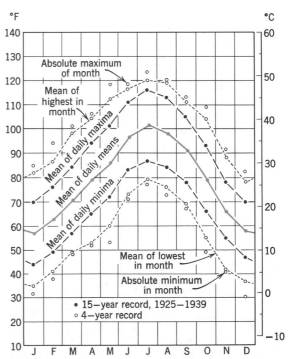

Figure 10.12   Air temperature data for Bou-Bernous, Algeria, 27°17′ N, 02°53′ W, elevation 1509 ft (460m). (Data from Meteorological Office of Great Britain.)

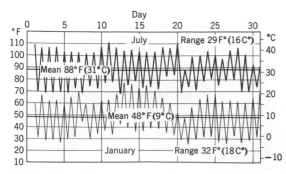

Figure 10.13 Daily temperatures for July and January at Phoenix, Arizona. (After Mark Jefferson, *Geographical Review*.)

A world record maximum of 136.4°F (58°C) was officially observed in the shade under standard shelter at Azizia, Tripoli. Figure 10.14 shows the highest temperatures ever observed in the western United States. The Sonoran Desert region of the lower Colorado River valley clearly surpasses all other areas for high temperatures.

In the extremely dry deserts (*BWh*), virtually the entire land surface appears to be free of vegetation and consists of bare rock, stream gravels and sands, or drifting dune sands. This does not mean

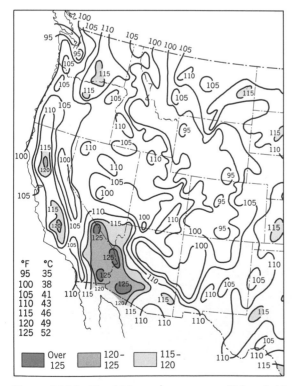

Figure 10.14 The highest temperatures (Fahrenheit) ever observed during the period 1899 to 1938. (After J. B. Kincer, U.S. Dept. of Agriculture, *Yearbook of American Agriculture*, 1941.)

that vegetation is wholly absent, rather, that the plants are thinly scattered over the surface and lack foliage to protect or obscure the bare ground. Desert plants are adapted to long drought periods by thick, fleshy leaves and stems, which store water for long periods of time but which do not permit loss of water through the surface. Representative of desert vegetation are the cacti and other shrubs seen in the Sonoran Desert of southwestern United States and northern Mexico (Figure 16.29). Chapter 15 treats the adaptation of plants to a dry (xerophytic) environment; Chapter 16 describes desert vegetation and its distribution.

Soils of the deserts are lacking in humus and are of a grayish or reddish color, depending upon the type of iron compound present to produce staining. These soils contain excessive amounts of calcium carbonate and other salts, which are left near the ground surface by evaporating water. In the centers of shallow lakes, the salts concentrate to form white saltflats, entirely sterile and almost perfectly smooth. Soil-forming processes of deserts are discussed in Chapter 13, the great soil groups of deserts in Chapter 14.

For a discussion of landforms of the mountainous desert topography, the student is referred to Chapter 22 on the fluvial cycle of landmass denudation.

Tropical steppe climate (*BSh*) borders the tropical deserts on both north and south, and in places on the east as well. Locally because of altitude, plateaus and high plains within what would otherwise be desert have the semi-arid steppe climate. Steppe zones lying equatorward of the deserts are transitional into the tropical wet-dry climate (*Aw*) and resemble it in many ways (Figure 10.15). Steppes on the poleward fringes of the tropical deserts grade into the Mediterranean climate (*Cs*) in many places. Steppes typically are grasslands of short grasses and other herbs, and with locally developed shrub and woodland (see Chapter 16). These areas are able to support limited numbers of grazing animals but are not generally moist enough for crop cultivation without irrigation. Soils are commonly of the *brown soil* and *chestnut soil* groups, containing some humus (Chapter 14).

4. West coast desert climate (BWk, BWh)

Again referring to the world rainfall map, Plate 1, it will be seen that all west coasts in latitudes 15° to 30° are extremely dry, generally with less than 10 in (25 cm) of rainfall annually. The Atacama desert of Chile and the Namib Desert of coastal southwest Africa are perhaps the most celebrated of these deserts, but they exist also in Lower California, the Morrocan coast of Africa, and the west coast of Australia. The arid belt

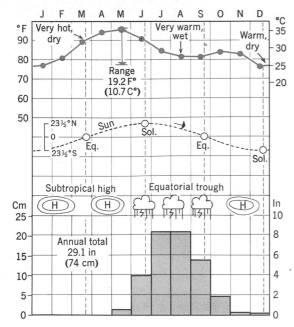

Figure 10.15 Climograph for Kayes, Mali, 14½° N lat. (Data from Trewartha.)

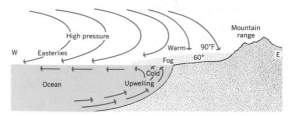

Figure 10.16 Subsiding air above a coastal temperature inversion. (From A. N. Strahler, *The Earth Sciences*, Harper and Row, New York.)

extends continuously eastward to inland continental tropical deserts.

Does it not seem strange that extreme dryness exists immediately along the shores of the oceans, close to possible sources of moist maritime air masses? The interior tropical deserts, already discussed, are logically explained by land-centered high-pressure cells into which moist air masses cannot easily drift, but the dry west coast strips are located between the oceanic and continental high-pressure cells where we might expect to find some development of fronts and convergence of air masses. The key to this problem seems to lie in the fact that the oceanic subtropical high-pressure cells are inherently dry on their east sides. The circulation in these cells is thought to be such that the air on the east sides is subsiding as it moves outward, hence, is adiabatically heated and its humidity reduced. The result is dry, stable air masses that bring an arid zone not only to the coast but extending far seaward as well. Strangely enough, there are dry deserts over the oceans in these tropical latitudes (Figure 10.16).

In what respects do the dry west coasts differ in climate from the interior continental deserts which they adjoin? The principal difference is in temperature. The coastal deserts are relatively cool, with annual average temperatures around 65°F (18°C), whereas the interior continental deserts average some 10 F° (5 C°) higher. The presence of cool upwelling and equatorward-flowing currents, such as the Humboldt and Benguela currents, explains the lowered temperatures. (See

Chapter 5.) The annual range of temperature in the coastal deserts is very low, as illustrated in Figure 10.17 by the temperature graph of Iquique, Chile (20° S), on the dry Atacama Desert coast of South America. The annual range is only 11 F° (6 C°). In contrast, the temperature cycle of Aswan, Egypt (24° N), shows both a higher average and a greater range.

Although the cool west coast desert climate is treated here as a separate climate from the tropical desert climate, not all climate classifications make the distinction. In the original Köppen classification, the cooler desert west coasts are designated by the symbol *BWn*, in which *n* means *frequent fog* (from the German, *nebel*, meaning fog). Persistent coastal fog banks form in the cool lower air layer overlying the cool ocean water (Figure 10.16). In the latest versions of the Köppen climate system, the symbols *BWh* and *BWk* are applied to the west coast deserts. The *BWk* climate is limited to the South American and southwest African coasts in latitudes 20° to 32° S, where the cool ocean currents are most influential.

Vegetation and soils of the cool west coast deserts are essentially similar to those of the interior deserts. An unusually high incidence of fog on these coasts leads to growth of some specialized

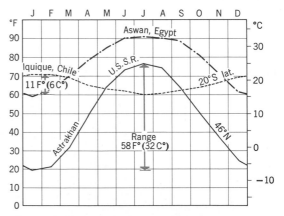

Figure 10.17 Temperature conditions compared for coastal and interior tropical desert locations. (Data from H. H. Clayton, Smithsonian Institution, and Trewartha.)

plants which can exist on condensed moisture close to the shore.

5. Tropical wet-dry climate (Aw, also Cwa)

We have thus far considered two extreme climate types. One, a wet climate lying in the equatorial zone, the other a desert climate arranged in two belts along the Tropics of Cancer and Capricorn. What of the intermediate zones where these two climate types come together? Knowing from previous study that the pressure and wind belts of the globe migrate northward at June solstice season and southward at December solstice season, we can infer that the intermediate zones in question will have climates that combine the characteristics of the first two types in a seasonal alternation. This results in the tropical wet-dry (*Aw*) climate, which has a wet season controlled by moist, warm equatorial and maritime tropical air masses at time of high sun and a dry season controlled by the continental tropical air masses at time of low sun.

The latitude belts in which tropical wet-dry climate is found lie roughly between 5° and 25° latitude, throughout Central and South America, Africa, and Australia. In southeast Asia this zone is pushed northward to latitudes 10° to 30° because the continental tropical air mass source region in summer is necessarily situated farther north to conform to the Asiatic landmass.

Climate characteristics of the tropical wet-dry climate can be judged from Figure 10.18, which gives temperature and rainfall data for a representa-

tive west African station. This can be studied in conjunction with Figure 10.19, which shows seasonal air mass and flow patterns. Timbo lies closer to the equator than Kayes, representing the tropical steppe (*BSh*) climate (Figure 10.15), hence has a longer wet season and a more even temperature cycle. Note especially the fact that maximum temperatures occur in March, April, or May, rather than in July, because the onset of the rains brings a cloud cover and cooler air temperatures. The water budget of an African tropical wet-dry station, Yambio, Sudan, is shown in Figure 9.9C. Here both a surplus and a deficiency occur in the cycle.

The Asiatic tropical wet-dry climate is somewhat different in that the monsoon controls are strong and bring an extreme contrast of wet and dry conditions. If we now look at the record of Allahabad, India, 25° N lat. (Figure 10.20), the basic similarity to Kayes and Timbo in western Africa (Figures 10.15 and 10.18), is very marked. The higher latitude of Allahabad results in much cooler January temperatures, and a greater annual range, but otherwise the temperature curves show the same dip during the rainy season. Köppen places Allahabad and a large belt of southeastern Asia in latitudes 20° to 25° in the *Cwa* climate (temperate, rainy climate, dry winter, hot summer). The *Cw* climate is a poleward extension of the *Aw* climate both here and in Africa and South and Central America. The two are here treated under the tropical wet-dry climate because tropical air masses dominate and the regime is basically of one pattern. Figure 8.8 shows the characteristic distorted figure-eight loop of the Indian wet-dry regime. Allahabad, India, is a *Cwa* station and shows a marked continental effect in its annual range. Paraná, a Brazilian *Aw* station, shows the thermal equability we associate with the wet equatorial climate but has a marked dry season.

The Köppen system provides an exact basis for the distinction between *Am* and *Aw* climates, both of which have a dry season. The *Am-Aw* boundary varies according to both total annual rainfall and driest-month rainfall, as shown by a graph (Figure 10.21). In the annual rainfall range between 40 in (100 cm) and 100 in (250 cm) the *Am* climate can exist with progressively lower values of driest-month rainfall, beginning with 2.4 in (6 cm) and declining to zero. Thus an *Am* station might have one month with no rainfall whatsoever, provided it has an annual total of more than 100 in (250 cm).

The contrast in wet and dry season rainfall in the tropical wet-dry climate is further illustrated in Figure 10.22. Here the rainfall of the one rainest month is compared with the total for the three consecutive driest months for a number of typical stations. Rainfall is not so reliable in the tropical

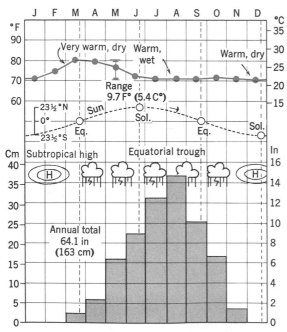

Figure 10.18 Climograph for Timbo, Republic of Guinea, 10°40′ N lat. (Data from Trewartha.)

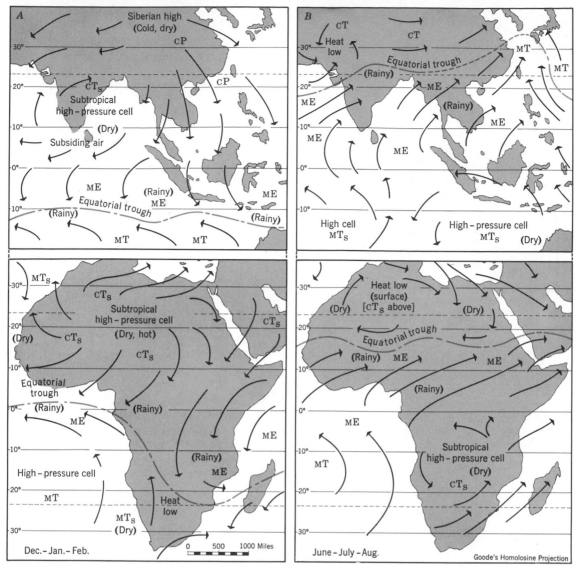

Figure 10.19 These maps of Africa and southern Asia illustrate one interpretation of the air-mass source regions and circulation patterns that govern the equatorial and tropical climates. (Modified after M. A. Garbell, *Tropical and Equatorial Meteorology*, New York, 1947. Based on a Goode Base Map. Copyright by the University of Chicago. Used by permission of the University of Chicago Press.)

wet-dry climate as in the wet equatorial climate but is more reliable than in the very dry tropical deserts. This can be seen from the record of Bombay, India, in Figure 10.9, and is also apparent from the world variability map, Figure 10.10.

Alternation of wet and dry seasons results in the growth of a distinctive vegetation known generally as the *tropical savanna*. This is characterized by open expanses of tall grasses, interspersed with hardy, drought-resistant shrubs and trees (Figure 16.20). Other areas have savanna woodland, monsoon forest, thornbush, and tropical scrub, all of which are formation classes grouped on Plate 4 under the heading "Raingreen." These formation classes are described in Chapter 16. Grasses in this climate are dried to straw and many tree species shed their leaves in the dry season. Other trees and shrubs have thorns and small leaves or hard, leathery leaves that resist water loss.

Soils of the wet-dry tropical climates are mostly yellowish or reddish latosols, similar to those described in connection with the wet equatorial climate. (See Chapter 14.) Excessive leaching is again the result of the heavy rainfall and high temperatures. In general, these residual soils of uplands are not fertile and are little cultivated in South America and Africa. Locally, rich floodplain alluvium is highly productive. Stream flow in these

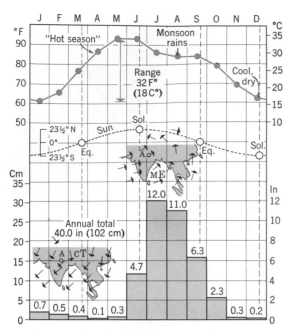

Figure 10.20   Climograph for Allahabad, India, 25° N lat. (Data from H. H. Clayton, Smithsonian Institution.)

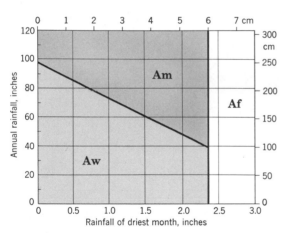

Figure 10.21   This graph shows how *Aw* climates are distinguished from *Am* climates in the Köppen system. (After Haurwitz and Austin.)

Figure 10.22   Rainfall (inches) of the one wettest month is shown contrasted with the total rainfall of the three consecutive driest months for five wet-dry tropical climate stations. (Data from Clayton and Trewartha.)

Africa are the natural home of such herbivores as wildebeest, gazelle, deer, antelope, buffalo, rhinoceros, zebra, giraffe, and elephant. On them feed the lion, leopard, hyena, and jackal. Some of the herbivores depend upon fleetness of foot to escape the predators. Others, such as the rhinoceros, buffalo, and elephant, defend themselves by their size, strength, or armor-thick hide. The giraffe is a perculiar adaptation to savanna woodlands; his long neck permits browsing upon the higher foliage of scattered trees.

The dry season brings a severe struggle for existence to animals of the African savanna. As streams and hollows dry up, the few muddy waterholes must supply all drinking water. Danger of attack by carnivores is greatly increased.

The Indian savanna, woodland, and thornforest have a somewhat similar assemblage. Deer and antelope are especially abundant, with some water buffalo and a few rhinoceroses. The tiger replaces the lion as the principal carnivore. The Indian elephant, however, is largely restricted in natural habitat to the rainforest coastal strips of Burma, the Malabar coast, and Ceylon.

REFERENCES FOR FURTHER STUDY

Gautier, E. F., tr. by D. F. Mayhen (1935), *Sahara, the great desert*, Columbia Univ. Press, New York, 264 pp.

Lee, D. H. K. (1957), *Climate and economic development in the tropics*, Harper and Row, New York, 182 pp.

Church, R. J. H. (1961), *West Africa, a study of environment and man's use of it*, John Wiley and Sons, New York, 547 pp.

Trewartha, G. T. (1961) *The earth's problem climates*, Univ. of Wisconsin Press, Madison, Methuen and Co., London, 334 pp.

Hills, E. S., ed. (1966), *Arid lands; a geographical appraisal*, Methuen and Co., London, 461 pp.

Meigs, P. (1966), *Geography of coastal deserts*, UNESCO Publications Center, New York, 140 pp.

regions contrasts greatly with that of the wet equatorial climate, in that the former has a very strong seasonal fluctuation. From flood conditions with extensive low-lying areas under water in the rainy season, the streams pass to a regime of little or no flow in the dry season, when channel bottoms of sand and gravel are exposed and mud flats dry out.

Closely related to vegetation and climate is the natural animal life of the savanna grasslands and woodlands. These are the regions of the carnivorous game animals and a vast multitude of grazing animals on which they feed. The grasslands of

REVIEW QUESTIONS

1. What conditions of temperature, pressure, and rainfall prevail in the equatorial belt? What air masses are present? Under what conditions does heavy precipitation occur? How does the equatorial trough change position throughout the year?

2. Describe the annual temperature curve for a typical station in the wet equatorial (*Af*) climate zone. What range is usually found? How does the daily temperature range compare with the annual range?

3. Approximately how great is the average annual rainfall of the wet equatorial climate? How is rainfall distributed throughout the year? Is variability of rainfall great or small?

4. What is rainforest? Of what types of vegetation is it composed? (See Chapter 16.)

5. What is a lateritic soil (latosol)? What processes are responsible for development of lateritic soils. Of what economic importance are laterites? (See Chapter 14.)

6. Is there much humus in lateritic soils (latosols)? Explain how temperature and bacterial activity affect the accumulation of decomposed vegetation in the hot equatorial regions as contrasted with cold regions.

7. Is the chemical decay of rock relatively small or great in the wet equatorial climate? Explain.

8. What can be said about stream flow in the wet equatorial climate? In what way are rivers of importance to native inhabitants of these regions?

9. In what way does the wet, monsoon-dominated climate of the southern Asiatic coasts (*Am*) differ from that close to the equator (*af*)?

10. Why is there heavy precipitation on east coast belts in latitudes 10° to 25°? What climate occurs here? What type of climate occurs at inland locations in this same latitude belt?

11. What causes the tropical deserts in latitudes 15° to 35°? Explain these deserts in terms of the global pattern of air circulation. What kind of air mass has its source over these areas?

12. Distinguish between desert and steppe climates. What Köppen symbols denote these climates?

13. Discuss the evaporation of moisture in the tropical deserts. How great is evaporation from a free water surface? What influence has this factor upon streams and the ground water table?

14. What type of rainfall occurs in the tropical deserts? Is it widespread or localized? Is rainfall variability in these deserts great or small?

15. What is the nature of the annual temperature cycle in the tropical deserts? Explain these characteristics. What is the nature of the daily cycle of temperature? How does it compare with that of the wet equatorial climate?

16. What kind of vegetation is native to the tropical deserts? To what extent does it afford cover to the ground? What kinds of soils are typical of these deserts?

17. Explain the west coast deserts in latitudes 15° to 30°. How are these related to the subtropical high pressure cells? Why is the air dry in these locations?

18. What is the outstanding feature of the annual temperature cycle of the west coast deserts? Why is the range small? Why are the temperatures abnormally low for this latitude? Why is fog abundant along these coasts?

19. Explain the characteristic seasonal features of the wet-dry tropical climate in terms of air masses and seasonal shifting of the pressure and wind belts of the globe.

20. How do annual precipitation and its distribution throughout the year change if we start with the wet equatorial climate zone and examine the rainfall records for stations progressively farther north until the Sahara Desert is reached?

21. Describe the annual temperature cycle of the tropical wet-dry climate. When is the maximum reached? Why does this not coincide with the time of highest sun?

22. In what way does the Asiatic wet-dry tropical climate differ from that of Africa or South America? Explain. How does Köppen distinguish the climate of north India from the climate of peninsular India?

23. What is the natural vegetation of the wet-dry tropical climate? What are the characteristics of the wild animal life of these regions? What type of soil is generally present? What effect has the seasonal contrast of dry and wet periods upon stream flow?

CHAPTER 11

Middle-Latitude Climates

CLIMATES of group II are those of middle latitudes occupying the polar front zone in which both tropical and polar air masses play an important part. The latitude belt in which these climates lie is subject to cyclonic storms; most of the precipitation in these climates occurs along fronts within these cyclones.

6. Humid subtropical climate (Cfa)

The moist nature of the west sides of the oceanic subtropical high-pressure cells has already been discussed in Chapter 6. Maritime tropical (мT) air in this part of the anticyclonic cell is engaged in a slight ascending motion combined with its flow toward higher latitudes. The air mass has a steep lapse rate. Moisture that enters the air mass by evaporation from the warm ocean surface is distributed to great heights. As this moist, unstable maritime tropical air mass moves over the eastern continental coasts in latitudes 25° to 35° and drifts inland, it brings the necessary moisture and latent heat energy for heavy precipitation. Lifting occurs along warm and cold fronts where the tropical air encounters polar air. This general pattern of climate, here called the *humid subtropical climate,* is exemplified by the southern Atlantic and Gulf Coast states of the United States; corresponding regions are found in the Argentina-Uruguay-southern Brazil area of South America, in eastern China and southern Japan, along a small part of the southeastern coast of Africa, and on the east Australian coast.

In the Köppen system, these areas lie in the *Cfa* climate, described as a temperate rainy climate with hot summers. Temperature specifications of the *C* climates have been stated in Chapter 8. The *Cfa* climate has no dry season, and even the driest summer month receives more than 1.2 in (3 cm) of rain. The hot summer specified by the letter *a* is one in which the average temperature of the warmest month is over 71.6°F (22°C).

For an analysis of the temperature and precipitation characteristics of this moist, subtropical climate, we may refer to the record for Charleston, South Carolina (Figure 11.1). Rainfall is ample at all times of the year, but is distinctly greater during the summer when the oceanic high strengthens and the flow of maritime tropical air is increased. Thunderstorms are especially frequent in summer (Figure 11.2). They may be of the thermal (heat) type, or of squall-line or cold front origin. An occasional tropical cyclone may strike the coastal area, bringing very heavy rains. Winter precipitation, some of it in the form of snow, is of frontal type in the frequent middle-latitude cyclones that sweep over these regions. Not all stations in the humid subtropical climate show a summer maximum. For example, Atlanta, Georgia, has a winter maximum, but a secondary peak in midsummer. Typically the annual precipitation totals more than 40 in (100 cm).

Temperatures show a moderately strong range, of very much the same magnitude as in the tropical deserts, but without the extreme heat in summer. Humidity is, of course, very high, in marked contrast to the deserts, and summer climate on the humid east coasts is at times similar to the wet equatorial climate in temperature, rainfall, and

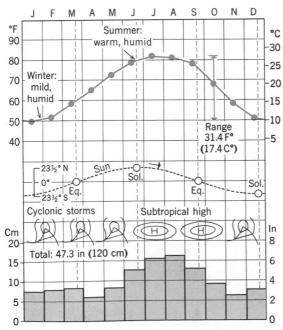

Figure 11.1   Climograph for Charleston, South Carolina, 33° N. lat. (Data from Trewartha.)

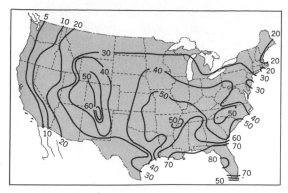

Figure 11.2 Average annual number of days with thunderstorms, based on records of the period 1899 to 1938. (After J. B. Kincer, U.S. Dept. of Agriculture.)

humidity. Winters show the influence of outbursts of polar air masses, which frequently penetrate into subtropical latitudes and bring below-freezing weather with killing frosts. We might say that this climate type is shared to some extent by both tropical and polar air masses, but that the tropical air masses prevail most of the time and dominate in summer.

In southeast Asia the humid subtropical climate is somewhat modified by intensive monsoon development. Winter air masses from interior Asia are very dry, and a winter scarcity of precipitation develops, whereas maritime air masses in summer, together with occasional typhoons, cause a strongly accentuated maximum in the summer. This shows clearly in the precipitation cycle of Shanghai, China (Figure 11.3), which has a winter both drier and cooler than that of Charleston.

With reference to climatic regime, the humid subtropical climate usually shows a pronounced continentality, as in the case of the thermohyet diagram for Shanghai, China, in Figure 8.10C.

Soils of the moister, warmer parts of the humid subtropical regions are strongly leached red-yellow soils related to the latosols of the humid tropical and equatorial climates. Rich in iron and aluminum oxides, these soils are poor in many of the plant nutrients essential for successful agricultural production.

Forest is the natural vegetation of most of the areas having the humid subtropical climate. Plate 4 shows temperate rainforest to be the dominant type, but with some areas of tropical rainforest occupying the coastal belts in lower latitudes. Much of the sandy coastal region of the southeastern United States today has a second growth forest of longleaf, loblolly, and slash pine, whereas the inland region has summer-green deciduous forest. (These forest types are treated in Chapter 16.) Toward higher latitudes forest gives way to tall-grass prairie such as the *Pampa* of Argentina

and Uruguay, and the prairies of Oklahoma and Missouri. Here the soils are of dark prairie and chernozem groups, typical of a dry continental climate. These grasslands occupy regions that are *subhumid*, i.e., best considered transitional to the semi-arid steppes, for they have less than 40 in (100 cm) annual precipitation and a marked dryness of the winter season.

7. Marine west coast climate (Cfb, Cfc)

Shifting attention now toward higher latitudes, and using the information about air masses, fronts, and cyclones presented in Chapter 7, we might expect to find that western coasts in the belt of cyclonic storms would receive ample rainfall from maritime polar air masses and would have rather moderate temperature variations because of the proximity to the oceans from which the air masses tend to drift landward.

These conditions are fulfilled in west coasts of landmasses lying between 40° or 45° and 60° latitude. Situated too far poleward to be dominated by the dry influence of the subtropical oceanic high-pressure cells, these climates lack a dry season. Because continental polar air masses tend to drift eastward, they rarely move westward to visit the west coasts; hence severe dry-cold conditions are uncommon.

Köppen classifies the marine west coast climate as *Cfb*, a temperate rainy climate with warm summers. The average temperature of the warmest month is under 71.6°F (22°C) and at least four

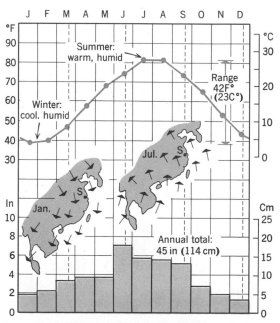

Figure 11.3 Climograph for Shanghai, China, 31° N lat. (Data from Meteorological Office of Great Britain.)

months average 50°F (10°C) or more. The cool summer variety *Cfc* has fewer than four months with averages over 50°F (10°C).

Climate of the middle-latitude west coasts is illustrated by the graphs for Brest, France (Figure 11.4), located at 49° N lat. on the Brittany coast. Precipitation is well distributed throughout the year, but shows a distinct reduction during the summer months. This feature is also found on rainfall graphs for coastal stations on the Pacific northwest coast of the United States. Why should this reduction occur? In summer, the oceanic subtropical high is most strongly developed and moves farthest north, bringing its arid influence to bear just enough to make a distinct decrease in summer rainfall. In other words, this is a manifestation of the same mechanism that causes a west coast desert in low latitudes.

Although the total rainfall for Brest and similar marine west coast stations over much of Europe is not great, judged by tropical or equatorial standards, the cooler air temperatures reduce evaporation and produce a very damp, humid climate with much cloud cover. Nearness to the ocean makes for a small annual temperature range, just as it does in lower latitudes on the west coast deserts. Mild winters and relatively cool summers are characteristic. Winters which are severely cold at the same latitudes in mid-continent and east-continent positions are by contrast surprisingly mild on the west coasts.

In terms of climatic regime, the marine west coast climate combines equability of temperature with a Mediterranean regime. Vancouver, B.C., has a thermohyet diagram with the typical Mediterranean shape, but is displaced to a colder and wetter position (Figure 8.9*C*). Auckland, New Zealand, is depicted in Figure 8.6*A* as equable in tendency, its annual temperature range being only 14.5 F° (8 C°).

The influence of coastal ranges upon rainfall is extremely marked in middle latitudes. Whereas low-lying coasts, such as in northern France or southern England, receive only 30 to 40 in (75–100 cm) of precipitation annually, mountainous coasts of British Columbia, Alaska, Norway, and Chile get 60 to 80 in (150–200 cm) and over. This has been a major factor in the development of fiords along the seacoasts. (See Chapter 23.) Heavy snows in the glacial period nourished vigorous valley glaciers which descended to the sea, scouring deep troughs below sea level at their lower ends.

Natural vegetation of the marine west coast climate is forest, but of widely different classes from one world locality to another. Western Europe has a summer-green deciduous forest whereas that of west coast North America is of the needleleaf class. On coast ranges of the Pacific northwest, Douglas fir, red cedar, and spruce grow in magnificent forests (Figure 16.14). Forests of the marine west coast climate belts of Chile and New Zealand are of the temperate rainforest class (Chapter 16).

Soils of the marine west coast climate regions bearing needleleaf forests are of a strongly leached type, the *podzols*, and are acid in nature. Under cool temperatures, bacterial activity is slow, in contrast to the warm tropics, so that vegetative matter is not consumed and forms a heavy surface deposit. Organic acids from the decomposing vegetation react with the soil compounds, resulting in removal of such bases as calcium, sodium, and potassium. Soils of western Europe are of the *gray-brown podzolic* group, typically associated with deciduous forests in middle latitudes. This soil is not so strongly leached and is suited to diversified forms of agriculture.

8. Mediterranean climate (dry summer subtropical climate) (Csa, Csb)

Situated on west coasts between latitudes 30° and 45° is a zone subject to alternate wet and dry seasons because it is located in the transitional zone between the dry west coast tropical desert (on the equatorward side) and the wet west coast climate (on the poleward side). Because these two climate types have been analyzed and described in previous pages, we need only substitute the information to derive a distinctive compound climate type.

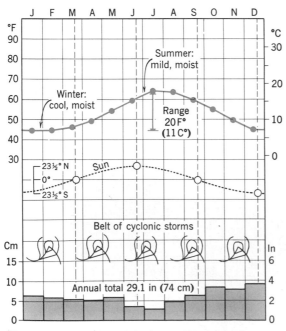

Figure 11.4 Climograph for Brest, France, on the coast of Brittany, 49° N lat. (Data from Blair.)

Consider the climate of Monterey, California, as an example (Figure 11.5). In summer, when the oceanic subtropical high is most powerfully developed and farthest north, the same desert conditions that prevail permanently farther south take over control of the climate and bring a severe drought. However, the proximity of the ocean with its cool current keeps summer temperatures to a mild 60°F (16°C) average. In winter, the humid regime of middle-latitude cyclones and moist maritime polar (mP) air masses is felt in the ample precipitation.

The dry-summer subtropical climate is particularly extensive in the Mediterranean lands. Hence, the name *Mediterranean climate* is commonly used for this climate. The unique regime of the Mediterranean climate has been described in Chapter 8. The thermohyet forms an elliptical figure slanting down to the right (Figure 8.9*A*, *B*).

Köppen classifies the climate of the Mediterranean lands as *Csa*, a temperate rainy climate with dry, hot summers. Those narrow coastal belts directly bordering the cool currents of the open Atlantic, Pacific, and Indian oceans, Köppen designates as *Csb*, distinguished by a markedly cooler summer than the *Csa* climate. Thus Monterey, California, represents a *Csb* climate.

An example of a *Csa* station in the Mediterranean lands is Naples, Italy (Figure 11.6). The annual temperature range here is 28.5 F° (16 C°), which is over twice that at Monterey. The summer is not rainless, but monthly averages are very low

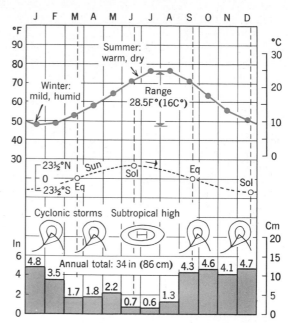

Figure 11.6   Climograph for Naples, Italy, 40½° N lat. (Data from Meteorological Office of Great Britain.)

in June and July. Annual precipitation is double that of Monterey. In short, the equable influence of the cool west coast desert is replaced by a continental influence.

It is apparent that if one travels from the Mediterranean shore into North Africa, he will pass from the Mediterranean climate into the tropical steppe climate of the Sahara. A transition climate type intermediate between the two is nicely shown by the record for Bengasi, Tripoli, at lat. 32° N (Figure 11.7). Note that the temperature record resembles that of the tropical dry climates, although not so hot, but the rainfall distribution is distinctly of the same type as the Mediterranean rainfall, though not so copious.

The Mediterranean climate, its drought coinciding with the high temperatures of summer, incurs a large water deficiency in middle and late summer (Figure 9.9*E*). Winter rains quickly restore the moisture deficiency and a surplus is developed by early spring.

The occurrence of a wet winter and a dry summer is unique among climate types and results in a distinctive natural vegetation of hardleaved evergreen trees and shrubs, known as *sclerophyll forest* (Chapter 16). Various forms of sclerophyll woodland and scrub are also typical (Figures 16.16, 16.17). Trees and shrubs must withstand the severe summer drought of two to four rainless months and intense evaporation.

Soils of the Mediterranean climate are not readily subject to simple classification. Reddish-chestnut and reddish-brown soils typical of semi-

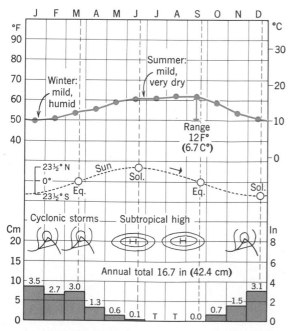

Figure 11.5   Climograph of Monterey, California, 36½° N lat. (Data from Blair.)

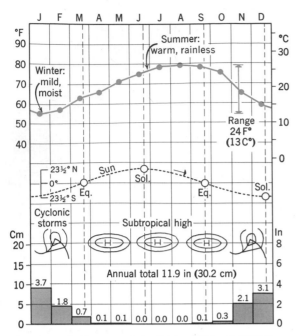

Figure 11.7 Climograph for Bengasi, Tripoli, 32° N lat., on the North African coast. (Data from Trewartha.)

arid climates are generally present (Plate 3). In the Mediterranean lands *terra rossa*, a red soil formed on limestone, occurs in various locations.

9. Middle-latitude desert and steppe climates (BWk, BSk)

The interiors of North America and Asia, in latitudes 35° to 50°, and, to a limited extent, that of southernmost South America have desert and steppe climates of somewhat complex origin. Three basic air mass controls operate. (1) In summer, when pressure and wind belts are shifted poleward, these regions temporarily become the source regions for continental tropical air masses, developed because of intense heating of the large continental interiors. (2) In winter, the intense development of the Siberian and Canadian highs, which are the source areas for continental polar air masses, causes frequent invasions of relatively dry continental air. (3) Mountain ranges separate these deserts from moist maritime polar and maritime tropical air masses which supply abundant rainfall to the west and southeast coasts. Through forced ascent over these ranges, followed by adiabatic heating upon descending the lee slopes (see Chapter 6), the maritime air masses are deprived of their moisture and raised in temperature as well. Thus the regions in question are poorly situated for obtaining precipitation. Because of the prevailing westerly winds in these latitudes, maritime tropical air masses cannot easily reach these areas from the east. In southern Asia, the northward flow of moist tropical

air from the Indian Ocean is blocked by the great Himalayan chain.

Under the Köppen system, two climate varieties are recognized, the semi-arid steppes, *BSk*, and the true desert, *BWk* (Plate 2). The letter *k* signifies a cool climate, with average annual temperature below 64.4°F (18°C). Where the letter *k'* is used, a cold climate is indicated, with the warmest month average below 64.4°F (18°C).

Only a small proportion of the area covered by dry middle-latitude climates is extremely dry: the Turkestan and Gobi deserts of central Asia and parts of the Great Basin in Nevada and Utah (Plate 1). The principal respect in which these deserts differ from the tropical deserts of lower latitudes is that their annual temperature range is much greater and the winter temperatures much lower. Astrakhan, USSR, at 46° N lat., illustrates this feature nicely (Figure 10.17). The enormous annual range of 58 F° (32 C°) is almost double that of Aswan, Egypt; the January mean temperature is a severe 20°F (−7°C), as compared with 60°F (16°C) for Aswan; and the July maximum is only 15 F° (8 C°) less than for Aswan.

Of considerably greater importance geographically than the deserts are the vast semi-arid steppe lands of the middle-latitude dry climates. Partly because of higher elevation, which increases the precipitation, or because of location nearer to maritime air mass invasions, these regions receive from 10 to 20 in (25 to 50 cm) of precipitation annually (Plate 1).

Continentality is the dominant climatic regime of the middle-latitude steppes and deserts. A repre-

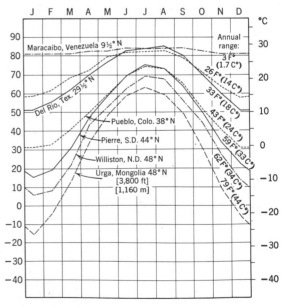

Figure 11.8 Temperature regimes of highland steppe climates. (Data from Trewartha.)

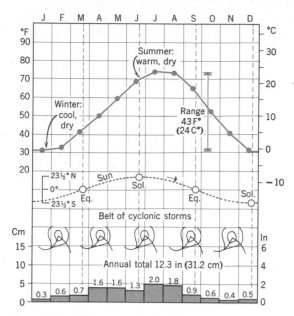

Figure 11.9 Climograph for Pueblo, Colorado, 38° N lat. (Data from Blair.)

sentative thermohyet diagram, that of Omaha, Nebraska (Figure 8.10A), has an elongate figure with a diagonal axis trending from upper right to lower left. As winter temperatures are lowered with increasing latitude the figure is drawn lower on the left side. This effect is seen when the semi-arid steppes are followed from subtropical highlands in Mexico to middle latitudes, the annual temperature cycles show not only progressively lower temperatures but also greatly increased annual ranges (Figure 11.8). A typical steppe climate of intermediate position is that of Pueblo, Colorado (Figure 11.9). Compared with the Mediterranean climate of Monterey, California (Figure 11.5), which lies at about the same latitude, the annual temperature range of Pueblo is very much greater, the precipitation cycle is just reversed, with the summer maximum clearly marked. In summer large highs drifting over the central United States produce a northerly return flow of air on their western sides, spreading moist tropical air from the Gulf of Mexico far northward and westward into the continental interior. This air is unstable and readily produces thunderstorms when lifted over a mountain range or along a cold front.

Steppes support a short-grass vegetation cover suited to grazing of cattle and sheep, but inadequate for farming without irrigation or special dry-farming methods. Special attention is given by geographers and climatologists to the boundary between dry and humid climates, particularly because the limits of agriculture without irrigation are drawn by nature as a fluctuating line over a hazardous frontier zone. Here, productivity may

in some years be high; in others, drought brings disastrous failures. Figure 11.10 is a map showing precipitation and temperature data for the boundary between dry and humid climates in the Great Plains region of the United States. Let us define a humid climate as one in which the precipitation on the average exceeds evaporation, so as to give permanently flowing streams and a generally moist soil; a dry climate as one in which evaporation on the average exceeds precipitation, so as to give ephemeral streams and a generally dry soil. The humid-arid boundary cannot be permanently located, but its average position runs somewhere along the dashed north-south line in Figure 11.10. Note that in North Dakota, the boundary cuts the 15-in (38-cm) annual isohyet but in southern Texas runs close to the 25-in (64-cm) isohyet. This shift is explained by the temperatures, which are higher

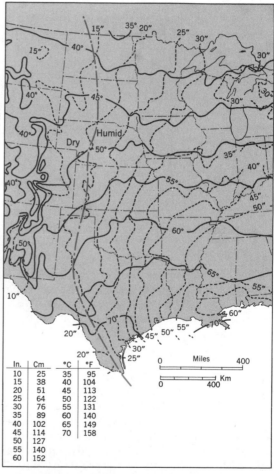

Figure 11.10 On this map, the boundary zone between semiarid and subhumid climates runs north-south. Superimposed are isotherms of average annual Fahrenheit temperature (solid lines) and lines of equal average annual precipitation in inches (broken lines). (After J. B. Kincer, U.S. Dept. of Agriculture.)

in the south and result in greater evaporation losses.

Middle-latitude steppes are characterized by short-grass prairie and by local occurrences of woodland and semi-desert shrubs (Plate 4, Figure 16.27). (This vegetation is described in Chapter 16.) Middle-latitude steppes now constitute the great sheep and cattle ranges of the world. On the vast expanses of the High Plains, the American bison lived in great numbers until almost exterminated by hunters. Likewise, the short-grass veldt of South Africa supported much game at one time. Steppe grasses do not form a complete sod cover, and loose, bare soil is exposed between grass clumps. For this reason, overgrazing or a series of dry years often reduces the hold of grasses enough to permit destructive soil erosion and gullying from heavy local downpours.

Soils of the steppe lands are deficient in humus. They belong to the group of brown soils in the less dry parts; elsewhere they are reddish or gray desert soils. Calcium carbonate is present in excess quantities and may form nodules in the soil. Encrustations of calcium carbonate make a hard, whitish crust termed *caliche* in the southwestern United States. These calcium-rich soils are highly fertile in terms of bases favorable to grain crops, such as wheat, but unless irrigation is employed the advantage is lost. Traced toward humid-climate zones which adjoin the steppes, the soils become darker brown in color, denoting the increasing amounts of humus from heavier growth of grasses. Thus the gray or red desert and steppe soils grade into the rich, dark brown and black chestnut and chernozem soils of the prairies (Plate 3).

10. Humid continental climate (Dfa, Dfb, Dwa, Dwb)

Consider a vast region located in a middle latitude, say between 40° and 55° N, extending from the continental interior to the east coast. North America and Asia are, of course, the areas that we have in mind. The location is intermediate between the source region of polar continental air masses on the north and maritime or continental tropical air masses on the south and southeast. In this polar-front zone, maximum interaction between polar and tropical air masses can be expected along warm and cold fronts associated with east-moving cyclones. In winter, the continental polar air masses dominate and much cold weather prevails; in summer, tropical air masses dominate and high temperatures prevail. Thus, strong seasonal temperature contrasts must be expected in this region. Precipitation is ample throughout the year, because the region lies in a frontal zone. Strong contrasts in air masses result in strong frontal

activity and highly changeable weather. This climate may be described as both humid and continental in properties.

The climate picture drawn above applies well to the north-central and northeastern United States and southeastern Canada as well as to northern China, southern Manchuria, Korea, and northern Japan. Very similar climatic conditions also hold for much of central and eastern Europe, the Balkan countries, and Russia, but this third region differs from the first two in that it is influenced by a great source region of continental-tropical air masses lying to the south and southeast, instead of the oceanic source regions of maritime tropical air masses.

Four varieties of climate, according to Köppen, are found in the regions that we are grouping together as having the humid continental climate. *Dfa* and *Dfb* are cold, snowy, forest climates, the first with a hot summer, the second with a warm summer. The precise definitions of *a* and *b* are as previously given. *Dwa* and *Dwb* climates, found in eastern Siberia, Manchuria, and northern Korea, are cold, snowy forest climates with a dry winter. Letters *a* and *b* refer to hot and warm summers, respectively. The winter dryness, typical of all climates of eastern Asia, is explained below.

We are also including here under the humid continental climate the northern part of Köppen's *Cf* climate region in the eastern United States. (Compare Plate 2 and Figure 8.5.) Recall that Köppen used as his *C-D* boundary the isotherm

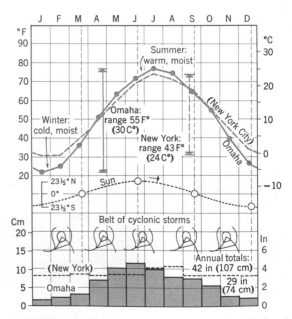

Figure 11.11 A comparison of climates at New York and Omaha, both in the humid continental climate zone. (Data from H. H. Clayton, Smithsonian Institution.)

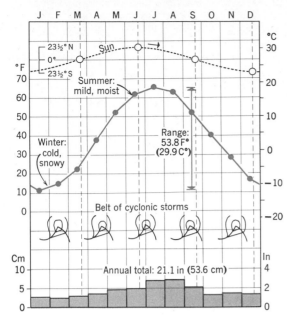

Figure 11.12 Climograph for Moscow, USSR, 56° N lat. (Data from Blair.)

of 26.6°F (−3°C) of the coldest month. Thus, for example, Köppen places New Haven and Cleveland in the same *Cfa* climate as New Orleans and Tampa, despite obvious contrasts in January mean temperatures, soil groups, and natural vegetation between these northern and southern zones.

Some characteristic features of the humid continental climate can be seen from the temperature-precipitation graphs of four stations: Omaha and New York City (Figure 11.11), Moscow, USSR (Figure 11.12), and Peiping, China (Figure 11.13). All four stations have similar temperature curves. Yearly range is near 55 F° (33 C°) for all but New York City, which is moderated by its near-ocean location. Omaha and Peiping, located at the same latitude (40° N), have almost identical temperatures; Moscow, being more than 1000 mi (620 km) farther north 56° N), has a temperature cycle which runs 10 F° (5.5 C°) lower than that of Omaha or Peiping.

Precipitation records of the four stations illustrated show certain marked dissimilarities which require explanation. A summer maximum is apparent in all four but is very weakly defined in New York City because maritime air masses, both polar and tropical, have ready access to the eastern seaboard at all times of the year. Omaha and Moscow have well-defined summer maxima and winter minima, reflecting the predominance of tropical air masses in summer and continental-polar air masses in the winter. Moscow, being farther north, near to the source region of polar continental air masses, has less precipitation than

the other stations. Peiping shows a very strong summer maximum and a winter drought. This contrast is characteristic of the east Asiatic middle-latitude stations and reflects the powerful monsoon control, whereby dry continental air masses dominate in winter.

Continentality is the dominant climatic regime expressed by the above stations. The thermohyet diagrams (Figure 8.10) are elongate in the vertical axis, but inclined to the right, reflecting a summer precipitation maximum.

The general temperature pattern of middle-latitude climates is well illustrated by daily temperature graphs of three stations at about the same latitude across North America (Figure 11.14). Victoria, British Columbia, on the west coast, occupies a windward position with respect to maritime polar air masses from the Pacific. Consequently, the contrasts between July and February (the two extreme months of the year) are small. Moreover, the daily ranges are small, especially in winter. The graph of Winnipeg, Manitoba, shows maximum continental influence with strong annual and daily contrasts. Note especially the wide fluctuations in winter, with outbursts of polar and arctic air bringing very low temperatures. St. Johns, Newfoundland, occupies a leeward position with respect to the continent and is accessible to maritime air masses from the Atlantic. It therefore has only a moderate annual temperature range, but the continental influence is nevertheless clearly marked in the sharply fluctuating daily temperature record.

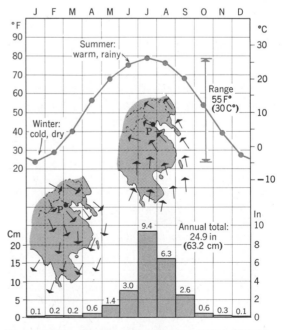

Figure 11.13 Climograph for Peiping, China, 40° N lat. (Data from Trewartha.)

10. Humid Continental Climate (Dfa, Dfb, Dwa, Dwb) | **169**

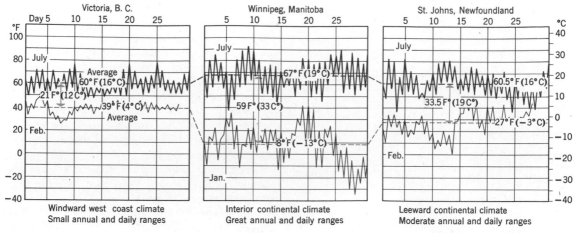

Figure 11.14 Along the 50th parallel in North America lie widely differing climate types. Averages are for months actually shown on graphs. (After Mark Jefferson, *Geographical Review.*)

Three major types of natural vegetation and their associated soils may be recognized in the humid continental climate. The distribution is well illustrated in North America. (See Plates 3 and 4.)

In the more humid eastern sections, including the warmer parts of the humid continental climate zone (*Dfa*), natural vegetation is summergreen deciduous forest (Figure 16.11). Here soils are of the *brown forest* and *gray-brown podzolic* types, rich in humus and moderately leached so as to have a distinct light-colored leached zone under the upper dark layer. In this region, diversified farming and dairying are the most successful uses of the land where topography is favorable.

A northern belt of needleleaf evergreen forest (Figure 16.12) extends along the entire length of the colder northern parts of the humid continental climate zone (*Dfb*). To this may be added the mountain regions of the Adirondacks and northern New England. Here soils are of the *podzol* type, strongly leached, but with an upper layer of humus (Chapter 14). Cool temperatures inhibit bacterial activity which would destroy this organic matter in tropical regions. Podzols are deficient in calcium, potassium, and magnesium, and are, in general, acid in chemical nature. Thus they are not highly productive for crop farming, even though adequate rainfall is generally assured. The podzols are, however, well suited to the growth of conifers.

The drier plains areas of the humid continental climate support a natural tall-grass prairie (Figure 16.26), which grades into the drier steppe regions of short grasses to the west. The prairie soils and chernozem soils, two major soil groups of these grasslands, are typically dark in color and consist of a single, thick upper layer grading into the parent soil material below (Chapter 14). These soils contain abundant calcium, magnesium, and potassium because rainfall is here distinctly less than farther eastward and leaching is less active as a soil-forming process.

REFERENCES FOR FURTHER STUDY

U.S. Dept. Agriculture (1941), *Yearbook of Agriculture, 1941*, U.S. Govt. Printing Office, Washington, D.C. See: C. O. Sauer, "The settlement of the humid East," pp. 157–176; C. W. Thornthwaite, "Climate and settlement in the Great Plains," pp. 177–196; J. Leighly, "Settlement and cultivation in the summer-dry climates," pp. 197–204.

White, G. F., ed. (1956), *The future of arid lands*, Am. Assoc. Advancement of Sci., Publ. 43, Washington, D.C., 453 pp.

Jaeger, E. C. (1957), *The North American deserts*, Stanford Univ. Press, Stanford, Calif., 308 pp.

Trewartha, G. T. (1961), *The earth's problem climates*, The Univ. of Wisconsin Press, Madison, Methuen and Co., London, 334 pp.

REVIEW QUESTIONS

1. Explain the occurrence of a moist climate in subtropical latitudes (25° to 35°) along the eastern continental margins. What air masses prevail over this region? What is the nature of these air masses?

2. Describe the annual cycle of temperature and precipitation at a typical humid subtropical climate (*Cfa*) station on the southeastern Atlantic Coast or eastern Gulf Coast of the United States. When is precipitation at a maximum? Why? Are thunderstorms common or rare?

3. What are the characteristics of climate on the southeastern Asiatic coast in the subtropical latitudes? How does the precipitation cycle differ from that in comparable localities in the southeastern United States? Why is there a difference?

4. What is the natural vegetation of the humid subtropical climate of east coasts in latitudes 25° to 35°? What is a characteristic feature of the soils in this climate zone? To what other groups of soils are they most closely related?

5. What are the important characteristics of the climates of west coasts in middle latitudes (40° to 60°)? What air masses are dominant? From what direction do most air masses and cyclonic storms approach these coasts? How is this reflected in the annual temperature cycle?

6. Why does the marine west coast climate (*Cfb*) show a marked reduction in rainfall during the summer, especially between 40° and 50° latitude?

7. What type of forest vegetation is developed on windward mountain slopes in the marine west coast belts. In what way is orographic rainfall related to the development of these forests and their associated soils?

8. Describe and explain the Mediterranean climate (*Csb*) which prevails on west coasts between 30° and 40° latitude. What is unique about the annual precipitation graph of this climate? How does it compare with graphs of the tropical savanna climate (*Aw*)?

9. What is the natural vegetation of the Mediterranean climate (*Csa, Csb*)?

10. Why are dry climates developed in middle latitudes in interior continental locations? Are they related to the presence of mountain barriers? Give examples. Explain the principle of the rainshadow desert.

11. What distinction is made between extremely arid regions and semi-arid regions or steppes? What is the natural vegetation of steppes? Of the very dry middle-latitude deserts (*BWk*)?

12. How does altitude influence the degree of aridity in the dry middle-latitude regions?

13. What is noteworthy about the annual temperature cycle of the middle-latitude steppes (*BSk*) and deserts (*BWk*)? How does this cycle differ from that of the tropical deserts?

14. How may the boundary between dry and humid climates be determined? Is it a fixed line, or does it fluctuate? Of what economic importance is this boundary?

15. Of what economic value are the middle-latitude steppe lands of the world? What types of soils underlie these regions? Is the soil suitable for grain crops? What is caliche? How does it form?

16. Describe the climate located in middle latitudes in the central and eastern parts of the continents. What air masses are involved in this climate? How does season of year determine which air masses are dominant?

17. Describe the annual temperature cycle of a typical station in the humid continental climate zone (*Dfa, Dfb*). Is the range great or small? Explain. Is the average temperature of the coldest month below freezing?

18. What can be said about the day-to-day temperature fluctuations in the humid continental climate? How do these variations reflect the conflict in air masses?

19. Discuss the natural vegetation and soils of the humid continental climate (*Dfa, Dfb*) in middle latitudes. How do southerly and northerly locations differ in this respect? How do easterly and westerly locations differ?

CHAPTER 12

Polar, Arctic, and Highland Climates

CLIMATES of group III, located at high latitudes, are controlled largely by polar and arctic air masses. They have low temperatures, usually low precipitation, and low evaporation.

11. Continental subarctic climate (Dfc, Dfd, Dwc, Dwd)

In the two great landmasses of North America and Eurasia, a vast expanse of interior continental area lies between 50° and 70° N lat. Here are the source regions for the continental polar air masses.

In winter, when excessive heat loss by radiation has resulted in the formation of the prevailing Siberian and Canadian highs, severely cold air temperatures develop over snow-covered surfaces, forming a cold, dense air mass. Typically, a severe temperature inversion prevails in winter, so that the air at, say, 5000 ft (1500 m) may be 10 F° (5 C°) warmer than air at the ground level. Low in moisture content, this air mass is stable and normally clear.

In summer, the source region is shifted farther north; air mass temperatures rise to moderately high levels, but the moisture content, although much greater in summer, is still small by comparison with that of maritime tropical air masses.

We might expect, therefore, a climate type showing very great seasonal temperature range, with extremely severe winters and a small annual total precipitation concentrated in the warm months. This climate, called here the *continental subarctic climate*, includes four of Köppen's climate types. The largest area, including belts from Alaska to Labrador and from Scandinavia to Siberia, is classified as *Dfc*, a cold, snowy forest climate, moist all year, with cool, short summers. Less than four months of the year have averages over 50°F (10°C). A still colder climate, *Dfd*, found in northern Siberia only, has very cold winters in which the average of the coldest months is below −36.4°F (−38°C). Climates *Dwc* and *Dwd* are also cold, snowy forest climates, but with a dry winter. They are found only in northeastern Asia. The letters *c* and *d* denote a cool summer and a very cold winter, respectively.

The continental subarctic climate (*Dfc*) is well illustrated by the graph of Ft. Vermilion, Alberta, at 58° N lat. (Figure 12.1). The annual range of 74 F° (41 C°) is remarkable enough, but it is greatly exceeded by that of Yakutsk, USSR, lying in the *Dwd* climate type. The annual temperature range of the continental subarctic climate is the greatest of any on earth, reaching 110 F° (61 C°) in Siberia. Even in central Antarctica, which has the lowest temperatures on earth, the annual range is not over 65 F° (36 C°).

Absolute minimum temperatures of −70° to −80°F (−57° to −62°C) are recorded in northwestern Canada, the lowest being −81.4°F (−63°C) at Snag, Yukon on February 3, 1947. Perhaps the coldest place in the northern hemisphere is Verkhoyansk, USSR, which lies in this climate zone and has a January mean of −59°F (−51°C) and an absolute minimum recorded temperature of −93°F (−69°C).

Summer in the subarctic climate regions is very short. The warmest month average may not greatly exceed 50°F (10°C), and frosts can occur at any time during the summer. Daily maximum temperatures, however, commonly reach 70°F (21°C). At these high latitudes the sun is above the horizon

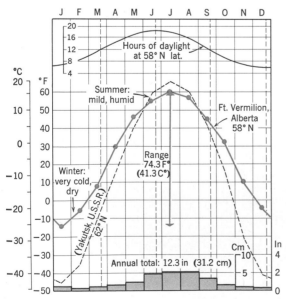

Figure 12.1 Climographs for Ft. Vermilion, Alberta, and Yakutsk, USSR. (Data from Trewartha.)

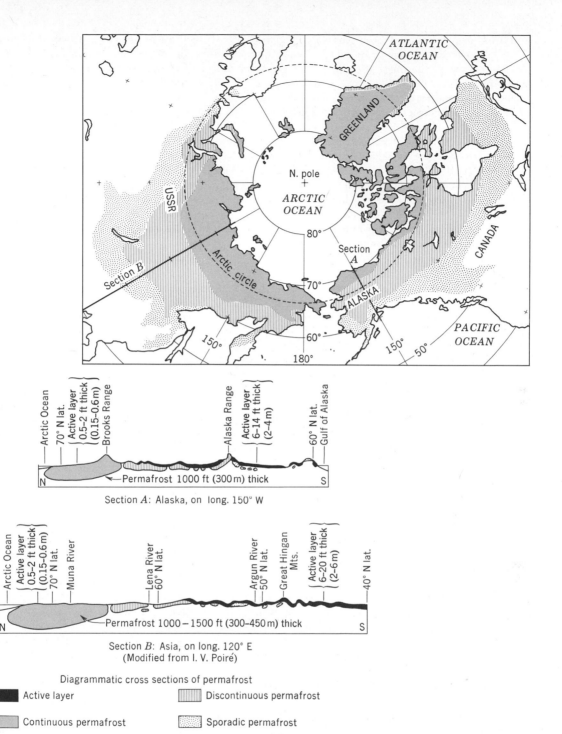

Figure 12.2 Distribution of permafrost in the northern hemisphere, and representative cross sections in Alaska and Asia. (From Robert F. Black, "Permafrost," Chapter 14 of P. D. Trask's *Applied Sedimentation*, John Wiley and Sons, New York, 1950.)

for sixteen to eighteen hours from May through August. (See graph on Figure 12.1.) The long hours of sunshine so accelerate the growth of plants that agriculture may be possible despite a very short growing season.

Winter is the dominant season of the continental subarctic climate. Because subfreezing monthly average temperatures occur for six or seven consecutive months, all moisture in the soil and subsoil is solidly frozen to depths of many feet. (See Figure 12.6.) Summer warmth is insufficient to thaw more than the upper few feet so that a condition of perennially frozen ground, or *perma-frost*, prevails over large parts of this and the tundra regions to the north. Seasonal thaw penetrates from 2 to 14 ft (0.6–4m), depending on location and nature of the ground. This shallow zone of alternate freeze and thaw is termed the *active zone*.

The distribution of permafrost in the northern hemisphere is shown in Figure 12.2. Three zones are recognized. Continuous permafrost, which extends without gaps or interruptions under all topographic features, coincides largely with the tundra climate (*ET*), but also includes a large part of the continental subarctic climate (*Dfc*, *Dfd*, *Dwd*) in Siberia. Discontinuous permafrost, which occurs in patches separated by frost-free zones under lakes and rivers, occupies much of the continental subarctic climate (*Dfc*) zone of North America and Eurasia. Sporadic occurrence of permafrost in small patches extends into the southern limits of the continental subarctic climate.

Depth of permafrost reaches 1000 to 1500 ft (300 to 450 m) in the continuous zone near latitude 70° (Figure 12.2). Much of this permanent frost is an inheritance from more severe conditions of the last ice age, but some permafrost bodies may be growing under existing climate conditions.

Permafrost presents problems of great concern in engineering and building construction in these cold regions. Buildings must be insulated underneath to protect the ground moisture beneath from melting; otherwise, the building might literally be engulfed in mud. Another serious problem is in the behavior of streams in winter. As the surface of streams or springs freezes over, the water beneath bursts out from place to place, freezing into huge accumulations of ice. Highways are thus made impassable. Scraping of insulating peat, forest litter, and vegetative cover from the frozen ground to make roads and air fields may result in dire consequences. The summer sun thaws the bare ground, which turns into a liquid mud, often growing into sizable lake basins by melting and sapping around the edges of the exposed areas.

Most of the source areas of the continental polar air masses are in regions of less than 20 in (50 cm) of precipitation annually, whereas the northerly portions have less than 10 in (25 cm). Although this amount would result in dry deserts at low latitudes, evaporation is greatly reduced by prevailingly low temperatures. Thus, essentially humid moisture conditions of air and soil prevail.

Precipitation is largely cyclonic in type and shows a very definite maximum in the summer months. Snowfall, although conspicuous in winter because it remains upon the ground, accounts for only a fraction of an inch of precipitation per month in the coldest months. Cyclonic storms crossing these areas bring little precipitation at this season. In summer, cyclonic rains are frequent, although thunderstorms are few.

The continental regime of climate modified by a polar influence is dominant in the continental subarctic climate. The thermohyet diagram of Yakutsk, Siberia (Figure 8.10*B*) is a good illustration.

The subarctic climate zone coincides with a great belt of needleleaf forest, often referred to as *boreal forest* (Figure 16.12), and open lichen woodland, the *taiga* (Figure 16.25). Trees tend to be small, so that they are economically of less value for lumber than for pulpwood.

Soils of the *podzol* group are associated with the arctic needleleaf forest. As explained before, these soils are strongly leached and of acid type. They are light gray and have a very distinct leached layer beneath the uppermost layer of humus and forest litter. These soils are extremely poor from the standpoint of agriculture. Added to the natural inadequacy of soils of this region is a great prevalence of swamps and lakes left by the departed ice sheets (Figure 23.16). Some rock surfaces were scoured by ice which stripped off the soil entirely. Elsewhere rock basins were formed, or previous stream courses dammed, making countless lakes. Insufficient geologic time has elapsed for good drainage even to begin to be reestablished over large areas.

12. Marine subarctic climate (EM)

In Chapter 8, the remarkable properties of the middle-latitude equable climatic regime were described. When traced farther poleward into the subarctic latitude zone, so as to span the latitude range 45° to 65°, the equable regime constitutes a distinct climate type, the *marine subarctic climate*, marked by unusually large total annual precipitation and an unusually small annual temperature range for such high latitudes.

The marine subarctic climate is dominated by maritime polar (MP) air masses throughout the year. The climate distribution actually coincides rather well with the MP air mass source regions in the North Atlantic, North Pacific, and the Southern Ocean. Important in defining the characteristics of this climate is the persistence of cloudy skies and strong winds, and the high percentage of days with precipitation.

The marine subarctic climate is not designated as a separate variety in the Köppen system, which includes it in the tundra climate (*ET*). Recent recognition by geographers of the marine subarctic climate has been coupled with the suggestion that it be entered in the Köppen system with the symbol *EM*, the letter "M" to mean an equable marine regime.

The marine subarctic climate is found on limited stretches of windward coasts, on islands, and over wide expanses of ocean in latitudes 50° to 60° in the Bering Sea, and latitudes 55° to 75° in the North Atlantic, touching points on the south Greenland coast, north Iceland, and extreme northern Norway. In the southern hemisphere, the marine subarctic climate occupies almost no land. The few representative stations are found in the southernmost tip of South America, the Falkland Islands, South Georgia Island, and a number of other small islands.

Representative stations in the marine subarctic climate qualify as belonging in the *E* climates of Köppen, for the warmest-month means are all below 50°F (10°C). In contrast to the typical tundra climate (*ET*) the annual ranges are all much smaller, the winters much milder, and the annual precipitation totals much larger. These contrasts are brought out strikingly in the comparison of two stations in Figure 12.3. Coldest month temperature mean is not below 20°F (−7°C).

Thermohyet diagrams of marine subarctic stations are shown in Figure 8.6 (*C,E*); they occupy a unique position well to the right of the cold climates and well below the rainy climates of lower latitudes. Combined with low evaporation rates, the large precipitation yields a large water surplus.

Soils and vegetation of the marine subarctic climate are essentially of the same tundra varieties associated with the tundra climate, treated below. The landscape is treeless and gives the impression of great barrenness, but it is nevertheless surprisingly rich in the variety of plant species that exist in the layer close to the ground.

13. Tundra climate (ET)

The northern continental fringes of North America and Eurasia from the Arctic Circle to about the 75th parallel lie within the outer zone of control of arctic-type air masses, whose source region covers the Arctic Ocean and Greenland. To the south lie the continental polar (cP) and maritime polar (mP) air mass source regions. The land fringes are thus in a frontal zone, which has been known as the *arctic front*, but which may be more simply identified as belonging to a widely shifting polar front, well developed over the northern Pacific and Atlantic oceans. Here many intense east-moving cyclonic storms are developed, and much bad weather may be expected.

These conditions produce a *tundra climate*, described by Köppen with the symbol *ET*. This is a polar climate in which the average temperature of the warmest month is below 50°F (10°C), but above 32°F (0°C).

The tundra climate is illustrated by the temperature-precipitation graph of Upernivik at 73° N lat.

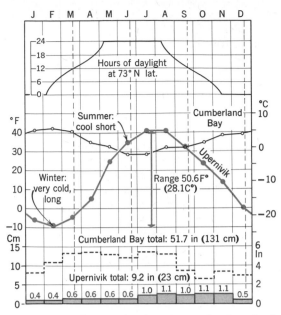

Figure 12.3 Climographs for Upernivik, Greenland, 73° N lat. and Cumberland Bay, South Georgia Island, 54° S lat. (Data from Blair and Meteorological Office of Great Britain.)

on the west Greenland coast (Figure 12.3). Note that (*a*) temperature range is large, but not nearly as great as in the continental subarctic climate; (*b*) the warmest month averages are just over 40°F (4°C) and the coldest month average is well below 0°F (−18°C); and (*c*) precipitation total is less than 10 in (25 cm), with an increased amount falling during and after the warm season. Nearness to the Arctic Ocean explains the somewhat more moderate temperature range and minimum, compared to the continental centers. Coolness of the summer is explained by the nearness to the large ocean body, keeping air temperatures down despite large receipts of solar energy at this latitude near summer solstice. Possibly a more persistent cloud cover is also a factor.

Thermohyet diagrams of the tundra climate combine the continental and polar regimes, as seen in the diagrams for Grønfjorden, Spitzbergen, and Barrow Point, Alaska (Figure 8.11*B,C*). Tundra climate of the southern hemisphere is more equable, lacking the continental influence, as seen in the thermohyet diagram for the South Orkney Islands (Figure 8.6*D*).

Vegetation of the treeless tundra consists of grasses, sedges, and lichens, along with shrubs of willow (Figure 16.28). Traced southward the vegetation changes into birch-lichen woodland, then into the needleleaf forest. In some places a distinct tree line separates the forest and tundra. Coinciding approximately with the 50° isotherm of the warmest month, this line has been used by Köppen as a boundary between *Df* and *ET* climates.

Tundra soils are noteworthy in that the soil particles are produced almost entirely by mechanical breakup of the parent rock and have suffered little or no chemical alteration. Grayish loam and blue-gray clay layers are present with much peat. Continual freezing and thawing of soil moisture has been responsible for disintegration of the soil particles. Like the soils of the northern continental interiors, soils of the tundra are affected by the permanently frozen, or permafrost, condition (Figure 12.2). The permafrost layer is more than 1000 ft (300 m) thick over most of this region; seasonal thaw reaches only 4 to 24 in (10–60 cm) below the surface.

Geomorphic processes have a somewhat distinctive pattern in the tundra regions, and a variety of curious landforms results. Under a protective sod of small plants, the soil water melts in summer, producing a thick mud which may flow downslope to create bulges and terraces without breaking through the surface. This process is known as *solifluction*, or *sludging*, and forms *solifluction terraces* and *lobes* on slopes (Figures 12.4 and 19.19). In the desert tundra, solifluction occurs without the confining cover and may be described as a layer of thick mud creeping down the slopes. Observers who have studied this flowage find that it is most rapid at midday, the mud flowing at a rate of several feet per hour and carrying along large blocks of rock (Figure 19.21).

The freeze and thaw of water in the soil gives

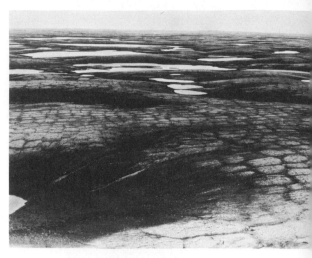

Figure 12.5   Ice-wedge polygons on Wollaston Peninsula, Victoria Island (lat. 70° N. long. 112° W), viewed from the air. (Photograph by A. L. Washburn, Arctic Institute of North America.)

rise to a curious system of polygonal cracks in flat ground (Figure 12.5). They may result from the shrinkage of the clay as water is withdrawn to form ice crystals in lenses or layers within the soil (Figure 12.6). The resulting pattern is termed *polygonal ground*. On hill and mountain summits the cracks are filled with stones, which seem to be gradually sorted and pushed to the sides of each

Figure 12.4   Solifluction, a slow soil flowage, has produced this lobate form on a tundra slope on Mt. Pelly, Victoria Island (lat. 69° N, long. 104° W). Surface of lobe stands about 6 ft (1.8 m) above surrounding slope. (Photograph by A. L. Washburn, Arctic Institute of North America.)

Figure 12.6   This vertical river-bank exposure near Livengood, Alaska, in the subarctic climate region reveals a V-shaped ice wedge surrounded by layered silt of alluvial origin. (Photograph by T. L. Péwé, U.S. Geological Survey.)

polygon during alternate freeze and thaw of soil water. These forms are called *stone rings*, or *stone polygons*. On slopes, the stone rings are drawn downslope to produce *stone stripes*, appearing from the air like giant hachures on the ground.

14. Icecap climate (EF)

Three vast regions of ice exist on the earth. These are the Greenland and Antarctic continental icecaps and the large area of floating sea ice in the Arctic Ocean (Figures 23.9 and 12.9). The continental icecaps differ in various ways, both physically and climatically, from the polar sea ice and can be separately treated. Glacial and topographic features of Greenland and Antarctica are treated in Chapter 23.

Icecap climate has by far the lowest average annual temperature of all global climates. Designated by Köppen *EF*, or polar climate of perpetual frost, this climate has no monthly average above freezing (32°F, 0°C). Available records indicate mean annual temperatures of −20° to −30°F (−30° to −35°C) for the Greenland icecap, but about −9°F (−23°C) for the Arctic Ocean. Compare this with the 10° to 25°F (−12° to −4°C) mean annual temperature of the tundra climate. The higher Arctic Ocean temperature is due to its sea-level altitude and the moderating influence of the ocean water, which can supply a continuous amount of heat conducted through the floating ice to the air.

The annual temperature graph for a Greenland icecap station, "Eismitte," is shown in Figure 12.7. This information was gathered by a German expedition under the direction of Alfred Wegener in the years 1929 through 1931. Note that only in three months of the year did the mean monthly temperatures rise above 0°F (−18°C), and then only to a July maximum of 12°F (−11°C). The coldest month was February, with −53°F (−47°C), making an annual range of 65F° (36 C°).

A shallow layer of air over the ice sheet is chilled severely and at times flows downslope under the influence of gravity toward the margins as a severe blizzard wind. It is reported that so great is the chill of the ice upon the air that tiny ice crystals sometimes form within a few feet of the ground. These may make a "snowstorm," above which the head and shoulders of a man rise clear. Driving blizzard winds pack the sandlike snow into a hard, smooth pavement.

Cyclonic storms frequently penetrate Greenland, bringing precipitation to the icecap. The principal nourishment of the ice sheet is from this source.

Climate of the Antarctic icecap was little understood until a number of weather stations were maintained in the International Geophysical Year of 1957–1958. Temperatures in the interior have

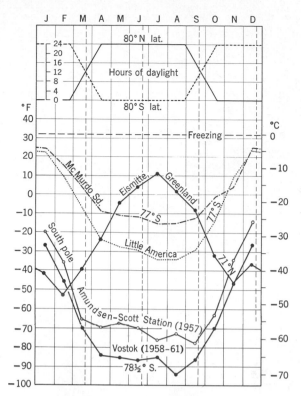

Figure 12.7 Temperature graphs for five icecap stations. (Data from Trewartha; *I. G. Y. Bulletin*; and Weatherwise, vol. 16, 1963.)

proved to be far lower than any place on earth. The Russian meteorological station "Vostok," located about 800 mi (1300 km) from the south pole at an elevation of 11,440 ft (3488 m), may be the world's coldest spot (Figure 12.7). Here a record low of −125.3°F (−87.4°C) was observed on August 25, 1958. Note that this minimum value occurred near the end of the long polar night. At the pole itself (Amundsen-Scott Station), July, August, and September of 1957 had averages about −76°F (−60°C) (Figure 12.7). Temperatures run roughly 40 F° (22 C°) higher, month for month, at Little America Station because it is located close to the Ross Sea and is at low elevation.

A remarkable feature of the high, antarctic interior is the intense chilling of air close to the snow surface. A strong temperature inversion develops in winter, so that the air near the surface may be 50 to 60 F° (28 to 33 C°) colder than air a few hundred feet higher. Downslope flow of this heavy, cold air layer causes blizzard winds to develop in favorable valley locations.

Sea ice

Greatly increased utilization and study of high latitudes by both military forces and civilian scientist research groups has brought to attention the

Figure 12.8   The U.S. Coast Guard icebreaker Northwind forces a passage through McClure Strait, Banks Island, Canadian Northwest Territory, in mid-August. To the left is an open lead. Cutting across from left to right is a rugged zone of pressure ridges. (Official U.S. Coast Guard photo.)

phenomenon of floating sea ice. Supply of arctic and antarctic outposts by ship, maintenance of observing stations on floating ice masses, and submarine operation in the polar sea are forms of activity influenced by sea ice.

The oceanographer distinguishes *sea ice*, formed by direct freezing of ocean water, from *icebergs*, and *ice islands*, which are bodies of land ice broken free from tide-level glaciers and continental ice shelves. Aside from differences in origin, a major difference between sea ice and floating masses of land ice is in thickness. Sea ice, which begins to form when the surface water is cooled to temperatures of about $28\frac{1}{2}°$F ($-2°$C) is limited in thickness to about 15 ft (5 m), because heat is supplied from the underlying water as rapidly as it is lost upward, once an insulating layer of floating ice has been formed.

Pack ice is the name given to ice that completely covers the sea surface (Figure 12.8). Under the forces of wind and currents, pack ice breaks up into individual patches, termed *ice floes*. The narrow strips of open water between such floes are *leads*. Where ice floes are forcibly brought together by winds, the ice margins buckle and turn upward into pressure ridges resembling walls or irregular hummocks (Figure 12.8). The difficulties of travel on foot across the polar sea ice are made extreme by the presence of such obstacles. The surface zone of sea ice is composed of fresh water, the salt having been excluded in the process of freezing.

The Arctic Ocean, which is surrounded by landmasses, is normally covered by pack ice throughout the year, although open leads are numerous in the summer (Figure 12.9). The relatively warm North Atlantic drift maintains an ice-free zone off the northern coast of Norway. The situation is quite different in the antarctic,

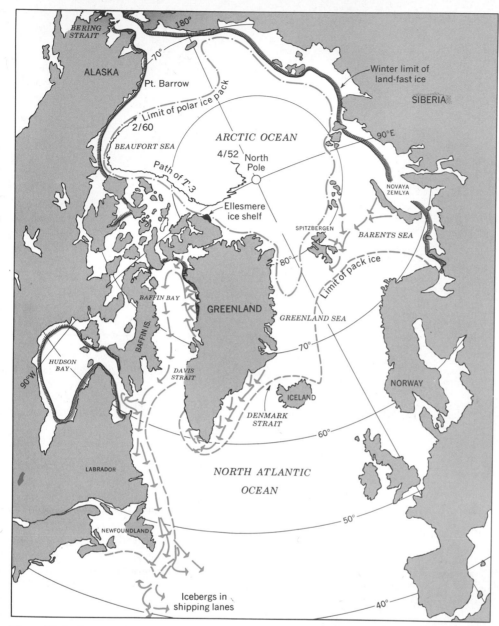

Figure 12.9 Sea ice of the Arctic Ocean. Common tracks of icebergs are shown by arrows. A tilted homolographic projection is used for this map. (Based on data of the National Research Council. From A. N. Strahler, *The Earth Sciences*, Harper and Row, New York.)

where a vast open ocean bounds the sea ice zone on the equatorward margin (Figure 12.10). Because the ice floes can drift freely north into warmer waters, the antarctic ice pack does not spread far beyond 60° S lat. in the cold season. In March, close to the end of the warm season, the ice margin shrinks to a narrow zone bordering the Antarctic continent.

Icebergs and ice islands

Icebergs, formed by the breaking off, or *calving*, of blocks from a valley glacier or tongue of an icecap, may be as thick as several hundred feet. Being only slightly less dense than sea water, the iceberg floats very low in the water, about five-sixths of its bulk lying below water level (Figure 12.11). The ice is fresh, of course, since it is formed of compacted and recrystallized snow.

In the northern hemisphere, icebergs are derived largely from glacier tongues of the Greenland icecap (Figure 12.9). They drift slowly south with the Labrador and Greenland currents and may find their way into the North Atlantic in the vicinity of the Grand Banks of Newfoundland. Icebergs

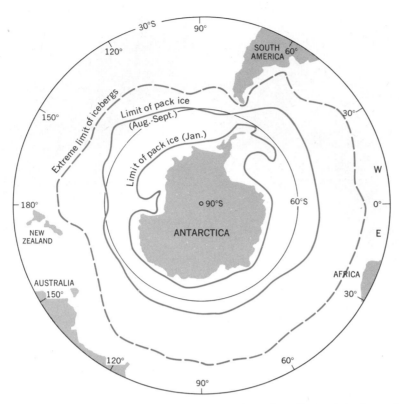

Figure 12.10   Sea ice of the antarctic region. (Data from National Academy of Sciences and American Geographical Society. From A. N. Strahler, *The Earth Sciences*, Harper and Row, New York.)

Figure 12.11   This great iceberg in the east Arctic Ocean dwarfs the U.S. Coast Guard icebreaker Eastwind. (Official U.S. Coast Guard photo.)

of the antarctic are distinctly different. Whereas those of the North Atlantic are irregular in shape and therefore present rather peaked outlines above water, the antarctic icebergs are commonly *tabular* in form, with flat tops and steep clifflike sides (Figure 12.12). This is because tabular bergs are parts of ice shelves, the great, floating platelike extensions of the continental icecap (Chapter 23). In dimensions, a large tabular berg of the antarctic may be tens of miles broad and over 2000 ft (600 m) thick, with an ice wall rising 200 to 300 ft (60 to 90 m) above sea level.

Somewhat related in origin to the tabular bergs of the antarctic are *ice islands* of the North Polar Sea. These huge plates of floating ice may be 20 mi (32 km) across and have an area of 300 sq mi (800 sq km). The bordering ice cliff, 20 to 30 ft (6 to 10 m) above the surrounding pack ice, indicates an ice thickness of 200 ft (60 m) or more. The few ice islands known are probably derived from a shelf of land-fast glacial ice attached to Ellesmere Island, about 83° N lat. (Figure 12.9). The ice islands move slowly with the water drift of the Arctic Ocean and a charting of their tracks reveals much about circulation in that ocean. (See path of T-3 in Figure 12.9). As permanent and sturdy platforms, ice islands serve as bases of

Figure 12.12   This tabular iceberg was observed near the Bay of Whales, Little America, Antarctica, in January 1947. (Official U.S. Coast Guard photo.)

scientific researches from which observations of oceanography, meteorology, and geophysics can be carried out over long periods.

Highland climates

As explained in earlier chapters, increasing elevation brings a great reduction in both pressure and temperature. Thus, climates change greatly within a vertical range of a few thousand feet. In a general way, a rise of altitude is equivalent to an increase in latitude, so that the tundra and icecap climate equivalents can be found among the glaciers of a mountain mass above timber line. In one major respect, however, the analogy is inadequate. Whereas intensity of insolation is progressively less toward the poles, it is increased at higher altitudes. Thus, daily temperature ranges are excessive at high altitudes in middle and low latitudes, but are much less pronounced in equivalent arctic climates.

Pressure and temperature

As stated in Chapter 3, for every 900 ft (275 m) of increased elevation the barometric pressure falls $\frac{1}{30}$ of its value (Figure 3.8).

The physiological effects of a pressure decrease are well known from the experiences of flying and mountain climbing. The principal influence is through an insufficient amount of oxygen to supply the blood through the lungs. At altitudes of 10,000 to 15,000 ft (3000 to 4500 m) mountain sickness (altitude sickness) occurs, characterized by weakness, headache, nosebleed, or nausea. Persons who remain at these altitudes for a day or two normally adjust to the conditions, but physical exertion is always accompanied by shortness of breath.

At reduced pressures the boiling point of water or other liquids is reduced so that cooking time of various foods is greatly lengthened.

The table gives some data on pressure and boiling point relationships. From these figures it is evident that the use of pressure cookers will be of great value about 5000 ft (1500 m) wherever the cooking involves boiling of water.

Elevation		Pressure		Boiling Temperature	
Ft	M	In	Cm	°F	°C
Sea level	0	29.9	76	212	100
1000	300	28.8	73	210	99
3000	900	26.8	68	206	97
5000	1500	24.9	63	203	95
10,000	3000	20.7	53	194	90

From the standpoint of weather and climate, reduced atmospheric pressure is principally effective in that the thinner atmosphere, with relatively less carbon dioxide, water vapor, and dust, absorbs and deflects less solar energy and thus permits a high intensity of insolation at the ground.

Increasing intensity of insolation at higher elevation has a profound influence upon temperature relations. Surfaces exposed to sunlight heat rapidly and intensely, shaded surfaces are quickly and severely cooled. This results in rapid air heating during the day and rapid cooling at night at high-mountain locations (Figure 12.13). Thus at mountain vacation resorts not only is the air pure and the landscape features sharply outlined as if washed clean, but the cool nights and warm days are stimulating physically.

The contrast between exposed and shaded surfaces is particularly noteworthy at high altitudes. It is said that temperatures of objects in the sun and in the shade differ by as much as 40 to 50 F° (22 to 28 C°).

Increased intensity of insolation is accompanied by an increase in intensity of violet and ultraviolet rays. Sunburn is very much more rapid above 5000 ft (1500 m) than at sea level, as many a person has learned by unfortunate experience. The red and infrared rays of the spectrum, on the other hand, are relatively less intensified by increased elevation because they are better able to pass through a dense atmosphere.

A general decrease in air temperature with elevation has been discussed in Chapter 3, where

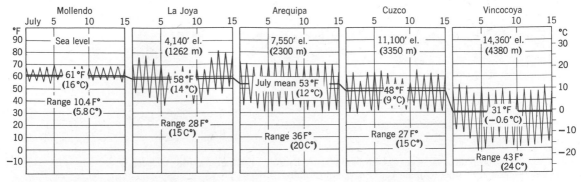

Figure 12.13 Daily maximum and minimum temperatures for mountain stations in Peru, 15° S lat. (After Mark Jefferson, *Geographical Review*.)

it was stated that the average lapse rate is about 3½ F° per 1000 ft (2 C° per 300 m). Thus, we might expect a station 10,000 ft (3000 m) in elevation to have a temperature about 35 F° (20 C°) below that of a nearby sea-level station. Actually it is somewhat less than this.

In high altitudes of the equatorial regions, annual range of temperature is very small, much as in the wet equatorial rainforest climates which surround these tropical mountains at low elevations (Figure 12.14). The range at Quito, Ecuador, is thus only 0.7 F° (0.4 C°) for the year. Compare this with Figure 10.1.

The effect of increasing altitude at low latitude is seen in the thermohyet diagrams of Figure 8.7. A high degree of thermal equability is maintained

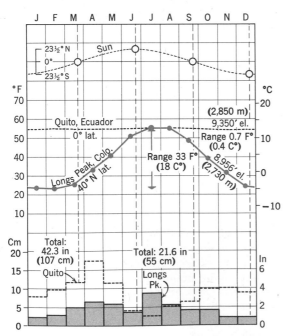

Figure 12.14 A comparison of rainfall and temperature characteristics of two mountain stations. (Data from Trewartha.)

as the diagram is shifted down and to the left.

In middle and high latitudes, on the other hand, a wide annual temperature variation is to be expected, following the marked variations of insolation from summer to winter, as the Longs Peak graph shows (Figure 12.14).

Precipitation

The general influence of increased elevation is first to bring an increase in precipitation, at least for the first few thousand feet of elevation. This is due to the production of orographic rainfall, generated by the forced ascent of air masses and the resultant cooling of the air. (See Chapter 6.) Above elevations of 6000 to 10,000 ft (1800 to 3000 m), which form the zone of heaviest rainfall in low latitudes, precipitation increase begins to slacken with elevation, owing to the inability of air at the lower temperatures to hold, and therefore to give up, as much moisture.

Generally speaking, mountains and plateaus are humid climate zones. This is particularly striking in an arid or semi-arid region, where the mountains form islands of humid climate surrounded by desert or steppe. Reduction in temperatures results in reduced evaporation, so that a humid condition prevails.

The influence of mountains upon precipitation is strikingly illustrated by the state of California (Figure 12.15), whose topography is greatly diversified. Note that in the Great Valley and Death Valley precipitation is less than 10 in (25 cm), but on the west slopes of the Sierra Nevada and Klamath Mountains it is over 70 in (175 cm).

From the standpoint of river flow and floods, mountain climates are of greatest importance in middle latitudes. The higher ranges serve as snow storage areas, keeping back the precipitation until early or midsummer, releasing it slowly through melting, and thus aiding in the maintenance of continuous river flow. As melting proceeds to successively higher levels, the meltwater is supplied

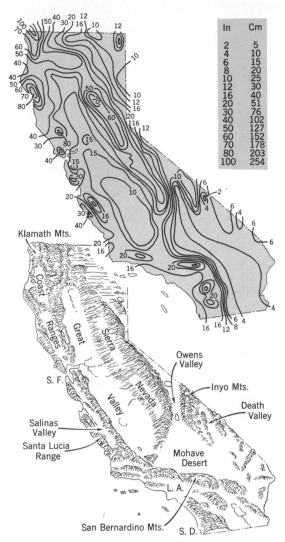

In	Cm
2	5
4	10
6	15
8	20
10	25
12	30
16	40
20	51
30	76
40	102
50	127
60	152
70	178
80	203
100	254

Figure 12.15 The effect of mountain topography on rainfall is well shown by the state of California. Isohyets in inches. (Rainfall map after U.S. Dept. of Agriculture, *Yearbook of American Agriculture*, 1941.)

to the drainage basin. Among the snow-fed rivers of the United States are the Columbia, Snake, Missouri, Platte, Arkansas, and Colorado. In the eastern United States, the Appalachians serve in a similar but less pronounced fashion for the Ohio and other rivers. Here, however, the snow is melted in late spring and has no influence in the middle and late summer.

Vegetation and life zones

It has been observed that increased elevation brings climatic conditions approximately equivalent to those encountered by increase in latitude. Consequently there exist distinct zones of natural vegetation which, in a general way, recapitulate the vegetation types of increasingly high latitudes.

Effects of increasing altitude on vegetation in wet equatorial lands are illustrated by an example from southern Peru (Figure 12.16*A*). Two other important regions besides the Andes of South America are the Ruwenzori Range of central Africa and the central mountain range of New Guinea. Where exposed to prevailing winds (trades) or to seasonal winds (wet monsoon) which bring moisture of maritime air masses (мT, мE) precipitation increases greatly with elevation.[1] As a result, rainforest extends high up the mountain slopes. Between 4000 and 6000 ft (1200 and 1800 m) the rainforest gradually changes into *montane forest*, which resembles the temperate rainforest found at low elevations farther poleward (Chapter 16). Montane forest is lower and less dense than equatorial rainforest and becomes still lower with increasing altitude. Tree ferns and bamboos are conspicuous. Epiphytes (air plants) are abundant. Mosses increase with altitude, giving what is termed *mossy forest* in the zone of persistent mists and high relative humidities. Near its upper limit the montane forest trees become dwarfed, constituting *elfin forest*, and are densely festooned with mosses. Above the forest limit, which occurs about at 12,000 ft (3600 m), there sets in a treeless vegetation which may be alpine tundra, scrub, grassland, or heath. (See Chapter 16 for a description of these vegetation forms.) Still higher, the alpine zone gives way to the zone of perpetual snow (Figure 12.16*A*). The snow line lies at about 16,000 ft (5000 m) in central Peru; about at 15,000 ft (4600 m) in the Ruwenzori Range.

In middle latitudes, where steppe or desert exist at low elevations, the zonation is particularly striking. Figure 12.16*B* shows the vegetation zones of the Colorado Plateau region of northern Arizona and adjacent states. Zone names, elevations, dominant forest trees, and annual precipitation data are given in the figure. Ecologists have set up a series of *life zones*, whose names suggest the similarities of these zones with latitude zones encountered in poleward travel on a meridian. The Hudsonian zone, 9500 to 11,500 ft (2900 to 3500 m), bears a needleleaf forest essentially similar to subarctic needleleaf (boreal) forest. Soils are similarly of podzolic type. As the limit of forest, or tree line, is approached the coniferous trees take on a stunted appearance and decrease in height to low shrublike forms.

A vegetation zone of *alpine meadows* (alpine tundra) lies above tree line and resembles in many ways the arctic tundra. The snow line is encountered at about 9000 to 10,000 ft (2750 to 3000 m)

[1] Altitude stations shown in the thermohyet diagrams of Figure 8.7 are in leeward positions and do not show this effect.

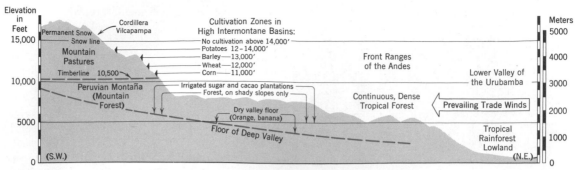

Figure 12.16A   Altitude zoning of climates in the equatorial Andes of southern Peru, 10°-15° S lat. (After Isaiah Bowman, *The Andes of Southern Peru*, 1916.)

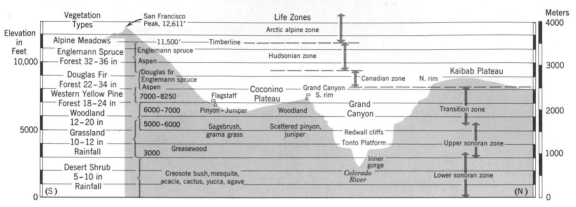

Figure 12.16B   Altitude zoning of mountain and plateau climates in the arid southwestern United States. Grand Canyon-San Francisco Mountain district of northern Arizona. (After G. A. Pearson, C. H. Merriam, and A. N. Strahler.)

in these middle latitudes, which is of course much lower than at the equator. Poleward the snow line decreases in altitude, eventually reaching sea level in the vicinity of the Arctic Circle.

REFERENCES FOR FURTHER STUDY

Bowman, I (1916), *The Andes of Peru*, Henry Holt and Co., New York, 336 pp.

Peattie, R. (1936), *Mountain geography. A critique and field study*, Harvard Univ. Press, Cambridge, Mass., 257 pp.

Kimble, G. H. T. and D. Good, eds. (1955), *Geography of the Northlands*, Amer. Geog. Soc., Special Publ. No. 32, John Wiley and Sons, New York, 534 pp.

Baird, P. D. (1964), *The polar world*, John Wiley and Sons, New York, 328 pp.

Bird, J. B. (1967), *The physiography of arctic Canada*, The Johns Hopkins Press, Baltimore, Md., 336 pp.

REVIEW QUESTIONS

1. Summarize and explain the characteristics of the continental subarctic climate. How do the Köppen symbols *Dfc*, *Dfd*, *Dwc*, and *Dwd* distinguish varieties of this climate? What is particularly noteworthy about the annual temperature cycle? Explain the excessively low winter temperatures. Where is the coldest place in the northern hemisphere?

2. What is perennially frozen ground, or permafrost? What is the origin of permafrost? How great a depth of ground is thawed during the summers in the continental subarctic climate? What are some of the engineering problems associated with permafrost?

3. How does the annual total of precipitation in the northern continental centers compare with that in the tropical deserts? How does evaporation compare in these two regions?

4. Describe the boreal forest and lichen woodland. What is the taiga? What types of soils are found in the continental subarctic climate regions? How has glaciation affected the surfaces of these regions?

5. What are the characteristics of the marine subarctic climate? How is the equability of this climate explained? Compare the marine subarctic climate with the tundra climate.

6. Describe the tundra climate along continental fringes between 65° and 75° N lat. What air masses interact in this zone? In what two regions is cyclonic activity strongly concentrated?

7. Compare the annual temperature cycle of the tundra climate with that of the continental subarctic climate? Compare ranges, warmest month averages, and coldest month averages.

8. What is the tundra? What types of plants are found on the tundra? What kinds of soils are developed here? Describe permafrost conditions in the tundra regions. To what depth is the soil moisture frozen over most of this area? To what depth does the annual summer thaw extend?

9. What is the tree line of the arctic regions? With what monthly isotherm is it approximately identified?

10. What are some of the unusual geomorphic processes active in the tundra regions? Explain the following terms: solifluction, solifluction terraces, polygonal ground, stone rings, stone stripes.

11. What are the essential characteristics of icecap climate? Describe temperature conditions throughout the year. With what air mass source regions is this climate associated?

12. Describe briefly the Greenland and Antarctic icecaps. How is the Greenland icecap nourished, that is, where does the moisture for the snow fields come from?

13. How is sea ice different from ice of icebergs and ice islands? What is the thickness of sea ice? Why is it limited? Describe pack ice, ice floes, leads, and pressure ridges.

14. Explain how the distribution of sea ice differs between the Arctic Ocean and the Antarctic Ocean.

15. How are icebergs formed? What proportion is submerged? Compare the form and source of bergs found in the North Atlantic with those of the antarctic region.

16. What is an ice island? What is its origin? How may it serve the purposes of scientific research?

17. Discuss the influence of increasing altitude upon air pressure. Describe the effects of high altitudes upon physiological processes and upon the boiling point of liquids.

18. What is the effect of increasing altitude upon air temperature? How are maximum and minimum daily temperatures affected? How is the mean annual temperature affected? Is the annual range significantly changed by increased elevation?

19. What influence has increased elevation upon precipitation? Explain this effect. Cite examples of variation of precipitation with elevation.

20. Explain how snow storage in high mountains affects the flow of rivers. Name some snow-fed rivers of the United States.

21. Describe the influence of increased altitude upon natural vegetation and crop cultivation in equatorial regions, using the Andes of Peru as an illustration. What is montane forest? Mossy forest?

22. Describe the influence of increased altitude upon vegetation in the southwestern United States, using the Grand Canyon-San Francisco Mountains region as an illustration. How does the elevation of the alpine meadow zone here compare with that of the upper limit of pastures in the Andes?

23. At what elevation is the snow line encountered near the equator? In middle latitudes? In high latitudes?

CHAPTER 13
Soils and Soil-Forming Processes

AN understanding of fundamental principles of soil science, or *pedology*, is indispensable to a geographer. Soils constitute a major geographical factor, influencing by their fertility and special qualities, not only whether a population can be fed, clothed, and housed but also the particular types of food and fiber or lumber products that can be obtained from a region. The systematic study of soils logically follows a study of climates because climate is a primary factor in soil making.

The soil as a dynamic body

Many persons think of the soil as a lifeless, residual layer, which has somehow accumulated over a long period of time and which merely holds a supply of things necessary for plant growth. As soil science has developed, however, it has become known that the soil is a dynamic layer in which many complex chemical, physical, and biological activities are going on constantly. Far from being a static, lifeless zone, it is a changing and developing body. We know now that soils become adjusted to conditions of climate, landform, and vegetation and will change internally when those controlling conditions change.

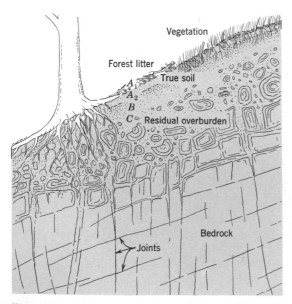

Figure 13.1 Bedrock, residual overburden, and true soil.

The soil scientist restricts the word *soil*, or *solum*, to the surface material, which, over a long period of years, has come to have distinctive layers, or horizons (Figure 13.1). A soil has certain distinctive physical, chemical, and biological qualities, which permit it to support plant growth and which set it off from the infertile substratum, which may consist of *overburden* or solid *bedrock* lying beneath. (See Chapter 17.) The true soil is composed both of mineral and organic particles, whereas the underlying material may be, and usually is, wholly mineral matter.

Soil is made up of substances existing in three states: solid, liquid, and gaseous. For plant growth a proper balance of all three states of matter is necessary.

The solid portion of soil is both inorganic and organic. Weathering of rock produces the inorganic particles that give a soil the main part of its weight and volume (Figure 13.1). These fragments range from gravel and sand down to tiny colloidal particles too small to be seen by an optical microscope. The organic solids consist of both living and decayed plant and animal materials, most being plant roots, fungi, bacteria, worms, insects, and rodents. Colloidal particles of organic matter share with inorganic colloidal particles an important function in soil chemistry.

The liquid portion of soil, the *soil solution*, is a complex chemical solution necessary for many important activities that go on in the soil. Soil without water cannot have these chemical reactions, or can it support life. Gases in the open pore spaces of the soil form the third essential component. They are principally the gases of the atmosphere, together with gases liberated by biological and chemical activity in the soil.

For an understanding of soils, information is needed about (1) the physical-chemical properties and materials of soils, and (2) processes that make and maintain soils.

Physical-chemical make-up of soils

Although a minor factor in itself, *soil color* is perhaps the characteristic that is first noticed about a soil. Color can tell much about how a soil is formed and what it is made of. Soil horizons are

TABLE 13.1   SOIL TEXTURE GRADES

(U.S. Department of Agriculture)

Name of Grade	Diameter. In	Diameter, Mm
Coarse gravel	Above 0.08	Above 2
Fine gravel	0.04–0.08	1–2
Coarse sand	0.02–0.04	0.5–1
Medium sand	0.01–0.02	0.25–0.5
Fine sand	0.004–0.01	0.1–0.25
Very fine sand	0.002–0.004	0.05–0.1
Silt	0.000,08–0.002	0.002–0.05
Clay	Below 0.000,08	Below 0.002

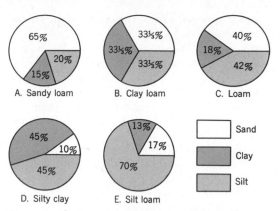

Figure 13.3 Typical compositions of five soil texture classes. These examples are shown as lettered points on Figure 13.2. (After U.S. Dept. of Agriculture, *Yearbook of American Agriculture*, 1938.)

usually distinguishable by color differences. One sequence of colors ranges from white, through brown, to black as a result of an increasing content of *humus* which is finely divided, partially decomposed organic matter. Abundance of humus depends in a general way on luxuriance of vegetation and upon intensity of microbial activity, which in turn depend on climate. Thus we find that in middle latitudes, soils range from black or dark brown in the cool, humid areas to light brown or gray in the semi-arid steppe lands and deserts.

Desert soils have little humus.

Reds and yellows are common colors in soils and are the result of small quantities of iron compounds. The red color is particularly associated with *sesquioxide of iron* (Fe_2O_3), whereas the yellow color may indicate the presence of this same iron compound combined with water (*hydrated iron oxide*). Red color indicates that the soil is well drained, but locally the color may be

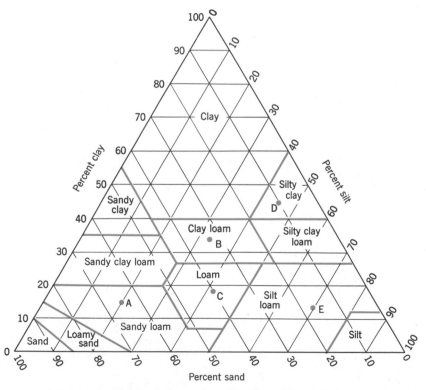

Figure 13.2 Texture classes shown as areas bounded by heavy lines on a triangular graph. (See also Figure 13.3.) (From U.S. Department of Agriculture and Millar, Turk, and Foth, *Fundamentals of Soil Science*, John Wiley and Sons, New York.)

derived from a red source rock such as a red shale or sandstone.

Grayish and bluish colors in soils of humid climates often mean the presence of reduced iron compounds (such as FeO) in the soil and indicate poor drainage or bog conditions. Grayish soils in dry climates mean a meager amount of humus; a white color may be a result of the deposit of salts in the soil. Although some recently formed soils retain the color of the parent overburden or bedrock, the color of fully developed soils is independent of what lies beneath.

Soil texture, a major characteristic of the soil, refers to particle sizes composing the soil. Particles are classified as various grades of gravel, sand, silt, and clay, in decreasing order of size (Table 13.1).

The U.S. Department of Agriculture has set up standard definitions of soil-texture classes in which the proportions of sand, silt and clay are given in percentages. Rather than to attempt to list these classes and give the limiting percentages for each, the information is given in a triangular diagram (Figure 13.2), which enables the percentages of all three components to be shown simultaneously. The corners of the triangle represent 100% of each of the three grades of particles—sand, silt, or clay. The word *loam* refers to a mixture in which no one of the three grades dominates over the other two. Loams therefore appear in the central region of the triangle. A particular soil whose components give it a position at Point *A* in the triangle has 65% sand, 20% silt, and 15% clay; it falls into a texture class known as *sandy loam*. Another soil, whose texture is represented by Point *B* has $33\frac{1}{3}\%$ sand, $33\frac{1}{3}\%$ silt, and $33\frac{1}{3}\%$ clay; it falls into the class of a *clay loam*. Figure 13.3 gives five examples of soil textures; their positions are shown on Figure 13.2.

Texture is important because it largely determines the water retention and transmission properties of the soil as explained later in this chapter. Sand may drain too rapidly; in a clay soil the individual pore spaces are too small for adequate drainage. Where clay and silt proportions are high, root penetration is difficult. Generally speaking, the loam textures are best for plant growth.

Included in the clay grade, smaller than 0.002 mm (0.000,08 in), are mineral soil colloids. Colloid particles are so small that they cannot be seen by optical microscope and will remain suspended indefinitely in water. Individual particles are in the form of thin flakes in the size range 0.001 to 1.0 microns[1] diameter. Finely divided humus provides another class of soil colloids, which may be referred to as *humus colloids*, or *organic colloids*.

Unusual chemical properties of colloids result from their very vast surface area for a given weight. Colloids have the property of being electrically charged and can therefore attract and hold *ions*, the unit chemical particles of dissolved substances. Ions of calcium, magnesium, and potassium are known in soil science as *bases*. These bases may be given up by the colloids to plants, which require them for growth, by a process known as *base*

[1] One micron = 0.001 mm = 0.00004 in.

TABLE 13.2   SOIL ACIDITY AND ALKALINITY[a]

pH	4.0	4.5	5.0	5.5	6.0	6.5	6.7	7.0	8.0	9.0	10.0	11.0
Acidity	Very strongly acid		Strongly acid	Moderately acid	Slightly acid		Neutral	Weakly alkaline	Alkaline	Strongly alkaline	Excessively alkaline	
Lime requirements	Lime needed except for crops requiring acid soil		Lime needed for all but acid-tolerant crops		Lime generally not required	No lime needed						
Occurrence	Rare	Frequent	Very common in cultivated soils of humid climates						Common in subhumid and arid climates		Limited areas in deserts	
Soil groups		Podzols	Gray-brown podzolic soils Tundra soils		Brown forest soils Prairie soils Latosols				Chestnut and brown soils Tropical black earths		Black alkali soils	

[a] Based on data of C. E. Millar, L. M. Turk, and H. D. Foth (1958), *Fundamentals of soil science*, third edition, John Wiley and Sons, New York, 526 pp. See Chart 4.

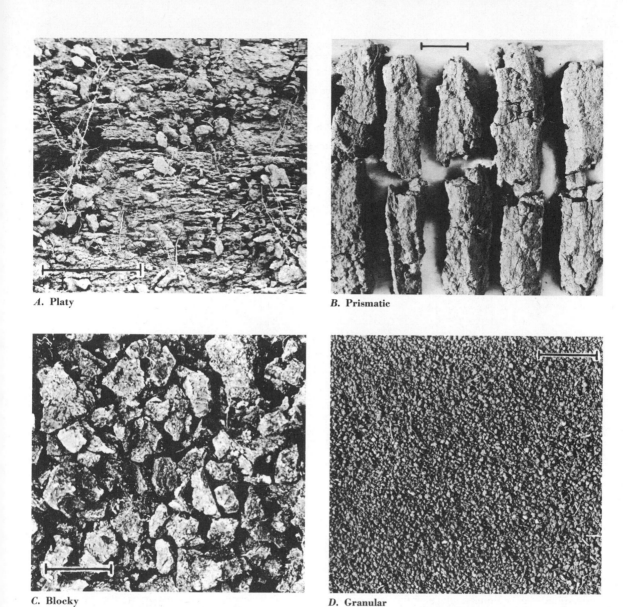

A. Platy

B. Prismatic

C. Blocky

D. Granular

Figure 13.4   Four basic soil structures are illustrated here. The black bar on each photograph represents one inch. (Photographs *A* and *C* by Roy W. Simonson; *B* and *D* by C. C. Nikiforoff. Courtesy of Division of Soil Survey, U.S. Dept. of Agriculture.)

exchange. On the other hand, the *hydrogen ion* in the soil solution makes for an *acid* condition. The concentration of hydrogen ions in the soil solution, relative to hydroxyl ions, is known as the pH of the soil, and is a measure of soil acidity or alkalinity. Hydrogen ions are held also by the soil colloids in exchange positions. In soils a value of 7.0 in the pH scale is *neutral*, whereas values are below 7.0 (4.0 to 7.0) in *acid soils* but above 7.0 (7.0 to 10.0) in *alkaline soils.* Table 13.2 will help to explain the scale of acidity and alkalinity as applied by agronomists to soils.

Soil colloids also are useful in holding water in the soil. When present in large quantities colloids

may make the soil sticky and tough so that it is difficult to cultivate.

Soil structure refers to the way in which soil grains are grouped together into larger pieces held together by soil colloids (Figure 13.4). Irregular pieces with sharp corners and edges give a *blocky* or nutlike structure. More or less spherical pieces make *granular* and *crumb* structure. Some soils have *columnar* and *prismatic* structure, made up of vertical columns or prisms 0.2 to 4 in (0.5 to 10 cm) across. *Platy* soil structure consists of plates, or flat pieces, in a horizontal position. Soil structure influences the rate at which water is absorbed by the soil, the susceptibility of the soil

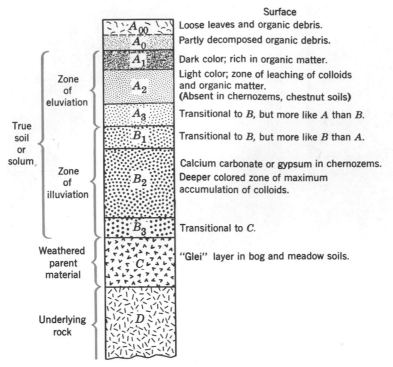

Figure 13.5 Horizons of the soil profile. (After U.S. Dept. of Agriculture, 1938.)

to erosion, and the ease of soil cultivation.

Another constituent of the soil is *soil air*, which occupies the pore spaces of the soil when it is not saturated with water. Soil air has been analyzed and found to contain an excess of carbon dioxide, but a deficiency of oxygen and nitrogen.

Soil water, the water temporarily held in the soil, is in reality a complex chemical solution. It is a dilute solution of such substances as bicarbonates, sulfates, chlorides, nitrates, phosphates, and silicates of calcium, magnesium, potassium, sodium, and iron.

The term *soil profile* denotes the arrangement of the soil into layerlike *horizons* of differing texture, color, and consistency (Figure 13.5). Soils are recognized and classified into broad groups on the basis of the parts of the profile that are present. Basically there are three parts to the soil profile. Horizons *A* and *B* represent the true soil, or *solum*; horizon *C* is the subsoil, or weathered parent body. Below this is the parent bedrock or other underlying rock, designated as horizon *D*. The *A* horizon in humid climates is composed of two very different parts. The upper, or A_1, horizon is rich in organic matter and is dark colored. The lower, or A_2, horizon is a zone of leaching. The *B* horizon is usually a zone of accumulation of soil colloids and is dark in contrast to the A_2 horizon above it. These processes and the horizons they produce are discussed more fully in later pages.

Soil-forming processes and factors

Many types of processes and influences, known altogether as *soil formers,* act together to develop a soil. Some of these are *passive* conditions; others are *active* agents. Five principal soil formers are (1) parent material, (2) landform, (3) time, (4) climate, and (5) biological activity.

The first of the passive soil formers is *parent material,* the residual or transported overburden of disintegrated rock making up the bulk of the soil. Certain of the original rock forming minerals have been thoroughly changed chemically into new compounds and reduced to colloidal size. Although many people think that the type of parent material alone determines the kind of soil that is present, this is an inaccurate concept. For example, the same granite forms the bedrock in the Piedmont region in both Maryland and Georgia, but because of climatic differences the soils of these two states are somewhat different. On the other hand, soils of the same major groups may be found to overlie two different types of overburden or bedrock.

An exception to the general rule that soil type is independent of parent material origin is found in young soils that have not had enough time to develop, and in some limestone areas where the influence of the rock is especially strong. Parent material may locally exert a strong control over the soil texture. For example, on the sandy cuestas of the New Jersey coastal plain, the soil consists largely of quartz sand and is very porous. On glacial clays of the Hudson Valley the soil is unusually sticky and dense, deterring rainfall from seeping through.

Another passive soil former is *landform*, or ground-surface configuration (Figure 13.6). Where slope is steep, surface erosion by runoff is more rapid and water penetration is less than on gentle slopes. As a result, the soil will be thinner on steeper slopes. Flat upland areas accumulate a thick soil that has a thick layer of dense clay (clay pan) and is excessively leached. There the products of weathering tend to remain in place. Flat bottom lands likewise have thick soils, but they are poorly drained and dark colored. Here, constant saturation retards decay of vegetation and allows organic matter to accumulate. Gentle slopes where drainage is good but erosion is slow are considered the norm for soil formation. Slow, continuous erosion is a normal geologic soil process whereby the removal balances the formation of new soil from the parent material. Only when this erosion is greatly accelerated does it become harmful to the soil.

Another influence of landform is the slope aspect or direction of exposure of the surface to the slanting rays of the sun. In middle latitudes it is common to find that south-facing slopes, exposed to the warming and drying effects of sunlight, have different conditions of vegetation and soils from north-facing slopes, which retain cold and moisture longer.

A third passive factor in soil formation is *time*. A soil is said to become *mature* when it has been acted upon by all soil-forming processes for a sufficient long time to have developed a profile that changes only imperceptibly with further passage of time. Soils that are evolving from recently deposited river alluvium or glacial till, for example, are considered *young*. In young soils the characteristic horizons are absent or poorly developed. No age in terms of years can be given to all mature soils because the rate at which a mature soil is developed depends on many other factors. Some soils of humid regions in sandy localities may require 100 to 200 years to develop, whereas more commonly, several thousand years may be needed to produce a mature soil. Some soils of tropical and equatorial regions are thought to be as old as one to six million years, or of Pliocene geologic age. As with development of landforms, age of soils is purely relative. Another way of defining a mature soil might be to say that it is in equilibrium with the many processes and forces acting upon it.

Climate and soil

Of the active soil formers, climate is perhaps the most important. Climatic elements involved in soil development are (1) *moisture conditions* affecting the soil (precipitation, evaporation, and humidity), (2) *temperature*, and (3) *wind*.

Precipitation provides the soil water, without which chemical and biological activities are not possible. When soluble chemicals are dissolved in water they ionize, or dissociate into positively and negatively charged particles. Without ionization the many complex chemical interchanges of elements necessary to soil development and plant growth cannot take place. An excess of precipita-

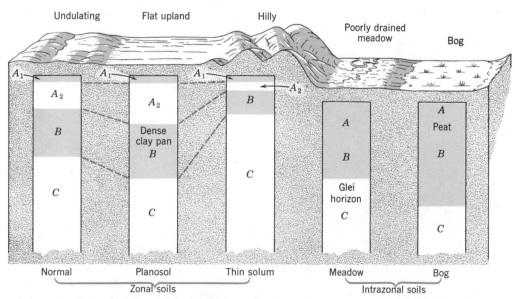

Figure 13.6 Topography as a factor in soil development. (After U.S. Dept. of Agriculture, *Yearbook of American Agriculture*, 1938.)

tion, however, tends to leach away the colloids and ions. This process of downward migration of soil components by waters percolating through the soil is known as *eluviation*. A distinctive leached horizon of the soil, the A_2 horizon, results from this process. (See Figure 13.5). The deposition of colloids and bases in the underlying B horizon is a process known as *illuviation*.

In warm climates where rainfall is extremely heavy, silica (SiO_2) is largely removed from the soil and carried off in streams. This process is termed *desilication*. Thus soils in the wet equatorial rain-forest belts are deficient in silica as well as in such bases as calcium, sodium, magnesium, and potassium, and are generally low in fertility.

In dry climates, evaporation exceeds precipitation and the soil is dry for long periods. Ground water is slowly brought to the surface by capillary attraction and evaporates in the soil, leaving behind the salts that were dissolved in the water. Calcium carbonate, the commonest of these deposits, forms a whitish crust, or *hardpan*, in the soil. In the southwestern United States this material is called *caliche*, and in places it makes the soil as hard and resistant to erosion as if it were a limestone. Gypsum (hydrous calcium sulfate) forms similar encrustations. In intermediate precipitation zones, such as the humid eastern border of the middle-latitude steppes, calcium carbonate appears as small nodules in the soil.

Rainfall and evaporation controls result in the formation of two major groups of soils (Figure 13.7). (1) *Pedalfer soils* show pronounced leaching and occur in the eastern United States where the rainfall is more than 25 in (60 cm) annually. (2) *Pedocal soils* have an excess of calcium carbonate and occur in the western United States where rainfall is less than 25 in (60 cm). The names are coined from the chemical content of the soil. The syllables *al* and *fer* in pedalfer refer to aluminum and iron, respectively, and were chosen because

excessive leaching leaves behind the aluminum and iron oxides as residual substances which thus become important in quantity. The syllable *cal* in pedocal refers, of course, to calcium, which is present in the carbonate form in all pedocal soils.

Temperature is another important climatic factor in soil formation. It acts in two ways. (1) Chemical activity is generally increased by higher temperatures but reduced by cold,[2] and it ceases when soil water is frozen. Thus, tropical soils have a parent material which is thoroughly altered chemically, whereas soils of the frozen tundra have a parent material which is composed largely of mechanically broken minerals. (2) Bacterial activity is increased by warmer soil temperatures. Where bacteria thrive, as in the humid tropics, they consume all dead plants that lie upon the ground. Thus there is no layer of decomposing vegetation on the ground and little humus within the soils of the humid tropics. In cold continental climates, bacterial action is reduced, and a generous layer of decomposing vegetation (leaf mold) covers the ground under forests. Hence, raw humus is preserved at the soil surface and is important in the upper part of the mature soil profile.

Wind is of minor importance as a climatic factor in soil development. Winds may increase the evaporation from soil surfaces and may remove surface soil in arid regions lacking a plant cover. Windblown dust may accumulate and thereby provide the parent material of a soil.

Biological soil formers

Both plants and animals profoundly influence soil development. The plant kingdom consists of the *macroflora* (trees, shrubs, and herbs) and *microflora* (bacteria and fungi).

Grasses and trees require somewhat different chemical substances for growth. Trees, particularly the conifers, use little calcium and magnesium. Hence they thrive well in the pedalfer soils from which these substances have been leached and which are usually acid. Grasses and small grains (wheat, oats, barley) need abundant calcium and magnesium and do well in the pedocal soils of the semi-arid and marginal lands. For grasses to grow well in acid pedalfers, calcium must be added to the soil in the form of lime or crushed limestone. Plants tend to maintain the fertility of soil by bringing the bases (calcium, magnesium, potassium) from lower layers of the soil into the plant stems and leaves, then releasing them to the soil surfaces as the plant decomposes.

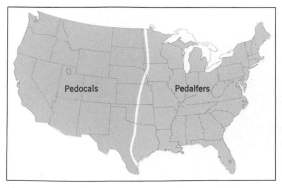

Figure 13.7 Major soil classes of the United States. (After C. F. Marbut.)

[2] The process of *carbonation*, or reaction of carbonic acid (H_2CO_3) upon minerals, is believed to be quite active at low temperatures because the concentration of carbonic acid is greater in cold water than in warm.

Dead plants provide *humus*, the finely-divided organic matter of the soil. Humus gives a dark brown or black color to the soil, as already noted. Humus particles of colloidal size act in the same way as mineral colloids in holding ions in the soil. The process of humus development, or *humification*, is essentially the slow oxidation, or burning, of the vegetative matter. Acids, known as the *organic acids*, are formed during humification. They aid in decomposing the minerals of the parent soil material. The hydrogen ions of the acid solution tend to replace the ions of potassium, calcium, magnesium, and sodium, which are removed in the leaching process. Soils of the cold humid climates are therefore deficient in the bases and are consequently of low fertility for crop farming.

Turning now to the microflora, or bacteria and fungi, we find that bacteria consume humus. In cold climates bacterial growth is slow, hence, humus may accumulate on and in the soil. Soils of the subarctic and tundra climates have much undecomposed organic matter, which locally forms layers of peat, but in humid tropical and equatorial climates, bacterial action is intense and all dead vegetation is rapidly oxidized by the bacteria. Here humus content of the soil is low. The organic acids formed by humus are therefore also lacking, and certain bases such as aluminum, iron, and manganese accumulate in a large proportion relative to silica. In this way the fundamental differences in soils of cold and warm climates can be traced back to intensity of bacterial activity.

Figure 13.8 shows the relation between mean annual air temperature and the relative rate of production and destruction of organic matter by macroflora and microflora, respectively. In the colder climate regimes production exceeds destruction and humus accumulates. Above a mean annual temperature of 77°F (25°C) destruction exceeds production and humus is absent.

Another function of some bacteria and other soil organisms (algae) is to take gaseous nitrogen from the air and convert it into a chemical form that can be used by plants. This process is known as *nitrogen fixation*. One kind of bacteria (*Rhizobium*) lives in the root nodules of leguminous plants and there fixes nitrogen beneficial to the plant host.

The influence of animals in the soil is largely mechanical, but nevertheless important. Earthworms are a particularly important agent in humid regions. They not only continually rework the soil by burrowing, but also change the texture and chemical composition of the soil as it passes through their digestive systems. Ants and termites bring large quantities of soil from lower horizons to the surface. Such burrowing animals as prairie

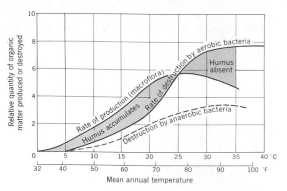

Figure 13.8 Relative rates of production and destruction of organic matter. (After M. W. Senstius, from A. N. Strahler, *The Earth Sciences*, Harper and Row, New York.)

dogs, gophers, ground squirrels, moles, and field mice disturb and rearrange the soil. Digging of burrows brings soil of lower horizons to the surface; collapse of burrows carries surface soil into lower horizons.

The pedogenic regimes

The foregoing analysis of soil-forming processes has been piecemeal in the sense that each factor and activity has been considered separately and in serial order. If the topics seemed in some cases to be out of order and some repetition occurred, it was because the soil-forming processes act in concert and affect the action of one another. To bring the study of soils into a more unified perspective we may concentrate upon several basic trends in soil development, each leading to formation of a distinctive major soil group under the control of a particular climatic regime. Such basic trends may be referred to as *pedogenic regimes*.

The regime of *podzolization* dominates in climates having sufficient cold to inhibit bacterial action, but sufficient moisture to permit larger green plants (macroflora) to thrive (Figure 13.9). Such conditions exist only in middle and high latitudes, and at high altitudes. The corresponding climatic regime may be equable, provided that it is prevailingly cool (marine west coast climates, poleward of 40° latitude), or it may be a continental regime which has cold winters and adequate precipitation distributed throughout the year (humid continental climate; continental subarctic climate). In its extreme development podzolization is associated with coniferous trees (spruce, fir, hemlock, pine). These plants do not require the bases (calcium, magnesium, and potassium) and hence do not restore them to the soil surface. The result is that humic acids, produced from the abundant leaf mold and humus, leach the upper soil strongly of bases, colloids, and the oxides of iron and aluminum, leaving a characteristic ashgray

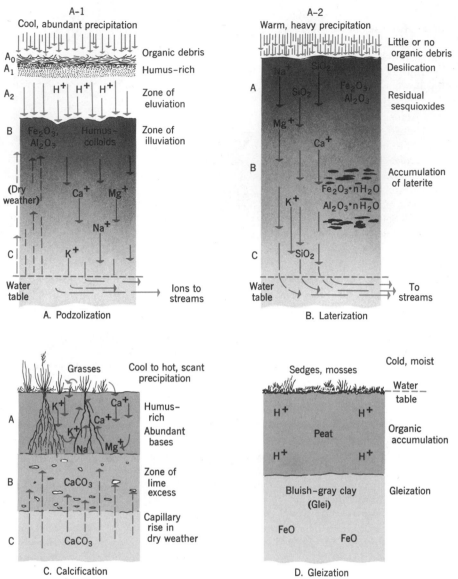

Figure 13.9 Soil profile development under four pedogenic regimes.

A_2 soil horizon composed largely of silica (SiO_2) (Figure 14.1). Colloids, humus, and oxides of iron carried out of the A_2 horizon accumulate in the B horizon, which may be dark in color, dense in structure, and in some cases hardened to rocklike consistency (*ortstein*).

The pedogenic regime of *laterization* is in some respects a warm-climate relative of podzolization, in that both are associated with climatic regimes of ample precipitation and with forests. Laterization takes place in a warm climate having copious rainfall well distributed throughout the year (equatorial rainforest climate; tropical wet-dry climate with long wet season; humid subtropical climate). A high mean annual temperature and a lack of severe winter season permit sustained bacterial action which destroys dead vegetation as rapidly as it is produced. Consequently little or no humus is found upon or in the soil (Figure 13.9B). In the absence of humic acids the sesquioxides of iron (Fe_2O_3) are insoluble and accumulate in the soil as red clays, nodules, and rocklike layers (*laterite*). Silica, on the other hand, is leached out of the soil and disposed of eventually by stream flow in the process of *desilication*. No distinctive soil horizons are developed. In the absence of silicate colloids the soil tends to be firm and porous rather than sticky and plastic, and will transmit water readily. Laterization results in very low soil fertility because bases are not held in the soil and humus is lacking.

Calcification is a pedogenic regime of climates

in which evaporation on the average exceeds precipitation. Calcification is associated with a continental climatic regime with low total annual precipitation (middle-latitude steppe climate) and with a tropical wet-dry climatic regime with a short wet season (tropical steppe climate). Rainfall is not enough to leach out the bases, so that calcium and magnesium ions remain in the soil (Figure 13.9C). Grasses, which use these bases, restore them to the soil surface. Colloids remain essentially in place and are not leached out, but are in a dense (flocculated) state and hold the soil into aggregate structures. Calcium carbonate, brought upward by capillary water films and evaporated in dry periods, is precipitated in the B horizon of the soil in the form of nodules, slabs, and even dense stony layers (caliche). Microbial activity is restricted and humus may be abundantly distributed throughout the A and B horizons. Humus occurs in progressively smaller amounts as one traces the soil into climate zones of increasing aridity. Calcification is characteristically associated with grasslands—the steppes and semi-deserts.

The pedogenic regime of *gleization* is characteristic of poorly drained (but not saline) environments under a moist and cool or cold climate. Gleization is thus associated with the climatic regime of polarization (tundra climate) but is also effective in bog environments of continental climates with cold winters. Low temperatures permit heavy accumulations of organic matter to form a surface layer of peaty material (Figure 13.9D). Beneath this is the *glei horizon*, a thick layer of compact, sticky, structureless clay of bluish-gray color. The glei horizon lies generally within the zone of ground water saturation; consequently the

iron is in a partially reduced condition and imparts the bluish-gray color.

Finally, there is the pedogenic regime of *salinization*, or accumulation of highly soluble salts in the soil. Salinization is associated with the desert climatic regime and takes place in poorly drained locations where surface runoff evaporates. Such locations are typically low-lying valley floors, flats, and basins in the continental interiors; and coastal flats in arid climates. Sulphates and chorides of calcium and sodium are common salts in such soils.

The pedogenic regimes form the basis for classifying the soils of the world into a number of great soil groups, discussed in Chapter 14. The pedogenic regimes also underlie, or are closely interrelated with the geography of plants and constitute one of the major classes of environmental influences upon plants.

REFERENCES FOR FURTHER STUDY

U.S. Dept. of Agriculture (1938), *Soils and men; yearbook of agriculture 1938*, U.S. Government Printing Office, Washington, D.C., 1232 pp.

Robinson, G. W. (1959), *Soils; their origin, constitution and classification*, third edition, John Wiley and Sons, New York, 573 pp.

Bunting, B. T. (1965), *The geography of soil*, Aldine Publ. Co., Chicago, 213 pp.

Millar, C. E., L. M. Turk, and H. D. Foth (1965), *Fundamentals of soil science*, fourth edition, John Wiley and Sons, New York, 491 pp.

Money, D. C. (1965), *Climate, soils, and vegetation*, University Tutorial Press, Ltd., London, 272 pp.

Eyre, S. R. (1968), *Vegetation and soils; a world picture*, second edition, Aldine Publishing Co., Chicago, 328 pp.

1. In what way is the soil a dynamic, rather than a lifeless, static body? To what general influences does the soil respond?

2. How is the term *soil* defined by the soil scientist? Of what three states of matter is soil made up? Are all three necessary for plant growth? Describe the types of substances that make up the soil.

3. What is the significance of soil color? What are some of the common soil colors, and what do they mean? In general, what relationship does soil color bear to climate?

4. What is meant by soil texture? What are the various soil textures recognized by the U.S. Department of Agriculture? What influence has soil texture upon agricultural use of the land?

5. What are some of the common soil structures? Describe four structures. How does soil structure influence the rate at which water is absorbed and transmitted by the soil?

6. What is meant by soil constitution? What properties are included in this term? Of what importance is soil constitution?

7. What are soil colloids? Explain how colloids hold ions. What are the bases commonly present in soils? What is base exchange? What is the relation of hydrogen ion concentration to acidity of the soil? What is meant by pH of the soil?

8. What is soil air? In what way is it different from normal air of the free atmosphere? What are some of the substances commonly found in soil water?

9. Describe an idealized soil profile containing *A*, *B*, and *C* horizons. Which of these horizons constitute the true soil?

10. List the five principal soil formers, including both passive conditions and active agents.

11. From what types of parent material can soils be derived? Does the type of parent material strongly control the characteristics of a fully developed soil profile? Give examples to illustrate. How may young soils strongly reflect the nature of the parent materials? Give examples.

12. How does landform influence development of the soil profile? Explain how various degrees of slope influence thickness of the profile. What conditions of slope are best for the development of soils suited to agriculture?

13. What is a mature soil? Does a mature soil profile continue to evolve through a series of changing forms? How long does it take for a mature soil to develop?

14. What influence has precipitation upon the soil profile? What is eluviation? What is illuviation? What horizons of the soil are affected by these processes? What is meant by desilication? Where is this process most intensively developed?

15. How do soils of dry climates differ fundamentally from soils of humid climates? What two major soil classes have been recognized on this basis?

16. How does temperature affect soil formation? Explain the control exerted by temperature over the type of weathering of the parent material and upon the accumulation of organic matter.

17. How are the needs of different plant groups suited to particular groups of soils? Are plants essential in maintaining the profile of a mature soil?

18. Explain the importance of humus in soil development. What is humification? What acids are formed during humification? What is the action of these acids in the soil? How does bacterial activity determine the amount of humus in the soil? How does this factor determine the fundamental difference between soils of equatorial and arctic regions?

19. What is meant by nitrogen fixation? How is it accomplished?

20. What important influence have earthworms, ants, termites, and burrowing animals upon the soil?

21. Name the basic pedogenic regimes and give the climatic regime in which each develops.

22. Describe and explain the soil profile characteristics associated with each pedogenic regime.

REPRESENTATIVE EXAMPLES OF PROFILES
OF THE GREAT SOIL GROUPS

Terminology and classification are according to the U.S. Department of Agriculture, 1938. Given in parentheses below are approximate equivalents from the U.S. Department of Agriculture, Soil Conservation Service, *Soil Classification, A Comprehensive System, 7th Approximation,* 1960 and 1967. Scale units are in feet. (Photographs were provided by the Soil Conservation Service, U.S. Department of Agriculture, courtesy of Dr. Guy D. Smith, Director, Soil Survey Investigations, Soil Conservation Service.)

ZONAL ORDER

   A *Podzol (Humod)* Humus podzol, Fontainbleu, France.

   B *Gray-Brown Podzolic Soil (Udalf)*, Minnesota.

   C *Yellow Podzolic Soil (Udult)*, Puerto Rico.

   D *Terra Rossa (Xeralf)*, Portugal.

   E *Latosol (Orthox)*, Puerto Rico.

   F *Prairie Soil* or *Brunizem Soil (Udoll)*, Story County, Iowa.

   G *Chernozem Soil (Udic Boroll)*, Spink County, South Dakota.

   H *Chestnut Soil (Typic Boroll)*, Raymon, Saskatchewan.

   I *Reddish Brown Soil (Ustalf)*, Vernon, Texas.

   J *Brown Soil (Aridic Boroll)*, Scobey, Montana.

   K *Sierozem* or *Gray Desert Soil (Calciorthid)*, Escalante, Utah.

   L *Red Desert Soil (Argid)*, Tucson, Arizona.

INTRAZONAL ORDER

   M *Bog Soil (Fibrist)*, Sphagnum peat, Minnesota.

   N *Meadow Soil* or *Humic Glei Soil (Aquoll)*, Union County, Iowa.

   O *Solonchak* or *Saline Soil (Salorthid)*, Virgin River, Nevada. White horizon is salt.

   P *Rendzina (Rendoll)*, Puerto Rico. Developed on marl.

SOIL PROFILES

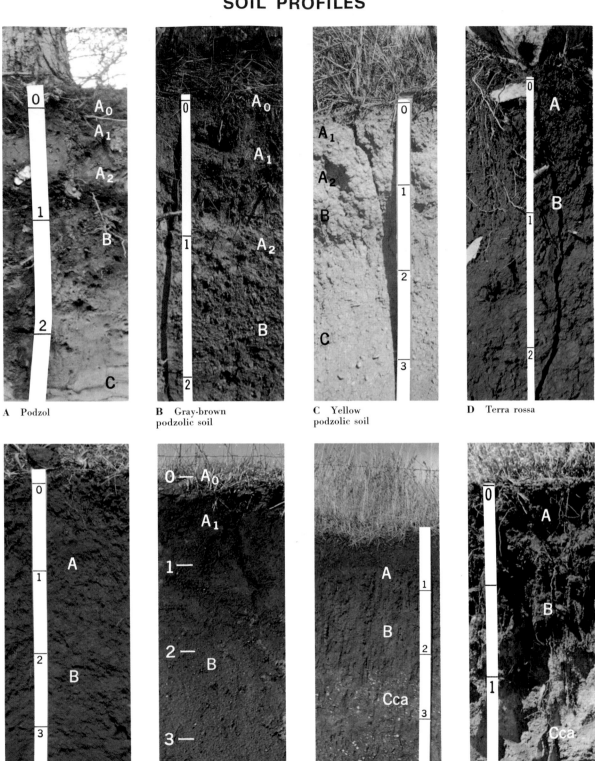

A Podzol

B Gray-brown
podzolic soil

C Yellow
podzolic soil

D Terra rossa

E Latosol

F Prairie soil

G Chernozem

H Chestnut soil

SOIL PROFILES

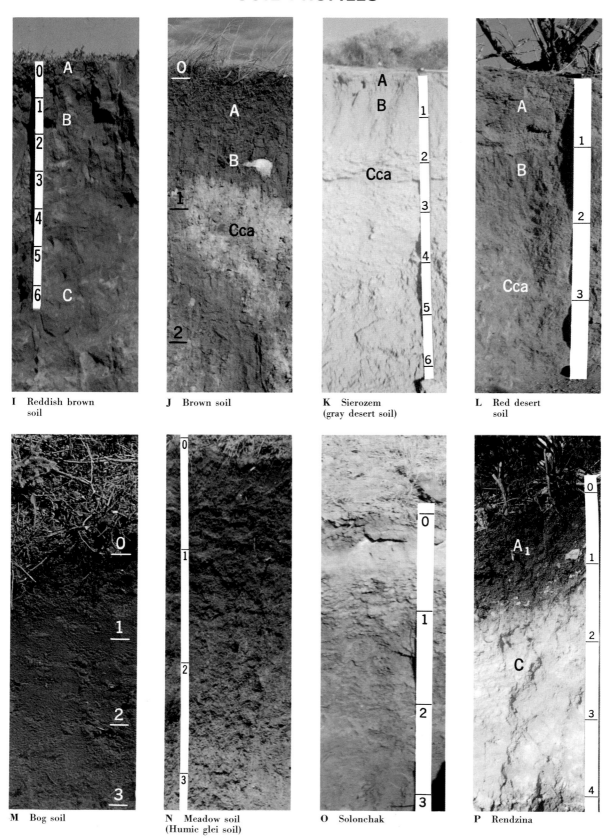

I Reddish brown
   soil

J Brown soil

K Sierozem
   (gray desert soil)

L Red desert
   soil

M Bog soil

N Meadow soil
   (Humic glei soil)

O Solonchak

P Rendzina

CHAPTER 14
The Great Soil Groups

THE soil scientist recognizes that all soils can be subdivided into three orders, known as *zonal*, *intrazonal*, and *azonal* orders. Zonal soils, formed under conditions of good soil drainage through the prolonged action of climate and vegetation, are by far the most important and widespread of the three orders. Intrazonal soils are simply those formed under conditions of very poor drainage (such as in bogs, flood-plain meadows, or in the playa lake basins of the deserts) or upon limestones whose influence is dominant.

Azonal soils have no well-developed profile characteristics, either because they have had insufficient time to develop or because they are on slopes too steep to allow profile development. Azonal soils include thin, stony mountain soils of the earth's mountain regions (*lithosols*), freshly laid alluvial materials, or dune sands (*regosols*). These usually have poorly developed profiles and cannot be easily classified, whereas the zonal and intrazonal soils have distinctive profile characteristics as the result of long development.

The great soil groups

A widely used classification of soils is illustrated in Table 14.1. The U.S. Department of Agriculture in 1938 recognized about 30 *great soil groups*, but these can be simplified into about 18 principal varieties, including both zonal and intrazonal orders, that have world-wide distributions under similar climatic and geomorphic conditions. Various zonal soil groups have already been mentioned and briefly discussed in earlier chapters describing climates. Combining knowledge of the pedogenic processes and regimes gained in Chapter 13 with an understanding of climatic elements, we can proceed to a systematic study of each of the great soil groups. Some of the soil groups bear Russian names, applied by soil scientists of Russia who pioneered in this field.

The founder of modern theories of soil origin and classification was V. V. Dokuchaiev, a Russian geologist; his studies between 1882 and 1900 led him to the concept that soil is an independent body whose character is determined primarily by climate and vegetation. A Russian follower of Dokuchaiev, K. D. Glinka, expanded the concepts of horizons in the soil profile. For the development of modern soil science in the United States during the 1920's and 1930's, much credit is given to C. F. Marbut, who served for many years as chief of the Soil Survey Division of the United States Department of Agriculture. Marbut translated Glinka's work into English, borrowed and modified the Russian pedologic views and ultimately created a comprehensive system of soil classification for the United States. It is Marbut's system, with further modifications, that is presented here.

Podzol soils

Of the zonal soils of cool humid climates the most widely distributed are the *podzol soils* (or simply, the *podzols*), found closely associated with the subarctic climate, the more northerly parts of the humid continental climate, and the cooler parts of the marine west coast climate. Podzol soils require a cold winter and adequate precipitation distributed throughout the year. (Plate 3 is a world soils map which may be referred to in conjunction with the world maps of climate and vegetation, Plates 2 and 4.) The pedogenic regime is that of podzolization (Chapter 13).

The podzol profile (Figure 14.1 and color plate, *A*) is the soil profile from which the various soil horizons were originally named. At the very top, lying on the soil surface, is a layer of leaf mold and acid humus termed the A_0 horizon. Below this is the first soil layer, designated as the A_1 horizon. It is a thin, acid layer, rich in humus and varying in color from gray through yellowish brown to a reddish brown. The A_1 horizon is rich in colloids and is a zone of interaction between acids and bases.

Below the A_1 horizon is a distinctive light-colored zone called the A_2 horizon. This is the strongly leached horizon from which colloids and bases have been carried down. It sometimes has a bleached ash-gray or whitish-gray color because coloring agents such as the iron oxides and colloidal humus have been removed. Leaching such as occurs in the A_2 horizon constitutes eluviation, explained in Chapter 13.

Below the A_2 horizon of the podzol profile is the *B* horizon, a brownish zone that is enriched by the colloids and bases brought down from the A_2 horizon. The process of accumulation is one of illuviation (Chapter 13). The colloids give a heavy, clayey consistency to the B horizon. Excessive deposit of oxides may cause the soil particles in this horizon to become strongly cemented into stony material known as *hardpan*, or, in European soil terminology, *ortstein*. Nodules of soil in the B horizon formed by the same process as hardpan are termed *concretions*. They may be composed of clay cemented by *limonite*, a hydrous iron oxide compound. Altogether the A and B horizons of the podzol profile total less than 3 ft (1 m) in thickness.

In the United States (Figure 14.2), the podzols are found in the northern Great Lakes states, the Adirondacks, and mountainous parts of New England. Here the ample precipitation combined with

TABLE 14.1   CLASSIFICATION OF SOILS[a]

ZONAL ORDER

Suborders	Great Soil Groups

Light-colored podzolized soils of forested regions	PODZOL SOILS
	   Brown podzolic soils
	GRAY-BROWN PODZOLIC SOILS
	RED-YELLOW PODZOLIC SOILS
	   (Incl. *terra rossa*)

Lateritic soils of warm, moist subtropical, tropical, and equatorial regions	LATOSOLS
	   Reddish-brown lateritic soils
	   Black and dark-gray tropical soils

Soils of the forest-grassland transition	Degraded chernozem soils
Dark-colored soils of the semi-arid, subhumid, and humid grasslands	PRAIRIE SOILS (BRUNIZEM SOILS) Reddish prairie soils CHERNOZEM SOILS CHESTNUT SOILS REDDISH-CHESTNUT AND REDDISH-BROWN SOILS
Light-colored soils of arid regions	BROWN SOILS GRAY DESERT SOILS (SIEROZEM SOILS) RED DESERT SOILS
Soils of the cold zone	TUNDRA SOILS Arctic brown forest soils

INTRAZONAL ORDER

Hydromorphic soils of marshes, swamps, bogs, and flat uplands	BOG SOILS MEADOW SOILS (WIESENBÖDEN) Alpine meadow soils PLANOSOLS
Halomorphic soils of poorly-drained arid regions and coastal deposits	SALINE SOILS (SOLONCHAK) ALKALI SOILS (SOLONETZ) Soloth
Calcimorphic soils	RENDZINA SOILS

AZONAL ORDER

LITHOSOLS	
REGOSOLS	Alluvial soils Sands (dry)

[a]Based on U.S. Department of Agriculture (1938), *Soils and men; Yearbook of Agriculture 1938*, U.S. Govt. Printing Office, Washington, D.C. See pp. 993–995.

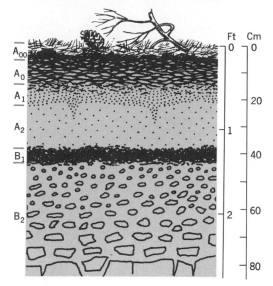

Figure 14.1 Podzol profile developed beneath pine forest. (From S. R. Eyre: *Vegetation and Soils*, Aldine Publishing Company. Copyright © 1963 by S. R. Eyre.)

long, cold winters favors surface accumulation of decomposed vegetation. Organic acids cause a strong leaching of the A_2 horizon. Islands of podzolic soil are found in the Allegheny Mountains, where high altitude gives a cooler climate, and in the sandy New Jersey and Long Island coastal plain belt, where leaching is intense in the porous sand. Podzols of the United States are but the small southern fringe of a great podzol belt corresponding with the needleleaf evergreen forest of the subarctic climate zones in Canada and Eurasia (Figure 14.3 and Plate 3). This forest is dominated by conifers which promote the process of podzolization.

Podzol soils are low in fertility. Leaching of important plant constituents is shown by the association of coniferous forest with this soil. Conifers need little of the calcium, magnesium, potassium, and phosphorus which many other plants require. Consequently the trees do not bring these bases to the ground surface from which they can be restored to the upper soil horizons. Certain species of oaks have a similar lack of use of the bases. The Pine Barrens on podzol soils of the New Jersey coastal plain are examples of adaptation of pine and oak forest to podzol soils.

The podzols cannot produce the crops to feed a large population. Addition of lime and fertilizers to the soil will largely correct soil acidity and replenish the leached bases, but the favorable areas for such treatment are limited by the effects of continental glaciation. Bouldery morainal topography interspersed with swamps and lakes still renders much of the podzol soil area unfit for farming.

Gray-brown podzolic soils

The second major great soil group of humid climates consists of the *gray-brown podzolic soils*. They differ from the podzols in that leaching is less intense and the soil color is brownish. The various soil horizons correspond with those of the podzols (Figures 14.4 and color plate, *B*). The A_1 horizon is a moderately acid humus layer. The A_2 horizon is a grayish-brown leached zone. It is less intensely leached than in the podzols and, consequently, is neither so light colored nor so distinctly limited. The *B* horizon is thick and yellowish brown to light reddish brown in color. Like the podzols, it has concentrated colloids and bases.

The gray-brown podzolic soils contain more of the important bases than the podzols but are nevertheless somewhat acid. Deciduous forests (maple, beech, oak) grow luxuriantly on the forest soils (Figure 14.3). These trees bring the bases up from the *B* horizon, returning them to the surface as dead leaves and branches. Thus the soil is replenished by these bases in a way not found in the podzols.

When treated with lime and fertilizers, the gray-brown podzolic soils make highly productive farms on which diversified crop farming and dairying are well developed. This is seen from the distribution of the gray-brown podzolic soils in the eastern-central United States where rainfall is 35 to 40 in (90 to 100 cm) yearly in the humid continental climate. Note that southern Wisconsin, southern Michigan, Indiana, Ohio, Kentucky, New York, Pennsylvania, and Maryland, and southern New England are largely underlain by these soils (Figure 14.2). These are states noted for the value of their diversified crop production. Gray-brown podzolic soils are also found in the Pacific northwest and on summit areas in the Rocky Mountains.

The gray-brown podzolic soils are also found over much of western Europe in both the marine west coast climate and humid continental climate. Smaller areas occur in the humid continental climate of northern China and northern Japan. Summer-green deciduous forest is associated with these areas of gray-brown podzolic soils.

Of relatively minor extent are the *brown podzolic soils* (Table 14.1), found in southern New England (Figure 14.2). These are transitional between podzols and gray-brown podzolic soils. The brown podzolic soils have only a very thin leached A_2 horizon. Although acid, they can be highly productive when heavily limed and fertilized.

Red-yellow podzolic soils

Farther south than the gray-brown podzolic soils, in a zone of increasingly warmer climate but

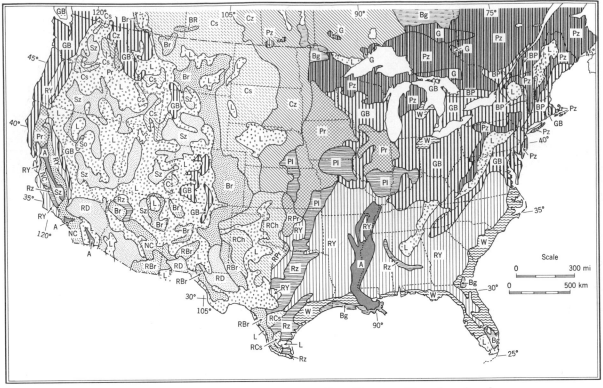

Sz   Sierozem (Gray desert) soils
RD   Red desert soils

Pz   Podzol soils
G   Gray wooded soils (Canada)
BP   Brown podzolic soils
GB   Gray-brown podzolic soils
RY   Red and yellow podzolic soils
Pr   Prairie soils
RPr   Reddish prairie soils
Cz   Chernozem soils
Cs   Chestnut soils
RCh   Reddish chestnut soils
Br   Brown soils
RBr   Reddish brown soils
NC   Noncalcic brown (Shantung brown) soils

INTRAZONAL SOILS

Pl   Planosols
Rz   Rendzina soils
So   Solonchak and solonetz soils
W   Wiesenböden, ground-water podzol, and half
         bog soils
Bg   Bog soils

AZONAL SOILS

L   Lithosols and shallow soils, sands, lava beds
A   Alluvial soils

Figure 14.2   Soils of the contiguous 48 United States and southern Canada. (Modified and simplified from map of soil associations of the United States by U.S. Department of Agriculture, Soil Survey Division, in *Yearbook of American Agriculture, 1938,* U.S. Government Printing Office, Washington, D.C., 1938, and *Atlas of Canada,* 1957, Plate No. 35, Canada Dept. of Mines and Technical Surveys.)

with equally abundant precipitation, lies a great area of *red-yellow podzolic soils* (Figure 14.2). These soils occupy the southern United States from Texas to the Atlantic, and coincide fairly well with the extent of the humid subtropical climate. A similar geographic relation holds in Japan, while in southern Brazil and southeastern Paraguay there is also a substantial area of these soils at comparable latitudes (Plate 3). Smaller coastal zones of red-yellow podzolic soils are found in South Africa, Australia, and New Zealand.

The red-yellow soils are of the podzolic type

and show the same characteristic leaching of the A_2 horizon (color plate, *C*). Warm summers and mild winters favor bacterial action. Humus content is low. Thus, both podzolization and laterization act in concert as pedogenic processes. The typical red and yellow colors are a staining in the form of hydroxides of iron. The yellow soils are the more strongly leached of the two and are found on sandy coastal plain belts. Aluminum hydroxides are also abundant in these soils, a condition typical of tropical soils in warm, humid regions.

Deciduous forest was the natural vegetation of

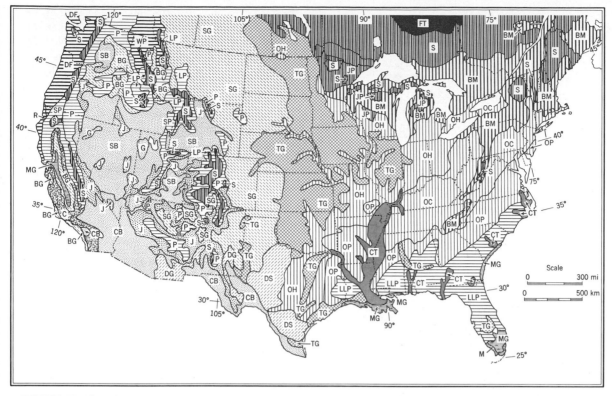

EASTERN FOREST VEGETATION

FT   Subarctic forest-tundra transition (Canada)
S   Spruce-fir (Northern coniferous forest)
JP   Jack, red, and white pines (Northeastern pine forest)
BM Birch-beech-maple-hemlock   (Northeastern hardwoods)
Oak forest (Southern hardwood forest):
     OC   Chestnut-chestnut oak-yellow poplar
     OH   Oak-hickory
     OP   Oak-pine
CT   Cypress-tupelo-red gum (River bottom forest)
LLP Longleaf-loblolly-slash pines (Southeastern pine forest)
M   Mangrove (Subtropical forest)

WESTERN FOREST VEGETATION

S   Spruce-fir (Northern coniferous forest)
Cedar-hemlock (Northwestern coniferous forest):
     WP Western larch-western white pine
     DF   Pacific Douglas fir
     R   Redwood

Yellow pine-Douglas fir (Western pine forest)
     Sp   Yellow pine-sugar pine
     P   Yellow pine-Douglas fir
     LP   Lodgepole pine
J   Pinon-Juniper (Southwestern coniferous woodland)
C   Chaparral (Southwestern broad-leaved woodland)

DESERT SHRUB VEGETATION

SB   Sagebrush (Northern desert shrub)
CB   Creosote bush (Southern desert shrub)
G   Greasewood (Salt desert shrub)

GRASS VEGETATION

TG   Tall grass (Prairie grassland)
SG   Short grass (Plains grassland)
DG   Mesquite-grass (Desert grassland)
DS   Mesquite and desert grass savanna (Desert savanna)
BG   Bunch grass (Pacific grassland)
MG   Marsh grass (Marsh grassland)
Alpine meadow (Not shown)

Figure 14.3 Vegetation of the contiguous 48 United States and southern Canada. (Modified and simplified from maps of H. L. Shantz and Raphael Zon, in *Atlas of American Agriculture*, 1929, U.S. Government Printing Office, Washington, D.C. and Canada Department of Forestry, Bulletin 123, 1963.)

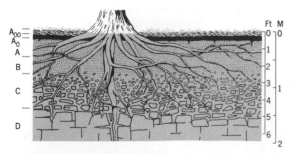

Figure 14.4 Brown forest soil profile formed beneath oak forest. (From S. R. Eyre: *Vegetation and Soils*, Aldine Publishing Company. Copyright © 1963 by S. R. Eyre.)

the northern part of the red soil belt in the United States. Soils of these southern states, though low in plant nutrients, respond well to fertilizers and have been important producers of tobacco, cotton, peanuts, soybeans, corn, sweet potatoes, cowpeas, and many other crops. Yellow soils, owing to their strong leaching, support forests of longleaf, loblolly, and slash pine (Figure 14.3). The pitch pine of the coastal belt is an important source of turpentine and resin; the slash pine is a good pulpwood tree.

In the other world regions of red-yellow podzolic soils noted above, natural vegetation is dominantly rainforest, of both tropical and temperate classes (Plate 4).

A great soil group related to the red podzolic soils is the *terra rossa* of Mediterranean lands (Plate 3). This red soil, poor in humus, is rich in sesquioxide of iron (Fe_2O_3) to which it largely owes its color. The origin of terra rossa is the source of much speculation. Perhaps it represents a soil that was once much richer in humus, but which has lost that humus because of the destruction of forest cover by man and his grazing animals over the course of many centuries. In some areas of its occurrence, terra rossa is associated with limestone as the parent matter, suggesting that its properties are inherited from a calcareous subsoil.

Latosols

Soils of the humid tropical and equatorial zones are called *latosols*, or *lateritic soils* (color plate, *E*). They are characterized as follows: (1) Chemical and mechanical decomposition of the parent rock is complete, owing to the favorable conditions of moisture and heat. (2) Silica has been almost entirely leached from the soil. (3) Sesquioxides of iron and aluminum have accumulated in the soil as abundant and permanent residual materials. (4) Humus is almost or entirely lacking because of the rapidity of bacterial action in the prevailingly warm temperatures. (5) The soil is distinctively reddish because of the presence of sesquioxides of

iron. The reasons for the development of these conditions have been explained in Chapter 13 under the pedogenic regime of *laterization*. Loss of silicate clay minerals renders the latosols relatively low in plasticity (stickiness) and remarkably porous. Consequently rainfall sinks readily into these soils.

True latosols are found only in warm, humid regions and, hence, correspond closely with the wet equatorial climate and the tropical wet-dry climate. Though the red-yellow podzolic soils show the effects of laterization, they are not true latosols.

Latosols quickly lose their fertility under crop cultivation because excessive leaching has removed the plant nutrients in all but a thin surface layer. However, the soil is favorable for the luxuriant growth of broadleaf evergreen rainforest (Plate 4). Other large areas have the raingreen forest and woodland associated with the tropical wet-dry climatic regime.

An interesting feature of latosols is the local development of accumulations of iron and aluminum sesquioxides into layers that can be cut out as building bricks. The material is termed *laterite*. On exposure to the drying effects of the air, these blocks become very hard. In Indochina, particularly, laterite bricks have been much used as a building material.

Valuable mineral deposits occur as laterites. These are thick layers of such minerals as *bauxite* (hydrous aluminum oxide), *limonite* (hydrous iron oxide), and *manganite* (manganese oxide). They are known as *residual ores* because they are not soluble in soil water and have continued to accumulate as the parent rock has weathered away and the silica and other soluble constituents have been removed. Important bauxite deposits of the Guianas of northern South America and of western India are of this type. (See Figure 10.5.) Valuable resources of manganese are in laterite deposits.

By no means all soils of the humid tropical and equatorial climates are latosols of the type described above. Variations of the latosols have been described as *yellowish-brown lateritic* and *reddish-brown lateritic* soils. Large upland areas of Africa and India within the tropical wet-dry climate regime have *black* and *dark-gray tropical soils*. The extent of these is given in Plate 3. These dark soils may be related to special conditions of underlying bedrock. For example, the dark soils of peninsular India shown in Plate 3 coincide rather well with the Deccan Plateau, underlain by basalts.

Hydromorphic soils (intrazonal)

Hydromorphic soils are soils associated with marshes, swamps, bogs, or poorly drained flat uplands. All are considered intrazonal soils because of poor drainage.

Bog soils are formed under bog vegetation in regions of cool, moist continental climates. Continental glaciation in North America and Europe left countless basins that have since been largely filled by a succession of water-loving plants (Chapter 15). Here the soil is saturated most of the time and plant decay is greatly retarded. The partly decomposed plants of the bog therefore accumulate into an upper peat layer 3 to 4 ft (1 m) thick (color plate, *M*). Below this is a horizon of sticky, structureless clay, known to soil scientists as the *glei* (also *gley*) horizon. It is of gray-blue color and is largely impervious to water seepage. The process of *gleization* was discussed in Chapter 13.

Meadow soils (*wiesenböden*) are formed on the flood plains of streams where drainage is somewhat better than in the bogs, but is nevertheless poor. These areas are extensively used in the humid middle-latitude climates as pastures because the grass grows rapidly and densely. A thick, humus-rich layer is developed, overlying a sticky gley horizon (color plate, *N*). A similar process of soil development takes place on poorly drained sites at the bases of hill slopes where there is a tendency for water to collect.

The name *humic-glei soils* has been proposed for the meadow soils in combination with half-bog soils.

At high altitudes, where the alpine tundra climate prevails, *alpine meadow soils* are found. Dark in color, these soils support a growth of grasses, sedges, and flowering plants.

Strongly leached soils developed on flat or gently sloping, elevated surfaces are known as *planosols*. The soil horizons are abnormally thick because of slow removal by erosion. Planosols have a thick, dense clay horizon in humid climates. In subhumid climates planosols have a dense, cemented horizon.

Tundra soils

Soils of the arctic tundra are so widespread (Plate 3) that they may be considered a zonal type along with the podzols, gray-brown forest soils, red-yellow soils, and latosols, but because they are poorly drained they are sometimes classified as intrazonal.

Climate conditions favorable for tundra soils have been described under tundra climates of the northern continental fringes, Chapter 12. Intensely cold, long winters cause soil moisture to be frozen during many months of the year. Under these cold conditions chemical alteration of the minerals is slow, and much of the parent material of the soil consists of mechanically broken particles. The slow rate of plant decomposition results in the presence of much raw humus, or peat. The tundra soils do not have simple, distinctive soil profiles, but consist of thin layers of sandy clay and raw humus. The surface may be covered with a sod of lichens, mosses, and herbaceous plants.

In the tundra regions of Siberia and North America, the condition of permanently frozen ground, or *permafrost*, is widespread. (See Chapter 12.) Curious lenses, layers, and vertical wedgelike bodies of ice are present under the soil.

In parts of central Alaska, principally in the valleys of the Yukon and Tanana rivers, are dark-colored soils which have been given great-soil group status as the *arctic brown forest soils* (Plate 3). The profiles of these soils have a thick, dark A_1 horizon rich in organic matter. Downward through lower horizons the soils grade into lighter brown colors and finally to gray in the C horizon. These soils may have originated in a surface layer of loess (wind-blown silt).

Chernozem soils

Of the zonal soils in a semi-arid climate among the most distinctive and widely distributed types are the *chernozem soils*, or *black earths* (color plate, *G*). A typical chernozem profile appears to consist essentially of two layers. Immediately beneath a grass sod is a black layer, the A horizon, 2 to 3 ft (0.6 to 1 m) thick and rich in humus. In this layer the structure is a crumb or nut. The A horizon grades downward into a B horizon of brown or yellowish-brown color, then, with a sharp line of demarcation, into a light-colored C horizon. As in the podzolic soils, in the B horizon colloids and bases accumulate by downward percolation through the A horizon, but unlike the podzols, the chernozem soils have no leached A_2 horizon.

Chernozem soils are rich in calcium, which appears in excess as calcium carbonate precipitated in the lower B horizon or just beneath the B horizon. It has been noted that chernozem soils develop in parent material rich in calcium carbonate. The origin and distribution of chernozem soils have long attracted the soil scientist. Russian investigators in particular have been intensely interested in chernozem soils because they occupy a large area in the Ukraine, surrounding the Black Sea on its west, north, and east sides (Plate 3) and continuing in a great belt eastward along the 55th parallel into the heart of Asia.

Chernozem soils are important in the United States and Canada, where they form a north-south belt starting in Alberta and Saskatchewan and running through the Great Plains of the United States to central Texas. A similar area lies in Argentina. Other areas are mapped in Australia and Manchuria (Plate 3).

Climate has long been thought to be a determining factor in the development of chernozem soils. Comparison of soil and climate maps shows that the middle-latitude chernozems, in the Americas and Europe, lie on the more arid western side of the humid continental climates and with decreasing latitude extend over into the middle-latitude steppe climates. Aridity is therefore a definite contributing cause. The continental location of chernozem areas makes for hot summers and cold winters. Drought periods with strong evaporation dry out the soil, and forests cannot exist. Instead, grasses, which can withstand drought readily and which are tolerant to soils with excesses of mineral salts, flourish on chernozem soils. Steppe grasslands and prairies are the natural vegetation of the middle-latitude chernozem soils.

An important factor in development of the middle-latitude chernozem soils is their occurrence on *loess*, the wind-transported dust so extensively deposited during the glacial period. (See Chapter 25.) Though chernozem soils are not limited to such areas, the texture and lime content of loess, together with the plains topography, have been especially favorable to chernozem development.

Geographically, perhaps the outstanding point of importance regarding the chernozem soils is their productivity for small grain crops—wheat, oats, barley, and rye. Great grain surpluses are exported from chernozem areas in the United States, Canada, and the Ukraine and Argentine, causing them to be described as breadbaskets of the world.

Where forests invade the chernozem soil grasslands, some influence of podzolization is felt and there develops a faint A_2 horizon. Such soils are known as *degraded chernozem soils*. They are transitional to the gray-brown podzolic soils and occupy a geographical position adjacent to them. One particularly large region of degraded chernozem soils lies northwest of the Black Sea and extends from the Danube River of Rumania to the southern Ukraine (Plate 3).

Prairie soils (brunizem soils)

Examination of the soil map (Figure 14.2) will show that between the chernozems and gray-brown podzolic soils in the United States lies a zone of *prairie soils* (or *brunizem soils*). Rainfall is 25 to 40 in (60 to 100 cm) and diminishes greatly across this belt (Figure 11.10). This soil group is similar to the chernozems in general profile and appearance, but differs in that it lacks the excess calcium carbonate of the chernozems (color plate, *F*). The prairie soil is, therefore, a transitional type between the major soil divisions: pedocal and pedalfer.

Of special interest in the United States has been the origin of natural tall-grass prairies of the upper Mississippi Valley and the Great Plains states: Illinois, Iowa, eastern Nebraska, southern Minnesota, northern Missouri, and eastern Kansas (Figure 14.3). Here forests were lacking over vast expanses at the time of the coming of white men. Although many explanations have been offered for the existence of the prairies, including the possibility of deforestation by burning, it seems likely that a major contributing factor has been that the prairie soils become almost completely dry between summer rains down to a depth of a foot or so (0.3 m) as a result of the dryness and heat of summer air masses. Although the prairie grasses can survive these conditions, deciduous forests, which border the prairies on the east, cannot.

Prairie soils are extremely productive, combining the fertility of the chernozems with somewhat moister climate. Perhaps the outstanding crop associated with the prairie-soil belt is corn. The corn belt is practically identical with the prairie-soil belt, although the corn belt actually extends eastward into the gray-brown forest soils of Indiana and Ohio. Corn requires not only high temperatures during the growing season but ample moisture as well. This last requirement is met by periods of summer thunderstorms interspersed with hot, dry periods.

South of the prairie soils in the United States lies a small area of *reddish prairie soils*, located south of the Arkansas River in Kansas, Oklahoma, and Texas (Figure 14.2). The reddish prairie soils are essentially like the prairie soils, but of reddish-brown color. They support a prairie grassland vegetation.

Chestnut soils and brown soils

To the arid side of the chernozem belt lies the belt of *chestnut soils* or *dark brown soils*. They occupy the semi-arid middle-latitude steppe lands, in North America and Asia (Plate 3). The chestnut-soil profile is generally similar to the chernozem, but contains less humus and hence is not so dark in color. The soil structure tends to be prismatic in the *B* horizon.

The chestnut soils are fertile under conditions of adequate rainfall or irrigation, but lie in a hazardous marginal belt in which years of drought and adequate rainfall are alternated. Bordering on the productive chernozem wheat belt of the Great Plains, the chestnut-soil belt of the United States offers a temptation for expanded wheat cultivation. Under special cultivation practices that conserve soil moisture, a period of moist years brings high grain yields to these marginal belts; but a series of drought years can bring repeated crop failures and poverty.

Toward still more arid regions the chestnut soils give way to the *brown soils*, generally similar, but with still less humus and consequently a lighter color (Figure 14.5 and color plate, *J*). In the western United States, the brown soils occupy basins in the central Rockies of Wyoming, the Colorado Piedmont, and parts of the Colorado Plateaus in Colorado, Utah, Arizona, and New Mexico.

The brown soils are typical of the middle-latitude steppes and support a light growth of grasses suitable for livestock grazing. With irrigation they are productive, but farming is not attempted without.

Reddish-chestnut and reddish-brown soils

Widespread in semi-arid and subhumid tropical and subtropical regions of the world are soils of reddish brown surface color which grade downward into a heavier dull-red or reddish-brown subsoil, then into a zone of lime accumulation. These are classified as *reddish-chestnut* and *reddish-brown soils* and grouped with the reddish prairie soils on the world soils map (Plate 3).

Comparison of the distribution of these reddish soils with the distribution of climate and natural vegetation (Plates 2 and 4) suggest that these soils are found in regions of tropical wet-dry climates with a short wet season and with their bordering tropical steppes. They also occupy regions of Asiatic monsoon wet-dry climate and regions of Mediterranean climates in Australia, North Africa, and the Middle East. Vegetation associated with reddish-chestnut and reddish-brown soils is varied, being tall-grass prairie, short-grass steppe, rain-green savanna and scrub, or monsoon forest. All of the areas of reddish-chestnut and reddish-brown soils have in common a marked degree of aridity of climate, whether associated with an annual cycle or drought in the winter (low-sun season) or in the summer (high-sun season). Very high air temperatures and a severe water deficiency occur at one season of the year. Otherwise, in terms of latitude, these reddish soils occupy the same global positions as areas of latosols and lateritic soils, into which the reddish soils grade in Central and South America, Africa, southern Asia, and Australia.

Aridity in the seasonal climatic cycle explains the occurrence of calcium carbonate in lower layers of the reddish-brown and reddish-chestnut soils, for this is a characteristic of the pedocal soils. Red color is evidence of an abundance of the sesquioxides of iron, which accumulate in warm climates where organic acids are not produced in quantity.

Gray desert soils (sierozem) and red desert soils

Soils of the middle-latitude deserts and tropical deserts fall into two great soil groups on the basis

Figure 14.5 Profile of a brown soil formed from loess in Colorado. The *B* horizon has good prismatic structure. Depths in feet. (Photograph by C. C. Nikiforoff. Courtesy of Soil Conservation Service, U.S. Dept. of Agriculture.)

of color. (1) *gray desert soils*, or *sierozem*, and (2) *red desert soils*.

The gray desert soils, or sierozem, are well developed in Wyoming and in the deserts of Nevada, western Utah, and southern portions of Oregon and Idaho (Figure 14.2). This latter region is sometimes referred to as the Great Basin because of the prevalence of interior drainage systems ending in evaporating basins. It corresponds approximately with the middle-latitude desert climate and has northward extensions into the middle-latitude steppe climate.

The gray desert soils contain little humus because of the sparse growth of vegetation, such as sagebrush and bunchgrass. Color ranges from light gray to grayish brown (color plate, *K*). Horizons are present, but being only slightly differentiated are not conspicuous. Excessive amounts of calcium carbonate are present at depths of less than one foot (0.3 m) in the form of *lime crust*, or *caliche*, a deposit of calcium carbonate or hydrous calcium sulfate. In places this deposit has the appearance of a hard rock layer and may even resist erosion in such a way as to produce small mesas or platforms. Gravels deposited by streams are often thus cemented into a conglomerate rock. Lime crust forms during prolonged dry periods when ground waters rise slowly surfaceword by capillary attraction and evaporate near the surface, leaving the salts behind in the soil.

In the more arid, hotter tropical deserts are found the *red desert soils* (Plate 3). These range from a pale reddish gray to a pronounced deep red (color plate, *L*). Humus is reduced to the minimum, there being only a scattered growth of desert shrubs. Thus the activity of plants as soil formers reaches its lowest point in the red desert soils, as does also the activity of animals. Color is derived from small amounts of oxides of iron. Horizons are poorly developed; texture is often coarse, with many fragments of parent rock throughout the soil. Deposits of lime carbonate are present, as with the gray desert soils.

The gray and red desert soils are suitable for cultivation only where they are fine textured, as along the floodplain terraces of exotic streams and on the outer slopes of alluvial fans. Irrigation is essential, whether water is diverted from a river or obtained from wells penetrating the ground water reserves in alluvial fans.

Altitude and soils

As explained in Chapter 12, climate zoning in mountains tends to reproduce world climates in a series of vertical zones, the effect of increasing altitude being much the same as that of increasing latitude. Because climate is a major determining factor in soils, we might also expect increasing altitude to result in a series of great soil groups. Such a series is illustrated by soils of the Big Horn Mountains in Wyoming (Figure 14.6). Starting with gray desert soils at the lowest elevation, the series progresses through the zonal soils of dry climates through prairie soils to podzols at the high elevations.

Compare these altitude changes with Figure 14.7, which summarizes the profile changes along an imaginary traverse line from the southwestern desert of the United States, eastward across the high plains region, then north into the Lake Superior region.

Halomorphic soils (intrazonal)

In the steppes and deserts, evaporation on the average greatly exceeds precipitation and there are many topographic depressions that have no outlet to external drainage systems. Here the products of rock weathering are brought by intermittent flowing streams in times of flood and left behind on the basin floor. Along with clay, silt and sand are the dissolved mineral salts that crystallize out as the water evaporates.

Shallow lake basins, or *playas*, of extreme flatness are often covered with only a thin film of water which evaporates rapidly, leaving its salts on the surface. Most persons are familiar with the salt flats of Great Salt Lake in Utah, on which so many automobile speed records have been set. Salts found in the various playa lakes of the southwestern United States are *soda* (Na_2CO_3), *borax* ($Na_2B_4O_7$), *calcium carbonate* ($CaCO_3$), various sulfates (Na_2SO_4, $MgSO_4$, K_2SO_4), chlorides ($NaCl$, $CaCl_2$, $MgCl_2$), and many others. Where salts are thick and pure in the inner parts of the playas there is no soil in the true sense of the word. The term *halomorphic soil* is properly applied to marginal areas where silts and clays make up a large proportion of the body of the soil.

The pedogenic process in the development of halomorphic soils is salinization (Chapter 13). Two major groups of halomorphic soils are recognized: *saline soils* (*solonchak*) and *alkali soils* (*solonetz*). Halomorphic soils are classed as intrazonal because of their poor drainage and limited extent.

Saline soils (solonchak, or *white alkali soils*) contain chlorides, sulfates, carbonates, and bicar-

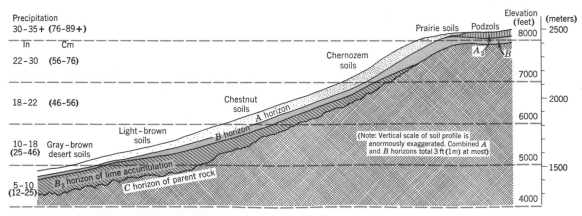

Figure 14.6 This schematic diagram shows the gradation of soils from a dry steppe-climate basin (left) to a cool, humid climate (right) as one ascends the west slopes of the Bighorn Mountains, Wyoming. (After J. Thorp, 1931.)

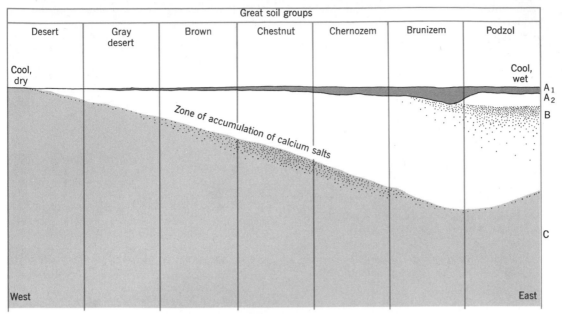

Figure 14.7 Schematic diagram of the relationships among profiles of the great soil groups, traced from dry to humid climates. (After C. E. Millar, L. M. Turk, and H. D. Foth, *Fundamentals of Soil Science*, John Wiley and Sons, New York, 1958.)

bonates of sodium, calcium, magnesium, and potassium. These soils are light colored and show poorly developed horizons (color plate, *O*). Although there are many species of plants adapted to saline soils the plant cover is at best sparse. Plants which are salt-tolerant are known as *halophytic* plants. They include grasses, shrubs, and some trees. Agriculture is not possible on saline soils unless they are flushed out with large quantities of irrigation water to remove the salts. This has been done to advantage to reclaim much good agricultural land in the southwestern United States.

In the second major group of halomorphic soils, the alkali soils (solonetz or *black alkali soils*), sodium salts predominate, especially sodium carbonate ($NaCO_3$). The soil profile is characterized by the presence of a dark, hard, columnar layer, i.e., a layer with prismatic structure (Figure 13.4*B*). Although salts of the alkali soils are of somewhat different chemical properties from salts of the saline soils, both soils occur in the same areas. Alkali soils occupy areas of somewhat better drainage than the saline soils and may be derived from the latter by the removal of more soluble salts (desalinization). Grass and shrub vegetation is found on alkali soils, but the species are particularly alkali-resistant.

With improved drainage—whether by man's intervention or by natural processes—solonetz is transformed by leaching of salts into *soloth*, a halomorphic soil of light-colored, slightly acid surface layer and a heavy-textured, dark brown *B* horizon with columnar structure.

Calcimorphic soils

Another class of intrazonal soils consists of the *calcimorphic soils*. These are soils whose characteristics are strongly related to the presence of lime-rich parent material. The process of *calcification* (introduction of calcium) is dominant in the formation of calcimorphic soils. An important example of the calcimorphic soils are the *rendzina soils*.

Rendzina soils have dark-gray or black surface layers overlying soft, light-gray or white material which is highly calcareous (lime-rich) (color plate, *P*). The parent matter of rendzina soils may be marl (a lime mud), soft limestone, or chalk, all of which are forms of calcium-carbonate. The soil profile is regarded as immature. The natural vegetation is typically grassland, which results in a well-distributed humus in the upper layers and imparts the dark color to the soils, just as in the chernozem.

Distribution of rendzina soils in the United States is fragmented into a number of geologically favorable areas. Substantial areas of these soils are found in undulating prairies of central and northeastern Texas and central-southern Oklahoma (Blackland Prairie and Grand Prairie). Another geographically important area is the Black Belt of Alabama and Mississippi, a lowland underlain by soft limestone. These areas are in the humid subtropical climate. Other rendzina areas occur on a high limestone plateau in northwestern Arizona, and in the Mediterranean climate regime of southern California.

Rendzina soils of the humid subtropical climate are productive agriculturally and yield cotton, corn, and alfalfa. In more arid grassland environments, uses are for grazing and some dry-farming.

A newer soil classification

Since publication of the 1938 soil classification, soil scientists of the Soil Survey, U.S. Department of Agriculture, have engaged in intensive review and modification of the 1938 system. Major changes in soil categories and nomenclature have been made. For students wishing to appraise these developments, there follows a general review of the new system, known as the *7th Approximation*. This review was written by Dr. Roy W. Simonson, Director, Soil Classification and Correlation, Soil Survey, U.S. Department of Agriculture, Soil Conservation Service. Excerpts from the review article are reprinted below by permission of both author and publisher.[1] Reprinted as Appendix I is a map of the United States showing patterns of soil orders and suborders, with explanatory legend and explanation of nomenclature.

History, purpose, and basic objectives of the 7th Approximation

Because it can be no better than the state of knowledge in the soil science of its day, any scheme must eventually be modified or replaced. This is the only way whereby new data or improved understanding of available data can be reflected in soil classification. The need for modifying or replacing old schemes was evident in the several reports on work in progress made to the Seventh International Congress of Soil Science in 1960.

The scheme presented by American pedologists to the 1960 congress has been carried through a succession of stages over a period of years. For purposes of identification, these stages have been numbered, and the stage presented to the 1960 congress is identified as the 7th Approximation of a comprehensive system of soil classification.[2] This approximation is currently being tested, as were earlier stages, and some modifications are likely to follow. These are expected to be modifications in details of the scheme, however, rather than in general structure. Hence, it seems that the scheme embodied in the 7th Approximation will be adopted, with minor changes, in this country within the next few years.

The scheme of soil classification now being developed in the United States differs from earlier schemes prepared in this country and elsewhere in several ways which are important. This scheme reflects evolution in the concept of soil itself. Basic to the scheme is the

[1] Simonson, Roy W., 1962, "Soil Classification in the United States," *Science*, Vol. 137, pp. 1027–1034. Copyright © by the American Association for the Advancement of Science.

[2] Soil Survey Staff, *Soil classification, a comprehensive system—7th Approximation*, U.S. Department of Agriculture Soil Conservation Service, Washington, D.C. 1960.

concept that soil comprises a continuum on the land surface, one which can be subdivided into classes in a variety of ways. Also basic to the scheme is an effort to achieve more quantitative definitions than have been devised heretofore. Definitions of classes at every categoric level are expressed in terms of properties that can be observed or measured. These are important departures from schemes developed earlier for classifying soils.

The basic objectives of the classification scheme are essentially the same as those of earlier schemes, despite the differences in approach. The scheme must first of all organize, define, and name classes in the lowest category, and it must group these classes into progressively broader classes in higher categories and provide names for these classes. Its general purpose is to make the characteristics of soils easier to remember, to bring out relationships among soils and between the soils and other elements of the environment, and to provide a basis for developing principles of soil genesis and soil behavior that have prediction value.

Concept of the pedon

The approach toward solution of this problem followed in the 7th Approximation differs from earlier approaches in several ways. An attempt is being made to define a small volume of soil as the basic entity, one for which the term *pedon* (plural, *pedons*) has been suggested. *Pedon* is proposed as a generic term for small volumes of soil, each large enough for the study of horizons and their interrelationships within the profile and having a roughly circular lateral cross section of between 1 and 10 square meters.

A group of contiguous pedons belonging to a single class of the lowest category (soil series) in the 7th Approximation is identified as a "soil individual" in the monograph. Since the monograph was prepared it has seemed that use of some other term to identify such groups of contiguous pedons would be desirable. Consequently, the term *polypedon* (plural, *polypedons*) has been proposed. A single polypedon is defined as a group of pedons contiguous within the soil continuum and having a range in characteristics within the limits of a single soil series.

Categories within the new system

Like other schemes developed earlier in this country, the 7th Approximation is a multiple-category system. Six categories are used in the scheme. These are identified, from top to bottom, as *orders, suborders, great groups, subgroups, families,* and *series*. Among these categories, that of the soil series has been used in the United States for a long time. The concept of the soil series has been changed over the years, but further change is not proposed in the 7th Approximation. In contrast to the category of the soil series, other categories in the scheme do not correspond exactly in level of generalization to any that have been used previously in this country or elsewhere, so far as is known. The "suborder" category in the 7th Approximation does approach in level of abstraction the category of the great soil group currently in use in this country, but the two are not fully equivalent.

Some measure of the span in properties permitted within classes is indicated by the numbers of classes in the categories. The 7th Approximation provides for the recognition of ten classes in the "order" category. The numbers of classes in the other categories (in rounded numbers) are as follows: suborders, 40; great groups, 120; subgroups, 400; families, 1500; and series, 7000. It should be mentioned that the totals for the three lowest categories cover only the soils of the United States. These totals would be appreciably larger if soils of other continents were included. On the other hand, the total number of each of the orders, suborders, and great groups is expected to remain the same, or virtually so, whether the scheme is applied to the United States or to the world as a whole. The intent has been to provide a place in the scheme for all known soils in the world, though this goal may not have been reached. In the construction of the scheme it is recognized that soils not known to the authors of the scheme may have been omitted. It is also recognized that modification of this scheme or its replacement will eventually become necessary as the knowledge and understanding of soils continue to grow.

Definitions of the classes in the different categories of the 7th Approximation are in terms of morphology and composition of the soils—that is in terms of soil characteristics themselves. Moreover, an attempt is made to have the definitions as nearly quantitative as currently available data will permit.

Nomenclature

The nomenclature proposed in the 7th Approximation represents a marked departure from past practice in soil classification. A new nomenclature is proposed for the classes in each of the four highest categories. The proposed names for classes in the order, suborder, great group, and subgroup categories consist of coined terms in which Greek and Latin roots are largely used. (See Appendix I.) The names are distinctive for the classes in each category, so that a name itself will indicate the category to which a given class belongs. Moreover, the names are designed so that each subgroup may be identified, by its name, with the great group, suborder, and order in which it is classified.

Several objectives have been kept in mind in developing the proposed nomenclature. Efforts were made to devise names that were distinctive and could be easily remembered, would suggest a few characteristics of the soils in each class, would identify the categorical level to which a class belonged, would fit into the existing pattern of language readily, and would provide convenient adjective as well as substantive forms. An effort to reach all of these objectives simultaneously is ambitious, and the several are not compatible in all cases. Hence, it is not surprising that some defects have already been noted in the proposed nomenclature.

The ten orders

The names of the ten orders consist of three or four syllables, and every name ends in the suffix *sol*. The names of the ten orders are "entisols," "vertisols,"

"inceptisols," "aridisols," "mollisols," "spodosols," "alfisols," "ultisols," "oxisols," and "histosols" (the letter s is added for the plural form). (See Appendix I.)

The ten orders are set apart on the basis of one or more of the following factors: gross composition, degree of horizonation, the presence or absence of certain horizons, and what is in effect a combined index of weathering and weatherability of minerals. The characteristics selected to distinguish the orders are believed to reflect major differences in paths of horizon differentiation, in stages reached in horizon differentiation, or in both. To state this another way, the intent has been to choose as differentiating characteristics properties that reflect major differences in genesis of the soils. Whether the selections have been successful for this purpose, and if they have been, to what degree, will become evident only after the scheme has been tested for a time.

The bases for distinguishing ten orders may be made clearer by some illustrations. Full definitions will not be given here for any of the orders, but the principal differentiae for the histosols, entisols, mollisols, and spodosols will be sketched briefly.

The histosols are organic soils, mainly those known as peats and mucks. these are distinguished from soils of the other nine orders by differences in gross composition: histosols are high in organic matter (20 percent or more). The balance among processes of horizon differentiation in soils so high in organic matter is very different from that in dominantly mineral soils.

The entisols are mineral soils with low degrees of horizonation, mainly those that have been identified as lithosols, regosols, and alluvial soils in the United States in recent years. The entisols have few and faint horizons in their profiles. These soils are in early stages of horizon differentiation. Some are forming in regoliths consisting of highly resistant minerals; others in areas where accretions of fresh materials keep pace with horizon differentiation; still others in areas where removal by erosion keeps pace with horizon differentiation.

The mollisols are mineral soils which have a characteristic known as a *mollic epipedon*. This is a darkened surface layer of considerable thickness, relatively high in organic matter, high in base saturation, and friable. For identification as a mollic epipedon, minimum requirements for thickness, color, base saturation, level of organic matter, and consistence are given. Unlike the histosols and entisols, mollisols tend to occur within certain geographic zones. Such soils are the major ones of the Corn Belt and Great Plains in the United States. Mollisols have been formed almost entirely under prairie vegetation in semiarid-to-subhumid climates. Included in the mollisols are the chernozems, studied almost a century ago by Dokuchaiev in Russia.

The spodosols are mineral soils which have a characteristic known as a spodic horizon. This is a subsurface horizon of illuvial accumulation of humus, usually in conjunction with accumulation of iron or aluminum, or both. The spodic horizon corresponds closely to the B horizon of podzols, as those soils have been described in North America and western Europe.

Like the mollisols, and spodosols tend to be associated with certain climatic and vegetation types. These soils are found in cool, humid regions, for the most part, and they are formed mainly under coniferous forest or under vegetation dominated by plants such as heather. The spodosols occur extensively in eastern Canada, in New England and the northern Lake States, and in the taiga region of the Soviet Union. Such soils often have strikingly different horizons in the same profile, and this may be why they were among the first to be studied with care.

Suborders

The basic approach followed in defining orders in the 7th Approximation is carried down to the suborder and great-group categories, though the same characteristics are not used as criteria. Additional characteristics of the soil are introduced as criteria for distinguishing classes at each level.

The kinds of characteristics used in differentiating suborders within orders are moisture regimes, temperature, mineralogy, and specific kinds of horizons. One suborder is set apart in each of eight orders because of the evidence of wetness in the morphology of the soils. In two of the orders, one pair of suborders is distinguished mainly on the basis of temperature. The mineralogy of the soil—for example, very high levels of quartz, dominance of allophone, or high proportions of calcium carbonate—provides criteria for recognizing at least one suborder in each of four orders. Characteristics such as an argillic horizon, a cambic horizon, and the tonguing of an albic horizon into an argillic horizon are definitive for at least one suborder in each of five orders.

The name of each suborder is a two-syllable rather than a three- or four-syllable term. Each name consists of a prefix syllable with a specific connotation plus a syllable from the name of the order in which the suborder is classified. Fourteen formative elements are used as prefixes in the construction of suborder names. Thus, for example, suborders in the order of entisols are identified as "aquents," "ustents," and "udents." (See Appendix I.)

Great groups and subgroups

The setting apart of great groups within suborders is based on the same kinds of criteria as is the distinguishing of suborders within orders. Great groups are distinguished within suborders by the presence or absence of characteristic horizons or other features, the occurrence of horizons extraneous to the sequence required for the suborder, and temperature. The range of definitive characteristics within individual classes has been reduced step by step in coming down the ladder from the order to the suborder to the great group. Thus, the soils of a great group are more homogeneous in their characteristics than are soils of classes in higher categories. For each great group, the soils have the same kinds of horizons in the same sequence within pedons, except for surface horizons, which may be obliterated by plowing or by erosion.

The approach in defining subgroups differs from that followed in defining classes in higher categories. As explained earlier, a typifying subgroup is first defined for each great group and identified by the term *orthic* preceding the great-group name. This subgroup has the median expression of the definitive characteristics of the great group. In addition to the orthic subgroup, other subgroups are set apart as intergrade or extragrade subgroups. Intergrade subgroups have some characteristics definitive of another great group, either in the same order or in some other order. Extragrade subgroups have some properties that are not definitive of any known great group. In both intergrade and extragrade subgroups, however, the soils are more like the orthic or central subgroup of the great group to which each belongs than to any other known kind of soil. Recognition of orthic, intergrade, and extragrade subgroups is one device for recognizing that the soil mantle forms a continuum in which changes are gradational rather than abrupt.

REFERENCES FOR FURTHER STUDY

U.S. Dept. of Agriculture (1938), *Soils and men; yearbook of agriculture 1938*, U.S. Govt. Printing Office, Washington, D.C., 1232 pp.

Robinson, G. W. (1959), *Soils, their origin, constitution and classification*, third edition, John Wiley and Sons, New York, 573 pp.

Bunting, B. T. (1965), *The geography of soil*, Aldine Publ. Co., Chicago, 213 pp.

Millar, C. E., L. M. Turk, and H. D. Foth (1965), *Fundamentals of soil science*, fourth edition, John Wiley and Sons, New York, 491 pp.

Money, D. C. (1965), *Climate, soils, and vegetation*, University Tutorial Press, Ltd., London, 272 pp.

Eyre, S. R. (1968), *Vegetation and soils; a world picture*, second edition, Aldine Publishing Co., Chicago, 328 pp.

REVIEW QUESTIONS

1. What are the three orders into which all soils can be classified? Under what conditions is each of these orders developed? What kinds of soils are included under the azonal order?

2. How did our present system of soil classification develop? What basic concepts underlie the classification? Name persons who made important contributions to this field.

3. Describe the profile of a typical podzol soil. What climatic conditions are required for the development of soils of this group? What is the natural vegetation of this great soil group? How do these soils rate in terms of agricultural productivity? In what ways do they require treatment?

4. How do the gray-brown podzolic soils differ from the podzol soils? How do the climatic factors differ? What geographic positions do the gray-brown podzolic soils have with respect to the podzols? How does natural vegetation of the two soil groups differ? Which soil group is better for agricultural purposes? Why?

5. What are the essential characteristics of the red-yellow soils? With what climate are they associated? Explain how temperature and precipitation determine the features of these soils. What types of forest are found on the red-yellow soils?

6. What is terra rossa? Where is it found? What ideas have been advanced as to the origin of terra rossa?

7. What are latosols (lateritic soils)? What are their essential characteristics? What is meant by laterization? How do latosols rate in terms of agricultural productivity?

8. How can laterite be used as a building material? What are some of the residual ores obtained from laterites?

9. Describe other soils of humid tropical and equatorial climates.

10. What are hydromorphic soils? Why are the bog and meadow soils classed as intrazonal? What is the appearance of the profiles of these soils? What is the glei horizon? How do meadow soils differ from bog soils?

11. What are planosols? To what order do they belong? Where do planosols develop, and what is different about their profiles as compared with associated zonal soils?

12. Describe a tundra soil. In what respect does it differ from the podzol soils? What effect has excessive cold upon the composition of the parent material of the soil? Where are the arctic brown forest soils found?

13. What are the chernozem soils? Where were these soils first studied? Where are they best developed? Describe the chernozem profile.

14. What is the essential difference between the chernozems and the podzolic soils? With what climatic characteristics are the chernozem regions associated?

15. What is the natural vegetation of chernozem soils? How is loess associated with the chernozem?

16. In what way are the prairie soils related to both the chernozems and the gray-brown podzolic soils? What is the nature of the geographic and climatic location of the prairie soils?

17. What are the tall-grass prairies? What theories have been advanced for the absence of forests in these areas? With what crops are the prairie regions now associated in the central United States? Where are reddish prairie soils found?

18. Where are the chestnut soils (dark-brown soils) located in relation to the chernozems? With what climatic conditions are they associated?

19. How do the brown soils differ from the chestnut soils? What is distinctive about the *B* horizon of a brown soil?

20. Describe reddish-chestnut and reddish-brown soils. Where are they found, and with what climates and natural vegetation types are they associated?

21. Describe the gray-desert soils and red-desert soils. How do these two groups differ in geographic location and in climatic controls? Are horizons well developed in these soils? What forms do excessive calcium carbonate and hydrous calcium sulfate take in these soils?

22. How does altitude influence the development of soil types? In an arid region in middle latitudes, what succession of soil groups might be encountered in passing from an intermontane basin to a lofty mountain summit?

23. What are halomorphic soils? Where do saline soils form? What are some of the salts found in playa lakes of the dry deserts? How do the alkali soils differ from the saline soils?

24. What are calcimorphic soils? Describe rendzina soils and give locations of three occurrences. Why are rendzina soils regarded as intrazonal?

CHAPTER 15

Structure and Environment of Vegetation

VEGETATION on the lands of the earth is of prime concern to the geographer. Plants, as stationary objects possessing distinctive physical properties, are an element of the landscape just as fully as landforms, soils, and hydrographic features. That the physical forms of individual plants and of their assemblages vary in some systematic way with latitude, elevation, and continental position excites the interest of the geographer and causes him to inquire more closely into the interrelations of vegetation, soils, landforms, and climate. Plants, as consumable and renewable sources of food, fuel, clothing, shelter and many other categories of life-essentials, form a great natural resource base essential to man. The ways in which man has used these resources to his advantage—or has been hindered by plants in his progress—throughout human history forms a vital part of historical geography; the present and future development and management of plant resources is a part of the fabric of many studies of economic and political geography.

Our concern here is with *natural vegetation*, that is, vegetation which develops without appreciable interference and modification by man. The widespread activities of man in agriculture, herding, forest-cropping, and urbanization guarantee that a considerable proportion of the earth's land surfaces does not bear a natural vegetation. Nevertheless, from the study of islandlike remnants of protected plant life and historical records, the plant geographer can achieve some success in the global mapping of broad units of natural vegetation. (See Chapter 16.) Such maps represent the vegetation that might be expected eventually to return through a series of stages, if man's interference with the natural regimen were to be suspended.

Floristic or structural approach?

Our selection of study topics is extremely limited in terms of the scope of the science of botany and even of plant geography as a specialized field within botany. Emphasis here is not upon a *floristic* approach, that is an approach primarily concerned with the *flora*. Should we describe the flora of a region, we would produce a list of the plant species found there. But such a list would tell us little about the relative abundance of the species; their physical forms; and their characteristic arrangements as landscape elements.

To be sure, where the trees of a forest consist largely of two or three dominant species, it will be desirable to name these species and to designate the forest accordingly. For example, one finds in the plateaus of Arizona and Utah a woodland in which the tree population consists almost exclusively of pinyon pine (*Pinus edulis*) and juniper (*Juniperus utahensis*). Through our knowledge of the characteristic height, branching habit, and foliage of these species we gain a true picture of the physical form of the woodland. Such a terse forest description by floristic means would be impossible in an equatorial forest composed of scores of different species. Instead, the forest must be described according to its *structure*, that is, by the way in which the living parts of the plants are shaped and distributed in space. Generally, in the classification of the world's natural vegetation, the structural approach is most meaningful to follow.

Bioclimatology and ecology

Our study of plant geography will not be concerned with the historical aspects, that is, with the evolution of plants and the spread and decline of floras through geologic time. On the other hand, we will be greatly interested in the *bioclimatological* aspects of plant geography in which the responses of plants to light, heat, and moisture are studied and the knowledge of climate elements and their global distribution is brought into play. If we choose, the bioclimatological approach can be narrowed to the response of a single species to a single meteorological element.

A quite different approach to the question of a plant's relation to its environmental influences is to consider together all the organisms—plant and animal—living in one place at one time in terms of their total environment and of their adjustments to both the environment and to each other. Such a whole dynamic system is referred to as an *ecosystem*; the study of an ecosystem is *plant ecology*. The ecosystem represents the combined use by plants and animals of the resources provided, in one place at one time, by the surroundings. Air, water, nutrients, heat, and light are variously utilized, transformed, stored, and returned to the nonliving state.

The biosphere and its subdivision

To fix the place of natural vegetation units in the scheme of life on earth we begin with the total system, or *biosphere*, which encompasses the entire near-surface zone of the earth favorable to life in one form or another. The biosphere may be subdivided into three environmental divisions, or *biocycles:* (1) *salt water* (oceans), (2) *fresh water* (streams, lakes, ponds), and (3) *land* (soil and the air in contact with it). We are interested primarily in the land biocycle, although the geographer's concern extends into the fresh water and salt water biocycles as well.

The land biocycle is divided into ecological systems of decreasing size and complexity, schematically represented in Figure 15.1*A*. Four great *biochores* comprise the land biocycle: (*a*) *forest*, (*b*) *savanna*, (*c*) *grassland*, and (*d*) *desert*. (See Figure 16.1). These four classes of vegetation are separated on the basis of the structure of the plant assemblages and represent the fundamental re-

sponse of vegetation to global climate controls, principally available moisture (precipitation and evaporation), but secondarily light, heat, and winds. The four biochores comprise systems too vast in area and having too wide a variation within each biochore to use in the mapping of world vegetation. Therefore, within the biochores are *formation classes*—altogether from 15 to 20 or more in number, depending upon the judgment of the plant geographer whose classification system is used. Thus there are clearly several varieties of forests (e.g., rainforest, deciduous forest, needleleaf forest) each responding to the climate regime to which it is subjected. Definitions of such terms as "forest," "woodland," and "grassland" are given in the discussion of the individual formation classes in Chapter 16. The distinction among the several biochores and their included formation classes is made on the basis of the plant structure, rather than species. Thus classification is structural, rather than floristic.

Plant habitats and communities

As we all know from direct experience, vegetation is strongly influenced by landforms and soil. Vegetation on an upland, that is, on relatively high ground of moderate surface slope and thick soil,

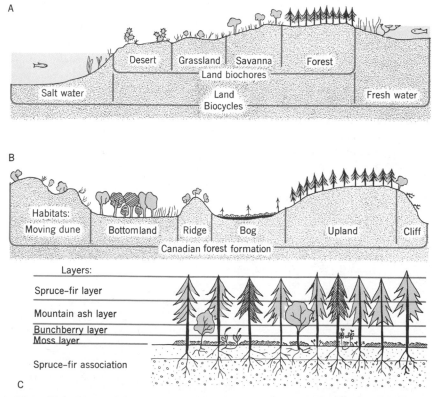

Figure 15.1 Dimensions of the environment. *A*, the biocycles and biochores. *B*, Habitats in the Canadian forest formation. *C*, Detail of spruce-fir upland association. (From Pierre Dansereau, 1951, *Ecology*, Vol. 32.)

is quite different from that on an adjacent valley floor where water lies near the surface much of the time. Vegetation is also strikingly different in form on rocky ridges and on steep cliffs where water drains away rapidly and soil is thin or largely absent.

Thus, each region which we assign to a given formation class is actually a mosaic of smaller units reflecting the inequalities in conditions of slope, elevation, and soil type. Such subdivisions of the plant environment are described as *habitats* (Figure 15.1*B*). In the example shown, the Canadian needleleaf forest (a formation class) actually comprises at least six habitats: upland, bog, bottomland, ridge, cliff, and active dune. Just where each habitat is located and how large an area it occupies depends largely upon the geologic history of the region, e.g., what processes of erosion and deposition have acted in the past to shape the landforms. The patterns of landforms from such processes are in turn influenced by the ancient geologic history which brought about varied combinations of arrangements of weak and resistant rocks. In the final part of this book, the evolution of landforms is explained in some detail.

Closely tied in with the effect of landform upon habitat is the distribution of water in the soil and rock, a topic also treated in later chapters. Finally, each habitat has characteristic soil properties which are determined not only by the conditions of slope and water, but in part by the plants themselves.

Adjacent habitats may possess strikingly different vegetation structures so that the question arises as to which habitat shall be selected as that for which the formation class is named. For this purpose the upland habitat, with its well-drained surface, moderate surface slopes, and well-developed soil, is usually taken as the reference standard.

The vegetation structure found in the upland habitat is thus what we have in mind when naming and describing a formation class (Figure 15.1*C*).

Within a given habitat will be found a smaller unit of ecosystem, the *plant community*, composed of rather definite proportions of organisms which are more or less interdependent and which use the resources of their habitat in such a way as to either maintain the habitat or modify it. No particular dimensions and boundary limits can be set for plant communities; these properties will depend on the changing habitat patterns. Within a given plant community can be recognized a structure of vegetation which is distinctive and a certain floristic composition, which may seem more or less fixed, or which may be gradually changing with time.

Structural description of vegetation

If the plant geographer is to describe vegetation according to its structure, or physical properties and outward forms, he must set up a number of categories of information, each designed to contribute an essential element to the description of the vegetation. Six categories are used in a system set up by Dr. Pierre Dansereau.[1] These six categories of information tell us about the growth-form of the plants, their size and stratification, the degree to which they cover the ground, their time functions, and their leaf forms.

1. *Life-form*. Plants can be classified according to their life-forms. The first two forms, *trees* and *shrubs* are erect, woody plants (Figure 15.2). The word "tree" means a perennial plant having a single upright main trunk, often with few branches

[1]Dansereau, Pierre (1957), *Biogeography, an ecological perspective*, The Ronald Press Co., New York. See pp. 147–154.

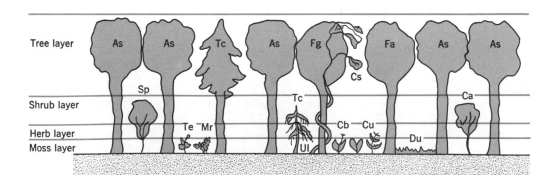

Figure 15.2   Schematic diagram of life forms in a beech-maple-hemlock forest. The tree layer consists of sugar maple (As), ash (Fa), beech (Fg), and hemlock (Tc) and includes a liana (Cs). The shrub layer includes elder (Sp), dogwood (Ca), and a young hemlock (Tc). An epiphyte (Ul) grows upon the hemlock. Plants designated Te, Mr, Cb, and Cu form the herb layer. Moss (Du) forms the lowest layer. (From Pierre Dansereau, 1951, *Ecology*, Vol. 32.)

Figure 15.3   A stand of mature sugar maples, 160 to 200 years old, in the Allegheny National Forest, Pennsylvania, 1939. (U.S. Forest Service photograph.)

Figure 15.4   Lianas of the equatorial rainforest, near Belém, Brazil. (Photograph by Otto Penner. Courtesy of Instituto Agronómico do Norte.)

in the lower part, but branching in the upper part to a crown which, in the mature individual, contributes to one of the upper layers of vegetation (Figure 15.3). The word "shrub" generally refers to a woody plant having several stems branching near the ground, so as to place the foliage in a mass starting close to ground level. Next are recognized woody vines which climb up on trees: the *lianas* (Figure 15.4). These may climb upon mature plants, or may be lifted during the growth of young trees to which they have become attached. Although the term "liana" is usually associated with woody vines of tropical forests, the meaning may be broadened to include woody vines of middle-latitude forests, such as the poison ivy and Virginia creeper. Thus it is apparent that the life-form classes cut across the taxonomic categories of the plant world. For example, palms and tree ferns look much alike outwardly but are of entirely different plant orders.

Fourth of the life-forms is the plant group known as the *herbs*. These are usually small tender plants, lacking woody stems. They occur in a wide variety of forms and leaf structures; including annuals and perennials, and broad-leaved plants as well as the grasses. Broad-leaved herbs are termed *forbs* in distinction with grasses. The adjective *herbaceous* is applied to this life-form. The herbaceous layer will normally occupy a low position in a layered or stratified plant community. Smaller, and lying in close contact with the ground or attached to tree trunks are the *bryoids*, a life-form named for a plant phylum which includes the mosses and liverworts.

Finally, among the life-forms are the *epiphytes*, plants which use other plants as supporting structures and thus live above the ground level, out of contact with soil (Figure 15.5). Familiar to all are the tropical orchids that live high upon tree limbs and are sometimes referred to as "air plants." Ferns also commonly exist as epiphytes (see Figure 16.5).

Not included in our list of life-forms are lower forms of plants, grouped under the name of *thallophytes*. These include bacteria, algae, molds, and fungi—all of which are plants lacking true roots, stems, and leaves. *Lichens*, plant forms in which algae and fungi live together to form a single plant structure, are included within the life-form classification of bryoids. Lichens occur as tough, leathery coatings or crusts and as leaflike masses attached to rocks and tree trunks (Figure 15.6). In arctic and alpine environments lichens may grow in profusion and dominate the vegetation in the near absence of more conspicuous plant forms.

Figure 15.5 An epiphyte, or air plant. This *Bromeliad* is attached to a limb of the host tree by a mass of fibrous roots. The plant leaves form an urn-shaped structure that collects and holds rainfall. About $\frac{1}{7}$ natural size. (After W. H. Brown, 1935, *The Plant Kingdom*, courtesy of Blaisdell Publishing Co., a Division of Ginn and Co.)

Figure 15.6 A lichen (*Stereocaulon vulcani*) growing on a rough lava surface, Waiakea Forest Reserve, Maua, Hawaii. (Photograph by Pierre Dansereau.)

2. Size and stratification. Each of the life-forms described above may be classified according to size. The words "tall," "medium," and "low" may be given definite limits for each life-form. For example, a tree higher than 25 meters (82 feet) is "tall"; from 10 to 25 meters (33 to 82 feet) is "medium"; 8 to 10 meters (26 to 33 feet), "low." For the smaller life-forms different limits are set (Table 15.1). Such standardization of size enables the plant geographer to make precise descriptions of vegetation. Further standardization is achieved by setting height limits for a series of layers, numbered successively from the ground up (Table 15.1).

3. Coverage. The degree to which the foliage of individual plants of a given life-form cover the ground beneath them is designated the *coverage*. We may use four terms descriptive of the coverage: barren or very sparse; discontinuous; in tufts or groups; and continuous. For example, the trees may be of discontinuous coverage whereas the herb layer is continuous, or vice versa.

4. Function, or periodicity. Of primary importance in the classification of forms of natural vegetation is the response of the plant foliage to the annual climatic cycle. *Deciduous* plants shed their leaves and become dormant in an unfavorable season, which is either too cold or too dry to permit growth. *Evergreen* plants retain green foliage year-around, although in some cases becoming almost dormant in a cold or dry season. Where the climate is equable (that is, moist and not cold throughout the year) evergreen plants grow continuously. A third class, the *semideciduous* plants, are those which shed their leaves at intervals not in phase with a season. Thus a semideciduous forest would not at any one time have all of its individuals devoid of foliage. As a fourth class we recognize *evergreen-succulent* and *evergreen-leafless* plants, those with very thick fleshy leaves, which retain their foliage year-around, and those with fleshy stems but no functional leaves, such as the cacti.

5. Leaf shape and size. Recognition of the leaf form of a plant constitutes an essential part of the structural description (Figure 15.7). One form is the *broadleaf*, familiar to us in such common trees as the maple, beech, and rhododendron. In contrast is the *needleleaf*, also familiar in the pine, spruce, fir, and hemlock. A similar form is the *spine*, which in some plants represents the transformation of the entire leaf. The slender, tapering leaves of grass are referred to as *graminoid* in form. We may also recognize the *small leaf-form*, as in the birch or holly; and the *compound leaf*, as in the hickory and ash.

6. Leaf texture. Leaf textures range widely according to the climate and habitat, because of the different degrees to which the water loss from the leaf into the air must be controlled. Leaves of average thickness are described as *membranous*; those which are thin and delicate (as in the maiden-hair fern) are described as *filmy*. Leaves which are hard, thick, and leathery are *sclerophyllous*; a forest dominated by trees and shrubs having such

TABLE 15.1 PLANT SIZE AND STRATIFICATION OF VEGETATION[a]

Plant Size

TALL:	Tree	Over 82 ft (25 m)
	Shrub	6.6 to 26 ft (2 to 8 m)
	Herb	Over 6.6 ft (2 m)
MEDIUM:	Tree	33 to 82 ft (10 to 25 m)
	Shrub	1.6 to 6.6 ft (0.5 to 2 m)
	Herb	1.6 to 6.6 ft (0.5 to 2 m)
	Bryoid	Over 4 in (10 cm)
LOW:	Tree	26 to 33 ft (8 to 10 m)
	Shrub	Under 1.6 ft (0.5 m)
	Herb	Under 1.6 ft (0.5 m)
	Bryoid	Under 4 in (10 cm)

Stratification

LAYER NUMBER	RANGE
7	Over 82 ft (25 m)
6	33 to 82 ft (10 to 25 m)
5	26 to 33 ft (8 to 10 m)
4	6.6 to 26 ft (2 to 8 m)
3	1.6 to 6.6 ft (0.5 to 2 m)
2	4 in to 1.6 ft (10 cm to 0.5 m)
1	0 to 4 in (0 to 10 cm)

[a] After Pierre Dansereau, 1958, Botanical Institute of the University of Montreal, *Contributions* No. 72, pp. 31–32.

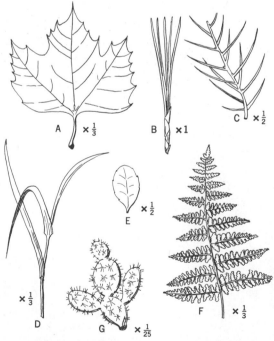

Figure 15.7   Leaf shapes. *A*, Large thin leaf (sycamore). *B*, Needle leaf (pine). *C*, Spine. *D*, Graminoid leaf (grass). *E*, Small leaf. *F*, Compound leaf (fern). *G*, Succulent stem (cactus).

leaves is termed a *sclerophyll forest.* Very greatly thickened leaves, capable of holding much water in their spongy structure, are described as *succulent.*

Environmental factors in plant ecology

With some information now at our disposal concerning the structure of vegetation and its organization into plant communities, associations, formation-classes, and biochores, we can study the environmental factors which require vegetation to assume such varied forms.

The four main classes of environmental factors are (1) *climatic,* (2) *geomorphic* (related to landform), (3) *edaphic* (related to soil), and (4) *biotic* (related to living organisms). Although we treat each of these classes of factors in turn and examine each factor separately, it must be kept in mind that the basic concept of ecology is that many factors act simultaneously and that factors affect one another in a most complex interrelationship. The very plants that are affected by environmental factors may react in such a way as to modify the environment and thus change the factors themselves.

In considering how various factors of the physical environment influence plant structures and distributions, two scale ranges can be treated. One is essentially the global scale and consists of such climatic factors as the seasonal and latitudinal patterns of insolation, light and darkness, temperature, precipitation, and prevailing winds. The other scale of consideration is that of variations of the physical environment found within a relatively small habitat and between adjacent habitats. Thus, although there are vast global climate patterns of deserts and humid regions, we may also find within a region of generally humid climate a few small habitats (such as a dune or cliff) which are extremely dry places for plants to live. We may also find in a large desert some small habitats which are extremely wet most of the time (such as a seep or spring).

Water needs of plants

The need for water is perhaps the dominating consideration in analyzing the physical environment of plants. In conjunction with their growth processes, leaf-bearing plants give off large quantities of water into the atmosphere, a process termed *transpiration,* which is essentially a form of evaporation from water films upon the exposed surfaces of plant cells. A relatively small amount of water is also required by green plants in the process of *photosynthesis* in which the energy of light is used together with carbon dioxide and water to produce carbohydrates. The dominant source of water needed for transpiration and photosynthesis is from the soil, where it is taken up into the plant roots.

The rate of transpiration varies greatly according to the type of plant and the prevailing atmospheric conditions. High temperatures, low humidities, and winds favor high rates of transpiration. The plant structure, particularly of the leaf, largely determines the rate of water loss. Plants with large total foliage surfaces composed of broad, thin leaves have higher rates of loss than plants bearing needle leaves, spines, or small thick leaves (sclerophylls). Under conditions of critical water supply but high rates of evaporation, only those plants can survive that minimize transpiration losses by their special leaf structures and by their small size.

The adaptation of plant structures to water budgets with large water deficiencies is of particular interest to the plant geographer. Transpiration occurs largely from specialized leaf pores, called *stomata*, which are openings in the *epidermis* (outer cell layer) and *cuticle* (an outermost protective layer) through which water vapor and other gases can pass into and out of the leaf (Figure 15.8). Surrounding the openings of the stomata are *guard cells* which can open and close the openings and thus to some extent regulate the flow of water vapor and other gases. Although most of the transpiration occurs through the stomata, some may pass through the cuticle. This latter form of loss is reduced in some plants by thickening of the outer layers of cells or by the deposition of wax or waxlike material on or near the leaf surface. Thus many desert plants have thickened cuticle

or wax-coated leaves, stems, or branches.

Another means of reducing transpiration is the development of stomata deeply sunken into the leaf surface that retard outward diffusion of water vapor into dry air, and the restriction in location of stomata to the shaded undersurfaces of leaves. A plant may also adapt to a desert environment by greatly reducing the leaf area, or by bearing no leaves at all. Thus needle-leaves and spines representing leaves greatly reduce loss from transpiration. In cacti the foliage leaf is not present and transpiration is limited to fleshy stems.

In addition to developing leaf structures that reduce water loss by transpiration, plants in a water-scarce environment improve their means of obtaining water and of storing it. Roots become greatly extended to reach soil moisture at increased depth. In cases where the roots reach to the ground water table, a steady supply of water is assured. Plants drawing from such a source are termed *phreatophytes* and may be found along dry channels (draws) and alluvial valley floors in desert regions. Other desert plants produce a widespread but shallow root system enabling them to absorb the maximum quantity of water from sporadic desert downpours which saturate only the uppermost soil layer. Stems of desert plants are commonly greatly thickened by a spongy tissue in which much water can be stored. As already noted, such plants would be described as *succulents*.

A quite different adaptation to extreme aridity is seen in many species of small desert plants that complete a very short cycle of germination, leafing,

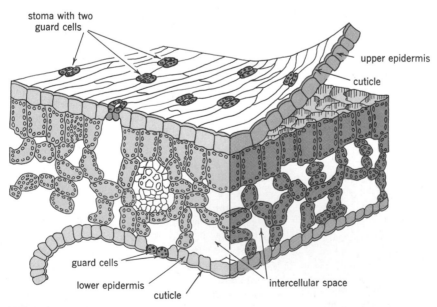

Figure 15.8 Cell structure of a foliage leaf. (After W. W. Robbins and T. E. Weier, 1950, *Botany*, John Wiley and Sons.)

flowering, fruiting, and seed dispersal immediately following a desert downpour.

Classification of plants by water need

Habitats of the land biocycle are differentiated primarily with reference to the degree of saturation of the soil by water. So important is this factor that plants may be classified according to their water requirements. The terminology associated with the water factor is built upon three simple prefixes of Greek roots: *xero-*, dry; *hygro-(hydro)*, wet; and *meso-*, intermediate or middle. Thus a habitat may be prevailing wet (*hygric*); prevailingly dry (*xeric*), or of intermediate degree of wetness (*mesic*). Those plants which grow in dry habitats are *xerophytes*; those which grow in water or in wet habitats are *hygrophytes* (*hydrophytes*); those of habitats of an intermediate degree of wetness and relatively uniform water availability are *mesophytes*.

The xerophytes are highly tolerant of drought and can survive in habitats which dry quickly following rapid drainage of precipitation, for example, on sand dunes, beaches, and bare rock surfaces. The plants typical of dry climates (deserts) are also xerophytes; cactus is an example. The hygrophytes are tolerant of excessive water and may be found in shallow streams, lakes, marshes, swamps, and bogs; an example is the water lily. The mesophytes are found in upland habitats in regions of ample rainfall. Here the drainage of precipitation is good and moisture penetrates deeply where it can later be used by the plants. On such upland locations the soil may be of intermediate texture—not too coarse or too fine—and is usually thickly and uniformly present.

Certain of the climatic regimes, such as the tropical wet-dry regime (including monsoon climates) and the continental regime where moist but not too cold, have a yearly cycle with one season in which water is unavailable to plants because of lack of precipitation or because the soil water is frozen. This season alternates with one in which there is abundant water. Plants adapted to such regimes are termed *tropophytes*, from the Greek word *trophos*, meaning change, or turn. Tropophytes may meet the impact of the season of unavailable water by dropping their leaves and becoming dormant. When water is again available, they leaf out and grow at a rapid rate. Trees and shrubs which seasonally shed their leaves are said to be *deciduous*; in distinction with *evergreen* trees which retain most of their leaves in a green state throughout the year.

Other climatic factors

The factor of light is of importance in plant ecology. Within the habitat of a given plant association or community the degree of light available will depend in large part upon position of the plant. Tree crowns of the upper layer receive maximum light, but correspondingly reduce the amount available to lower layers. In extreme cases forest trees so effectively cut off light that the forest floor is almost free of shrubs and herbaceous plants. In certain deciduous forests of middle latitudes, the period of early spring, before the trees are in leaf, is one of high light intensity at ground level, permitting certain herbaceous plants to go through a rapid growth cycle. In summer these plants will largely disappear as the leaf canopy is completed. Other herbaceous plants in the same habitat require shade and do not appear until later in the summer.

Treated on a global basis, the factor of light available for plant growth is varied by latitude. Duration of daylight in summer increases rapidly with higher latitude and reaches its maximum poleward of the arctic and antarctic circles, where the sun may be above the horizon for 24 hours (Chapter 2). Thus, although the growing season for plants is greatly shortened at high latitudes by frost, the rate of plant growth in the short frost-free summer is greatly accelerated by the prolonged daylight. But in still higher, subarctic latitudes plant growth is greatly slowed by the low heat budget, despite perpetual summer daylight.

In middle latitudes, where vegetation is of a deciduous type, the annual rhythm of increasing and decreasing periods of daylight determines the timing of budding, flowering, fruiting, leaf-shedding, and other vegetation activities. As to the importance of light intensity itself, it is generally believed that even on overcast days there is sufficient light to permit plants to carry out photosynthesis at their maximum rates and that direct sunlight constitutes a considerable excess of light.

Temperature, another of the important climatic factors in plant ecology, acts directly upon plants through its influence upon the rates at which the physiological processes take place. In general, we can say that each plant species has an optimum temperature associated with each of its functions, such as photosynthesis, flowering, fruiting, or seed germination, and that there exist some overall optimum yearly temperature conditions for its growth in terms of size and numbers of individuals. There are also limiting lower and upper temperatures for the individual functions of the plant as well as for its total survival. Temperature acts as an indirect factor in many other ways. Higher air temperatures increase the water-vapor capacity of the air and thus induce greater plant transpiration as well as greater rates of evaporative loss of moisture from the soil.

In general, the colder the climate, the fewer

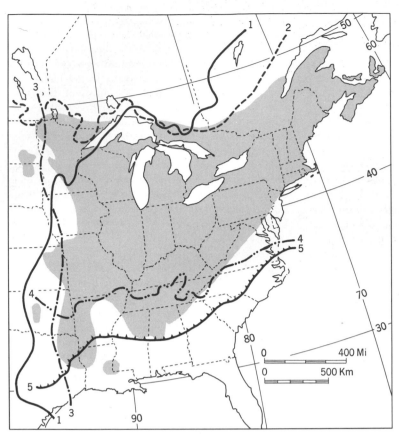

Figure 15.9 Bioclimatic limits of the sugar maple (*Acer saccharophorum*) in eastern North America. The shaded area represents the distribution of the sugar maple. Line 1, 30 in (76 cm) annual precipitation. Line 2, −40°F (−40°C) mean annual minimum temperature. Line 3, eastern limit of yearly boundary between arid and humid climates. Line 4, 10 in (25 cm) mean annual snowfall. Line 5, 16°F (−10°C) mean annual minimum temperature. (Based on *Biogeography—An Ecological Perspective*, by Pierre Dansereau. Copyright © 1957, The Ronald Press Company.)

numbers of species that are capable of surviving. A large number of tropical plant species cannot survive below-freezing temperatures. In the severely cold arctic and alpine environments of high latitudes and high altitudes only a few species can survive. Application of this principle explains why a forest in the equatorial zone has many species of trees, whereas a forest of the subarctic zone may be dominantly of one, two, or three tree species. Tolerance to cold is closely tied up with the ability of the plant to withstand the physical disruption that accompanies the freezing of water. If the plant has no means of disposing of the excess water in its tissues, the freezing of that water will damage the cell tissue.

On the basis of preference for, or tolerance to, temperatures, plant geographers distinguish the following: *megatherms*, plants favoring warm regions; *microtherms*, favoring cold regions; and *mesotherms*, favoring regions of intermediate temperatures.

It is a law of bioclimatology that there is a critical level of climatic stress beyond which a plant species cannot survive; hence that there will exist a geographical boundary that will mark the limits of its distribution. Such a boundary may also be referred to as a *frontier*. Although the frontier is determined by a complex of climatic elements, it is sometimes possible to single out one climatological element that coincides with the plant frontier. An example is seen in the limits of growth of the sugar maple (*Acer saccharophorum*) in North America (Figure 15.9). Here the boundaries on the north, west, and south are found to coincide roughly with selected values of annual precipitation, mean annual minimum temperature, and mean annual snowfall. Another example is seen in the distribution of the yellow pine (*Pinus ponderosa*) of western North America (Figure 15.10). In this mountainous region annual rainfall varies sharply with elevation. The 20 in (50 cm) isohyet of annual total precipitation encloses most of the upland areas having the yellow pine. It is the parallelism of the isohyet with forest boundary that is significant, rather than actual degree of coincidence.

Wind is seen as an important environmental factor in the structure of vegetation in highly exposed positions. Close to timber line in high mountains and along the northern limits of tree growth in the arctic zone trees will be found to be deformed so that the branches project from the lee side of the trunk only (flag shape), or the trunk and branches are bent to near-horizontal attitude, facing away from the prevailing wind direction (Figure 15.11). In such habitats the effect of wind is to cause excessive drying on the exposed side of the plant. The tree limit on mountainsides thus varies in elevation with degree of exposure of the slope to strong prevailing winds and will extend higher on lee slopes and in sheltered pockets.

Geomorphic factors

Geomorphic, or landform, factors influencing plant forms include such elements as *slope steepness* (the angle which the ground surface makes

Figure 15.10   Areas of yellow pine (*Pinus ponderosa*) in western North America are shown in solid black. Edge of the shaded area represents the isohyet of 20 in (50 cm) annual precipitation. (Based on *Biogeography—An Ecological Perspective*, by Pierre Dansereau. Copyright © 1957, The Ronald Press Company.)

with respect to the horizontal), *slope aspect* (the orientation of a sloping ground surface with respect to geographic north), and *relief* (the difference in elevation of divides and adjacent valley bottoms). In a much broader sense, the geomorphic factor includes the entire sculpturing of the landforms of a region by processes of erosion, transportation and deposition by streams, waves, wind, and ice, and by forces of vulcanism and mountain building. The unique landform assemblage found in any one region is understood in terms of geomorphic processes; these are treated in Chapters 19 through 28. An infinite variety of plant habitats can be ascribed to the geomorphic processes and their individual landforms.

Slope steepness acts indirectly by its influence upon the rate at which precipitation is drained from the surface. On steep slopes surface runoff is rapid and the water does not long remain available to plants. On gentle slopes much of the precipitation can penetrate the soil and become available for prolonged plant use. More rapid erosion on steep slopes may result in thin soil, whereas that on gentler slopes is thicker. Slope aspect has a direct influence upon plants by increasing or decreasing the exposure to sunlight and to prevailing winds. Slopes facing the sun have a warmer, drier environment than slopes facing away from the sun and therefore lying in shade for much longer periods of the day. In middle latitudes these slope-aspect contrasts may be so strong as to produce quite different plant formations on north-facing and south-facing slopes (Figure 15.12.)

Geomorphic factors are in part responsible for the dryness or wetness of the plant habitat within a region having essentially the same overall climate. Each plant community has its own microclimate. Upon divides, peaks, and ridge crests the soil tends to dryness because of rapid drainage of water away from such places and because the surfaces are more exposed to sunlight and to drying winds. By contrast, the valley floors tend to wetness because surface runoff over the ground and into streams causes water to converge there. In humid climates the ground water table may lie close to the ground in the valley floors or may actually coincide with the ground surface to produce marshes and swamps. Watertable ponds and bogs may also occur. Hygrophytic plants thus form distinctive plant communities in the valley floors in humid climates while at the same time mesophytic or xerophytic communities occupy the intervalley surfaces.

Edaphic factors

Edaphic factors are those related to the soil. In Chapter 13 principles of soil development (pedogenesis) have been taken up systematically. In

Figure 15.11 Trees deformed by the effects of cold, dry winds. Timberline scene in Arapaho National Forest, Colorado, 1946. (U.S. Forest Service photograph.)

terms of plant geography we can look at soils in two perspectives. One views broad patterns of the great soil groups, reflecting the pedogenic regimes of podzolization, laterization, calcification, gleization, and salinization. These pedogenic regimes are largely controlled by the climatic regimes and will be found closely correlated with the global patterns of the vegetation formation classes. These broad relationships will be taken up together in Chapter 16. A second view point is in terms of plant habitats—the small-scale mosaic of place-to-place variations of the earth's surface. Here the edaphic factors also act as important controls.

Among the edaphic factors treated in Chapter 13 are: soil texture and structure, humus content; presence or absence of soil horizons; soil alkalinity, acidity, or salinity; and the activity of bacteria and animals in the soil.

Although this book treats the systematic principles of soil science ahead of those of natural vegetation, a good argument might be made for reversing this order of treatment on the grounds that vegetation plays a leading role in the development of soil characteristics. Given a barren habitat, recently formed by some geologic event such as the outpouring of lava or the emergence of a

Figure 15.12 Contrasts in vegetation on opposing valley-side slopes. The heavily wooded slope on the left faces northeast and is in shade during the afternoon. The sparsely wooded slope on the right, facing southwest, receives intense insolation when air temperatures are highest. Long Canyon of the Rio Hondo, Carson National Forest, New Mexico. (U.S. Forest Service photograph.)

coastal zone from beneath the sea, the gradual evolution of a soil profile goes hand in hand with the occupance of the habitat by a succession of plant communities. The plants profoundly alter the soil by such processes as contributing organic matter, or by producing acids which act upon the mineral matter. Animal life, feeding upon the plant life, also makes its contribution to physical and chemical processes of soil evolution. We will give further attention to this topic under the subject of dynamics of vegetation.

Biotic factors

The sustained activity within a particular plant community functioning as a stable ecosystem and the gradual modification of the vegetation at a given place with time throughout a succession of stages require that the plants and animals within that community contribute to its operation through their own physical and chemical processes and through their life cycles of growth and subsequent decay. So vast and complex are these biotic factors that we can do no more here than to suggest by examples what is meant by biotic influence.

Among the biotic factors may be mentioned the activity of bacteria in consuming the dead tissue of larger plants; of earthworms in altering and aerating the soil; of insects and herbivorous animals which attack and consume plants; and of insects and birds which perform such functions as pollination of plants and the dispersal of seeds. A whole realm of biotic influence lies in the diseases of plants, the spread of which may drastically alter the plant associations over wide areas. The geographer is particularly interested in the way in which large herds of grazing and browsing animals, whether wild or domesticated, have contributed to the development of certain formation classes of vegetation. To what extent, say, have herds of American bison been responsible for the maintenance of grasslands against the spread of woodland and forest?

Man is himself perhaps the most potent biotic factor influencing vegetation over the globe today. His role has been dominantly destructive with respect to the plant associations and formation classes that might otherwise be expected to cover the lands in response to the various factors of climate, soil, and geomorphology that we have thus far reviewed.

Dynamics of vegetation

One of the main themes of plant geography is that the vegetation at a particular place evolves with time, usually starting with very simple plant communities, then leading gradually to more complex communities, and ultimately to the establish-ment of a relatively stable plant community—the *climax*. Starting with a newly formed ground surface, or one that has been denuded of vegetation, the process of *succession* takes place, in which one plant community invades the area and is followed in turn by other plant communities in an orderly sequence, or *sere*, culminating in the vegetation climax.

A new site for the development of vegetation may have one of several origins: a sand dune, a sand beach, the surface of a new lava flow or of a freshly fallen layer of volcanic ash, or the deposits of silt on the inside of a river bend which is gradually shifting. Such a site will not have a true soil with horizons; rather it may be a lithosol—perhaps little more than a deposit of coarse mineral fragments. In other cases, such as that of the floodplain silt deposits, the surface layer may represent redeposited soil endowed with substantial proportions of soil colloids and bases. Ground surfaces from which vegetation has been destroyed by fire will have the soil profile largely intact.

The first stage of a succession is a *pioneer stage*, consisting of a few plant species unusually well adapted to adverse conditions of rapid water drainage and drying of soil, and to excessive exposure to sunlight, wind, and extreme ground and lower air temperatures. As these plants grow, their roots penetrate the soil; their subsequent decay adds humus to the soil. Fallen leaves and stems add an organic layer to the ground surface. Bacteria and animals begin to live in the soil in large numbers. Soon conditions are favorable for other plant species which *invade* the area and displace the pioneers. The new arrivals may be larger plant forms providing more extensive cover of foliage over the ground. In this case the climate near the ground, or *microclimate*, is considerably altered toward one of less extreme air and soil temperatures, higher humidities, and less intense insolation. Now still other species can invade and thrive in the modified environment.

When the succession has finally run its course, there will exist a stable community consisting of certain definite proportions of the various species, each of which contributes to the total structure of the vegetation. This climax stage represents an ideal model for the so-called natural vegetation of a region. Doubt exists as to whether a climax, unchanging with time, can truly be maintained. Some plant geographers consider that the climax must be followed by disturbance leading to a *regression*, or return to one of the earlier stages of the succession, so that a type of self-repeating cycle is the rule. There is also the possibility that climatic change prevents the climax from being maintained in one place.

One type of succession is that in which the

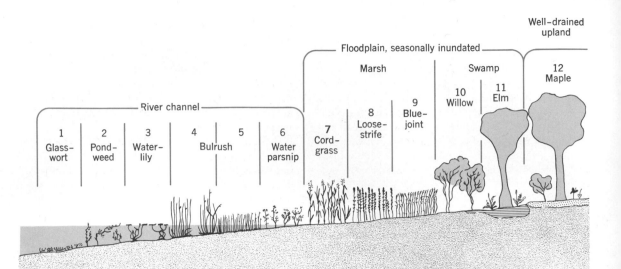

Figure 15.13 Allogenic succession on the bank of the St. Lawrence River. Twelve vegetation belts, each characterized by a plant association, shift gradually from right to left as bank deposition continues. (After Pierre Dansereau, 1956, *Revue Canadienne de Biologie*, Vol. 15.)

normal geomorphic processes build new ground continuously, as on a floodplain, delta, or sandspit. The succession associated with such continuous accretion of ground is described as *allogenic*. An example is shown in Figure 15.13, representing the zonation of vegetation along the St. Lawrence River. The twelve zones shift gradually to the left as silting by the river raises the level of the ground surface and shifts the water line to the left. Here we see the hygrophytes, living largely submerged, giving way to plants which thrive under conditions of intense sunlight on ground that is alternately exposed and inundated. Higher upon the bank are

Figure 15.14 Bog succession in Emmet County, Michigan. A bog mat lies at the edge of the lake, a black spruce forest in background. (Photograph by Pierre Dansereau.)

successive forest zones: willow, elm, and finally maple, representing the climax of mesophytic trees.

Another form of succession, described as *autogenic*, results from the alteration of the environment by the plants themselves, and not by outside agencies (such as the silt accumulation from flood stages of a river). The gradual covering of a sand dune by small plants, then by forest, would illustrate autogenic succession. Another illustration is provided by the evolution of a shallow pond or lake of glacial origin in the regions of cold continental climate such as prevail in Canada and northern Europe (Figure 15.14). This *bog succession* is illustrated by stages beginning with the lake as it was left following disappearance of glacial ice, perhaps 10,000 to 15,000 years ago. A remarkable feature of the bog succession is that the organic matter produced by the growth and partial decay of the plants accumulates to such thickness that the open water is actually replaced by an organic mass of a substance known as *peat*. At the water's edge is a zone of sedges, followed by rushes (Figure 15.15). These construct a floating layer that encroaches upon the open water. There follows a zone of sphagnum (peat moss), which eventually completely fills the lake. Now the peat deposit supports hygrophytic trees (largely spruce) which produce a woody peat. This community may in turn be replaced by mesophytic trees, marking the climax stage. In the shallower upland ponds, shown on either side of the profile, the mesophytic growth is achieved much earlier than over the larger water body.

In terms of the vast scope and depth of the

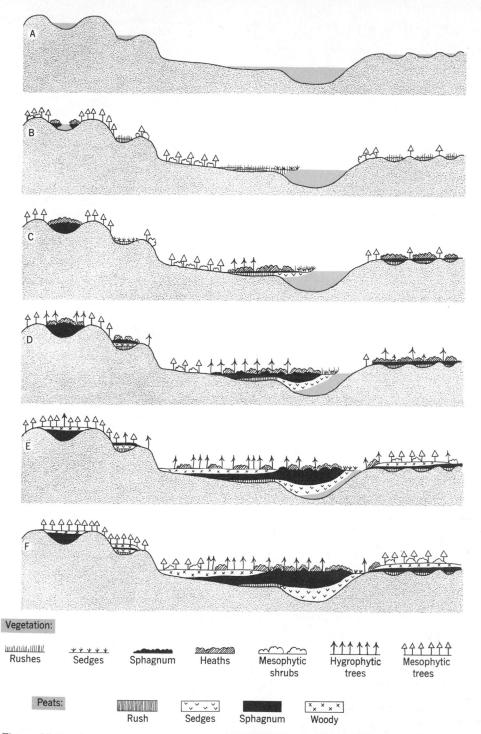

Vegetation:

| Rushes | Sedges | Sphagnum | Heaths | Mesophytic shrubs | Hygrophytic trees | Mesophytic trees |

Peats:

| Rush | Sedges | Sphagnum | Woody |

Figure 15.15 Autogenic bog succession typical of the Laurentian shield area of Canada. (*After Dansereau and Segadas-Vianna, 1952, Canadian Journal of Botany, Vol. 30.*)

science of plant geography, this chapter must be taken as a greatly simplified review touching upon only a few essential topics. As with all other subjects which we call collectively "physical geography," plant geography can be seriously ap-

proached only through intensive study of principles of botany, including the classification of plants (taxonomy), their evolution and floristic distribution, plant morphology, and the chemistry and physics of plant physiology.

REFERENCES FOR FURTHER STUDY

Dansereau, Pierre (1957), *Biogeography; an ecological perspective*, The Ronald Press, New York, 394 pp.

Daubenmire, R. F. (1959), *Plants and environment: a textbook of plant autecology*, second edition, John Wiley and Sons, New York, 422 pp.

Oosting, H. J. (1956), *The study of plant communities; an introduction to plant ecology*, second edition, W. H. Freeman, San Francisco, 440 pp.

Polunin, Nicholas (1960), *Introduction to plant geography and some related sciences*, McGraw-Hill Book Co., New York, 640 pp.

Robbins, W. W., T. E. Weier, and C. R. Stocking, (1964), *Botany: an introduction to plant science*, third edition, John Wiley and Sons, New York, 480 pp.

REVIEW QUESTIONS

1. How does vegetation act as a geographic factor? To what other branches of physical geography is plant geography most closely related?

2. What is meant by natural vegetation? What does a map of natural vegetation attempt to portray?

3. Compare floristic with structural approach in the study of natural vegetation. Which approach is most useful in physical geography and why?

4. What is bioclimatology? Define an ecosystem. What is the subject of plant ecology?

5. Name the three environmental divisions of the biosphere. Into what four great biochores is the land biocycle divided? Explain the basis of this subdivision.

6. What are formation-classes? How many formation classes are commonly recognized? On what basis are formation classes set up?

7. How does habitat influence the structure of vegetation? What six habitats can be recognized within the Canadian needleleaf forest? What is the cause of differing habitats? On the basis of what type of habitat are the formation classes defined? What is a plant community?

8. What six categories have been used to form a system of description of the structure of vegetation? Define each category and give the units of subdivision within each.

9. How is a tree distinguished from a shrub? What are lianas? What kind of plants fall within the category of herbs? What are forbs? Bryoids? Epiphytes? Lichens? What kinds of plants are included within the thallophytes?

10. Distinguish deciduous from evergreen plants. What are semideciduous plants? Name the principal leaf forms and give examples from common plants.

11. What are the varieties of leaf textures? What is a sclerophyll forest? What are succulent leaves?

12. List the four main classes of environmental factors in plant ecology. What two scale ranges of phenomena are included in the environment?

13. How do plants use water? What factors affect the rate of transpiration? How are plants adapted to reduce water loss? How do some plants improve their means of obtaining and storing water? What are phreatophytes?

14. Explain how plants are classified according to water need. Name and define the three plant groups so classified. What are tropophytes?

15. Discuss the climatic factors of light, air temperature, and wind as they apply to plant ecology. How do numbers of plant species in a region depend upon temperature conditions? Distinguish megatherms, microtherms, and mesotherms.

16. What bioclimatological law relates a plant species to the level of climatic stress imposed upon it? Give an example of a plant frontier.

17. Discuss the geomorphic factors in plant ecology and show how each factor acts to influence vegetation.

18. What are edaphic factors? What two points of view can be taken in treating the edaphic controls? How do plants themselves modify the soils upon which they grow?

19. Discuss the biotic factors in plant ecology. Give specific examples of the way in which animal life affects vegetation. What is the role of man as a biotic factor in our present day?

20. What is the process of succession in development of vegetation? Define sere and climax. What kinds of new sites may become available for plant successions? What is the pioneer stage? What is meant by a regression following climax?

21. Distinguish allogenic from autogenic succession and give one example of each type.

CHAPTER 16

Distribution of Natural Vegetation

THE principles of description of vegetation in terms of its structure, and the organization of vegetation into plant assemblages of various orders of magnitude (biochore, formation class, association, community), have been treated in Chapter 15. Using these principles in combination with our understanding of the climatic regimes, pedogenic regimes, and the soil-water budget, we are now prepared to analyze the world-wide distribution of vegetation and to explain its variations with latitude, continental position, and altitude.

The great biochores

All natural vegetation of the lands falls into four major structural subdivisions, the *biochores*, illustrated schematically in Figure 16.1. First is the *forest biochore*. We define a *forest* as a plant formation consisting of trees growing close together and forming a layer of foliage that largely shades the ground. Forests often show stratification, with more than one layer. Shading of the ground gives a distinctly different microclimate than that which would be found over open ground. Forests require a relatively large annual precipitation, but it does not need to be uniformly distributed throughout the year. No single value of precipitation can be stated because the effectiveness of the precipitation depends upon the water loss by evapotranspiration, and this in turn depends upon air temperature and humidity. Consequently, the forest biochore spans a great climatic range, from wet equatorial to cold subarctic.

The *savanna biochore* is a formation consisting of a combination of trees and grassland in various proportions. The appearance of the vegetation can be described as parklike, with trees spaced singly or in small groups and surrounded by, or interspersed with, surfaces covered by grasses, or by some other plant life form, such as shrubs or annuals in a low layer. The savanna biochore indicates a climate of limited total annual precipitation with an uneven distribution throughout the year.

The *grassland biochore* consists of an upland vegetation largely or entirely of herbs, which may include grasses, grasslike plants, and forbs (broadleaf herbs). Degree of coverage may range from continuous to discontinuous and there may be stratification. The grassland biochore may include trees in moister habitats of valley floors and along stream courses where ground water is available. The grassland biochore is typical of a climate which has small total annual precipitation, but otherwise the climate may range from one of extreme heat to one of extreme cold.

The *desert biochore*, associated with climates of extreme aridity, has thinly dispersed plants and hence a high percentage of bare ground exposed to direct insolation and to the forces of wind and water erosion or to freeze-thaw action. Although essentially treeless, the desert biochore may have scattered woody plants. Typically, however, the plants are small, e.g., herbs, bryoids, lichens. Because the desert biochore includes climates ranging from extremely hot tropical desert to extremely cold arctic desert, a great range in plant communities and habitats is spanned by the biochore.

In describing the four great biochores, emphasis has been placed on the vast range of climates spanned by each. Essentially, the biochores are determined by the degree to which moisture is available to plants in a scale ranging from abundant (forest biochore) to almost none (desert biochore). But within each biochore conditions of temperature are vastly different from low to high latitudes and from low to high altitudes. Consequently, there is need to subdivide each biochore into a number of formation classes.

The formation classes

The description of world vegetation in terms of formation classes was first developed fully by Professor A. F. W. Schimper whose monumental work in two volumes entitled *Plant Geography on a Physiological Basis* was first published in German in 1903. This work was subsequently revised by Professor F. C. von Faber and published in 1935. A somewhat similar approach to structural classification of world vegetation was taken by Professor Eduard Rübel who published a major volume in 1930. The classification system described below follows in many respects the recent work of Professor Pierre Dansereau [1] and is based

[1] Dansereau, Pierre (1957), *Biogeography, an ecological perspective*, The Ronald Press Co., New York, 394 pp.

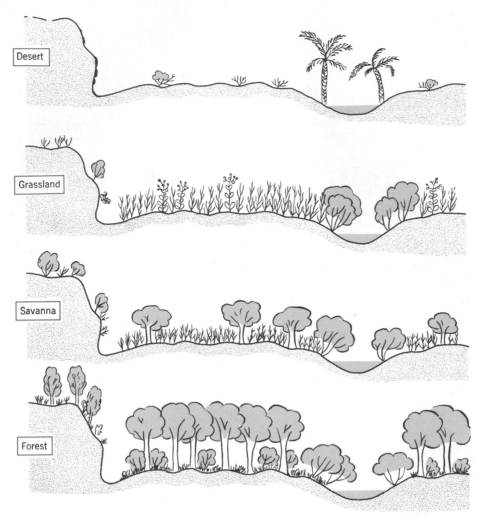

Figure 16.1 Diagrammatic representation of the four great biochores. Well-drained upland is shown in the middle of each profile; steep cliffs on the left; wet or poorly drained ground on the right. Note that the vertical scale of the grassland biochore is exaggerated in comparison with the other three. (From *Biogeography—An Ecological Perspective*, by Pierre Dansereau. Copyright © 1957, The Ronald Press Company.)

essentially on Schimper's and Rübel's principles. The world map of vegetation, Plate 4, follows Rübel's system in showing the distribution of ten map units, designated here by the letters *A* through *J*. The eighteen formation classes listed in this chapter are all included within Rübel's ten map units so that in certain cases two to four formation classes are combined into a single map unit. The map is thus simplified in appearance and brings out broad world patterns that are obviously related to the climate regimes.

Table 16.1 lists eighteen formation classes within the four biochores and gives the corresponding world map units of Plate 4. On the map legend is a similar table in reverse, giving the ten map units and listing the formation classes which are included in each.

Forest biochore

1. *Equatorial rainforest*[2] consists of tall, closely set trees whose crowns form a continuous canopy of foliage and provide dense shade for the ground and lower layers (Figure 16.2). The trees are characteristically smooth barked and unbranched in the lower two-thirds. Trunks commonly are buttressed at the base by radiating, wall-like roots (Figure 16.3). Tree leaves are large and evergreen; from this characteristic the equatorial rainforest is often described as "broadleaf evergreen forest." Crowns of the trees tend to form into two or three layers, or strata, of which the highest layer consists of scattered emergent crowns rising to 130 ft (40 m) and protruding conspicuously above a second

[2] Called *tropical rainforest* by Dansereau and others.

TABLE 16.1 THE FORMATION CLASSES

I Forest Biochore	Equivalent map units (Plate 4)	
1. Equatorial rainforest		
2. Tropical rainforest	A.	Equatorial and tropical rainforest
3. Monsoon forest	C.	Raingreen forest, woodland, scrub, and savanna.
4. Temperate rainforest	B.	Temperate rainforest or Laurel forest
5. Summergreen deciduous forest	E.	Summergreen deciduous forest
6. Needleleaf forest	F.	Needleleaf forest
7. Evergreen-hardwood forest (Sclerophyll forest)	D.	Evergreen-hardwood forest (Sclerophyll forest)

II Savanna Biochore		
8. Savanna woodland		
9. Thornbush and tropical scrub	C.	Raingreen forest, woodland, scrub, and savanna
10. Savanna		
11. Semidesert	H.	Dry desert and semidesert
12. Heath		(Heath not shown on map)
13. Cold woodland	I.	Tundra

III Grassland Biochore		
14. Prairie		
15. Steppe	G.	Steppe and Prairie grasslands
16. Grassy tundra	I.	Tundra

IV Desert Biochore		
17. Dry desert	H.	Dry desert and Semidesert
18. Arctic fell field	I.	Tundra

layer, 50 to 100 ft (15 to 30 m) which is continuous (Figure 16.4). A third, lower layer consists of small, slender trees 15 to 50 ft (5 to 15 m) high with narrow crowns (Figure 16.4).

Typical of the equatorial rainforest are lianas, thick woody vines supported by the trunks and branches of trees. Some are slender, like ropes, others reach thicknesses of 8 in (20 cm). They rise to heights of the upper tree levels where light is available and may have profusely branched crowns. Lianas may depend upon a growing tree to be carried upward, where the liana has no devices with which to climb by itself. Other woody climbers rise by winding about the tree trunk. Epiphytes are numerous in the equatorial rainforest. These plants are attached to the trunk, branches and foliage of trees and lianas, using the "host" solely as a means of physical support. Epiphytes are of many plant classes and include ferns, orchids, mosses, and lichens (Figure 16.5). Some epiphytes are *stranglers*. These woody vines send down their roots to the soil and may eventually surround the tree, perhaps ultimately replacing it. The *strangling fig* (*Ficus*) is an example (Figure 16.6). Other stranglers begin as lianas.

A particularly important botanical characteristic of the equatorial rainforest is the large number of species of trees that coexist. It is said that as many as 3000 species may be found in a square mile. Individuals of a species are thus widely separated. Consequently, if a particular tree species is to be extracted from the forest for commercial uses, considerable labor is involved in seeking out the trees and transporting them from their isolated positions. Representative trees of the rainforest of the Amazon valley are the Brazilnut (*Bertholletia excelsa*) and the silk-cotton tree (species of *Bombax*).

The floor of the equatorial rainforest is usually so densely shaded that plant foliage is sparse close to the ground and gives the forest an open aspect, making it easy to traverse. The ground surface is covered only by a thin litter of leaves. Rapid consumption of dead plant matter by bacterial action results in the absence of humus upon the soil surface and within the soil profile. As explained in Chapter 13, these conditions are typical of the pedogenic process of laterization, with which the rainforest is identified.

Equatorial rainforest is a response to an equable climatic regime which is continuously warm, frost-free, and has abundant precipitation in all months of the year (or, at most, only one or two dry months). A large water surplus characterizes the annual water budget (see Figure 9.9A), so that soil moisture is adequate at all times and the export of large amounts of stream runoff allows permanent removal of bases and silica from the soils of the region. In the absence of a cold or dry season, plant growth goes on continuously throughout the year. Individual species have their own seasons of leaf shedding, possibly caused by slight changes in the light period.

Variations in the equatorial rainforest structure are found in specialized habitats and where man has disturbed the vegetation. Where the forest has been cleared by cutting and burning (as for small-plot agriculture or highways), the returning plant growth is low and dense and may be described as *jungle*. Jungle can consist of a tangled growth of lianas, bamboo scrub, thorny palms, and thickly branching shrubs, constituting an impenetrable

Figure 16.3 Buttress roots at the base of a large tree (*Bombacopsis fendleri*) of the rainforest on Barro Colorodi Island, Canal Zone. (Courtesy of the American Museum of Natural History.)

barrier to travel, in contrast to the open floor of the climax rainforest.

Coastal vegetation in areas of equatorial rainforest is highly specialized. Coasts which receive suspended sediment (mud) from river mouths, and where water depths are shallow, typically have *mangrove swamp forest*, consisting of stilted trees (Figure 16.7). The mangrove prop-roots serve to trap sediment from ebb and flood tidal currents, so that the land is gradually extended seaward. Mangroves commonly consist of several shoreward belts of the red (*Rhizophora*), the black (*Avicennia*), and the white (*Laguncularia*) mangrove. Another common coastal salt-marsh plant of the humid tropics is the screw-pine (*Pandanus*). Typical of recently formed coastal deposits are belts of palms, such as the cocoanut palm (*Cocos nucifera*) (Figure 16.8).

World distribution of equatorial rainforest is shown on the vegetation map, Plate 4, by those areas of rainforest (Class *A*) located close to the equator. The principal world areas are: the Amazon lowland of South America; the Congo lowland of Africa and a coastal zone extending westward from Nigeria to Guinea; and the East Indian region, from Sumatra on the west to the islands of the western Pacific on the east. Poleward borders of these equatorial rainforest regions are, of course, transitional into rainforest of higher latitudes, particularly that found on tropical windward coasts and that of coastal monsoon rainforest belts of south and southeast Asia.

2. *Tropical rainforest* is in many respects structurally similar to equatorial rainforest but has distinct differences imposed upon it by its location, which is on windward coasts, roughly from 10° latitude to the tropics of Cancer and Capricorn

Figure 16.2 Equatorial rainforest near Belém, Brazil. This forest of tall broad-leaved evergreen trees and numerous lianas is on the relatively high ground (*terra firme*) of the Amazon lowland. (Photograph by Otto Penner, courtesy of Instituto Agronómico do Norte.)

Figure 16.4   Diagram of the structure of tropical rainforest in Trinidad, British West Indies. A representative species of the tallest trees is *Mora exelsa*. (After J. S. Beard, 1946, *The Natural Vegetation of Trinidad*, Clarendon Press, Oxford.)

Figure 16.5   A tree limb covered by epiphytic ferns. Upland rainforest of Kenya, Africa. (Courtesy of the American Museum of Natural History.)

Figure 16.6   This strangler fig has surrounded the trunk of a cabbage palm in southernmost Florida. (Photograph by Bob Haugen, Everglades National Park.)

(23½°). Here we find a distinct annual precipitation cycle, consisting of a long wet season alternating with a season of reduced rainfall, if not actual drought. There is also a marked annual temperature cycle resulting from the variations in height of the sun's path in the sky in tropical latitudes. The cooler temperatures, coinciding approximately with the period of reduced rainfall, impose some stress upon the plants. As a result, there are fewer species and fewer lianas. Epiphytes are, however, abundant because of continued exposure to humid air and cloudiness of the maritime tropical air masses which impinge upon the coastal hill and mountain slopes.

In terms of global distribution, the tropical rainforest is represented on the world map (Plate 4) by those areas of rainforest (Class *A*) which lie between 10° and 25° latitude. The Carribean lands represent one important area of tropical rainforest, although the rainforest is predominantly limited to windward locations. Of interest to American students is the fact that the Everglades of southern-most Florida contains small isolated communities of tropical rainforest, termed *hammocks*, in which one finds the mahogany tree and strangler fig, along with abundant epiphytes. The coast of this same area has extensive mangrove forest and scrub.

In southern and southeastern Asia tropical rain-forest is extensive in coastal zones and highlands which have heavy monsoon rainfall and a very short dry season. The Western Ghats of India and the coastal zone of Burma have tropical rainforest supported by orographic rains of the southwest monsoon. In the zone of combined northeast trades and Asiatic summer monsoon are rainforests of the coasts of Vietnam and the Philippine Islands. In the southern hemisphere belts of tropical rain-forest extend down the eastern Brazilian coast, the Madagascar coast, and the coast of northeastern Australia.

As is the equatorial rainforest, the tropical rainforest is associated with a pedogenic regime of laterization under an equable climatic regime. It is common practice among plant geographers to combine the equatorial and tropical rainforests into a single formation class, because their points of similarity greatly outweigh their points of difference.

Figure 16.8 A grove of cocoanut palms (*Cocos nucifera*) along a sandy beach in the Solomon Islands. Wave action is undermining the trees. (Courtesy of the American Museum of Natural History.)

3. *Monsoon forest* presents a more open tree growth than the equatorial and tropical rainforests. Consequently, there is less competition among trees for light but a greater development of vegetation in the lower layers. Maximum tree heights range from 40 to 100 ft (12 to 35 m), which is less than in the equatorial rainforest. Many tree species are present and may number 30 to 40 species in a small tract. Tree trunks are massive; the bark is often thick and rough. Branching starts at comparatively low level and produces large round crowns. Perhaps the most important feature of the monsoon forest is the deciduousness of most of the tree species present, e.g., the abundance of trophophytes. The shedding of leaves results from the stress of a long dry season which occurs at time of low sun and cooler temperatures. Thus the forest in the dry season has somewhat the dormant winter aspect of deciduous forests of middle latitudes (Figure 16.9). Some writers use the name "tropical deciduous forest" for monsoon forest, emphasizing the deciduousness rather than the climate regime. A representative example of a monsoon forest tree is the teakwood tree. (*Tectona grandis*).

Lianas and epiphytes are locally abundant in monsoon rainforest but are fewer and smaller than in the equatorial rainforest. Undergrowth is often a dense shrub thicket. Where second-growth vegetation has formed, it is typically jungle. Clumps of bamboo are an important part of the vegetation in climax teakwood forest.

As already noted, monsoon forest is a response to a wet-dry tropical climate regime in which a long rainy season with a large total precipitation alternates with a dry, rather cool season. Such

Figure 16.7 Mangrove growing in salt water, Harney's River, Florida. (Courtesy of the American Museum of Natural History.)

conditions are most strongly developed in the Asiatic monsoon climate, but are not limited to that area. Perhaps the type regions of monsoon forest are in Burma (inland from the coastal tropical rainforest belt) and in Thailand and Cambodia. Large areas of deciduous and semideciduous tropical forest occur in west Africa and in central and South America, bordering the equatorial and tropical rainforests into which there is a gradation. Areas of monsoon forest or related types are also described in Indonesia (especially Java and Celebes), in northern Australia, and in western Madagascar. World vegetation maps and standard reference works do not agree closely on the location and extent of the monsoon, or deciduous tropical forest areas. In Plate 4, monsoon forests are included in Class *C*, the raingreen vegetation of the wet-dry tropics, which also represents other formation classes.

The prevailing pedogenic regime of monsoon forest areas is that of laterization. Despite the dry season, a substantial water surplus is developed during the warm rainy season. Humus does not accumulate; leaching of bases and silica is the dominant soil-forming process.

4. *Temperate rainforest*, also referred to as *temperate evergreen forest* and *laurel forest*, differs from the equatorial and tropical rainforests in having relatively few species of trees, and hence large populations of individuals of a species. Trees are not as tall as in the low-latitude rainforests; the leaves tend to be smaller and more leathery; the leaf canopy less dense. Among the trees commonly found in the temperate rainforests of southern Japan and the southeastern United States are evergreen oaks (such as *Quercus virginiana*) and members of the laurel and magnolia families (such as *Magnolia grandiflora*). A quite different temperate rainforest flora is found in New Zealand and consists of large tree ferns, large conifers such as the kauri tree (*Agathis australis*), podocarp trees (*Podocarpus*), and small-leaved southern beeches (*Nothofagus*) (Figure 16.10). Another important type of temperate rainforest, found in the Azores and Canary island groups, is the Canary laurel forest, which formerly covered Europe in the Miocene geological epoch.

Temperate rainforests tend to have a well-developed lower stratum of vegetation that, in different places, may include tree ferns, small palms, bamboos, shrubs, and herbaceous plants. Lianas and epiphytes are abundant. Particularly striking at higher elevations where fog and cloud are persistent is the sheathing of tree trunks and branches by mosses. An example of conspicuous epiphyte accumulation at low elevation is the Spanish "moss" (*Tillandsia usneoides*) which festoons the Evangeline oak, bald cypress, and other trees of

Figure 16.9 Monsoon forest in Chieng Mai Province, northern Thailand. Scale is indicated by a line of porters crossing the clearing. (Photograph by Robert L. Pendleton, courtesy of the American Geographical Society.)

the gulf coast of the southeastern United States.

Temperate rainforest represents a response to an equable climatic regime in which the annual range of temperature is small or moderate and rainfall is abundant and well distributed throughout the year. Such conditions are met in three quite different geographical locations: (1) at higher altitudes in the equatorial and tropical zones; (2) along eastern continental margins and on islands in the latitude belt 25° to 35° or 40°; (3) on west coasts from 35° to 55°. In the first location the effect of higher elevation is to reduce temperatures and evaporation and thereby to increase the moisture available to plants.

Figure 16.10 This podocarp forest illustrates a variety of temperate rainforest in a marine west-coast climate, Hari Hari, western coast of South Island of New Zealand. (Photograph by Pierre Dansereau.)

A word about the vegetation of the coastal region of the southeastern United States may save misunderstanding. On forest maps of the United States (Figure 14.3), and on a number of maps of world vegetation, this coastal zone is shown as having needleleaf evergreen or coniferous forest, whereas on Plate 4, it is shown as temperate rainforest. It is true that large areas of sandy upland bear forests of loblolly and slash pine and that bald-cypress is a dominant tree in swamps, but such vegetation represents xerophytic and hydrophytic forms in excessively dry or wet habitats, or the second-growth forest following fire and deforestation. The climax vegetation of mesophytic habitats is nevertheless the evergreen-oak and magnolia forest.

Temperate rainforest spans two pedogenic regimes: laterization and podzolization. In lower latitudes laterization is characteristic, and this grades through regimes in which both laterization and podzolization are effective (red-yellow soils), to the cool higher latitude regions of podzolization.

5. *Summergreen deciduous forest*, sometimes called *temperate deciduous forest*, is familiar to inhabitants of eastern North America and western Europe as a native forest type. It is dominated by tall, broadleaf trees which provide a continuous and dense canopy in summer but shed their leaves completely in the winter (Figure 16.11). Lower layers of small trees and shrubs are weakly developed. In spring a luxuriant low layer of herbs quickly develops, but this is greatly reduced after the trees have reached full foliage and shaded the ground.

Summergreen deciduous forest is almost entirely limited to the middle-latitude landmasses of the northern Hemisphere (Plate 4, Class *E*). Common trees of the deciduous forests of eastern North America, southeastern Europe, and eastern Asia (all in the humid-continental climate) are oak (*Quercus*), beech (*Fagus*), birch (*Betula*), hickory (*Carya*), walnut (*Juglans*), maple (*Acer*), basswood (*Tilia*), elm (*Ulmus*), ash (*Fraximus*), tulip tree (*Liriodendron*), sweet chestnut (*Castanea*), and hornbeam (*Carpinus*). In western and central Europe, under a marine west coast climate dominant trees are mostly oak and ash, with beech (*Fagus sylvatica*) in cooler and moister areas.

In poorly drained habitats, the deciduous forest consists of such trees as alder, willow, ash, and elm, and many hygrophytic shrubs. Where the deciduous forests have been cleared in lumbering, pines readily develop as second-growth vegetation.

The summergreen deciduous forest represents a response to a continental climatic regime which at the same time receives adequate precipitation in all months. There is a strong annual temperature cycle with a cold winter season and a warm summer. Precipitation is markedly greater in the summer months (especially so in eastern Asia) and thus increases at the time of year when evapotranspiration is great and the moisture demands are high. Only a small water deficit is incurred in the summer, while a large surplus normally develops in spring. In eastern Asia the winter is exceptionally dry, but this factor is compensated for by cold.

The pedogenic process associated with summergreen deciduous forest is that of podzolization, but moderated by the warm wet summers. As a result, the soils are characteristically of the gray-brown forest group. Toward lower latitudes the tendency to laterization becomes stronger and red-yellow podzolic soils are encountered; toward the continental interiors the tendency to calcification sets in and the deciduous forest extends into regions of darker soils of the grasslands (prairie and cher-

Figure 16.11 This stand of beech and hemlock in the Allegheny National Forest, Pennsylvania, illustrates mixed summergreen deciduous forest and needleleaf forest in the northeastern United States. (U.S. Forest Service photograph.)

nozem soils). In the summergreen deciduous forests a thick layer of fallen leaves covers the ground and there is abundant humus in the soil.

Because the regions of summergreen deciduous forest have for centuries been the areas of dense populations, few remnants of a primeval forest have survived. Instead, most existing forests are modified by practices of tree farming. Large areas are completely and permanently removed from forest growth by farming, urban development, and roadways.

6. *Needleleaf forest* is composed largely of straight-trunked, conical trees with relatively short branches, and small, narrow, needlelike leaves. These trees are conifers (Figure 16.12). Where evergreen, the needleleaf forest provides continuous and deep shade to the ground so that lower layers of vegetation are sparse or absent, except for a thick carpet of mosses in many places. Species are few and large tracts of forest consist almost entirely of but one or two species.

Reference to the world vegetation map (Plate 4, Class *F*) will show that needleleaf forest is predominantly in two great continental belts, one in North America and one in Eurasia, which span the landmasses from west to east in latitudes 45° to 75°. The needleleaf forest of North America, Europe, and western Siberia is composed of evergreen conifers, such as spruce (*Picea*), fir (*Abies*), and pine (*Pinus*), whereas that of north-central and eastern Siberia is dominantly of larch (*Larix*) which sheds its needles in winter and comprises a deciduous forest (Figure 16.13). Associated with the needleleaf trees is mountain ash. Aspen and balsam poplar (*Populus*), willow, and birch tend to take over rapidly in areas of needleleaf forest

Figure 16.13 Woodland of larch (*Larix dahurica*), Tompo River region, about latitude 64° N, Yakutsk, A.S.S.R. This scene shows the larch tree near the northern limit of its growth. (Photograph by I. D. Kild'ushevsky, courtesy of Professor B. A. Tikhomirov, Komarov Botanical Institute, Leningrad.)

which have been burned over, or can be found bordering streams and in open places.

Needleleaf evergreen forest extends into lower latitudes wherever mountain ranges and high plateaus exist. Thus in western North America this formation class extends southward into the United States on the Cascade, Sierra Nevada, and Rocky Mountain ranges and over parts of the higher plateaus of the southwestern states. In Europe, needleleaf evergreen forests flourish on all of the higher mountain ranges, and well into Scandinavia.

The needleleaf evergreen forests of British Columbia and California are of particular interest. Here, under a regime of heavy orographic precipitation and prevailing high humidities, are perhaps the densest of all coniferous forests, with the world's largest trees. Forests of coastal redwood (*Sequoia sempervirens*), big tree (*Sequoiadendron giganteum*) and Douglas fir (*Pseudotsuga taxifolia*) are particularly remarkable (Figure 16.14). Individuals of redwood and big tree attain heights of over 325 ft (100 m) and girths of over 65 ft (20 m) (Figure 16.15).

As much of the area of needleleaf evergreen forest in North America and Europe was subjected to glaciation in the Wisconsin stage of the Pleistocene epoch, there abound lakes and poorly drained depressions. These bear the hygrophytic vegetation, described in Chapter 15 as forming a bog succession, and leading to large, thick peat accumulations known in Canada as *muskeg*.

Podzolization, the pedogenic regime of the needleleaf forests, is a direct result of a climatic

Figure 16.12 Spruce forest, Upper Peribonka River, Quebec, Canada. (Photograph by Pierre Dansereau.)

regime of continentality with a strong polar influence. As explained in Chapter 13, the conifers require little of the bases and these are leached from the soil, which is characteristically acidic.

7. *Evergreen-hardwood forest*, also termed *sclerophyll forest*, consists of low trees with small, hard, leathery leaves. Typically the trees are low-branched and gnarled, with thick bark. The formation class includes much *woodland*, an open forest in which the canopy coverage is only 25 to 60%. Also included are extensive areas of *scrub*, a plant formation type consisting of shrubs having

Figure 16.15　A grove of Sequoia trees, Sequoia National Park, California. The largest tree measures 51 ft (15m) in circumference. A man can be seen standing at the right of this tree (look for hat). (Courtesy of the American Museum of Natural History.)

Figure 16.14　This mixed stand of white pine. Douglas fir, and Englemann spruce is on the western slopes of the Cascade Range in Washington. (U.S. Forest Service photograph.)

a canopy coverage of perhaps 50% (Figure 16.16). The trees and shrubs are evergreen, their thickened leaves being retained despite a severe annual drought. There is little stratification in the sclerophyll forest and scrub, although there may be a spring herb layer.

Evergreen hardwood forest is closely associated with the dry-summer subtropical (Mediterranean) climate and hence quite narrowly limited in geographical extent—primarily to west coasts between 30° and 40° or 45° latitude (Plate 4, Class *D*). In the Mediterranean lands the hardwood forest forms a narrow peripheral coastal belt. Here the woodland consists of such trees as cork oak (*Quercus suber*), live oak (*Quercus ilex*), Aleppo pine (*Pinus halepensis*), stone pine (*Pinus pinea*), and olive (*Olea europaea*). What may have once been luxuriant forests of such trees were greatly disturbed by man over the centuries and reduced to woodland or entirely destroyed. Instead, large areas consist of dense scrub termed *maquis* which includes many species, some of them very spiny. The other northern hemisphere region of evergreen hardwood forest is that of the California coast ranges. Some of this is a woodland composed largely of the live oak (*Quercus agrifolia*) and white oak (*Quercus lobata*). Much of the vegetation is a scrub or "dwarf forest" known as *chaparral*, which varies in composition with elevation and exposure (Figure 16.17). Chaparral may contain wild lilac (*Ceanothus*), manzanita (*Arctostaphylos*), mountain mahogany (*Cercocarpus*), "poison oak" (*Rhus diversiloba*), and live oak. The evergreen hardwood forest is represented in central Chile and in the Cape region of South Africa by a maquislike scrub vegetation which is, however, of quite different flora from those of the northern hemisphere. Important areas of sclerophyll forest, woodland, and scrub are found in southeast, southcentral, and southwest Australia, including several species of eucalypts and acacias (Figure 16.18).

The dry-summer subtropical (Mediterranean) regime of the evergreen hardwood forest is one of great environmental stress because the severe drought season coincides with high air temperatures. Thus a large water deficit occurs in the summer (see Figure 9.9*E*). The wet, mild winter is, by contrast, highly favorable to rapid plant growth. The pedogenic regime is one associated with semi-aridity, namely calcification, and results in soils with some precipitated calcium carbonate in the *B* horizon. Much of the area of the evergreen hardwood forest, woodland, and scrub is classified as having reddish-chestnut, reddish-prairie, and reddish-brown soils, and in the Mediterranean lands, *terra rossa*.

Savanna biochore

8. *Savanna woodland* consists of trees spaced rather widely apart, permitting development of a dense lower layer, which may be of grasses or shrubs. This formation class is sometimes referred to as "parkland" because of the open, parklike appearance of the vegetation. Savanna woodland is associated with a climate regime in which aridity is sufficiently developed to prevent the tree growth from forming a closed canopy.

Many geographers associate savanna woodland closely with the tropical wet-dry climate. It is this tropical variety of woodland that is implied in the world vegetation map (Plate 4) by the inclusion of savanna woodland with Class *C* vegetation. However, from the standpoint of the vegetation structure itself, the definition of savanna woodland can be broadened to include woodlands of middle latitudes, such as the open stands of yellow pine and of pinyon pine and juniper which, in the western United States, occur in an elevation zone

Figure 16.16 Chaparral on steep mountain slopes of the San Dimas Experimental Forest, near Glendora, California, 1953. This vegetation has been protected from burning for many years. (Photograph by A. N. Strahler.)

above that of the sagebrush scrub and below that of the needleleaf forests. One might also include in this formation class the Eucalyptus woodlands of southern Australia, which we have elsewhere placed in the formation class of evergreen-hardwood forest.

Figure 16.17 Detail of chaparral vegetation, San Dimas Experimental Forest (see Figure 16.16). A tilted rain gauge, 8 in (20 cm) in diameter, rests on bare ground at the left. (Photograph by A. N. Strahler.)

Figure 16.18 Eucalyptus woodland, Darling Range, Western Australia, about 1930. (Photograph by Agent General for Western Australia, courtesy of the American Geographical Society.)

Referring now to the tropical savanna woodland, the trees are of medium height, the crowns flattened or umbrella-shaped, and the trunks have thick, rough bark (see Figure 16.20). The trees tend toward xerophytic forms with small leaves and thorns, or may be deciduous, shedding their leaves in the dry season. In this respect, tropical savanna woodland is closely akin to monsoon forest, into which it grades. The trees are of species capable of withstanding fires which sweep through the lower layer in dry season; thus many rainforest tree species which might otherwise grow in the wet-dry climate regime are prevented by fires from invading.

Tropical savanna woodland is found widely throughout Africa, South America, southeast Asia, northern Australia, and in Central America and the Carribean Islands (Plate 4, Class C). The pedogenic process usually dominant is laterization.

9. *Thornbush and tropical scrub* (which may be treated as two different formation classes) consist of xerophytic trees and shrubs responding to a climate with a very long dry season and only a short, but intense, rainy season. Thornbush, also referred to as "thorn forest," and "thornwoods," consists of tall, closely spaced woody shrubs commonly bearing thorns and largely deciduous (Figure 16.19). Cacti may also be present. The lower layer of herbs may consist of annuals which largely disappear in the dry season, or of grasses. In the drier areas the lower layer may consist only of scattered grass clumps with much bare ground

between. One example of the thornbush is the *caatinga* of northeastern Brazil, an open thorn forest of the dry highlands. Another is the *dornveld* of South Africa.

Tropical scrub is a dense growth of low woody shrubs which may occur in patches or clumps separated by barren ground. Scrub may develop in stony, sandy, or gravelly sites in areas of thornbush.

Thornbush and tropical scrub have been included in Class C of the world vegetation map (Plate 4) as formation classes of the wet-dry tropical climate regime. Thornbush and tropical scrub are found in many parts of the world where the wet-dry tropical climatic regime is transitional into the tropical desert regime. Soils show the influence of aridity and are subject to calcification, and (in poorly drained sites) of salinization. Reddish-brown and reddish-chestnut soils occur in much of the area of these formation classes.

10. *Savanna* is a vegetation consisting of widely scattered trees rising from a more or less continuous lower layer dominated typically by grasses. Although the term "savanna" most commonly refers to a vegetation in tropical and subtropical latitudes, structurally similar vegetation can be found in the subarctic lands of the northern hemisphere—the *taiga*. (See "cold woodland.") As discussed here, savanna is that of low latitudes, included in Class C in the world vegetation map (Plate 4) as a relative of the tropical savanna woodland, monsoon forest, and thornbush and tropical scrub. Like its relatives, the tropical sa-

Figure 16.19 Thornbush in the region of the Tona River, central Africa. (Photograph of the Akeley Expedition, courtesy of the American Museum of Natural History.)

vanna is a response to a wet-dry tropical climate regime in which the severe drought period is one of relatively cooler temperature but which experiences great heat just preceding the onset of the rains.

As with the savanna woodland, the tropical savanna supports trees and shrubs which are xerophytic or deciduous. In some localities the trees are palms, giving a "palm savanna." Grasses in the tropical savanna are characteristically tall, with stiff coarse blades, commonly higher than the height of a man, and even up to 12 ft (4 m) high. In the dry season these grasses form a yellowish straw mat that is highly flammable and subject to periodic burning. It is widely held that, to a greater or lesser degree, periodic burning of the savanna grasses is responsible for the maintenance of the grassland against the invasion of forest. Fire does not kill the underground parts of grass plants, but limits tree growth to a few individuals of fire-resistant species. The browsing of animals, which kills many young trees, is also a factor in maintaining grassland at the expense of forest.

The African savanna is perhaps the most celebrated of the tropical grasslands, spanning the continent from west to east in two great belts, centered about on the 10th parallels of latitude north and south, and connected in east Africa by a broad north-south belt (Figure 16.20). In equatorial latitudes the African savanna replaces the rainforest because of the aridity associated with highlands in Sudan, Kenya, and Tanzania. Characteristic are the flat-topped acacia trees (*Acacia*), and the grotesque baobab (*Adansonia digitata*), which has a large barrel-shaped water-storing trunk (Figure 16.21). Elephant grass (*Pennisetum pur-*

Figure 16.21 Baobab tree (*Adansonia digitata*), near Nairobi, Kenya, Africa. (Photograph of the Akeley Expedition, courtesy of the American Museum of Natural History.)

pureum), forming almost inpenetrable thickets, may grow to heights of 16 ft (5 m).

Savanna in South America is exemplified in the *campo cerrado* of the interior Brazilian Highlands (Figure 16.22). Here the trees are largely evergreen, deep-rooted, and capable of tapping lower levels of soil moisture not available to the grasses during the dry season.

Other important savanna areas occur in northern Australia, India, and southeast Asia. The mesquite savanna of Texas provides an American example. Of interest to American students is the small patch

Figure 16.20 African tall-grass savanna in Kenya. (Photograph by Richard U. Light, courtesy of American Geographical Society.)

Figure 16.22 The *campo cerrado* at Pirassununga, Sao Paulo, Brazil. (Photograph by Pierre Dansereau.)

of rather specialized (edaphic) tropical savanna in southern Florida—the Everglades. Underlain by a limestone formation, the extremely low, flat plain of the Everglades is flooded by a shallow layer of runoff from summer rains and becomes a swamp. In winter the area becomes extremely dry. Coarse saw grass (*Cladium effusum*) covers much of the plain. Scattered trees are represented by palms and, on higher ground, by pines.

The pedogenic process most closely associated with tropical savanna is laterization, promoted by the high temperatures associated with the rainy season. However, laterization gives way to calcification as the savanna is traced toward higher latitudes where thornbush and, ultimately, steppe grasslands are encountered.

11. *Semidesert* (also called *half desert*) is a xerophytic shrub vegetation with a poorly developed herbaceous lower layer. Trees are generally absent. Semidesert shrub vegetation is well developed in subtropical and middle-latitude dry climates having a small annual total rainfall and high summer temperatures. An example is the sagebrush (*Artemisia tridentata*) vegetation of the middle and southern Rocky Mountain region and Colorado Plateau (Figure 16.23). Semidesert shrub vegetation seems recently to have expanded widely into areas of the western United States that were formerly steppe grasslands, as a result of overgrazing and trampling by livestock.

On the world map of vegetation, Plate 4, semidesert is included with dry-desert vegetation. The xerophytic shrub vegetation is to be expected along the less arid margins of the deserts and in favorable highland locations within the desert. Pedogenic processes are calcification, producing brown soils,

Figure 16.23 Sage-brush semidesert, Vermilion Cliffs, near Kanab, Utah, 1906. (Photograph by Douglas Johnson.)

Figure 16.24 Lowland heath, mostly heather (*Calluna vulgaris*), on moraine, North Norfolk. (Photograph by S. R. Eyre.)

and gray desert soils. Salinization is found in poorly drained sites.

12. *Heath* is a low, dense layer of shrubs commonly not more than 10 in (25 cm) in height, which occurs in regions of highly equable but cool climate in middle and high latitudes (Figure 16.24). The shrubs are usually dominated by members of the Heath family (*Ericaceae*); mosses are also important. A common plant of heath vegetation is heather (*Calluna vulgaris*). In very cold climates heath includes small shrubby birches and willows.

Distribution of heath is not shown on the world vegetation map (Plate 4) but is found in small areas in Classes *E* and *I*. A cool, marine west coast climate is favorable to development of heath. Well-distributed precipitation and a temperature regime with small annual range are required. The pedogenic process is podzolization and soils are acidic. Examples of heath vegetation are seen in Ireland and other exposed west coasts of the British Isles (*moors*) and in western and north-central Europe.

13. *Cold woodland*, the last on our list of formation classes in the savanna biochore, is a form of vegetation limited to very cold subarctic and tundra climates. Trees are low in height and well spaced apart; a shrub layer may be well developed. The ground cover of lichens and mosses is distinctive. Cold woodland is essentially equivalent to what is widely referred to as *taiga*,[3] and occurs along the northern fringes of the high-latitude, needleleaf forests. Cold woodland is thus transitional into the treeless tundra and arctic heath.

In North America representative trees of the

[3]The term "taiga" is commonly used to include also needleleaf forest, needleleaf woodland, and needleleaf savanna.

cold woodland are black spruce (*Picea mariana*) and tamarack (*Larix laricina*). In northern Scandinavia a scrubby birch (*Betula odorata*) forms a woodland with a low layer of lichens, such as reindeer "moss" (*Cladonia rangiferina*) (Figure 16.25). In Siberia larch (*Larix*) is the woodland tree. Birch-lichen woodland is also found in the subarctic regions of northwestern Canada and Alaska.

Cold woodland is dominated by the polar climatic regime with its long and severe winters in which all soil moisture is frozen for many months of the year. A shallow depth of thaw accompanies the short "summer," when insolation continues throughout much of the day. The pedogenic process is toward gleization in the poorly drained sites; toward podzolization on upland sites.

Cold woodland is not shown separately on the world vegetation map (Plate 4) but can be found in the tundra areas (Class *I*) along the boundary with needleleaf forest (Class *F*).

Grassland biochore

14. *Prairie* consists of tall grasses, comprising the dominant herbs, and subdominant forbs (broad-leaved herbs) (Figure 16.26). Trees and shrubs are almost totally absent but may occur in the same region as forest or woodland patches in valleys and other topographic depressions. The grasses are deeply rooted and form a continuous and dense sward. The grasses flower in spring and early

Figure 16.26 Kalsow Prairie, Iowa, has been set aside as an example of virgin prairie. (Photograph by courtesy of State Conservation Commission of Iowa.)

summer; the forbs, in late summer. In Iowa, a representative region of tall-grass prairie, typical grasses are big bluestem (*Andropogon gerardi*) and little bluestem (*A. scoparius*), a typical forb is black-eyed Susan (*Rudbeckia nitida*).

The tall-grass prairies are typically associated with continental, middle-latitude climates described as *subhumid*, that is, in which evapo-transpiration and precipitation are almost balanced on an average yearly basis and range between 20 and 40 in (50–100 cm). In summers air and soil temperatures are high, so that on uplands soil moisture is not adequate for tree growth and deeper water sources are beyond reach of tree roots. The North American Prairies are found in a broad belt extending from Illinois, northwestward to southern Alberta and Saskatchewan. On the world map of vegetation (Plate 4) the North American prairie includes the eastern and northern parts of Class *G*. Areas of forest are mixed with areas of prairies in a transitional belt between forest and prairie regions.

Because the tall-grass prairies grade into short-grass prairies and eventually into steppe grasslands in the direction of increasing aridity, it is not practical to try to list all world regions of tall-grass prairies. In Europe, a typical region of tall-grass prairie is the *puszta* of Hungary. The Argentine *pampa* is often cited as a region of prairie vegetation, as are areas in north China.

The pedogenic process associated with prairie vegetation is calcification and results in development of prairie and chernozem soil profiles.

15. *Steppe*, sometimes called *short-grass prairie*, is a formation class consisting of short grasses tending to be bunched and sparsely distributed (Figure 16.27). Scattered shrubs and low trees may be found in the steppe and there exist all grada-

Figure 16.25 Lichen woodland (open boreal forest) on a sand plain of the Hamilton River delta, Labrador. Black spruce and a few small white birch are widely spaced. The floor is covered by lichen (*Cladonia*). (Photograph by F. Kenneth Hare, courtesy of the American Geographical Society.)

Figure 16.27   Short-grass vegetation of the Great Plains, South Dakota. (Photograph by Douglas Johnson.)

tions into semidesert and woodland formation classes. Ground coverage is small and much bare soil is exposed. Many species of grasses and other herbs occur; a typical grass of the American steppe is buffalo grass (*Buchloe dactyloides*), other typical plants are the sunflower (*Helianthus rigidus*) and loco weed (*Oxytropis lambertii*).

The world distribution of steppe vegetation is extremely wide in terms of latitude, ranging in location from the equator to 55° N and 45° S (Plate 4, Class *G*). Steppes of low latitudes are transitional from vegetation of the wet-dry tropical climates into the dry deserts. Steppes of North Africa are transitional from Mediterranean climate, with its sclerophyll forest, to the African desert. Steppes of middle latitudes are associated with a semi-arid continental climatic regime in which, despite a summer rainfall maximum, evaporation exceeds precipitation on the average. Winters in the middle-latitude steppes are cold and dry; the summers warm to hot. In this climatic regime the dominant pedogenic process is calcification, with salinization in poorly drained sites. Soils contain a large excess of precipitated calcium carbonate and are very rich in bases. Brown soils are typical. Humus content is relatively small because of the sparseness of the vegetation.

16. *Grassy tundra* (including *alpine meadow*) is a grassland biochore formation limited to very cold climates having ample available moisture and often saturated soils. Grassy tundra of the arctic regions flourishes under a regime of long summer days during which time the ground ice melts only in a shallow surface layer. The frozen ground beneath (permafrost) remains impermeable and melt water cannot readily escape. Consequently, in summer a marshy condition prevails for at least a short time over wide areas. Humus accumulates in a well-developed layer.

Plants of the arctic grassy tundra are low and mostly herbaceous, although dwarf willow (*Salix herbacea*) occurs in places. Sedges, grasses, mosses, and lichens dominate the tundra in a low layer (Figure 16.28). Typical species are ridge sedge (*Carex bigelowii*), arctic meadow grass (*Poa arctica*), cottongrasses (*Eriophorum*), and snow lichen (*Cetraria nivalis*). There are also many species of forbs which flower brightly in the summer. Considerable variations in composition of the tundra are seen in the range from wet to well-drained habitats. One form of tundra consists of sturdy hummocks of plants with low, water-covered ground between. In the regions of grassy tundra will also be found areas of arctic scrub vegetation composed of willows and birches.

Under the subarctic climate and a condition of poor soil-water drainage, the pedogenic regime tends toward gleization. Size of plants is in part limited by the mechanical rupture of roots during freeze and thaw of the surface layer of soil, producing shallow-rooted plants. In winter drying winds and mechanical abrasion by wind-driven snow tend to reduce any portions of a plant that project above the snow.

In all latitudes, where altitude is sufficiently high, an alpine tundra is developed above the limit of tree growth and below the vegetation-free zone of barren rock and perpetual snow. Alpine tundra resembles arctic tundra in many physical respects.

Desert biochore

17. *Dry desert* is a formation class of xerophytic plants widely dispersed and providing almost negligible ground cover. In dry periods (which are the rule) the visible vegetation consists of small hard-leaved or spiny shrubs, succulent plants (cacti), or hard grasses. Many species of small annuals may be present, but appear only after a rare but heavy rain has saturated the soil.

Desert floras differ greatly from one part of the

Figure 16.28   Cotton-grass meadows, Arctic coastal plain of Alaska. (Photograph by William R. Farrand.)

Figure 16.29 Shrub vegetation of the Sonoran Desert, southwestern Arizona. Ocotillo plant in the left foreground; saguaro cactus plants in left distance. (Photograph by A. N. Strahler.)

world to another. In the Mohave-Sonoran deserts of the southwestern United States, plants are often large and in places give a near-woodland appearance (Figure 16.29). Well known are the treelike saguaro cactus (*Carnegiea gigantea*), the prickly-pear cactus (*Opuntia imbricata*), the ocotillo (*Fouquiera splendens*), creosote bush (*Larrea tridentata*) and smoke tree (*Dalea spinosa*). Much of the "desert" of the southwestern United States is in fact scrub, thorn scrub, savanna, or steppe grassland (Figure 14.2). In the Sahara desert (most of it very much drier than in the American desert) a typical plant is a hard grass (*Stipa*); another, found along the dry beds of watercourses, is tamarisk (*Tamarix*). The coastal desert of southwest Africa is known for the strange tumboa plant (*Welwitschia mirabilis*) with strap-shaped leaves radiating from a central tap root which penetrates deep into the ground (Figure 16.30).

Much of the area assigned to dry-desert vegetation has no plants of visible dimensions, being composed of shifting dune sands, or almost sterile salt flats.

Distribution of dry deserts is shown in the world vegetation map, Plate 4, as Class *H*. As a study of climate will show, three variations of the desert trend exist. The true tropical-continental deserts have not only extreme aridity but also extremely high air and soil temperatures; the middle-latitude deserts (30° to 35° lat.) have both aridity and a great annual temperature range including extremes of winter cold; the tropical west coast deserts have

remarkable uniformity and relative coolness of temperature along with persistent coastal fog. A dominant pedogenic process of dry deserts is salinization, locally producing areas of salt crust where only salt-loving (halophytic) plants can survive. Calcification is conspicuous on well-drained uplands; encrustations and deposits of calcium carbonate (caliche) are commonplace. Humus is lacking and soils are pale gray or red in color, e.g., sierozems and red-desert soils.

18. *Arctic fell field* is the arctic equivalent of the dry desert, being found in the extremely cold tundra and icecap climates. The arctic fell field consists of rocky ground surfaces, produced by

Figure 16.30 The tumboa plant (*Weltwitschia mirabilis*) on a sandy plain in the Kalahari Desert of southwestern Africa. (Photograph by Robert J. Rodin.)

Figure 16.31 Arctic fell-field vegetation, Etah, Greenland. On the left is *Dryas integrifolia*; on the right, *Cerastium alpinium*. (Photograph of Crockerland Expedition, courtesy of the American Museum of Natural History.)

distribution by means of a map can be only partly successful, at best. The infinite variations in structure of vegetation defy a simple categorical system. Moreover, no two authorities in the field of plant geography will set up the same array of classes or produce closely corresponding maps. An appreciation of broad systems of plant formation classes in response to the restraints and freedoms of the spectrum of climatic regimes should be foremost in the mind of the geographer and he must stress the underlying principles and distributional aspects that most systems show in common, rather than be diverted by the seeming contradictions and complexities he frequently encounters.

REFERENCES FOR FURTHER STUDY

Shantz, H. L., and R. Zon (1924), *Natural vegetation,* Atlas of American Agriculture, Section E, U.S. Department of Agriculture, U.S. Govt. Printing Office, Washington, D.C., 29 pp.

Newbigin, Marion I. (1948), *Plant and animal geography,* second edition, E. P. Dutton and Co., New York, 298 pp.

Brockmann-Jerosch, H. (1951), *Vegetation of the earth* (wall map), Justus Perthes, Gotha.

American Geographical Society (1956), *A world geography of forest resources,* edited by Haden-Guest, Wright, and Teclaff, The Ronald Press Co., New York, 736 pp.

Dansereau, Pierre (1957), *Biogeography; an ecological perspective,* The Ronald Press Co., New York, 394 pp.

Polunin, Nicholas (1960), *Introduction to plant geography,* McGraw-Hill Book Co., New York, 640 pp.

Küchler, A. W. (1964), Potential natural vegetation of the conterminous United States (map and manual), Amer. Geographical Soc., Special Publ. No. 36, 116 pp.

Eyre, S. R. (1968), *Vegetation and soils; a world picture,* second edition, Aldine Publ. Co., Chicago, 328 pp.

intense frost shattering and having only small patches of finetextured mineral soil. Vegetation is very sparse and consists mostly of lichens, mosses, and a few small shrubs (Figure 16.31). For example, in northern Baffin Island, typical plants are a small polar willow (*Salix polaris*), purple saxifrage (*Saxifraga oppositifolia*), and woolly moss (*Rhacomitrium lanuginosum*).

Arctic fell field can be found as far north as the extreme limits of land, namely up to 84° N latitude and southward to ice-free parts of Antarctica. Soil-forming processes consist of little more than rock disintegration by frost action, manifested in the formation of stone polygons in which finer particles are sorted from the coarse into patches. On the world vegetation map, Plate 4, arctic fell field is included in Class *I* along with grassy tundra and arctic scrub.

This attempt to classify world vegetation into a number of formation classes and to show their

REVIEW QUESTIONS

1. What are the four great biochores? Describe each briefly and give its general climatic associations.

2. What investigators were responsible for developing the system of vegetation formation classes commonly followed today? Discuss some of the problems that might be connected with preparing a useful world map of natural vegetation.

3. Describe the equatorial rainforest in terms of structure. What is the importance of lianas in this rainforest? Of epiphytes? Comment on the numbers of species of trees found in the equatorial rainforest. Name two representative trees. What climatic and pedogenic regimes are associated with the equatorial rainforest?

4. What coastal vegetation forms may be expected in regions of equatorial rainforest? Describe particularly the mangrove coasts.

5. How does tropical rainforest differ from equatorial rainforest and in what respects is it similar? In what geographic regions is tropical rainforest found? Explain.

6. What are the unique characteristics of the monsoon forest as compared with other low-latitude forests? What accounts for the deciduousness of the trees? Relate the monsoon forest to a particular climatic regime and describe its world distribution pattern.

7. How does temperate rainforest differ in structure and geographic location from equatorial and tropical rainforests? Describe temperate rainforests of America, Asia, and New Zealand. What is particularly noteworthy of the development of epiphytes in the temperate rainforest?

8. What three types of world locations favor temperate rainforest? What climatic and pedogenic regimes are associated with temperate rainforest?

9. Describe the summergreen deciduous forest as to structure and seasonal development. Name common trees of the deciduous forests of eastern North America and Europe. In what ways does the summergreen deciduous forest represent a response to the continental climatic regime? What is the associated pedogenic process?

10. Describe a needleleaf forest in terms of vegetation structure. Define the major world regions of occurrence of needleleaf forest and name common trees of these belts.

11. How does topography influence the extension of needleleaf forests into lower latitudes? Describe the needleleaf forests of the Pacific coastal regions of North America. What is the dominant pedogenic regime in areas of needleleaf forest?

12. Describe the evergreen-hardwood (sclerophyll) forest in terms of vegetation structure. With what climatic regime is this formation class closely associated and how is this association seen in the world geographical distribution of sclerophyll forest, woodland, and scrub? Name some representative sclerophyll forest plants of the Mediterranean region, California, and Australia.

13. What is savanna woodland? With what climatic regime is it most commonly associated? What are the life-form characteristics of trees and shrubs of the tropical savanna woodland?

14. Describe thornbush and tropical scrub as vegetation classes. What is the *caatinga* of Brazil? The *dornveld* of South Africa?

15. How does savanna differ from savanna woodland? Describe the vegetation soils, climate, and animal life associated with tropical savanna, with particular reference to Africa. What is the *campo cerrado* of Brazil?

16. What is semidesert vegetation? Describe semidesert in the southwestern United States.

17. What is heath? What common plants are associated with heath? Describe the conditions of climate and geographic location favorable to occurrence of heath.

18. Describe cold-woodland vegetation. Is *taiga* an equivalent designation for this formation class? What are some representative plants of the cold woodland of North America? Of Scandinavia? Of Siberia?

19. What plants comprise prairie as a vegetation formation class? Name representative examples from the tall-grass prairie of Iowa. With what climatic and pedogenic regimes is prairie vegetation associated? Explain. What is the *puszta* of Hungary? The *pampa* of Argentina?

20. Compare steppe (short-grass prairie) with tall-grass prairie in terms of vegetation structure and climatic regime.

21. Describe the grassy tundra of cold climates, giving names of typical plants. With what conditions of climate, drainage, and pedogenic process is the grassy tundra associated? Compare alpine tundra with grassy tundra of low elevations.

22. Describe dry desert as a plant formation class. Name some typical desert plants of the southwestern United States, the Sahara, and southwest Africa. What is the dominant pedogenic process of the dry desert?

23. What relation does arctic fell field bear to other formation classes of the cold lands in high latitudes? Explain how freeze and thaw of soil water plays a dominant role in the landscape associated with arctic fell field.

CHAPTER 17
Landforms and Earth Materials

LANDFORMS, the distinctive geometric configurations of the earth's land surface, are of primary concern to the geographer because they exert far-reaching and fundamental influence on the patterns of human activity. The direct influences are obvious to any thoughtful person. A mountain chain is an effective barrier between groups of people who live in adjacent lowlands. A plain, on the other hand, may be densely populated, rich in agricultural resources, and unified culturally and politically by a network of good roads and railroads that permit people with common interests to intercommunicate freely. One coast line, deeply indented with excellent natural harbors but bordered by a rocky rugged coastal belt, may produce a community of seafaring humans, adapted to fishing, ocean commerce, and shipbuilding. Another coast line, whose simple plan and shallow bottom provide not a single good natural harbor, may be bordered by a low, fertile coastal plain. Here human activity turns naturally to agriculture.

Examples of the direct effect of landforms, or topographic relief features, on human beings could be cited almost without end, but there are, in addition, indirect geographical influences to be considered. As we have seen in the study of rainfall distribution, a high mountain range profoundly influences climate in adjacent areas. If it shields a nearby lowland from prevailing moisture-bearing winds, a desert results. On mountain slopes, climates become cooler and moister with increased altitude, bringing a changing succession of agricultural and forest conditions, which, in turn, determine to what extent occupation is possible and what natural products can be obtained. Steepness of slopes is closely associated with fertility of soils. Hill and mountain slopes have thin, relatively poor soils, readily subject to devastating soil erosion when exposed by axe or plow. Plains may have thick, rich soils, not easily eroded even under poor farming practices.

Geomorphology

The systematic study of topographic relief features is known as geomorphology (*geo*, earth; *morph*, form; *ology*, science), but for simplicity may well be called "the study of landforms." Landscape features must be sorted out into classes or groups of those essentially similar both in outward form and in origin. The geomorphologist is equally interested in forms and in the processes and stages of development of those forms. This book will treat landforms in terms of how they came about. Landscape features pass through an orderly series of changes, just as do human beings in their life span. Once these life stages are known, any landscape feature can be related to a particular event in the life cycle of landforms, thereby lending order and natural law to our concepts of landform development.

Genetic landform description

It is possible, of course, to describe all landscape features by tabulating their size, shape, angles of slope, and orientation without thought to their origin or development. This is an *empirical* approach to natural science. Volumes of figures and other factual data would be required to convey the proper description of even the simplest assemblages of landforms.

If, on the other hand, the development of landforms is carefully examined, it is seen that the same series of forms is repeated with remarkable similarity over and over again in nature. In order to classify and describe a complex assemblage of forms in terms of orderly sequences of development, we need only a single brief statement giving (*a*) the *structure* of the rock mass beneath, (*b*) the *process* which sculptured the landform and (*c*) the *stage* of its development. Such a description is *genetic* because it emphasizes genesis, or origin. The person who hears or reads such a description, knowing what the ideal forms are like, can relate any particular landform to its proper place in the natural scheme.

The systematic study of landforms according to their origin and stage of development was introduced by Professor William Morris Davis, of Harvard University, about 1890. His influence has been so profound that many English-speaking geomorphologists follow the basic plans that he laid down.

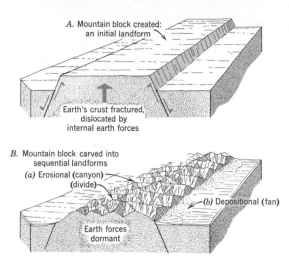

A. Mountain block created:
an initial landform

Earth's crust fractured,
dislocated by
internal earth forces

B. Mountain block carved into
sequential landforms
(a) Erosional (canyon)
(divide)
(b) Depositional (fan)

Earth forces
dormant

Figure 17.1 Initial and sequential landforms.

Initial and sequential landforms

In terms of the grand scale of geological processes there are two fundamental classes of landforms. First, there are the original crustal masses raised by internal earth forces and by volcanic eruptions. These comprise the *initial landforms* (Figure 17.1*A*). Second, there are the landforms made by agents of denudation. Because these follow the initial forms and occur in orderly sequences, they are called collectively the *sequential landforms* (Figure 17.1*B*).

Any landscape is really nothing more than the existing stage in a great struggle or contest. The internal earth forces spasmodically elevate parts of the crust to create initial landforms. The external agents patiently keep wearing these masses down and carving them into vast numbers of smaller sequential landforms.

All stages of this struggle can be seen in various parts of the world. Where high, rugged mountains exist, the internal earth forces have recently dominated. Where certain kinds of low plains exist today, denudational forces have finally triumphed. All intermediate stages can be found. Because the internal earth forces act repeatedly, new landmasses keep coming into existence as old ones are subdued. Judging from conditions in various periods of the geologic past, we are now in a time when continents stand relatively high above sea level. This suggests that internal forces were active relatively recently, geologically speaking.

Agents of land sculpture

Sequential landforms are products of one or more of the land-sculpturing agents, running water, waves, ice, and wind. These erosional agents, aided by processes of rock decay and downslope movements of soil and rock under gravitational force, attack from the outside all continental rock masses that become elevated by mountain-making upheavals or warping of the earth's solid crust. No part of the land surface is immune from attack. As soon as any rock mass comes to be exposed to the air or to wave attack it is set upon by these denudational agents and processes. They work to one ultimate goal—wearing away the landmass until it becomes a low plain, which is then slowly consumed by waves and perhaps finally covered by ocean waters. The disintegration products are spread over the sea floors surrounding the continents. The processes act with extreme slowness, to be sure, but geologic time is enormously great. Streams and waves seen in action today have had millions of years to do their work. Geologists believe that all sequential landforms can be explained as results of processes that can be seen acting at the present time.

In the wearing down of landmasses, a great variety of sequential landforms results. Where rock is eroded away, valleys or topographic depressions of one sort or another are formed. Between the eroded depressions are ridges, hills, or mountains representing unconsumed parts of landmasses. All such sequential landforms shaped by progressive removal of the bedrock mass are designated *erosional landforms*.

Rock and soil fragments that were removed are deposited elsewhere to make an entirely different set of topographic features, the *depositional landforms*. Figure 17.2 illustrates these two groups of landforms. The ravine, canyon, peak, spur, col and bluffs are erosional landforms; the fan, built of rock fragments below the mouth of the ravine, is a depositional landform. The floodplain, built of material transported by a stream, is also a depositional landform.

Bedrock, soil, and residual overburden

Examination of a freshly cut cliff, such as that in a new highway cut or quarry wall, will reveal several kinds of earth materials, shown in Figure 17.3. Solid, hard rock which is still in place and

Figure 17.2 Erosional and depositional landforms.

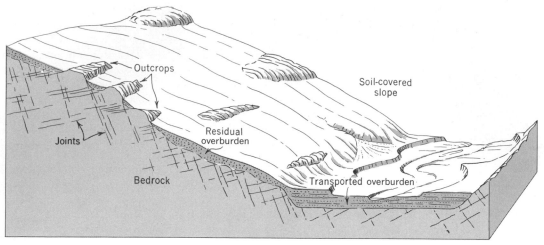

Figure 17.3 Residual and transported overburden.

relatively unchanged is called *bedrock*. It grades upward into a zone where the rock has become decayed and has disintegrated into clay, silt, and sand particles. This may be called the *weathered overburden*, or *residual overburden*. At the top is a layer of true *soil*, often called *topsoil* by farmers and gardeners. Soil properties, soil-forming processes, and soil classification were treated in Chapters 13 and 14. Over the soil may be a protective layer of grass, shrubs, or trees.

One or more of these zones may be missing. Sometimes everything is stripped off down to the bedrock, which then appears at the surface as an *outcrop*. Sometimes following cultivation or forest fires the true soil only is eroded away, leaving exposed the overburden, which is infertile and may become scored by deep gullies. The thickness of soil and overburden is quite variable. Although the true soil is rarely more than a few feet thick, the residual overburden of decayed and fragmented rock may extend down tens or even hundreds of feet. Formation of the overburden is greatly aided by the presence of innumerable bedrock cracks termed *joints* (Figure 17.3), along which water can move easily to promote rock decay.

Transported overburden

Another variety of overburden that may be found covering the bedrock is *transported overburden*. It consists of such materials as stream-laid gravels and sands, floodplain silts, clays of lake bottoms, beach and dune sands, or rubble left by a melting glacier. All types have in common a history of having been transported by streams, ice, waves, or wind.

Whereas residual overburden, formed in place by disintegration of bedrock below it, is of local origin, transported overburden consists of rock and mineral varieties from distant sources and may be quite unlike the underlying minerals and rocks. Figure 17.3 shows stream valley deposits, called *alluvium*, which would be designated as transported overburden in contrast to residual overburden of the adjacent hill slope. Once deposited, transported overburden may remain undisturbed for many thousands of years, in which case a true soil is formed in its uppermost layer.

In a broad sense all of the materials of which the depositional landforms are built up are of transported overburden. In later chapters many kinds of depositional landforms are described and explained in terms of the processes by which the particles are carried.

Influence of rocks on landforms

Bedrock strongly influences shape, size, and development of the erosional landforms. In some places, rock takes the form of thin layers, lying horizontally, tilted, folded, or broken, as the case may be. Elsewhere it consists of thick, irregular masses extending down to great depths. Some varieties of rock are soft and are readily washed away by streams and waves; others are extremely resistant to all agents of weathering and erosion. To a considerable degree the weakness or resistance of rocks is determined by their origin and age. When varied rocks lie side by side near the surface of the earth's crust the agents of denudation etch them out according to their degree of resistance, the weak rock tending to form valleys or other types of depressions, the resistant ones standing out in bold relief as hills, mountains, or plateaus. Consequently, landforms reflect closely the shape and arrangement of the original rock masses and will show certain distinctive qualities whereby they may be referred to an orderly classification.

The first step in approaching geomorphology is,

therefore, to learn the fundamental facts about rocks of the earth's crust, their composition, physical and chemical properties, characteristic forms of occurrence, processes of origin, and geologic age. A knowledge of common rocks and the predominant minerals which comprise them is of further value to the geographer because rocks and minerals have intrinsic value to man as ores, fuels, and construction materials. An evaluation of the natural resources of a region is thus in part a question of geology.

Igneous rocks

Rocks may be divided into three major classes: (1) *igneous*, (2) *sedimentary*, and (3) *metamorphic*, according to their origin. *Igneous rocks* solidify from a molten state. The highly heated fluid rock, termed *magma*, originates at considerable depths beneath the earth's surface and is forced by great internal pressures to penetrate the solid, brittle outer crust of the earth (Figure 17.4). It should not be supposed that the whole interior of the earth is molten. It seems more likely that local pockets of magma, lying at depths of 20 mi (30 km) or more, are formed in various places and eventually are forced toward the surface.

The nature and origin of molten rocks bodies and the forces that control their behavior comprise a highly speculative phase of geology. Unlike the study of landforms, in which both the landform and the erosional process can be observed in their entirety, the study of deep-seated igneous rock masses cannot be made during their formation, but can only be interpreted in the light of the appearance and composition of the rock after it has cooled and become solidified, then exposed to the surface by long ages of landmass denudation.

Intrusive and extrusive igneous rocks

Molten masses that do not reach the surface but solidify in spaces that they have made by pushing the surrounding rock apart, or melting, or dissolving it are termed *intrusive* igneous rocks (Figure 17.4). They do, in fact, intrude or invade the

previously formed rock. Where the intrusion has been violent the surrounding rocks may show breaking, crushing, folding, or uparching around the newly formed rock body. Where the intrusion has been gradual and the surrounding rock slowly melted or dissolved, no disturbance of the adjoining rock may result.

A second class of igneous rock, termed *extrusive*, is that which reaches the ground surface, issuing from a pipe or crack and pouring out upon the ground, then quickly solidifying into a hard rock. Volcanoes and lava flows are the principal products of igneous rock extrusion (Figure 17.4).

Intrusive rocks differ from extrusive rocks considerably in both gross outward form and details of internal structure and texture. Intrusive rocks commonly occur in enormous masses, termed *batholiths*, covering areas as great as an entire state and extending many miles down (Figure 17.4). The adjective *plutonic* is commonly used for this and other deep-seated intrusive rock bodies. Such large masses cooled very slowly; consequently, the individual mineral crystals of which the rock is composed are relatively large (Figure 17.5), commonly measuring 0.1 to 0.5 in (0.25 to 1.5 cm) across. Crystal grains of intrusive igneous rock interlock closely so as to produce a dense, strong rock, free of cavities.

Minerals

A *mineral* is a naturally occurring, inorganic substance that has a definite or fairly definite chemical composition and usually exhibits a characteristic crystal form, together with certain distinctive qualities of color, hardness, luster, and fracture habit. Although a few kinds of rocks consist almost wholly of a single mineral variety, most rock types are aggregates or physical mixtures of two or more minerals. Both the mineral varieties present and the proportions in which they are mixed determine the name and properties of a given rock variety.

Mineralogy, the study of minerals, and *petrology*, the study of rocks, are important branches of geology. Several thousand mineral varieties have been identified and named; the list of rock varieties runs into the hundreds. Fortunately, only a few minerals are so abundant in volume and worldwide distribution as to combine into major rock types. This introductory study goes little further than to describe a few of the commonest igneous rocks and to name the principal constituent minerals.

Common varieties of intrusive rocks

One of the most abundant and best known of the intrusive rocks is *granite*, composed of a

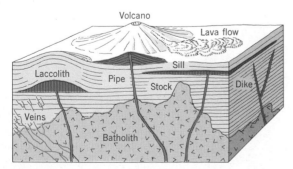

Figure 17.4   Igneous rock bodies.

Figure 17.5 Seen close up, this granite, a coarse-grained intrusive rock, proves to be made of tightly interlocking crystals of a few kinds of minerals.

mixture of these minerals: *quartz* (silicon dioxide), *potash feldspar* (an aluminum silicate of potassium), *hornblende* (an aluminum silicate of iron and magnesium), and *biotite mica* (a complex aluminum silicate with manganese and iron) (Figure 17.6). Because quartz and feldspar, which together may comprise three-quarters of the rock, are light-colored, whereas hornblende and biotite mica are black, granite has a speckled appearance when closely examined (Figure 17.5).

The potash feldspar grains will be identified by their opacity, blocklike outlines, and shiny luster on cleavage surfaces. In some granites this feldspar is almost white, producing gray granite; in others, the feldspar is salmon pink, producing pink granite. The quartz grains are clear and appear glassy beside the feldspar grains. Quartz grains tend to have irregular outlines and break irregularly, after the manner of glass. Both hornblende and biotite crystals are black, but the biotite mica is soft and can be split into thin flakes with the point of a knife or needle.

A second variety of intrusive igneous rock is

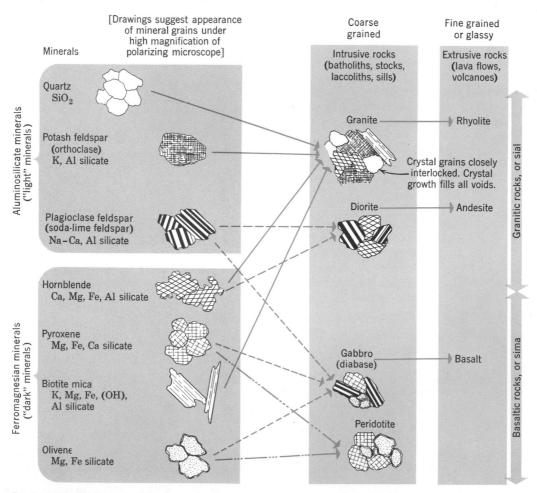

Figure 17.6 Various combinations of light and dark minerals produce the igneous rocks.

diorite. Not nearly so widespread in occurrence as granite, it illustrates a rock of more limited mineral composition, in that quartz is entirely lacking and only two minerals dominate: (1) *plagioclase feldspar* (soda-lime feldspar), which is an aluminum silicate mineral with varying proportions of sodium and calcium, and (2) a dark mineral, which may be either hornblende or *pyroxene*. The plagioclase feldspar, which is milky white to gray in color, with vitreous luster on its cleavage surfaces, makes up about three-fourths of the rock. Both hornblende and pyroxene are commonly black and are extremely difficult to tell apart without the aid of a microscope. Pyroxene, like hornblende, is a silicate mineral with magnesium, iron, and calcium, but pyroxene is often lacking in aluminum.

A third intrusive igneous rock of importance is *gabbro*, composed primarily of two minerals, plagioclase feldspar and pyroxene. The difference between this rock and diorite is largely in the proportions of the two minerals. Feldspar dominates in diorite, but makes up only about 50 percent in gabbro. Another mineral, *olivine*, is commonly present in gabbro. Olivine, a silicate of magnesium and iron, is commonly green in color. Gabbro is a coarse-grained, plutonic rock like granite or diorite and also occurs in great masses.

A fourth intrusive rock variety is *peridotite*, a dense, dark rock composed largely of pyroxene and olivine. Little or no feldspar is present. Peridotite is found usually in small masses or dikes, not in vast batholiths such as granite.

Granitic and basaltic rock groups

As Figure 17.6 shows, the four examples of intrusive igneous rocks have been placed in an orderly arrangement, such that the light-colored minerals, quartz and feldspar, decrease in proportion while the dark-colored minerals increase in proportion. In granite and diorite there is a preponderance of quartz and feldspar, which are light-colored minerals of low density. A cube of granite will weigh about 2.6 times as much as a similar volume of water; its relative density, or *specific gravity*, is therefore stated as 2.6. Igneous rocks containing largely light-colored minerals of low density, rich in aluminum silicate combined with potassium, sodium, and calcium (as in the feldspars), are said to belong to the *granitic group*. Geologists often refer to these rocks collectively as *sial*, a word coined from *si*, for silica, and *al*, for alumina.

Gabbro and peridotite, on the other hand, are composed predominately of minerals rich in magnesium and iron, the so-called *ferromagnesian minerals*. These minerals are relatively dense, so that peridotite, for example, may have a specific gravity over 3.2, which is roughly 25 percent greater than granite. Rocks composed largely of the dense ferromagnesian minerals belong to the *basaltic group* (or *basic* group). The word *sima*, coined for these rocks collectively in distinction to sial, is taken from *si* for silica and *ma* for magnesium. The importance of granitic and basaltic (or sialic and simatic) groups of rocks in the great layers of the earth's crust is discussed in Chapter 18.

Extrusive igneous rocks

The extrusive igneous rocks take two forms: (*a*) fluid magma flows in the form of thin tongues or sheets, called *lava flows*, and (*b*) solid or near-solid pieces blown violently from volcanic vents and collectively termed *volcanic ejecta*. (See Figure 28.9.) Some lava flows may have a rough, blocky surface; others a smooth billowy surface. Usually flows are dark red or black in surface color. Because of the rapid cooling, mineral crystals have little or no opportunity to develop during the

Figure 17.7 Above: The outer surface of lava has wrinkles produced by flowage. Below: A frothy, gaseous lava solidifies into a light, porous scoria (left). Rapidly cooled lava may form a dark volcanic glass (right).

solidification and cannot be detected with the unaided eye. Expanding gases may fill the rock with countless bubblelike cavities, producing a texture of rock called *scoria* (Figure 17.7). Where solidification is extremely rapid, a glass is formed. This glass goes by the name *obsidian*, or simply *volcanic glass* (Figure 17.7).

Lavas may be similar in composition to the various kinds of intrusive igneous rocks, inasmuch as the same magma may reach the surface in one place but not in another. The lava that corresponds to granite is *rhyolite* (Figure 17.6). That which corresponds to diorite is *andesite*. These rocks are both rather light gray or pink on freshly broken surfaces. Even close study will show only a dull rock surface, the crystals being too fine to see with the naked eye. In some lavas, however, there is a sprinkling of large crystals of feldspar easily seen in outline against the fine-grained background.

The extrusive equivalent of diabase is *basalt*, a dense black rock which is perhaps the commonest and most widespread of the lavas. Areas of huge size, such as the Columbia River plateaus in Washington, Oregon, and Idaho and the Deccan region of western India, are underlain by thousands of feet of thickness of basalt lavas, poured out in tremendous upwellings through long cracks known as *fissures*. A horizontal layer of basalt, upon cooling, shrinks somewhat and forms into four-, five-, or six-sided columns which stand vertically. The structure is termed *columnar jointing* and gives a most striking appearance to cliffs of basalt (Figure 17.8).

Volcanic ejecta come in various sizes and shapes. Very fine material is termed *volcanic dust* or *volcanic ash*. Coarser particles, up to 2 or 3 in (5 or 8 cm) across, are termed *cinder* and resemble the clinkers from a coal furnace. Still larger masses are *volcanic bombs* (Figure 28.9). These may be still plastic when blown from a volcanic vent and become shaped into rounded masses resembling a loaf of bread. Volcanic ejecta result from the presence of large amounts of gas under high pressure in the magma. As it approaches the surface the gas is released from confining pressure and expands, frothing the magma into a bubbly foam and producing the violent explosions that blow the fragments into the air. One particularly spongy, foamlike variety of lava is *pumice*, which is so light that it floats on water.

Sedimentary rocks

Those rocks that are composed of particles derived from previously existing rock, then laid down after transportation by streams, ocean or wave currents, wind, or ice are *sedimentary rocks*. Rocks of any origin may be the parent material for sediment. In addition, sediment may result

Figure 17.8 This cliff of old lava has weathered away, revealing the columnar jointing system formed during cooling. (Photograph by Gifford.)

from chemical reaction and precipitation. As the solid particles settle out, a layer is formed. Should the size or composition of the particles change repeatedly, the result will be a series of layers of varied type. Thus a series of muds or clays, sands, or gravels might be deposited upon an ocean or lake bottom, or upon land if the material is deposited from streams, winds, or glaciers. In their original state, these sediments are soft, but in time they become hardened into rocks as we are accustomed to seeing them. Hardening occurs both because of pressure compaction driving out water, and because chemical deposits of calcium carbonate or silicon dioxide form between the particles, causing them to be cemented together.

Sedimentary rocks are distinguished from igneous rocks by the usual presence of layers resulting from changing size and composition of sediment. The layers are termed *strata*, or simply *beds*. The planes of separation between successive layers are *planes of stratification*, or *bedding planes*. The rock is said to be *stratified* or *bedded*. (See Figures 26.14, 26.15, 26.16.) Bedding planes in their original condition are nearly horizontal but may be tilted to any angle by subsequent movements of the earth's crust (Figure 17.9).

Groups of contiguous rock strata having some common properties of rock variety or geologic age or both are designated as *formations* and are usually named for the geographic locality at which they are well displayed for study, or were first described by the geologist. For example, the *Mesa Verde Formation* is a sandstone named for its striking cliff exposures at Mesa Verde, in southwestern Colorado, yet it is found in several states and is believed to have been once a vast, continuous layer of sands laid down in the floor of an ancient shallow sea.

Clastic sediments

One great class of sedimentary rocks is described as *clastic*, meaning that the particles are produced by disintegration of previous rocks through the combined weathering processes (Chapter 19). Processes of transportation in water or wind tend to sort the particles into various size grades, from which the clastic sedimentary rocks are named as follows:

Rock Name	Particle Name	Diameter or Fineness of grains
Conglomerate	Boulders Cobbles Pebbles (gravel)	Over 10 in (25 cm) 2½ to 10 in (6 to 25 cm) $\frac{1}{16}$ to 2½ in (2 mm to 6 cm)
Sandstone	Sand	10 to 1000 per inch (0.02 to 2 mm)
Siltstone	Silt	1000 to 10,000 per inch (0.002 to .02 mm)
Shale	Clay	Finer than 10,000 per inch (Under 0.002 mm)

Conglomerate, although named for the presence of boulders, cobbles, or pebbles, has a large proportion of sand filling the interstices. The larger particles are usually well rounded from tumbling on stream beds or in the surf zone of beaches (Figure 17.10). In fact, conglomerates may be thought of as fossil gravel bars or beaches of ancient rivers or shorelines. Where the large fragments are angular, instead of rounded, the rock is called a *breccia*, rather than a conglomerate. Breccia might be formed of rock fragments rolled

Figure 17.9 These limestone strata in Oklahoma were steeply tilted long after being laid down. (Photograph by Lofman. Courtesy of Standard Oil Co., New Jersey.)

from a cliff, crushed within a landslide, or swept from a steep ravine in a brief torrential flood.

Sandstone and *siltstone* grade into one another and are distinguished much as coarse sandpaper is distinguished from very fine. The grains of sandstone may be of any durable mineral, such as those described in the foregoing section on igneous rocks. Nevertheless, the long distance of travel of the grains in rivers or along shores eliminates the soft, easily powdered minerals and allows chemical decay to turn others, such as the feldspars and ferromagnesian minerals, into soft clay. Therefore, the commonest mineral of sandstones is quartz, whose resistance to mechanical and chemical attack is phenomenal. A few grains of dark minerals may be present, as well as tiny gleaming mica flakes.

Sandstone, siltstone, and conglomerate owe their hardness to cementation following deposition.

Figure 17.10 Conglomerate is a mixture of pebbles and sand cemented into a hard rock.

Gradually, in the vast spans of geologic time that follow deposition and burial, deposits of silica (SiO_2) or calcium carbonate ($CaCO_3$) accumulate in the pore spaces between sand grains, producing a solid, strong rock. The cementing substances are carried in solution by slowly moving underground water. Cementing material rarely fills all of the openings of a sandstone, so that the rock can hold large stores of fluids such as water or oil. We depend on sandstones as reservoirs of our natural gas, petroleum, and in many cases of ground water supplies as well (Chapters 20 and 26).

Shale is compacted clay or mud, greatly increased in density and hardness by the application of pressure from overlying strata. Clays commonly consist of submicroscopic flakes of mineral matter. Such particles behave as colloids, which have a strong affinity for water. Under sufficient pressure, the excess water is largely driven off; an extremely dense rock results. Whereas a clay or mud will soften and swell when soaked for a short time in water, shale will not. Shales are gray, earth-red, or black. The red color is due to presence of iron oxides; black may indicate the presence of finely divided organic matter in the form of carbon compounds. Shales have a soft, smooth feel in contrast to the gritty surface texture of sandstones. Furthermore, shale is mechanically weak and can easily be shattered into thin flakes and plates. For this reason shale plays the geomorphic role of a weak rock, easily carved away by the agents of land erosion to produce valleys and broad lowlands.

Pyroclastic rocks

A special class of clastic rocks results from the violent eruption of volcanoes, where fine ash is belched from the vent in great clouds. Sediment settling from the air during such explosions is described as *pyroclastic* material. An important variety of pyroclastic rock is *tuff*, a fine-grained, light-gray rock composed of compacted volcanic ash. Scattered small crystals of igneous minerals, blown from the vent, are sometimes visible in the tuff. If the ash has been picked up by streams and carried into a body of standing water, the tuff may be reformed into a distinctly stratified sedimentary rock. Other volcanic ash outbursts were carried by gases of such high temperature that the ash was fused into a hard rock known as *welded tuff*.

Organic and chemical sedimentary rocks

A second important class of sedimentary rocks, described as *organic*, consists of rock matter produced by the growth activities of plants and animals. A sedimentary rock variety of major importance is *limestone* (Figure 17.9), much of which is formed from the limy parts of such organisms as corals, algae, foraminifera, clams, and snails. The organic origin is obvious where large shell fragments have been cemented into a variety of limestone termed *coquina*.

Limestone in its pure state consists of *calcite* (calcium carbonate), a mineral easily scratched with the sharp point of a knife and responding by vigorous effervescence when a drop of dilute hydrochloric acid is placed on its surface. Limestones may be pure white, as in the case of a very soft variety termed *chalk*, or gray to black in the more common dense varieties. Limestone may also form by direct chemical precipitation from waters of lakes or oceans heavily charged with lime in solution. The soft lime mud thus formed is termed *marl*. Because of its susceptibility to weak acids present in the rain water, soil water, and streams, limestone is rapidly eaten away in regions of humid climates, forming valleys and broad lowlands.

Many limestones have considerable amounts of the mineral *dolomite* (calcium-magnesium carbonate) incorporated with the calcite. Where the mineral dolomite predominates, the rock itself is called *dolomite*.

Quite different from limestone is another organic type of sedimentary rock, *coal*, formed of vast bog accumulations of partly decayed plant matter. In the early stages of its formation, coal passes through the stage of *peat*, a soft brown, combustible mass of plant fragments. Through later burial and compaction, water and certain volatile compounds are slowly driven off, yielding dense, black, rocklike layers of coal which contain a high proportion of fixed carbon.

Of great importance economically, though not abundant in comparison with limestone, are a group of chemical sediments termed *evaporites*. These are salts that have been precipitated from waters of shallow desert lakes or constricted bays of the ocean, where evaporation of the water is rapid. One evaporite which is found in rock layers of wide extent is *anhydrite* (calcium sulfate). Closely related is *gypsum* (hydrous calcium sulfate), a soft white mineral. Gypsum is often found interlayered with shales and limestones. Another evaporite familiar to all is rock salt, occurring as the mineral *halite* (sodium chloride), which forms thick sedimentary rock layers along with shales in a few localities.

Sedimentary rocks as landform controls

Figure 17.11 shows five types of sedimentary rock together with a mass of much older igneous rock upon which the sediments were laid. Their usual landform habit, whether to form valleys or mountains, is indicated, together with the conventional symbols used on cross sections by geologists. These rock strata have been strongly tilted and deeply eroded, so that there is maximum oppor-

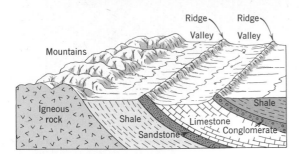

Figure 17.11 Many landscape features originate through the slow erosional removal of weaker rock, leaving the more resistant rock standing as ridges or mountains.

tunity for the development of relief as a consequence of the different degrees of resistance offered to denudational processes.

Metamorphic rocks

Any of the types of igneous or sedimentary rocks may be altered by the tremendous pressures and high temperatures that accompany mountain-building movements of the earth's crust. The result is a rock so greatly changed in appearance and composition as to be classified as a *metamorphic rock*. Generally speaking, metamorphic rocks are harder and more compact than their original types, except when the latter are igneous rocks. Moreover, the kneading action and baking that metamorphic rocks have undergone produces new structures and even new minerals. Each sedimentary and igneous rock has an equivalent metamorphic rock. The term *metasediment* conveniently covers all metamorphic rocks derived from sedimentary strata.

Shale, on being squeezed and sheared under mountain-making forces, turns into *slate*, a gray or brick-red rock that splits neatly into thin plates so familiar to all as roofing shingles and flagstones of patios and walks. The planes of splitting form a structure called *slaty cleavage*. This is a new structure imposed upon the rock during the process of internal slippage during metamorphism, not merely stratification or bedding. Slate is fine textured and of rather dull surface texture. It can be distinguished from shale by the fact that a thin slab of slate rings when struck sharply.

With continued application of pressure and internal shearing, slate changes into *schist*, the most advanced grade of metamorphic rock. Schist has a structure termed *foliation*, consisting of thin but rough and irregularly curved planes of parting in the rock. Schist is set apart from slate and phyllite by the coarse texture of the mineral grains, the abundance of mica, and the presence of scattered large crystals of new minerals such as *garnet* and *staurolite*, which have grown during the proc-

ess of internal shearing of the rock (Figure 17.12).

Slates and schists are relatively resistant to the processes of denudation and tend to form hills and uplands. In comparison with granite, however, these rocks are less resistant, so that granites will usually form markedly higher mountain masses.

The metamorphic equivalent of conglomerate, sandstone, and siltstone is *quartzite*, formed by the addition of silica (SiO_2) to fill completely the interstices between grains, most of which are normally quartz (also silica). This process is carried out by the slow movement of underground waters carrying the silica into the sandstone, where it is deposited. Pressure and kneading of the rock is not essential in producing a quartzite, but may deform the quartz grains. When a quartzite is fractured, as with a hammer blow, the break will cut across sand grains and pebbles in the rock. In this way, quartzite can be distinguished from a sandstone, which usually breaks around the grains, leaving them mostly intact. The extreme hardness of quartzite, combined with its high immunity to chemical decay, makes it the most resistant of all rocks. Many prominent ridge crests and peaks in a region of metamorphic rock will be found to be composed of quartzite.

Limestone, upon metamorphism, becomes *marble*, commonly a white rock of sugary texture when freshly broken. During the process of internal shearing, the calcite mineral of the limestone has reformed into larger, more uniform crystals than before. Bedding planes are obscured and masses of mineral impurities are drawn out into swirling streaks and bands. Like limestone, marble is easily decomposed by weak acids in the soil water and streams, hence, is usually found occupying valleys and lowlands.

Finally, the important metamorphic rock,

Figure 17.12 This fragment of schist, 6 in (15 cm) long, has a glistening, undulating surface (above) consisting largely of mica flakes. An edgewise view (below) of the same specimen shows the wavy foliation planes.

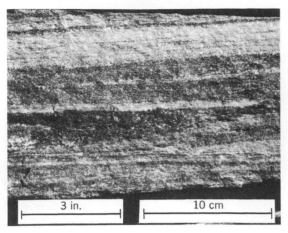

Figure 17.13 Gneiss shows a banded surface.

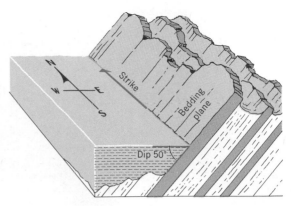

Figure 17.15 Strike and dip.

gneiss, may be formed either from intrusive igneous rocks, or as a metasediment from strata that have been in close contact with intrusive magmas.

A single description will not fit all gneisses, which vary considerably in appearance, mineral composition, and structure. One common variety is *granite gneiss*, formed directly by flowage of granite in a somewhat plastic state. Granite gneiss resembles granite in its massiveness, general texture, and mineral components, but possesses a streaked appearance, called *lineation*, produced by parallelism of dark minerals which have been drawn out into long, pencil-like shapes in the direction of flow.

Still other gneisses are strongly banded into light and dark layers or lenses (Figure 17.13), which may be contorted into wavy folds. It is possible that in some instances these bands, which have differing mineral compositions, may be the relics of sedimentary strata such as shale and sandstone to which new mineral matter has been added from nearby intrusive rocks. Gneisses are strong, resistant rocks, which, like granite, generally form bold highlands or mountain chains. Figure 17.14 shows the topographic expressions we would expect in regions where various metamorphic rocks rest side by side and have been acted upon for vast spans of time by the agents of landmass denudation.

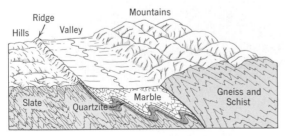

Figure 17.14 Metamorphic rocks tend to form elongate, parallel belts of valleys and mountains.

Dip and strike

Because natural planes are characteristic of the structure of each type of rock, the geologist requires a system of geometry to enable him to measure and describe the attitude of these natural planes and to indicate them on maps. Examples of such planes are the bedding layers of sedimentary strata, the sides of a dike, the upper and lower surfaces of a sill, slaty cleavage of slates, and the joints in a granite. Rarely are these planes truly horizontal.

The acute angle formed between a natural rock plane and an imaginary horizontal plane is termed the *dip*, and is stated in degrees ranging from 0° for a horizontal plane to 90° for a vertical plane. Figure 17.15 shows the dip angle for an outcropping layer of sandstone, against which rests a horizontal water surface.

The compass direction, or bearing, of the line of intersection between the inclined rock plane and an imaginary horizontal plane is the *strike*. In Figure 17.15 the strike is north.

REFERENCES FOR FURTHER STUDY

Dunbar, C. O. and J. Rodgers (1957), *Principles of stratigraphy*, John Wiley and Sons, New York, 356 pp.

Pettijohn, F. J. (1957), *Sedimentary rocks*, second edition, Harper and Row, New York, 718 pp.

Spock, L. E. (1962), *Guide to the study of rocks*, Harper and Row, New York, 298 pp.

Leet, L. D. and S. Judson (1965), *Physical geology*, third edition, Prentice-Hall, Inc., Englewood Cliffs, N.J., 406 pp. See Chapters 4, 5, 6, 8, and 18.

Gilluly, J., A. C. Waters, and A. O. Woodford (1968), *Principles of geology*, third edition, W. H. Freeman and Co., San Francisco, 687 pp. See Chapters 2, 3, 17, and 18.

Longwell, C. R., R. F. Flint, and J. E. Sanders (1969), *Physical geology*, John Wiley and Sons, New York, 685 pp. See Chapters 3, 4, 16, and 20.

1. What are landforms? What is geomorphology? Of what importance are landforms to the geographer? Cite several examples.

2. What is a genetic system of description? How does it differ from an empirical description?

3. What advantage is there to describing and classifying landforms according to structure, process, and stage? Explain these terms.

4. What is the basis for dividing all landforms into initial and sequential groups?

5. What are the agents of land sculpture? Toward what goal do they work?

6. Explain what is meant by the terms *erosional landforms* and *depositional landforms.*

7. Explain how a landscape reflects the work of both internal and external earth forces and processes.

8. Define bedrock, overburden, and soil. What is an outcrop?

9. What is the distinction between transported and residual overburden? What is the relation between transported overburden and the depositional landforms?

10. In what way is an understanding of rocks and their structures necessary in explaining landforms?

11. What are igneous rocks? What is a magma? What is the difference between intrusive and extrusive igneous rocks? How does each type occur? What difference is there in the texture of extrusive and intrusive rocks? Why?

12. What is a mineral? How does a mineral differ from a rock? What name is given to the science of minerals? Of rocks?

13. What is granite? Of what minerals is it composed? Give the names and mineral components of at least two other intrusive igneous rocks.

14. What minerals tend to make an igneous rock light in color and low in specific gravity? What minerals tend to make an igneous rock dark and high in specific gravity? How do the words *sial* and *sima* serve to distinguish two major divisions of igneous rocks?

15. What forms do extrusive igneous rocks take? What is lava? How does the texture of lava differ from that of the intrusive igneous rocks? Why? Through what kind of passages do extrusive rocks reach the earth's surface? What is columnar jointing?

16. How are volcanic ejecta classified according to the sizes of the fragments? How are pyroclastic sedimentary rocks related to volcanic ejecta?

17. What is a sedimentary rock? How are the sedimentary materials hardened into rock? What are the sources of sediment? Define strata; stratification planes.

18. On what basis are formations set apart as units in sedimentary rock?

19. Name in order of size the particles that make up clastic sediments and the rock name for each grade. In what way is conglomerate different from breccia? What is the principal mineral of sandstone? What characterizes the shape and properties of particles which make up shale? Name a pyroclastic rock and explain how it is formed.

20. How does limestone differ from the clastic sedimentary rocks? In what ways may limestone be formed? What is chalk? What is marl? What mineral composes pure limestone? How does dolomite differ from limestone in chemical composition?

21. How are peat and coal produced? What class of sedimentary rocks do they illustrate? Explain how an evaporite forms. Name two common evaporites and give their compositions.

22. How do the various kinds of sedimentary rocks influence the development of landforms? Which form valleys? Which form mountains and ridges? How do the sedimentary rocks compare with the igneous rocks in this respect?

23. What are metamorphic rocks? How are they formed? From what other rocks may they be derived? Describe slate, schist, quartzite, marble, and gneiss. Which of these rock types generally form valleys? Which form ridges or mountains?

24. Explain how dip and strike tell the orientation in space of an inclined rock plane.

CHAPTER 18
The Earth's Crust and Its Relief Forms

THE geographer concerned largely with man's occupation of the earth's surface may justifiably feel little need to study those branches of geology dealing with the earth's interior, the subsurface structures of the continents, the relief features of the ocean floor, and the events of ancient geologic history. It is true that the remoteness of these things—whether in place or in time—relegate them to indirect or remote roles in man's economic, social, and cultural processes, inasmuch as few immediate or direct environmental effects can be established. Nevertheless, it is worthwhile for the student of physical geography to place the surficial elements of the lands, atmosphere, and oceans in the broader setting of global geology.

Internal structure of the earth

Man's direct observation of the composition and physical properties of the earth's interior is limited by mining and drilling operations to depths of a few miles at best, so that he must turn to indirect means of obtaining information. The science of *geophysics* is concerned largely with obtaining information about the physical properties of the earth by means of instruments that measure earthquake waves, earth magnetism, and the force of gravity. Interpretation of these data through established laws of physics has yielded surprisingly detailed knowledge of the earth's structure and properties.

Figure 18.1 is a cutaway diagram of the earth to show its major parts. The earth is an almost spherical body approximately 3960 mi (6370 km) in equatorial radius. The center is occupied by the *core*, a spherical zone about 2160 mi (3475 km) in radius. Because of the sudden change in behavior of earthquake waves upon reaching this zone, it has been concluded that the outer core has the properties of a liquid, in abrupt contrast to a solid mass which surrounds it. However, the innermost part of the core with a radius of 780 mi (1255 km) may be solid, or crystalline.

Through astronomical calculations it can be shown that the earth has a specific gravity of about $5\frac{1}{2}$, whereas the surface rocks average 3 or less (Chapter 17). This observation must mean that specific gravity increases greatly toward the interior, where it may be about from 10 to 15. Iron, with a small proportion of nickel, is considered as the probable substance comprising the liquid core. This conclusion is supported by the fact that many meteorites, representing disrupted fragments of our solar system, are of iron-nickel composition. Temperatures in the earth's core may lie between 4000° and 5000°F (2200° and 2750°C); pressures are as high as three to four million times the pressure of the atmosphere at sea level.

Outside of the core lies the *mantle*, a layer about 1800 mi (2895 km) thick, composed of mineral matter in a solid state. Judging from the behavior of earthquake waves, the mantle is probably composed largely of the mineral olivine (magnesium iron silicate), which comprises a basic rock variety called *dunite*. This rock, which may be in a glassy state, exhibits qualities of great rigidity and high density in response to sudden stresses of earthquake waves which pass through it. On the other hand, the mantle rock can also adjust by slow flowage to unequal forces which act over great periods of time. In this respect it is somewhat like cold tar, which is hard and shatters easily if struck, but which will slowly flow downhill if left undisturbed for a long time.

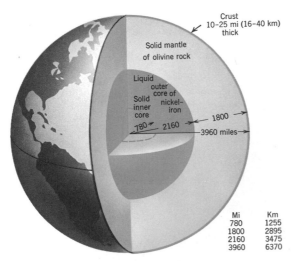

Figure 18.1 Concentric zones make up the earth's interior.

The crust

Outermost and thinnest of the earth zones is the *crust,* a layer some 5 to 25 mi (8 to 40 km) thick. Formed of crystalline rocks, largely igneous in classification, the crust exhibits properties of brittleness when subjected to mountain-building forces and therefore may break along great faults. The base of the crust, where it contacts the mantle, is sharply defined, a fact known because earthquake waves change velocity abruptly at that level (Figure 18.2*A*). The surface of separation between crust and mantle is called the *Moho,* a simplification of Mohoroviçić, the name of the seismologist who discovered it.

From a study of earthquake waves it is concluded that the crust consists of two layers: (1) a lower, continuous layer of basaltic rock otherwise termed *sima* (see Chapter 17); (2) an upper layer of granitic rock, otherwise termed *sial,* which constitutes the continents. The granitic layer is therefore discontinuous in areal extent, being absent over the ocean basins. Figure 18.2*B* shows schematically a small part of the crust near the margin of a continent. The sedimentary strata of the continents are on the average merely a thin skin, not shown on this diagram, although locally they are several thousand feet thick.

Those parts of the crust forming the continents are much thicker than the crust under the ocean basins, and may be likened to vast icebergs floating in the sea with only a small part visible above the water, but with a great bulk deeply submerged. The glassy rock of the earth's mantle has yielded by slow flowage, much like a very viscous fluid; this has permitted the lighter, rigid, continental plates of the crust to come to rest in the manner of the floating iceberg.

Distribution of continents and ocean basins

The first-order relief features of the earth are the continents and oceans basins. From a globe or atlas we can compute that about 29% of the globe is land; 71% oceans. If, however, the seas were to drain away, it would become apparent that broad areas lying close to the continental shores are actually covered by shallow water, less than 600 ft (180 m) deep. From these relatively shallow *continental shelves* the ocean floor drops rapidly to depths of thousands of feet. In a sense, then, the ocean basins are brimful of water and have even spread over the margins of ground that would more reasonably be assigned to the continents. If the ocean level were to drop by 600 ft (180 m) the surface area of continents would increase to 35%; the ocean basins decrease to 65%; figures which we may regard as representative of the true relative proportions.

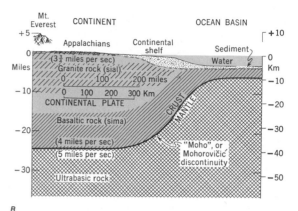

Figure 18.2   The earth's crust is much thicker under continents than beneath the ocean basins.

Figure 18.3 shows graphically the percentage distribution of the earth's surface area with respect to elevation both above and below sea level. Note that most of the land surface of the continents is less than 3300 ft (1 km) above sea level. There is a rapid drop off from about −3000 to −10,000 ft (−1 to −3 km) until the ocean floor is reached. A predominant part of the ocean floor lies between 10,000 and 20,000 ft (3 and 6 km) below sea level. Disregarding the earth's curvature, the continents can be visualized as platformlike masses; the oceans as broad, flat-floored basins.

Scale of the earth's landforms

Before turning to a description of the major subdivisions of the continents and ocean basins, it is revealing to consider the true scale of the earth's landforms in comparison with the earth as a sphere. Most of the relief globes and pictorial relief maps seen commonly in magazines and atlases are greatly exaggerated in vertical scale. For a true-scale profile around the earth we might draw a chalk-line circle 21 ft (6.4 m) in diameter, representing the earth's circumference on a scale of 1:2,000,000. A chalk line $\frac{3}{8}$ in (0.15 cm) wide would include within its limits not only the highest point on the earth, Mt. Everest, +29,000 feet (+8840 m) but also the deepest known ocean trenches, below −35,000 feet (−10,700 m).

Figure 18.4 shows profiles correctly curved and scaled to fit a globe whose diameter is 21 ft (6.4 m). The topographic profile is drawn to natural scale, without vertical exaggeration. Although the most imposing landforms of Asia and North

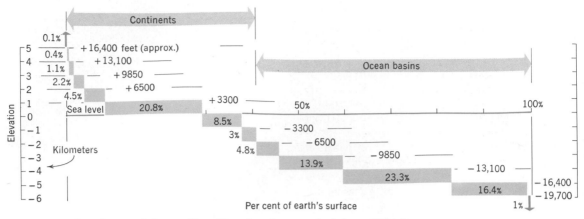

Figure 18.3 Distribution of the earth's solid surface in successively lower altitude zones.

America are shown, they seem little more than trivial irregularities on the great global circle.

Second-order relief features of the ocean basins

The North Atlantic Ocean illustrates certain typical features of the ocean basins and the continental margins. Three major units of the North Atlantic are (1) the *continental margin*, (2) the *ocean basin floor*, and (3) the *mid-oceanic ridge* (Figure 18.5). These units may be regarded as second-order relief features of the earth's surface.

Along the eastern margin of North America lies the *continental shelf*, a fairly smooth, sloping plain 75 to 100 mi (120 to 160 km) wide and reaching a depth of 600 ft (180 m) at the outer edge. This shelf, a part of the continental margin, is essentially a zone of deposition of sedimentary rock layers built of material brought from the eastern United

States by streams and spread over the sea floor by currents. At its outer edge, the shelf abruptly gives way to a descending *continental slope* leading down to the true ocean basin floor at a depth of about 12,000 ft (3700 m).

The slope is scored by strange *submarine canyons* (Figure 18.6), whose origin has been strongly debated. The canyons seem to be the work of eroding streams, very likely flows of muddy water, called *turbidity currents*, which are produced when storms or earthquake shocks disturb soft sediment at the canyon heads. These flows travel swiftly down the continental slope because their density is greater than that of the surrounding sea water. Spreading out upon the deep sea floor, turbidity currents come slowly to rest. The sediment is spread in broad layers which have accumulated over millions of years, gradually burying the irregular topographic features of the sea floor and

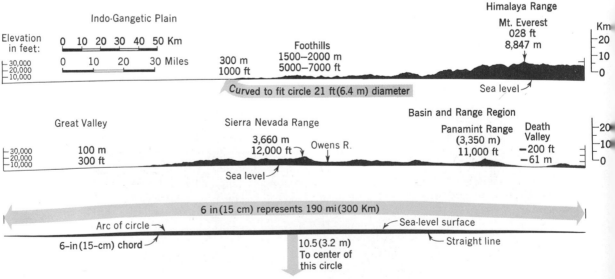

Figure 18.4 These profiles show the earth's great relief features in true scale with sea-level curvature fitted to a globe 21 ft (6.4 m) in diameter.

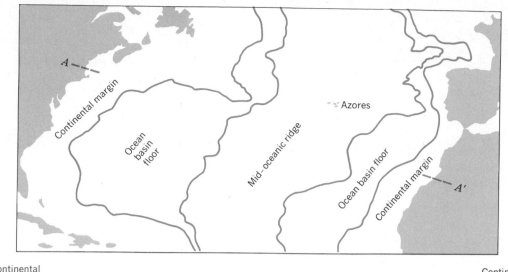

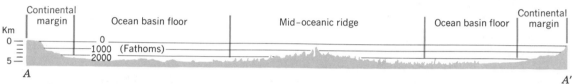

Figure 18.5 Major divisions of the North Atlantic Ocean basin (above), and a representative profile from New England to the coast of Africa (below). Profile exaggeration is about 40 times. (Data of B. C. Heezen, M. Tharp, and M. Ewing, 1959, from A. N. Strahler, *The Earth Sciences*, Harper and Row, New York.)

producing vast, flat *abyssal plains* lying within the ocean basin floor (Figure 18.7). The broad basin of the North Atlantic thus has a remarkably smooth floor over large areas at a depth of about 18,000 ft (5500 m). Rising abruptly from the floor are isolated submarine mountains, named *seamounts*, some of which may be ancient volcanoes.

In the center of the Atlantic lies a great submarine mountain range, the *Mid-Atlantic Ridge*, comparable with the Rocky Mountain chain in size and relief but entirely submerged except for the

Azores Islands. This ridge is merely a part of a single, continuous mid-ocean ridge traced through the South Atlantic, Indian, South and East Pacific, and Arctic oceans. It is the expression of a major fracture system of the earth's crust (Figure 18.8). The oceanic crust is in process of being pulled apart along the mid-ocean ridge, causing a *rift valley* to lie along the center line of the ridge (see profile of Figure 18.5). At the same time mantle rock is being pushed up to elevate the ridge.

Evidence in support of crustal spreading in the

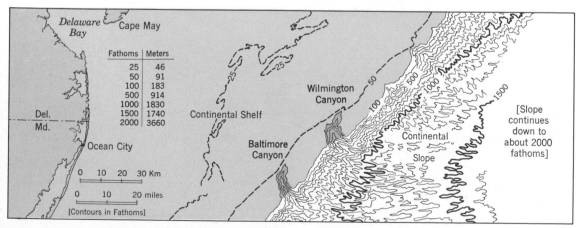

Figure 18.6 Two submarine canyons notch the outer edge of the Atlantic continental shelf off the Delaware-Maryland coast. (After Veatch and Smith.)

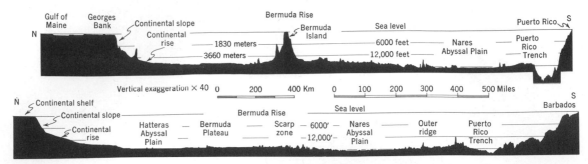

Figure 18.7 Profiles across the Atlantic Ocean basin reveal continental shelf and slope, abyssal plain, trench, and various minor irregularities. (After Bruce C. Heezen, Lamont Geological Observatory of Columbia University.)

mid-ocean ridge zone has come from studies of magnetic properties of the basalt rock in belts adjacent to the ridge. Rate of separation of the crustal masses on either side of the mid-ocean ridge has been estimated to be about 0.8 in (2 cm) per year in the North Atlantic region. Separation affects the continents lying east and west of the Atlantic ocean basin and constitutes a geologic process known as *continental drift.* Present-day scientific opinion strongly supports the hypothesis that North and South America were formerly joined to the African and Eurasian landmasses. If so, the basaltic rocks underlying the floor of the Atlantic Ocean basin are much younger than the

bulk of rocks comprising the crust of the drifted continents.

Other features of the ocean floor are *trenches* or *foredeeps*—long, narrow depressions whose bottoms reach depths of 24,000 to 30,000 ft (7500 to 10,000 m) or more (Figure 18.7 and 18.9). The trenches are thought by geologists to represent downfolded zones associated with recent mountain-building movements. Accumulation of sediment on the trench floors has been so slow that these depressions are not filled in, as they would be if above sea level.

Geologists and geophysicists have long been convinced that the continental crustal masses have

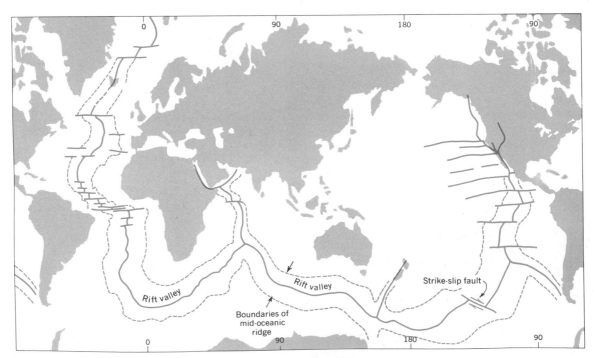

Figure 18.8 World map of the mid-ocean ridge system and its transverse fracture zones. (Based on data of B. C. Heezen, 1962, from A. N. Strahler, *The Earth Sciences,* Harper and Row, New York.)

endured throughout all recorded geologic time and have not exchanged places with the ocean basins by crustal warping. Evidence for this belief lies in the fact that most sedimentary rocks of past geologic ages seen on the continents are of types deposited in fairly shallow depths, whereas the very fine-grained deep-sea clays, or *oozes*, which settle out upon seamounts and isolated submarine

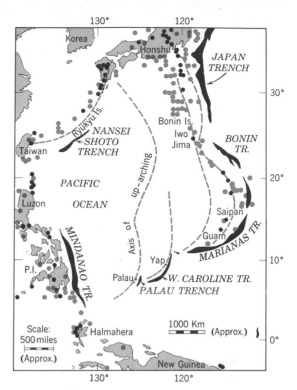

Figure 18.10 This map of the western Pacific shows trenches (solid black), island arcs (dashed lines), active volcanoes (black dots), and deep-focus earthquake centers (circles). (After H. H. Hess.)

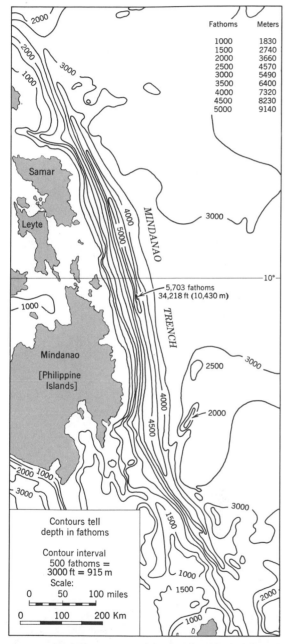

Fathoms	Meters
1000	1830
1500	2740
2000	3660
2500	4570
3000	5490
3500	6400
4000	7320
4500	8230
5000	9140

5,703 fathoms
34,218 ft (10,430 m)

Contours tell depth in fathoms

Contour interval
500 fathoms =
3000 ft = 915 m
Scale:

Figure 18.9 The great Mindanao Trench, lying east of the Philippines, reaches a maximum depth of 34,218 ft (10,430 m) which is considerably deeper than Mt. Everest is high. (After H. H. Hess, Hydrographic Office Chart 5485.)

ridges, are absent from geological strata of the continents. Sample submarine cores, penetrating more than 50 ft (15 m) into such oozes, show thin clay layers dating back as far as 100 million years and indicating that the rain of sediment from the overlying ocean water has been extremely slow as compared to deposition on continental shelves and in shallow inland seas. The lower, deeper parts of the ocean basins have, however, received the deposition of thousands of feet of layered clay and sand carried by turbidity currents from continental margins.

Present-day mountain building near the margins of the ocean basins is distributed along long, narrow, curving zones, known as *island arcs* (Figure 18.10). Each arc represents a zone of sharp crustal folding or compression and is associated with active volcanoes and earthquakes. The great ocean deeps generally lie along the outer side of the island arcs.

Second-order relief features of the continents

Just as with the ocean basins, the continents can be subdivided into second-order relief units. Essentially, the continents consist of two fundamental kinds of geologic units: (1) *shields* and (2) *moun-*

tain belts, or *orogenic belts*. The shields are the heartlands of the continental plates and consist of extremely ancient rocks. The age of the shield rocks is *Precambrian*, that is, older than 600 million years. Most of the shield rock is much older than one billion years and some has been dated as 3½ billion years old. Several periods of mountain-making, or *orogeny*, affected the shields throughout the Precambrian era of geologic time and there was much metamorphism of ancient sedimentary rocks as well as many large intrusions of granitic rock in the form of batholiths (Chapter 17). Whereas great mountain ranges existed at various times during the Precambrian era, processes of erosion effectively reduced those mountains to their roots and leveled the ancient continents to nearly featureless plains. The shields today are therefore largely plains and low plateaus and are highly stable parts of the earth's continental crust.

Crustal movements of the shields in later geologic time have been of a type known as *epeirogenic* movements; that is, rising or sinking of the crust over broad areas without appreciably breaking or bending the rocks. Epeirogenic movements reflect crustal stability generally, in contrast to compressional and tensional orogenic movements affecting unstable zones.

When epeirogenic movements were of a negative, or sinking, type, large parts of the shields became submerged as shallow seas and continental shelves. This inundation gave opportunity for sedimentary strata to be deposited on the old metamorphic and igneous rocks. Later, positive, or rising, epeirogenic movements brought the sedimentary cover above sea level where it has since been carved up by streams into hills and plateaus.

Shields of the northern hemisphere are shown in Figure 18.11. North America's geological heartland is the *Canadian Shield*; that of Europe is the *Russian-Baltic Shield*, also called the *Fenno-Scandian Shield*. Rocks in both of these regions date back to the oldest era of geologic time, from one to three billion years ago. In the southern hemisphere, similar shield areas occupy parts of Australia, South Africa, and the Antarctic.

The mountain belts of the continents are long, narrow orogenic zones in which the crust has been compressed and forced to buckle into tight folds and at the same time to be strongly elevated. Two basic ages of mountain belts are distinguished. The first includes mountains produced in the Paleozoic and early Mesozoic (200 million to 600 million years ago). Now inactive, these ancient mountain ranges in some places still possess rugged relief and moderately high elevations; elsewhere, they are reduced to hill belts. The second age group of mountain belts belongs to the late Mesozoic and Cenozoic eras of geologic time (younger than about 150 million years) and are characterized by bold relief and great altitude. Those ranges produced in the last, or Cenozoic, era are in many places still active zones of crustal deformation and volcanic activity; they rise as spectacular alpine mountains.

The geologic time scale

To discuss the development of relief features of the continents requires reference to events of the geologic past. Even the explanation of landforms produced in the more recent of the geological periods requires some knowledge of the length and sequence of geologic time units. The economic geographer interested in the occurrence and distribution of ores and mineral fuels will find that a knowledge of historical geology will be extremely helpful in gaining understanding of the occurrence of economic mineral deposits in various parts of the world. Table 18.1, therefore, outlines the important events and subdivisions of geologic history. Note that the upper part of the table, representing the younger or more recent events, subdivides geologic time into small intervals, whereas with increasing age the units are larger and more generalized. This is because the geologic record is most fragmentary for the oldest ages but becomes generally more detailed as recent time is approached.

Absolute ages given in the table have been verified by means of chemical analyses of radioactive mineral substances and are generally accepted by geologists, subject to a small percentage of error. In every period of geologic time, there were widespread accumulations of sedimentary strata; in fact, the strata comprise the record itself and constitute sole evidence of the climatic and geomorphic conditions of the time, as well as containing the fossil remains of plant and animal life.

Generally speaking, each major time unit was brought to a close by orogeny, also called *revolution*, disrupting the sequence of sediment deposition. As already noted, the largest time unit is the *era* of which the last three are *Cenozoic*, *Mesozoic*, and *Paleozoic*, in order of increasing age. All time before this is designated as belonging to the *Precambrian* time. Eras have been recognized within the Precambrian, but the record tends to be fragmentary and confused. The second order of time unit is the *period*, of which the Paleozoic era has seven; the Mesozoic era, three; and the Cenozoic, two. *Epochs* are still shorter units of time, listed on the table only for the Cenozoic era.

Figure 18.11 Shields are bounded by mountain belts (dashed lines) of Paleozoic and younger age. Shield areas include sedimentary covers. (After A. J. Eardley, *Structural Geology of North America*.)

Scheme of major geologic events

It is beyond the scope of this volume to treat physical and historical geology. But some brief insight into a general scheme of geologic events which has been repeated many times in the past may be had from the diagrams of Figure 18.12. These represent certain inferred events in the development of the Hudson Valley region throughout the Paleozoic, Mesozoic, and Cenozoic eras encompassing approximately the last 600 million years.

In block *A* the region is shown as a Paleozoic seaway in which thousands of feet of sedimentary

strata had accumulated. An inland seaway deposit of this nature is called a *geosyncline*. The source of sediment was in large part from a chain of volcanic islands lying to the east. The Paleozoic era was brought to a close by a great orogeny, the *Appalachian Revolution*. Sedimentary strata of the geosyncline were severely crumpled, as well as broken into slices which slid over one another (block *B*). As a result, a great mountain range stood where formerly there had been a seaway. The bending of the strata is generally referred to as *folding*, and the corrugated structures thus produced are simply termed *folds*. The slanting surfaces upon which sliding occurred are termed

TABLE 18.1 TABLE OF GEOLOGICAL HISTORY

ERA	PERIOD	EPOCH	Absolute age in years before present	Major geologic events in United States given in order of increasing age	Distinctive features of plant and animal life	
CENOZOIC	QUATERNARY	Recent (HOLOCENE)	10,000	Minor changes in land forms by work of streams, waves, wind	Rise of civilizations	Age of Man
		PLEISTOCENE		Four stages of spread of continental ice sheets and mountain glaciers	Development of man; extinction of large mammals	
			1,000,000	Cascadian orogeny: Cascade and Sierra Nevada ranges uplifted; volcanoes built		
	TERTIARY	PLIOCENE			Early evolution of man; dominance of elephants, horses, and large carnivores	Age of Mammals
		MIOCENE	13,000,000	Marine sediments deposited on Atlantic and Gulf coastal plain; stream deposits spread over Great Plains and Rocky Mountain basins; thick marine sediments deposited in Pacific coastal region	Development of whales, bats, monkeys	
		OLIGOCENE	25,000,000		Rise of anthropoids	
			36,000,000			
		EOCENE			Development of primitive mammals; rise of grasses, cereals, fruits	
		PALEOCENE	58,000,000		Earliest horses	
			63,000,000	Laramide orogeny: Rocky Mountains formed		
MESOZOIC	CRETACEOUS			Marine sediment deposition over Atlantic and Gulf coastal plain and in geosyncline of Rocky Mountain region	Extinction of dinosaurs; development of flowering plants	Age of Reptiles
			135,000,000	Nevadian orogeny: Intrusion of batholith of Sierra Nevada region		
	JURASSIC			Marine sediment deposition in seas of western United States; desert sands deposited in Colorado Plateau	Culmination of dinosaurs; first birds appear	
			180,000,000	Palisadian disturbance: Block faulting in eastern United States		
	TRIASSIC			Deposition of red beds in fault basins of eastern United States and in shallow basins of western United States	First dinosaurs; first primitive mammals; spread of cycads and conifers	
			230,000,000	Appalachian orogeny: Folding of Paleozoic strata of Appalachian geosyncline		
PALEOZOIC	PERMIAN			Deposition of red shales and limestones in southwestern United States; much salt and gypsum (glaciation of southern hemisphere continents)	Conifers abundant; reptiles developed; spread of insects and amphibians; trilobites become extinct.	Age of Amphibians
	CARBONIFEROUS — PENNSYLVANIAN		280,000,000	Deposition of coal-bearing strata in eastern and central United States	Widespread forests of coal-forming spore-bearing plants; first reptiles; abundant insects	
	CARBONIFEROUS — MISSISSIPPIAN		310,000,000	Deposition of limy, shaly sediments in widespread, shallow seas of central and eastern United States	Spread of sharks; culmination of crinoids	
			345,000,000	Acadian orogeny: Folding and igneous rock intrusion in New England		
	DEVONIAN			Deposition of thick marine strata in geosynclines of eastern and western United States	First amphibians; many corals; earliest forests spread over lands	Age of Fishes
	SILURIAN		405,000,000		First land plants and air-breathing animals; development of fishes	
			425,000,000	Taconian orogeny: Folding of rocks in eastern United States, Nevada, and Utah		
	ORDOVICIAN			Deposition of thick marine strata in geosynclines of eastern and western United States	Life only in seas; spread of molluscs; culmination of trilobites	Age of Marine Invertebrates
			500,000,000			
	CAMBRIAN				Trilobites predominant; many marine invertebrates	
	Precambrian time; age goes back to over four billion years		600,000,000	Many periods of sediment deposition alternating with orogeny	Earliest known forms of life; few fossils known	

overthrust faults, the process being known as *overthrusting.* Mountain topography thus produced by folding and thrusting is composed of initial landforms, as explained in Chapter 17. The agents of denudation are nevertheless shown to be dissecting the initial mountain forms even before the folding and thrusting have ceased.

After the Appalachian Revolution, the region shown here remained essentially stable and quiescent for many millions of years. The mountains were reduced by the denudational processes to a land surface of very faint relief termed a *peneplain.* The peneplain is shown in block *C.* Notice that the oldest rock, a gneiss of Precambrian age, is exposed in the core of the mountain belt, whereas the youngest sedimentary layers, of Paleozoic age, remain in the zone of least disturbance, at the left end of the diagram.

In Figure 18.12*D*, The region is shown to have been again subjected to crustal movement, but of a quite different nature from previous movements. This was the *Palisadian disturbance,* which began in the Triassic period. The region has been broken into a series of blocks, each tilted with respect to its neighbor. The fractures are a type of fault, but are different from overthrust faults by having nearly vertical inclination of the breaks and by the absence of any pronounced crustal compression and shortening. The entire breakage scheme may be termed *block faulting.* The peneplain made before faulting forms the smooth, sloping surfaces of the fault blocks.

In block *E,* the region has again been reduced to a peneplain, indicating another prolonged period of relative stability of the earth's crust lasting into the Tertiary period. The more resistant gneiss, granite, conglomerate, and sandstone rise as low hills between broad, flat lowlands underlain by shale.

The region next experienced still another crustal movement, of epeirogenic nature. This was a very simple rising, or upwarping. No faulting or folding occurred. Consequently, the peneplain of Figure 18.12*E* was merely uplifted about 2000 ft (600 m). Streams and other agents of land denudation again set to work and excavated the weaker rocks to form valleys. The resistant sandstones, conglomerates, granites, and gneisses were left as ridges and mountain masses, as shown in diagram *F,* representing the present. The topography of today is solely the result of different rates of downwasting of the ground surface upon complex rock structures formed through a long series of geologic events.

Chapters 26 to 28 treat the individual peculiarities of landforms developed on flat-lying strata, folded strata, domed strata, block faults, and the extrusive and intrusive igneous rocks.

A world-wide system of landform classification

World-wide classification systems for climates, great soil groups, and natural vegetation types are developed in earlier chapters. Plates 2 to 4 present the results of those classifications. It is essential for the geographer to add a fourth classification system—that of landforms—to complete his analysis of global distribution of the four basic elements of physical geography: climates, soils, vegetation, and landforms.

A classification system which is based in part upon the genesis (origin) of landforms and in part upon the actual configuration of the land surface is both appropriate and meaningful in terms of the explanatory-descriptive system of landform analysis described in Chapter 17. Such a classification system has been devised and applied by Dr. Richard E. Murphy[1] and has led to the world map of landforms reproduced as Plate 5 (folding plate inside back cover). Description of the system as given below conforms with definitions set forth by Dr. Murphy in 1967.[2]

The Murphy system of landform classification uses three levels, or categories, of information in successive application to identify a landform type in terms of its geologic origin and rock composition (structural regions), the configuration of its surface (topography), and finally the nature of the geomorphic process by which it has been shaped (erosional or depositional landscapes). The threefold basis of the genetic approach to landform study—structure, process, and stage—is included in the first and third categories of the classification system. A particularly valuable attribute of the genetic approach is that a highly trained and experienced geomorphologist can use his background of knowledge to predict or anticipate many characteristic details of the landforms that are not implicitly stated in the definitions of the various classes.

The empirical ingredient of the Murphy system is found in the second level of classification, namely topography. Here subdivision of geometrical properties of the land surface follows strict numerical definitions. Elevation above sea level and local relief (difference in elevation between highest and lowest points in adjacent locations) form the basis for defining classes. The empirical

[1] Richard E. Murphy, 1968, *Landforms of the world,* Annals Map Supplement Number Nine, Annals, Association of American Geographers, **58(1).**
[2] Richard E. Murphy, 1967, *A spatial classification of landforms based on both genetic and empirical factors—a revision,* Annals, Association of American Geographers, Vol. 57, No. 1, pp. 185–186. Material from the text of Dr. Murphy's paper is quoted here with the author's permission.

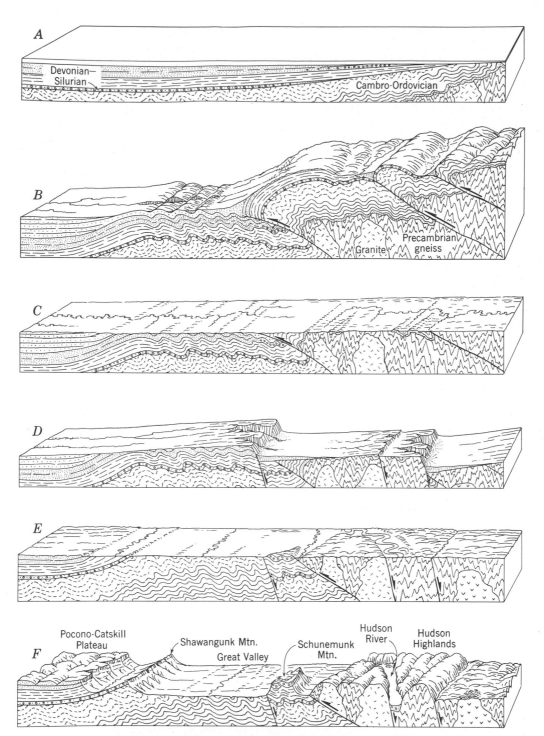

Figure 18.12   A sequence of events which has been generally repeated throughout geologic time over the continents of the globe is well illustrated in the Hudson Valley region. In these diagrams northwest is to the left, southeast to the right. *A,* A shallow inland seaway accumulated thousands of feet of sediments during the Paleozoic era. *B,* Mountain-making at the end of the Paleozoic era produced a series of folds and thrust faults. A general uplifting brought a large mass above sea level. *C,* Following a long period of erosion, a peneplain was produced. *D,* Faulting in the Triassic period produced these gently tilted blocks. *E,* A second long period of erosion resulted in another peneplain. *F,* The region today owes its relief to different rates of removal of the various kinds of rocks (From A. N. Strahler, *The Earth Sciences,* Harper and Row, New York.)

approach to description of the topography lends an element of useful and unambiguous information to the classification system.

The entire system uses three sets of letter symbols, the first to represent structural regions, the second to represent topographical classes, and the third to indicate the kinds of erosional or depositional landscapes.

Structural regions

Seven structural regions are recognized and are designated by the capital letters *A, C, G, L, R, S,* and *V,* defined as follows.

A   *Alpine system.* World-girdling system of mountain chains formed since the Jurassic period. Faulted areas, plateaus, basins, and coastal plains enclosed by such ranges are included in the system.

C   *Caledonian (or Hercynian or Appalachian) remnants.* Remains of mountain chains and ranges formed during the Paleozoic and Mesozoic eras, prior to the Cretaceous period and experiencing no orogeny since then, although epeirogenic movements may (and often have) occurred. In some cases, only worn-down roots remain. Faulted areas, plateaus, basins, and coastal plains enclosed by these remnants are included with them.

G   *Gondwana shields.* Areas of stable, massive blocks of the earth's sialic crust, lying south of the great east-west portion of the Alpine system, where Precambrian rocks form either the entire surface rock or where Precambrian rocks form an encircling enclosure with no gap of more than 200 mi (320 km) between outcroppings or covering extrusives and within which crystalline rocks form more than 50 percent of the surface rock. These shields have not been subject to orogeny since the Cambrian period.

L   *Laurasian shields.* Areas of stable, massive blocks of the earth's sialic crust lying north of the great east-west portion of the Alpine system. (Remainder of definition same as in *G,* above.)

R   *Rifted shield areas.* Block-faulted areas of shields forming grabens together with associated horsts and volcanic features.

S   *Sedimentary covers.* Areas of sedimentary layers which have not been subjected to orogeny and which lie outside either the crystalline rock enclosures of the shields or the enclosing mountains and hills of the Alpine or older orogenic systems. These areas of sedimentary rock form continuous covers over underlying structures.

V   *Isolated volcanic areas.* Areas of volcanoes, active or extinct, with associated volcanic features, lying outside the Alpine or older mountain systems and outside the rifted shield areas.

Figure 18.13 shows the distribution of the seven structural regions, together with major oceanic rift and fault lines, undersea axial connections of the Alpine system, and continental shelf areas.

Topographical regions

Six classes of topography, represented by the capital letters *P, H, T, M, W,* and *D,* are defined as follows (metric equivalents are approximate).

P   *Plains.* Surfaces with local relief less than 325 ft (100 m). On the marine side the surface slopes gently to the sea. Plains rising continuously inland may attain elevations of high plains, over 2000 ft (600 m).

H   *Hills and low tablelands.* Hill areas have local relief more than 325 ft (100 m) but less than 2000 ft (600 m). At the oceanic shoreline, however, local relief may be as low as 200 ft (60 m). A low tableland is an area less than 5000 ft (1500 m) in elevation, with local relief less than 325 ft (100 m), but which (unlike plains) either does not reach the sea or, where it does, terminates in a bluff at least 200 ft (60 m) high. A tableland may also terminate in a similar bluff overlooking a low coastal plain.

T   *High tablelands.* Upland surfaces over 5000 ft (1500 m) in elevation having local relief less than 1000 ft (300 m), except where cut by widely separated canyons.

M   *Mountains.* Areas of steep slopes with local relief more than 2000 ft (600 m).

W   *Widely spaced mountains.* Mountains which are discontinuous and stand in isolation with intervening areas having local relief less than 500 ft (150 m).

D   *Depressions.* Basins surrounded by mountains, hills, or tablelands which abruptly delimit the basins.

Figure 18.14 shows the distribution of the six topographical regions.

Erosional and depositional landforms

The kind of geomorphic process acting currently, or relatively recently in geologic time, to shape the landscape into its present form provides the basis for five classes of areas, indicated by lower-case letters *h, d, g, w,* and *i,* and defined as follows.

h   *Humid landform areas.* Areas in which the pattern of permanent streams has a density of at least one stream in every 10 mi (16 km) traverse distance, and which have not been subject to glaciation since the beginning of the Pleistocene epoch.

d   *Dry landform areas.* Areas in which the pattern of stream density is more sparse than one stream in 10 mi (16 km), and which have not been subject to glaciation since the beginning of the Pleistocene epoch. Some karst as well as arid areas are included in this category.

g   *Glaciated areas.* Areas covered by glacial ice at some time since the beginning of the Pleistocene epoch, but earlier than the Wisconsin and Würm glaciations. The symbol *g* is also used for undifferentiated glaciated areas.

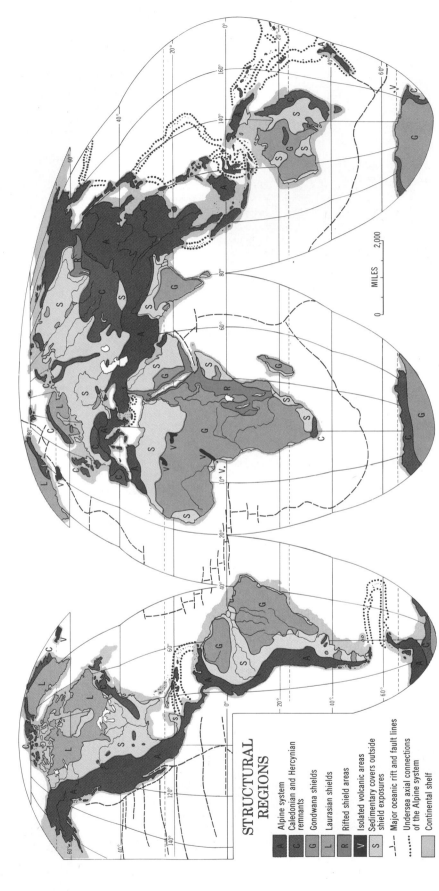

STRUCTURAL REGIONS

A	Alpine system
C	Caledonian and Hercynian remnants
G	Gondwana shields
L	Laurasian shields
R	Rifted shield areas
V	Isolated volcanic areas
S	Sedimentary covers outside shield exposures
	Major oceanic rift and fault lines
	Undersea axial connections of the Alpine system
	Continental shelf

MILES
0   2,000

Figure 18.13 World structural regions. (From R. E. Murphy, 1968, Annals, A.A.G., Map Supplement No. 9. Based on Goode Base Map. Copyright by the University of Chicago. Used by permission of the University of Chicago Press.)

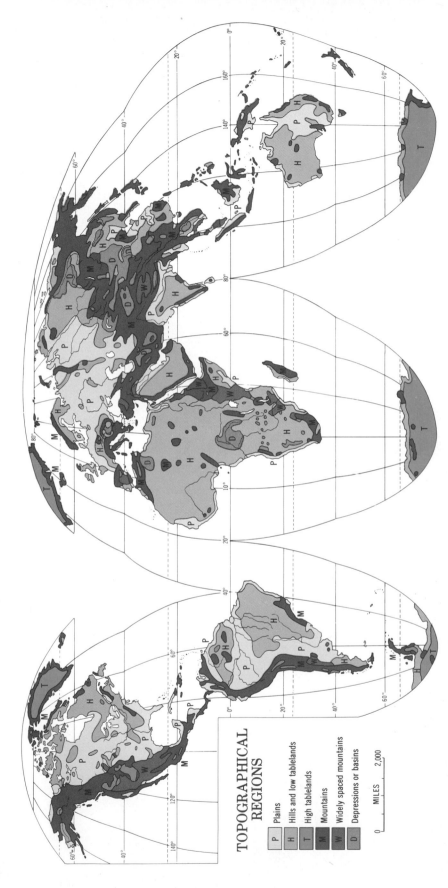

TOPOGRAPHICAL REGIONS

P Plains
H Hills and low tablelands
T High tablelands
M Mountains
W Widely spaced mountains
D Depressions or basins

0   1,000   2,000
|——————————|——————————|
MILES

Figure 18.14 World topographical regions. (From R. E. Murphy, 1968, Annals, A.A.G., Map Supplement No. 9. Based on Goode Base Map. Copyright by the University of Chicago. Used by permission of the University of Chicago Press.)

271

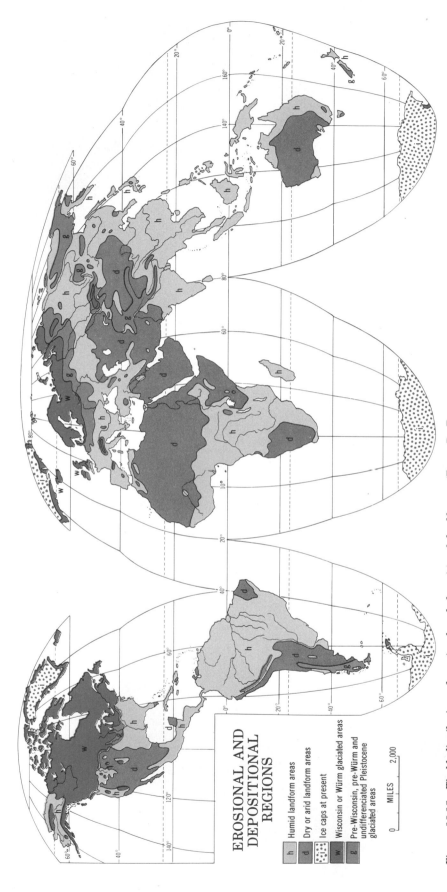

EROSIONAL AND DEPOSITIONAL REGIONS

h	Humid landform areas
d	Dry or arid landform areas
	Ice caps at present
w	Wisconsin or Würm glaciated areas
g	Pre-Wisconsin, pre-Würm and undifferenciated Pleistocene glaciated areas

MILES

0   2,000

Figure 18.15 World distribution of erosional and depositional landforms. (From R. E. Murphy, 1968, Annals, A.A.G., Map Supplement No. 9. Based on Goode Base Map. Copyright by the University of Chicago. Used by permission of the University of Chicago Press.)

w Wisconsin and Würm glaciated areas. Areas covered by glacial ice during or since the Wisconsin and Würm glaciations but now free of glacial ice.

i Icecaps. Areas covered by glacial ice at present.

Figure 18.15 shows the distribution of the five classes of erosional and depositional landforms.

Combined landform classes

Under the Murphy system, the three categories of classification are superimposed to yield a complete world landform map (Plate 5). To designate a particular area within the complete system, the code letter of each category is stated in sequence. For example, the Colorado Plateau is symbolized as *ATd*, meaning that it is an elevated tableland over 5000 ft (1500 m) in elevation, enclosed within the Alpine system, and characterized by landforms developed under conditions more arid than humid. The Congo basin is symbolized as *GDh*, indicating a depression in the Gondwana shield subject to geomorphic processes of a humid climate. Central Poland is designated as *SPg*, a sedimentary plain which has been subjected to Pleistocene glaciation.

As with all attempted classifications of climates, soils, and natural vegetation types, any classification of landforms is difficult to implement in the form of a world map because of lack of sufficient information. Over vast areas of the continents information on bedrock geology and surface topography is only generalized and of questionable reliability because of the reconnaissance nature of field surveys. Consequently, as new information of increasing detail of scale and accuracy becomes available, the world map must be modified. In the light of new information, minor changes in the categories and their definitions may also be desirable. Nevertheless, the existing world landform map (Plate 5) represents a vast quantity of soundly established geologic and topographic information, organized into a logical and meaningful classification system.

REFERENCES FOR FURTHER STUDY

Guilcher, A. (1958), *Coastal and submarine morphology*, John Wiley and Sons, New York, 274 pp. See Part II.

Heezen, B. C., M. Tharp, and M. Ewing (1959), *The floors of the oceans*, Geological Society of America, Special Paper 65, 122 pp.

Dunbar, C. (1960), *Historical geology*, second edition, John Wiley and Sons, New York, 500 pp.

Shepard, F. P. (1963), *Submarine geology*, second edition, Harper and Row, New York, 557 pp.

Kay, M. and E. H. Colbert (1965), *Stratigraphy and life history*, John Wiley and Sons, New York, 736 pp.

Woodford, A. O. (1965), *Historical geology*, W. H. Freeman and Co., San Francisco, 512 pp.

1. Describe the earth's core and mantle, giving dimensions, mineral composition, and physical properties. What type of evidence is used to obtain this information? What temperatures and pressures may be expected at the earth's center?

2. What is the earth's crust? How thick is it? How can it be distinguished in properties from the underlying mantle? Of what two rock layers does the crust consist? What is the general distribution of sima and sial over the earth?

3. Describe the general form of the continents and ocean basins as regards the levels of concentration of surface areas. If sea level were lowered by 600 ft (180 m), what percentage of the earth would be land?

4. On a globe 21 ft (6.4 m) in diameter, how far would the greatest relief features of the earth depart from a perfect circle drawn to represent sea level?

5. What are the principal second-order relief features of the ocean basins?

6. What is a continental shelf? By what type of rock material is it underlain? What is a continental slope? What relation do submarine canyons bear to these features?

7. What are turbidity currents? What work do they perform? What deposits do they build?

8. What kinds of relief features are found on the floors of the ocean basins? Describe seamounts, trenches (foredeeps), and submarine mountain ledges.

9. Describe the mid-ocean ridge system. What feature marks the central line of the ridge? How do geologists account for the mid-ocean ridge system?

10. With what sort of crustal deformation are the island arcs and deep trenches of the ocean basins associated?

11. Why are the ocean basins thought to have remained as permanent features throughout known geologic history? What is the evidence?

12. What are the second-order relief features of the continents? What are shields? What are orogenic belts?

13. What are epeirogenic crustal movements? Contrast orogenic movements with epeirogenic movements.

14. Name the eras of geologic time and give the total duration of each in years. For each of the eras, name the periods of geologic time. Into what epochs is the Tertiary period subdivided? What great events occurred during the Pleistocene epoch?

15. What type of geologic event has brought to a close each era and many of the periods? What is the known duration of Precambrian time? How much longer is Precambrian time than all of post-Cambrian time?

16. Explain the general scheme of geologic events in which sedimentary rocks are deposited, then deformed, and finally reduced by erosion. What is a geosyncline? What is a revolution?

17. What is meant by folding and overthrusting of strata? Do these deformations require compression or tension of the earth's outer crust?

18. What is a peneplain? At what level do peneplains form? How long a period of time is required to reduce a mountain range to a peneplain?

19. What is faulting? How does block faulting differ from overthrust faulting? Why would you not expect both types to occur simultaneously in the same region?

20. When a peneplain is upwarped, what type of landform development follows? What happens to the peneplain?

21. Describe the system of landform classification devised by Dr. Richard E. Murphy. Compare this system with the Köppen system of climate classification. Compare it with systems of great soil groups and structural classes of vegetation used in this book.

22. What are the seven structural regions that comprise the first category of the Murphy system? Describe the broad patterns of worldwide distribution of the mountain and shield classes.

23. How is the empirical approach used by Murphy in defining six classes of topography? What advantages does an empirical approach offer?

24. How are geomorphic processes reflected in the third category of classification in the Murphy system? Of what value is this type of information to the geographer?

CHAPTER 19
The Wasting of Slopes

THE term *slope*, as used throughout the science of geomorphology, designates some small element or area of the land surface which is inclined from the horizontal. Thus, we speak of "mountain slopes," "hill slopes," or "valley-side slopes" with reference to the inclined ground surfaces extending from divides and summits down to valley bottoms.[1]

Slopes are required for the flow of surface water under the influence of gravity. Therefore, slopes are fitted together to form drainage systems in which surface water flow converges into stream channels, these, in turn, conduct the water and rock waste to the oceans to complete the hydrologic cycle. Nature has so completely provided the earth's land surfaces with slopes that perfectly horizontal or vertical surfaces are extremely rare.

Our concern in this chapter is with the wasting of land slopes under the dominant influence of water acting in conjunction with gravity. Emphasis is on the slow processes whereby bedrock is transformed into residual overburden. This material in turn moves down the slopes to channels where it can be taken by stream flow to still more distant, lower areas. Slopes are also shaped by other processes—glaciers, winds, and waves—which are treated in later chapters.

Weathering and mass wasting

Weathering is the combined action of all processes whereby rock is decomposed and disintegrated because of exposure at or near the earth's surface. Weathering normally changes hard, massive rock into finely fragmented, soft residual overburden, the parent matter of the soil. For this reason, weathering is often described as the preparation of rock materials for transportation by the agents of land erosion—flowing water, glacial ice, waves, and wind. Because gravity exerts its force on all matter, both bedrock and the products of weathering tend to slide, roll, flow, or creep down all slopes in a variety of types of earth and rock movements grouped under the term *mass wasting*.

[1] *Slope* is also used to mean inclination from the horizontal, measured as in *dip* (Chapter 17); or it is used as a verb *to slope*, meaning *to incline*.

Weathering processes may be subdivided into two large groups, *physical (mechanical) weathering* and *chemical weathering*. Although these processes are extremely complex and act in combinations that are hard to separate into simple concepts, we shall attempt to identify the most important individual changes and to show what landforms or surface features of the rock and soil are caused by each.

Geometry of rock breakup

Before examining weathering processes, it is well to introduce four terms applied to the geometrical manner in which bedrock breaks into smaller pieces. In so doing we are not considering the possible forces involved, merely the shapes of the rock fragments as they appear to the eye.

Rocks composed of rather coarse mineral grains (intrusive igneous rocks of granitoid texture and coarse clastic sedimentary rocks) commonly fall apart grain by grain, a form of breakup termed *granular disintegration* (Figure 19.1). The product is a gravel or sand in which each grain consists of a single mineral particle separated from its fellows along the original crystal or grain boundaries. *Exfoliation* is the formation of curved rock shells which separate in succession from the original rock mass, leaving behind successively smaller spheroidal bodies (Figure 19.11). This type of breakup is also called *spalling*.

Where a rock has numerous joints produced previously by mountain-making pressures or by shrinkage during cooling from a magma, the common form of breakup is by *block separation* (Figure 19.1). Obviously, comparatively weak forces can separate such blocks, whereas great forces are required to make fresh fractures through solid rock. In sedimentary rocks the planes of stratification, or bedding planes, comprise one set of planes of weakness commonly cutting at right angles to the joints. Figures 19.7 and 19.9 show joint blocks being separated by weathering forces. Of course, it is quite possible that a single, solid joint block will later break up either by granular disintegration or by exfoliation.

Shattering is the disintegration of rock along new surfaces of breakage in otherwise massive,

Granular disintegration

Exfoliation

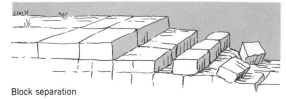
Block separation

Shattering

Figure 19.1 Rock breakup takes various forms.

strong rock, to produce highly angular pieces with sharp corners and edges (Figure 19.1). The surface of fracture may pass between individual mineral crystals or grains, or may cut through them. Blocks seen in Figure 19.2 are joint blocks, many of which have been shattered into smaller pieces.

Figure 19.2 A felsenmeer atop Medicine Bow Peak, Snowy Range, Wyoming, at 12,000 ft (3650 m) elevation. The rock is quartzite. (Photograph by A. N. Strahler.)

Physical weathering processes and forms

The physical, or mechanical, processes of weathering produce fine particles from massive rock by the exertion of stresses sufficient to fracture the rock, but do not change its chemical composition. One of the most important physical weathering processes in cold climates is *frost action*, the repeated growth and melting of ice crystals in the pore spaces or fractures of soil and rock. As water in joints freezes, it forms needlelike ice crystals extending across the openings. As these ice needles grow, they exert tremendous force against the confining walls and can easily pry apart the joint blocks. Even massive rocks can be shattered by the growth of ice crystals created from water that has previously soaked into the rock. Where soil water freezes, it tends to form ice layers parallel with the ground surface, *heaving* the soil upward in an uneven manner.

Freezing water strongly affects soil and rock in all middle- and high-latitude regions having a cold winter season, but its effects are most striking in high mountains, above the timberline. Here the separation and shattering of joint blocks may produce an extensive ground surface littered with angular blocks (Figure 19.2). Such a surface is termed a *felsenmeer* (rock sea), or *boulder field*. Where cliffs of bare rock exist at high altitudes, fragments fall from the cliff face, building up piles of loose blocks into conical forms, termed *talus cones* (Figure 19.27).

Closely related to the growth of ice crystals is the weathering process of rock disintegration by growth of salt crystals. This process operates extensively in dry climates and is responsible for many of the niches, shallow caves, rock arches, and pits in sandstone formations. During long drought periods, ground water is drawn to the surface of the rock by capillary force. As evaporation of the water takes place in the porous outer zone of the sandstone, tiny crystals of salts are left behind. The growth force of these crystals is capable of producing granular disintegration of the sandstone, which crumbles into a sand and is swept away by wind and rain. Especially susceptible are zones of rock lying close to the base of a cliff, for here the ground water tends to seep outward, perhaps prevented from further downward percolation by impervious layers below (Figure 19.3). In the southwestern United States, many of the deep niches thus formed were occupied by Indians, whose cliff dwellings obtained protection from the elements as well as safety from armed attack (Figure 19.4).

An important but little appreciated process of physical weathering is the continual swelling and shrinking of soils as the particles of fine silt and

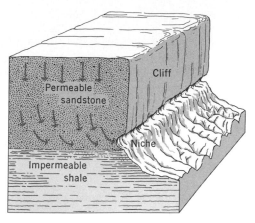

Figure 19.3 Seepage of water from the cliff base localizes development of niches through rock weathering.

clay absorb or give up soil water in alternate periods of rain and drought. Shrinkage forms soil cracks in dry periods, making the infiltration of rainfall much more rapid in early stages of an ensuing rain. In clay-rich sedimentary rocks such as shales, the swelling is largely responsible for a spontaneous breakup known as *slaking*, in which the shale crumbles into small chips or pencil-like fragments when exposed to the air.

Most crystalline solids, such as the minerals of rocks, tend to expand when heated and to contract when cooled. Where rock surfaces are exposed daily to the intense heating of the sun alternating with nightly cooling, the resulting expansion and contraction exerts powerful forces upon the rock. Given sufficient time (tens of thousands of such daily alternations), even the strongest rocks may develop fractures. Breakage can take the form of exfoliation or granular disintegration.

A curious but widespread process related to physical weathering results from *unloading*, the relief of confining pressure, as rock is brought nearer to the earth's surface through the erosional removal of overlying rock. Geologists think that rock formed at great depth beneath the earth's surface (particularly igneous and metamorphic rock) is in a slightly contracted state because of the tremendous pressures applied during mountain-making crustal deformations. On being brought to the surface, the rock expands slightly in volume and, in so doing, great shells of rock break free from the parent mass below. The new surfaces of fracture are a form of joint termed *sheeting structure* and show best in massive rocks such as granite and marble, because in a closely jointed rock the expansion would be taken up among the blocks. The rock sheets or shells produced by unloading generally parallel the ground surface and therefore tend to dip valleyward. On granite coasts the shells are found to dip seaward at all points along the shore. Sheeting structure is well seen in quarries, where it greatly facilitates the removal of rock (Figure 19.5).

Where sheeting structure has formed over the top of a single large body of massive rock, an *exfoliation dome* is produced (Figure 19.6). These are among the largest of the landforms due primarily to weathering. In the Yosemite Valley region, California, where domes are spectacularly displayed, the individual rock shells may be as thick as 20 to 50 ft (6 to 15 m).

Other large, smooth-sided rock domes lacking in shells are not true exfoliation domes, but are formed by granular disintegration of a single body of hard, coarse-grained intrusive igneous rock lacking in joints. Examples are the Sugar Loaf of Rio de Janeiro and Stone Mountain, Georgia (Figure 22.4), which rise prominently above surrounding areas of weaker rock.

Finally, in this list of physical weathering processes, the wedging of plant roots deserves consideration as a possible mechanism whereby joint blocks may be separated. We have all seen at one time or another a tree whose lower trunk and roots are firmly wedged between two great joint blocks of massive rock (Figure 19.7). Whether the tree has actually been able to spread the blocks farther apart is doubtful at best. However, it is certain that the growth of tiny rootlets in joint fractures must be of great importance in loosening countless small rock scales and grains, particularly when a rock has already been softened by decay or fractured by frost action.

Figure 19.4 White House Ruin occupies a deep niche in the sandstone wall of Canyon de Chelly, Arizona. (Photograph by Ray Atkeson.)

Figure 19.5 Sheeting of granite, a large scale form of exfoliation, facilitates quarrying operations. (Photograph by courtesy of Smith Quarry Division of Rock of Ages Corporation, Barre, Vermont.)

Chemical weathering processes and forms

Chemical weathering denotes changes in chemical compositions of rock-forming minerals to produce new minerals better suited to existing in equilibrium with the relatively low temperatures and pressures found at the earth's surface. One group of these changes involves addition of oxygen and water, both of which are abundantly available in the soil and bedrock. A second group of changes is the reaction of natural acids of the soil solution with rock-forming minerals to yield salts which are readily carried away by ground water movement. Third, certain salts of the type known as *evaporites* (see Chapter 17), found in layers in sedimentary strata, are readily dissolved without change in chemical composition and are carried away in ground water flow. The first two groups of chemical processes are considered in somewhat more detail.

Water enters into a permanent chemical union with many common rock-forming minerals in a chemical change termed *hydrolysis*. The igneous rocks are particularly susceptible. Potash feldspar commonly turns to a clay mineral known as *kaolin*. When this happens in a granite, in which potash feldspar is an important constituent, the granite will undergo granular disintegration. Not only is the kaolin a soft mineral, but it tends to swell and thus to burst the grains apart. Similarly, the iron-magnesium minerals such as hornblende and biotite decompose and soften, causing disintegration of basic rocks such as gabbro or basalt. The iron thus released unites with oxygen and water to form *limonite*, a soft, yellowish mineral that is responsible for much of the staining of rock surfaces with earth-red colors.

Figure 19.6 North Dome and Basket Dome in Yosemite National Park, California, are exfoliation domes developed from huge masses of solid igneous rock. (Photograph by Douglas Johnson.)

Figure 19.7 Jointing in sandstone resembles pavement blocks at Artists View, Catskill Mountains, New York. (Photograph by A. N. Strahler.)

Figure 19.8 Weathering converts rectangular blocks into rounded forms. (Drawn by E. Raisz.)

Figure 19.9 Egg-shaped granite boulders are produced from joint blocks by granular disintegration in a semiarid climate near Prescott, Arizona. (Photograph by A. N. Strahler.)

The hydrolysis of granite, with accompanying granular disintegration and some exfoliation of thin scales, produces many interesting boulder and pinnacle forms by rounding of angular joint blocks (Figures 19.8, 19.9, 19.10). These forms are particularly conspicuous in arid regions because of the absence of any thick cover of soil and vegetation. There is ample moisture in most deserts for hydrolysis to act, given sufficient time. Hydrolysis in fine-grained basic igneous rocks, such as basalt, commonly gives small-scale exfoliation of a type called *spheroidal weathering* (Figure 19.11).

In warm, humid climates, hydrolysis of susceptible rocks goes on below the soil and may result in the deep decay or rotting of igneous and metamorphic rocks to depths as much as 100 to 300 ft (30 to 90 m). Geologists who first studied this deep rock decay in the Southern Appalachian region termed the rotted layer *saprolite*. To the engineer, occurrences of deep weathering are of major importance in construction of highways, dams, or other heavy structures. Advantageous as is the property of softness of the saprolite, so as to be removable by power shovels with little blasting, there is serious danger in the weakness of the material in bearing heavy loads, as well as undesirable plastic properties because of a high content of clay minerals.

Of the acid reactions affecting rock-forming minerals, perhaps the most important is that caused by *carbonic acid,* a weak acid formed when carbon dioxide gas of the atmosphere is dissolved in soil and ground water. Particularly susceptible is limestone, formed of the mineral calcite (calcium carbonate). The action of carbonic acid on limestone produces a salt (calcium bicarbonate) which is readily carried off in the flow of ground water and streams. Because limestone is a common rock, this process is of major importance in land-

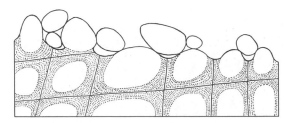

Figure 19.10 Stages in the development of egg-shaped boulders from rectangular joint blocks. (After W. M. Davis.)

Figure 19.11 Spheroidal weathering, shown here, has produced many thin concentric shells in a basaltic igneous rock. Lucchetti Dam, Puerto Rico. (Photograph by C. A. Kaye, U.S. Geological Survey.)

form development. Surfaces of limestones are commonly deeply pitted and grooved (Figure 19.12). More important is the removal of vast amounts of rock in underground locations to produce cavern systems into which surface water disappears. The landforms associated with cavern development are explained in Chapter 26.

Figure 19.12 Solution rills in limestone, west of Las Vegas, Nevada. Scale is indicated by pocket knife in center. (Photograph by John S. Shelton.)

Those who have studied the rate at which carbonic acid acts have estimated that in a humid climate, such as that of the eastern United States, the ground surface of a limestone region may be lowered at the average rate of 1 ft (0.3 m) in 10,000 years through this process alone.

In any soil rich in decaying plant matter, a variety of organic acids is formed in the soil solution, and these also react with mineral surfaces to produce chemical weathering. The salts that are products of such reactions are carried down through the soil into the ground water zone, then eventually to streams.

The weathering processes reviewed above, both physical and chemical, work universally but produce few distinctive large landforms or spectular activities that would draw the attention of the average person. Nevertheless, these processes are of enormous importance in slope development in that they prepare the bedrock for soil formation and for erosional removal by the agents of land sculpture. Without the weathering processes, vegetation could not thrive as we know it today, or could the great continental land masses be easily reduced by the agents of denudation.

Mass wasting

Everywhere on the earth's surface, gravity pulls continually downward on all materials. Bedrock is usually so strong and well supported that it remains fixed in place, but should a mountain slope become too steep through removal of rock at the base, bedrock masses break free, falling or sliding to new positions of rest. In cases where huge masses of bedrock are involved, the result may be catastrophic in loss to life and property in towns and villages in the path of the slide. Soil and overburden, being poorly held together, are much more susceptible to gravity movements. There is abundant evidence that on most slopes at least a small amount of downhill movement is going on at all times. Much of this is imperceptible, but sometimes the overburden slides or flows rapidly.

Taken altogether, the various kinds of downslope movements occurring under the pull of gravity, which we have collectively termed mass wasting, constitute an important process in slope wasting and denudation of the lands. A few of the commoner kinds of gravity movements and resulting landforms are described here.

Soil creep

On almost any moderately steep, soil-covered slope, some evidence may be found of extremely slow downslope movement of soil and overburden, a process called *soil creep*. Figure 19.13 shows some of the evidence that the process is going on.

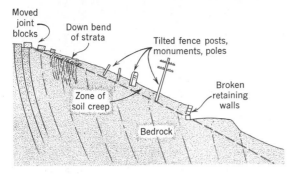

Figure 19.13 Slow, downhill creep of soil and weathered overburden. (After C. F. S. Sharpe.)

Joint blocks of distinctive rock types are found moved far downslope from the outcrop. In some layered rocks such as shales or slates, edges of the strata seem to bend in the downhill direction. This is not true plastic bending, but is the result of slight movement on many small joint cracks (Figure 19.14). Fence posts and telephone poles lean downslope and even shift measurably out of line. Retaining walls of road cuts lean and break outward under pressure of soil creep from above.

What causes soil creep? Heating and cooling of the soil, growth of frost needles, alternate drying and wetting of the soil, trampling and burrowing by animals, and shaking by earthquakes all produce some disturbance of the soil and mantle. Because gravity exerts a downhill pull on every such rearrangement that takes place, the particles are urged progressively downslope.

Creep affects rock masses enclosed in the soil or lying upon bare bedrock. Huge boulders which have gradually crept down a mountain side in large numbers may accumulate at the mountain base to produce a boulder field containing blocks the size of a house. Creep also affects the rock pieces in a talus slope, causing the angle of the talus surface gradually to become flatter. In some alpine mountains, high above timberline, sheets of frost-shattered rock fragments creep slowly down the valleys, making curious tonguelike bodies which in many ways resemble a glacier of ice. These forms are called *rock glaciers* (Figure 19.15). Among the best examples are those from mountain ranges in Alaska, where a single rock glacier may be 1 to 2 mi (1.5 to 3 km) long and 0.25 mi (0.4 km) wide.

Earthflow

In humid climate regions, if slopes are steep, masses of water-saturated soil, overburden, or weak

Figure 19.14 Slow creep has caused this down hill bending of steeply dipping sandstone layers. (Photograph by courtesy of Ward's Natural Science Establishment.)

Figure 19.15 Its wrinkled surface suggesting internal flowage, this rock glacier descends a steep mountain slope in the Copper River Region of Alaska. (Photograph by Bradford Washburn.)

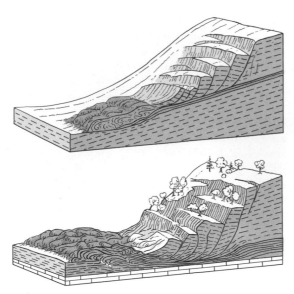

Figure 19.16 Two varieties of earthflow. (After E. Raisz.)

bedrock may slide downslope during a period of a few hours in the form of *earthflows*. Figures 19.16 and 19.17 are sketches of earthflows showing how the material slumps away from the top, leaving steplike terraces bounded by arcuate scarps, and flows down to form a bulging "toe" in which wrinkles are curved convexly downslope.

Shallow earthflows, affecting only the soil and residual overburden, are common on sod-covered slopes that have been saturated by heavy rains. An earthflow may affect a few square yards, or it may cover an area of several acres. If the bedrock is rich in clay (shale or deeply weathered igneous rocks), earthflow sometimes include millions of tons of bedrock, moving by plastic flowage like a great mass of thick mud.

A special variety of earth flowage characteristic of arctic regions is *solifluction* (from Latin words meaning *soil* and *to flow*. In late spring and early summer, when thawing has penetrated the upper few feet, soil is fully saturated with water which cannot escape downward because of the underlying impermeable frozen mass (permafrost). Flowing almost imperceptibly, this saturated soil forms terraces and lobes that give the mountain slope a stepped appearance (Figure 19.19).

Figure 19.17 Earthflows in a mountainous region. (After W. M. Davis.)

Mudflow

One of the most spectacular forms of mass wasting is the *mudflow*, a mud stream of fluid consistency which pours down canyons in mountainous regions (Figure 19.20). In deserts, where vegetation does not protect the mountain soils, violent local storms produce rain much faster than it can be absorbed by the soil. As the water runs down the slopes it forms a thin mud, which flows down to the canyon floors. Following stream courses, the mud continues to flow until it becomes so thickened that it must stop. Great boulders are carried along, buoyed up in the mud. Roads, bridges, and houses in the canyon floor are engulfed and destroyed. If the mudflow emerges from the canyon and spreads across a piedmont plain, property damage and loss of life can result, because in desert regions the plains lying at the foot of a mountain range which supplies irrigation water may be heavily populated.

Mudflows also occur on the slopes of erupting volcanoes. Freshly fallen volcanic ash and dust is turned into mud by heavy rains and flows down the slopes of the volcano. Herculaneum, a city at the base of Mt. Vesuvius, was destroyed by a mudflow during the eruption of 79 A.D., when the neighboring city of Pompeii was buried under volcanic ash.

Still other mudflows, usually of small size, occur in high mountains and in arctic tundra regions during periods of thaw when excess water is produced by melting of snow and soil ice (Figure 19.21).

Landslide

Landslide is the rapid sliding of large masses of rock with little or no flowage of the materials as in the previous types. Two basic forms of landslide are (*a*) *rockslide*, in which the bedrock mass slips on a relatively flat inclined rock plane, such as a fault or bedding plane, and (*b*) *slump*, in which there is backward rotation on a curved up-concave slip plane (Figure 19.22).

Wherever steep mountain slopes occur, there is a possibility of great and disastrous rockslides. In Switzerland, Norway, or the Canadian Rockies, for example, villages built on the floors of steep-sided valleys are sometimes destroyed and their inhabitants killed by the sliding of millions of cubic yards of rock, set loose without any warning (Figures 19.23 and 19.24). Certain kinds of artificial excavations, made in connection with the building of dams, railroads, or highways, may undermine rock masses, causing troublesome landslides. Aside from occasional great catastrophes, rockslides do not have strong geographical influences because of their sporadic occurrence in thinly populated

Figure 19.18   The great earthflow in Slumgullion Gulch, in the San Juan Mountains of Colorado, dammed a river to produce Lake San Cristobal. (Photograph by C. W. Cross, U.S. Geological Survey.)

Figure 19.19   Solifluction lobes cover this Alaskan mountain slope in the tundra climate region. (Photograph by P. S. Smith, U.S. Geological Survey.)

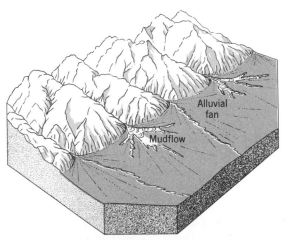

Figure 19.20   Thin streamlike mudflows commonly issue from canyon mouths in arid regions, spreading out upon the piedmont alluvial fan slopes.

Figure 19.21 A small mudflow resulting from summer thaw in a tundra climate, De Salis Bay, Banks Island (lat. 71½° N, long. 122° W). The inner channel is about 2 ft (0.6 m) wide. (Photograph by A. L. Washburn, Arctic Institute of North America.)

Figure 19.22 Landslides may involve (*A*) slip on a nearly plane surface or (*B*) slump with rotation on a curved plane.

mountainous regions. Small slides may, however, repeatedly block or break an important mountain highway or railway line.

Because rockslides are characteristic of over-steepened topography formed by glaciers, streams, and waves, and of regions having certain geologic characteristics, the basic reasons for the existence of steep slopes favorable to landsliding will be clarified in later chapters.

The second form of landslide produces *slump blocks*, great masses of bedrock or overburden that slide downward from a cliff, at the same time rotating backward on a horizontal axis (Figure 19.25). Wherever massive sedimentary strata, usually sandstones or limestones, or lava beds, rest upon weak clay or shale formations, a steep cliff

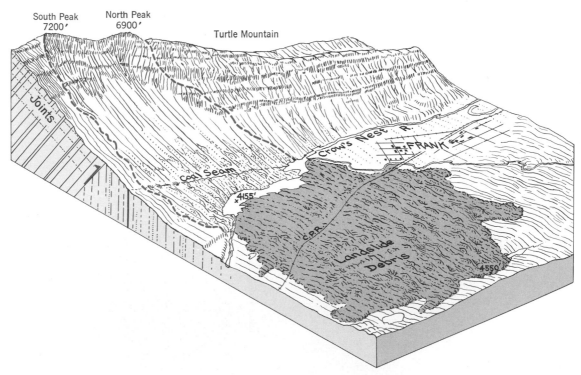

Figure 19.23 A classic example of a great, disastrous landslide is the Turtle Mountain slide, which took place at Frank, Alberta, in 1903. A huge mass of limestone slid from the face of Turtle Mountain between South and North peaks, descended to the valley, then continued up the low slope of the opposite valley side until it came to rest as a great sheet of bouldery rock debris. (After Canadian Geological Survey, Dept. of Mines.)

Figure 19.24 The debris of a landslide in Yellowstone Park, Wyoming, consists of rock masses of all sizes tumbled in a disordered fashion. (Photograph by Douglas Johnson.)

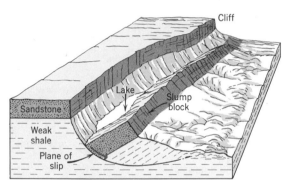

Figure 19.25 Slump blocks rotate backward as they slide from a cliff.

tends to be formed by erosion. As the weak rock is eroded from the cliff base, the cap rock is undermined. When a point of failure is reached, a large block breaks off, sliding down and tilting back along a curving plane of slip. Slump blocks may be as much as 1 to 2 mi (1.5 to 3 km) long and 500 ft (800 m) thick. A single block appears as a ridge at the base of the cliff. A closed depression or lake basin may lie between the block and the cliff.

Slumping commonly occurs on a small scale wherever weak overburden, such as floodplain alluvium or a glacial deposit, is cut away, hence it is seen in caving river banks or along a sea cliff.

Rockfall and talus

Most rapid of all mass wasting processes is *rockfall*, the free falling or rolling of single masses of rock from a steep cliff. Individual fragments may be as small as sand grains, or as large as a

city block, depending upon the overall scale of the cliff and the manner in which the rock breaks up. Large blocks disintegrate upon falling, strewing the slope below with rubble and leaving a conspicuous scar on the upper cliff face (Figure 19.26).

Rockfall goes on continuously over long periods of time, yielding a rain of countless individual small fragments at the cliff base. These accumulate in a distinctive landform, the *talus cone*, already referred to as a by-product of rapid frost disintegration on exposed cliff faces at high altitudes or in arctic regions (Figure 19.27). A talus slope, or *scree slope*, as it is often called, has a remarkably constant slope angle of about 34° or 35°. So long as the talus slope is freshly formed and contains little very fine material mixed in with the coarse, the angle is constant within one or two degrees of variation, regardless of the rock type or the shape of the blocks.

Most cliffs are notched by narrow ravines which funnel the fragments into individual tracks, so as to produce conelike talus bodies arranged side by side along the cliff. Where a large range of sizes of particles is supplied, the larger pieces, by reason of their greater momentum and ease of rolling, travel to the base of the cone, whereas the tiny

Figure 19.26 This rockfall on the Palisades of the Hudson River, just north of the George Washington Bridge, took place in November 1955. About 1200 tons of broken rock fragments resulted from the fall of a large columnar joint block of diabase. (Photograph by Bergen Evening Record.)

Figure 19.27 Talus cones at the base of a frost-shattered cirque headwall. Moraine Lake in the Canadian Rockies. (Photograph by Ray Atkeson.)

grains lodge in the apex. This tends to sort the fragments by size, progressively finer from base to apex (Figure 19.28).

Most fresh talus slopes are unstable, so that the disturbance created by walking across the slope, or dropping of a large rock fragment from the cliff above, will easily set off a sliding of the surface layer of particles. The upper limiting angle to which coarse, hard, well-sorted rock fragments will stand is termed the *angle of repose*. Other examples of this critical angle of slope are seen in the leeward surfaces (slip faces) of sand dunes and on the side slopes of small volcanic cones.

REFERENCES FOR FURTHER STUDY

Matthes, F. E. (1930), Geologic history of Yosemite Valley, *U.S. Geol. Prof. Paper 160*, 137 pp. See pp. 114–116.

Sharpe, C. F. S. (1938), *Landslides and related phenomena*, Columbia University Press, New York, 137 pp.

Lobeck, A. K. (1939), *Geomorphology*, McGraw-Hill Book Co., New York, 731 pp. See Chapter 3.

Reiche, P. (1950), *A survey of weathering processes and products*, University of New Mexico Press, 95 pp.

Highway Research Board (1958), *Landslides and engineering practice*, Special Report 29, N.A.S.—N.R.C. Publication 544. Washington, D.C., 232 pp.

Thornbury, W. D. (1969), *Principles of geomorphology*, second edition, John Wiley and Sons, New York, 594 pp. See chapter 4.

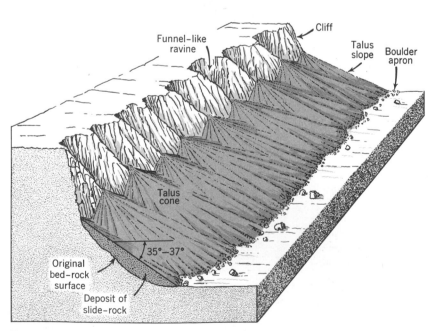

Figure 19.28 Idealized diagram of talus cones formed at the base of a cliff, which might be 200 to 500 ft (60 to 150 m) high.

REVIEW QUESTIONS

1. What is the meaning of the term slope? In what way do slopes contribute to the functioning of the hydrologic cycle?

2. Define and distinguish between weathering and mass wasting. Into what two large groups can the weathering processes be divided?

3. Describe four common geometrical forms of rock breakup. For each, name two or three common rock types in which the particular form of breakup might be expected.

4. List the physical processes of weathering. Of these, which require water to be present? Which do not? Which are controlled by changes in air temperature? Which can act deep in the soil or bedrock? Which cannot?

5. What are felsenmeer and talus? Under what climatic environment would they be formed? What process of weathering dominates in such places?

6. How can the growth of salt crystals cause rock disintegration? What climatic conditions favor this process? What forms result?

7. Under what conditions can temperature changes alone cause rock disintegration? What evidence is available that the process is effective?

8. Explain how sheeting structure and exfoliation domes result from unloading. What rock type is most likely to exhibit such structure? Name a locality famous for the display of exfoliation domes.

9. Comment on the effectiveness of plant growth forces in producing the disintegration of rock.

10. What three groups of changes come under the general heading of chemical weathering processes?

11. What is the nature of changes in minerals when hydrolysis takes place? How does hydrolysis promote weathering? What rocks are most susceptible to this type of decay? What are some of the visible forms resulting from hydrolysis in granitic rocks? In basic rocks? How deep in the ground do the effects of hydrolysis extend?

12. How is carbonic acid formed and how does it act on rock? What type of rock is most susceptible? What surface forms are produced? What other acids are commonly found in soil water?

13. What role does mass wasting play in the denudation of the lands? List the evidences of soil creep. What general type of mechanism causes soil creep? Are large rock masses also affected by creep? Explain.

14. Describe an earthflow. How large are earthflows? Under what conditions of season and climate might earthflows be expected to take place? Describe and explain the special features of solifluction.

15. Under what topographic and climatic conditions do mudflows occur? Point out the similarities and differences between a mudflow and a stream.

16. To what kinds of movement and earth material is landslide restricted? Distinguish between rockslide and slump as two basic types of landslide. Describe the topographic and geologic conditions favorable to each of these types.

17. What is rockfall? What sizes of rock masses may be included? How is a talus cone produced and what angle of slope does it normally have? How may the particles of a talus cone be arranged in order of size from apex to base? Explain.

CHAPTER *20*
Runoff and Ground Water

IN earlier chapters we followed the hydrologic cycle from the water vapor phase in air masses through precipitation and evapotranspiration to infiltration as subsurface water. Chapter 9 treated in detail the part of the cycle dealing with soil moisture and the water budget. Study of the cycle is now resumed by a consideration of *runoff*, which consists of all surface water flow, both over the land slopes and in streams. Runoff may be derived directly from excessive precipitation which cannot infiltrate the soil, or it may originate as the outflow of ground water along lines where the water table intersects the earth's surface.

Seeking to escape to progressively lower levels and eventually to the sea, runoff becomes organized into *drainage systems*, which we may describe as more or less pear-shaped areas bounded by divides, within which ground slopes and branching stream networks are adjusted to dispose as efficiently as possible of the runoff and its contained load of mineral particles, and thus to lower the land surfaces progressively toward the ultimate goal of the processes of denudation—reduction to a nearly featureless plain close to sea level. Most drainage systems possess a constricted exit, normally the mouth of the master stream, where it meets a large body of water. Thus, a drainage system is a converging mechanism funneling and integrating the weaker and more diffuse forms of runoff into progressively deeper and more intense paths of activity.

The study of drainage systems brings us in contact with two fields of science, hydrology and geology. Much of the study of the water itself, particularly as to the quantities of water involved in runoff and their variations in response to precipitation, is done by hydrologists, who are affiliated with the profession of civil engineering. The study of streams in eroding and transporting rock materials, so as to shape the landforms of drainage systems, is done by geologists. In the United States, both groups pool their efforts in the analysis and solution of problems of both runoff and ground water under the Water Resources Division of the U.S. Geological Survey, to which is given the responsibility for assessing the surface and ground water resources of the nation. Many facets of the

work of that agency will be touched upon in this chapter; also the work of the Forest Service, which studies problems of runoff in the nation's forests, and the Soil Conservation Service, which is concerned with the effects of runoff in causing land erosion and related agricultural problems. Runoff problems concerned with the improvement of irrigation works and navigable waterways are dealt with by engineers and scientists of the Bureau of Reclamation and the U.S. Army Corps of Engineers.

Runoff, ground water, and geography

For the geography student, runoff and ground water are subjects of vital concern as comprising a basic natural resource upon which agricultural and industrial development are heavily dependent. Runoff held in reservoirs behind dams provides water supplies for great urban centers, such as New York City and Los Angeles; diverted from large rivers, it provides irrigation water for highly productive lowlands in arid lands, such as the Imperial Valley of California and the Nile Valley of Egypt. To these uses are added hydroelectric power, where the drop of the river is steep; or routes of inland navigation, where the drop is gentle.

Available sources of both surface and ground water are fast being fully exploited in heavily populated areas. More and more attention is being paid to reducing various forms of waste of useful water, making a larger proportion available for man's productive use. In underdeveloped lands, many proposed improvements in the agricultural and industrial economy require development of surface and ground water resources. A geographer analyzing the potential of such regions needs to understand certain principles of hydrology and geology if he is to make realistic appraisals of the available water resources.

Forms of overland flow

Runoff that flows down the slopes of the land in more or less broadly distributed films, sheets, or rills is referred to as *overland flow* in distinction with *channel flow*, or *stream flow*, in which the water occupies a narrow trough confined by lateral banks. Within this broad definition, overland flow can take many forms. It may be a continuous thin film, called *sheet flow*, where the soil or rock surface is extremely smooth, or a series of tiny rivulets connecting one water-filled hollow with another, where the ground is rough or pitted. On a grass-covered slope, overland flow is subdivided into countless tiny threads of water, passing around the stems. Even in a heavy and prolonged rain, overland flow in full progress on a sloping lawn may not be visible to the casual observer. On heavily forested slopes bearing a thick mat of decaying leaves and many fallen branches and tree trunks, overland flow may pass almost entirely concealed beneath this cover.

Intermediate in classification between overland flow and channel flow is flow in shallow *shoestring rills*, which may score the hillside surface as a system of long, parallel lines (Figure 20.4). In some cases, shoestring rills are merely seasonal features, developed during periods of torrential rain in the spring and summer, but healing over as ground frost heaves the soil during the winter season. Again, shoestring rills may actually represent a permanent change brought on by deforestation or cultivation, in which new stream channels are in the process of formation.

Progress of overland flow

Imagine that a hillside slope which has been thoroughly drained of moisture in a period of drought is subjected to a period of rain. What are the successive stages in the production of overland flow? If heavy vegetation such as forest is present, much of the rain at the beginning is held in droplets on the leaves and plant stems, a process termed *interception*. This water may be returned directly to the atmosphere by evaporation, so that if the rainfall is brief, little water reaches the ground.

As explained in Chapter 9, the ground surface is capable of absorbing by infiltration even a heavy rain in the early stages of a rain period. Therefore, unless the rain continues for an hour or so, no overland flow may be expected. Soon, however, the soil passages are sealed or obstructed, dropping the infiltration rate to a low, but constant value. Any excessive precipitation now remains on the ground surface, first accumulating in small puddles or pools which occupy natural hollows in the rough ground surface or are held behind tiny check dams formed by fallen leaves and twigs (Figure 20.1). *Surface detention* is the term applied to the holding of water on a slope by such small natural containers. Assuming that the rain continues to fall with sufficient intensity, water then overflows from hollow to hollow, becoming true overland flow.

Because any given square unit of ground on a hill side must receive the overland flow from the

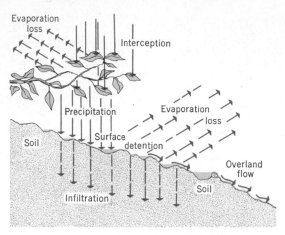

Figure 20.1 Precipitation and overland flow.

entire strip of ground of that width lying upslope of it, we may expect the rate of discharge (volume of water passing across a given line in a given unit of time) to increase in direct proportion to the length of the total path of flow. Depth of the flowing layer might therefore be expected to increase the farther downslope it progresses, but this increase may be small because the flow velocity will also be increasing down the slope. Figure 20.2 shows heavy runoff at the base of a long slope, where the accumulated overland flow has converged into broad shallow streams spreading across the slope.

At the base of a hill slope, overland flow is disposed of by passing into a stream channel or lake, or by sinking into the ground, should a highly permeable layer of sand, gravel, or blocky slide rock be encountered.

Overland flow is measured in inches or centimeters of water per hour, just as for precipitation and infiltration. Therefore, a simple formula expresses the rate at which overland flow will be produced by a given unit of ground surface as follows:

Rate of production of overland flow = rate of precipitation − rate of infiltration

For example, if the rate of infiltration became constant at a value of 0.4 inches per hour, and the rate of rainfall was a steady 0.6 inches per hour (a heavy rain), the runoff would be produced at a rate of 0.2 inches per hour, assuming none to be returned to the atmosphere by evaporation.

Accelerated land erosion

Overland flow, by exerting a dragging force over the soil surface, picks up particles of mineral matter ranging in size from fine clay to coarse sand or gravel, depending on the speed of the flow and the degree to which the particles are bound by

Figure 20.2   Overland flow running down an 8 percent slope following a heavy thunderstorm. The ditch in the foreground receives the runoff and conducts it away as channel flow. (Soil Conservation Service photograph.)

plant rootlets or held down by a mat of leaves. Added to this solid matter is dissolved mineral matter produced by acid reactions or direct solution. Such slow removal of soil is part of the natural geological process of landmass denudation and is both inevitable and universal. Under stable, natural conditions, the erosion rate in a humid climate is slow enough that a soil with distinct horizons is formed and maintained, enabling vegetation to maintain itself. Soil scientists refer to this state of activity as the *geologic norm.*

By contrast, the rate of soil erosion may be enormously speeded up through man-made activities or rare natural events to result in a state of *accelerated erosion,* removing the soil much faster than it can be formed. This condition comes about most commonly from a change in the conditions of vegetative cover and physical state of the ground surface. Destruction of vegetation by clearing of

land for cultivation, or by forest fires, directly causes great changes in the relative proportions of infiltration to runoff. Interception of rain by foliage is ended; protection afforded by a ground cover of fallen leaves and stems is removed. Consequently the rain falls directly upon the mineral soil.

Direct force of falling drops (Figure 20.3) causes a geyserlike splashing in which soil particles are lifted and then dropped into new positions, a process termed *splash erosion.* It is estimated that a violent rainstorm has the ability to disturb as much as 100 tons of soil per acre (225 metric tons per hectare). On a sloping ground surface, splash erosion tends to shift the soil slowly downhill. A more important effect is to cause the soil surface to become much less able to infiltrate water because the natural soil openings become sealed by particles shifted by raindrop splash. Reduced infiltration permits a much greater proportion of overland flow to occur from rain of given intensity and duration. The depth and velocity of overland flow then increase greatly, intensifying the rate of soil removal.

Another effect of destruction of vegetation is to reduce greatly the resistance of the ground surface to the force of erosion under overland flow. On a slope covered by grass sod, even a deep layer of overland flow causes little soil erosion because the energy of the moving water is dissipated in friction with the grass stems, which are tough and elastic. Similarly on a heavily forested slope, countless check dams made by leaves, twigs, roots, and fallen tree trunks take up the force of overland flow. Without such vegetative cover the eroding force is applied directly to the bare soil surface, easily dislodging the grains and sweeping them downslope.

Summarizing these things, we may state that the eroding capacity of overland flow is directly proportional to the rate of precipitation and length of slope, but inversely proportional to both the infiltration capacity of the soil and the resistance of the surface. To complete this equation, we need only to add the effect of the steepness of ground slope. Obviously, the steeper the slope of ground, the faster is the flow and the more intense the erosion. We therefore add that the eroding capacity of overland flow increases directly with angle of slope. As the slope angle approaches the vertical, however, erosion will become less intense from overland flow because the ground surface intercepts much less of the vertically falling rain.

Forms of accelerated erosion

When a plot of ground is first cleared of forest and ploughed for cultivation, little erosion will occur until the action of rain splash has broken down the soil aggregates and sealed the larger opening. Following this, overland flow begins to remove the soil in rather uniform thin layers, a process termed *sheet erosion.* Because of seasonal cultivation, the effects of sheet erosion are often little noticed until the upper horizons of the soil (*A* and *B* horizons) are removed or greatly thinned. Reaching the base of the slope, where the angle of surface is rapidly reduced to meet the valley bottom, soil particles come to rest and accumulate in a thickening layer termed *colluvium,* or simply *slope wash.* This, too, has a sheetlike distribution and may be little noticed, except where it can be seen that fence posts or tree trunks are being slowly buried.

Material that continues to be carried by overland flow to reach a stream in the valley axis is then carried further down valley and may be built up into layers on the valley floor, where it becomes

Figure 20.3 A large raindrop (above) lands on a wet soil surface, producing a miniature crater (below). Grains of clay and silt are thrown into the air and the soil surface is disturbed. (Official U.S. Navy photograph.)

alluvium, a word applied generally to any stream-laid deposits. Colluvium and alluvium together are described as products of *sedimentation,* the opposite process from erosion. In many ways, sedimentation at the base of slopes and in valley bottoms is a process equally serious to erosion from the agricultural standpoint, because it results in burial of soil horizons under relatively infertile, sandy layers and may choke the valleys of small streams, causing the water to flood broadly over the valley bottoms.

Where slopes are exceptionally steep and runoff from storms is exceptionally heavy, sheet erosion progresses into a more intense activity, that of *rill erosion,* or *rilling* (Figure 20.4), in which innumerable, closely spaced channels, already referred to as *shoestring rills,* are scored into the soil and subsoil. If these rills are not destroyed by soil tillage, they may soon begin to integrate into still larger channels, termed *gullies.* This transformation comes about as the more active rills deepen more rapidly than their neighbors and incorporate

Figure 20.4 Shoestring rills on a barren slope. (Soil Conservation Service photograph.)

the adjacent drainage areas. Erosive action thus is concentrated into a few large channels which deepen into steep-walled, canyonlike trenches whose upper ends grow progressively upslope (Figure 20.5).

Ultimately, a rugged, barren topography, resembling the badland forms of the arid climates, may result from accelerated soil erosion allowed to proceed unchecked. Curative measures developed by the Soil Conservation Service have proved effective in stopping accelerated soil erosion and permitting the return to slow erosion rates approaching the geologic norm. These measures

Figure 20.5 This great gully, eroded into deeply weathered overburden, was typical of certain parts of the Piedmont region of South Carolina and Georgia before remedial measures were applied. (Soil Conservation Service photograph.)

include construction of terraces to reduce slope angle and distance of overland flow, permanent restoration of overly steep slope belts to dense vegetative cover, and the healing of gullies by placing check dams in the gully floors.

Stream channels

The channel of a stream may be thought of as a long, narrow trough, shaped by the forces of flowing water to be most effective in moving the quantities of water and sediment supplied from the drainage basin or watershed. Channels may be so narrow that a person can jump across them, or as wide as one mile (1.5 km) for great rivers such as the Mississippi. Taking the entire range of natural channel widths as between one foot and one mile, a 5000-fold difference in size can exist.

Hydraulic engineers who must measure stream dimensions and flow rates have adopted a set of terms to describe channel geometry (Figure 20.6). *Depth*, in feet or meters, is measured at any specified point in the stream as the vertical distance from surface to bed. *Width* is the distance across the stream from one water's edge to the other. *Cross-sectional area*, A, is the area in square feet or square meters of a vertical slice across the stream at any specified place. *Wetted perimeter*, P, is the length of the line of contact between the water and the channel, as measured from the cross section. An important characteristic of streams is the *hydraulic radius*, R, which is defined as the cross-sectional area, A, divided by the wetted perimeter, P, or $R = A/P$.

Another important ratio expressing channel geometry is *form ratio*, defined as depth, d, divided by width, w, or d/w. Form ratio is commonly stated as a simple fraction such as $1/100$ or $1:100$, meaning that the stream channel is 100 times as wide as it is deep. Finally, a most important measure is *slope*, S (or *gradient*), which is the angle between the water surface and the horizontal plane. Slope can be stated in feet per mile or meters per kilometer. Thus a slope of five feet per mile means that the stream surface undergoes a vertical drop of five feet for each mile of horizontal distance downstream. Slope can also be given in terms of *per cent grade,* a common practice in engineering. A grade of 3 percent, or 0.03, means that the stream drops 3 feet for every 100 feet of horizontal distance.

Stream flow

Gravity acts upon the water of a stream to exert pressure against the confining walls. A small part of the gravitational force is aimed downstream parallel with the surface and bed, causing flow. Resisting the force of downstream flow is the force

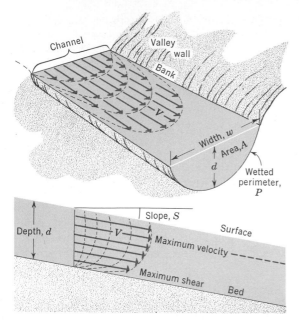

Figure 20.6   Geometry of a stream channel.

of resistance, or friction, between the water and the floor and sides of the channel. As a result, water close to the bed and banks moves slowly; that in the deepest and most centrally located zone flows fastest. Figure 20.6 indicates by dotted lines the manner in which flow takes place, or the *velocity distribution*. We can imagine that each dot is a given drop of water and that we observe its subsequent positions at equal time intervals. The single line of highest velocity is located in mid-stream, if the channel is straight and symmetrical, but about one-third of the distance down from surface to bed.

The above statements about velocity need to be qualified. Actually, in all but the most sluggish streams, the water is affected by *turbulence*, a system of innumerable eddies that are continually forming and dissolving. Therefore, a particular molecule of water, if we could keep track of it, would actually describe a highly irregular, cork-screw path as it is swept downstream. Motions would include upward, downward, and sideward directions. Turbulence in streams is extremely important because of the upward elements of flow that lift and support fine particles of sediment. The murky, turbid appearance of streams in flood is ample evidence of turbulence, without which sediment would remain near the bed. Only if we measure the water velocity at a certain fixed point for a long period of time, say several minutes, will the average motion at that point be downstream and in a line parallel with the surface and bed. It is such average values that are shown by the arrows in Figure 20.6.

Because the average velocity at a given point in a stream differs greatly according to whether it is being measured close to the banks and bed, or out in the middle line, a single figure, the *mean velocity*, is computed for the entire cross section to express the activity of the stream as a whole. Mean velocity in streams is commonly equal to about six-tenths of the maximum velocity, but depends on the relative depth of the stream.

The last and most important measure of stream flow is *discharge*, *Q*, defined as the volume of water passing through a given cross section of the stream in a given unit of time. Commonly, discharge is stated in cubic feet per second, abbreviated to *cfs*. Sometimes the hydraulic engineer simply states this quantity as *second feet*. In metric units discharge is stated in cubic meters per second (cms). Discharge may be obtained by taking the mean velocity V, and multiplying it by cross-sectional area, A. This relationship is stated by the important equation, $Q = AV$, sometimes referred to as "the equation of continuity of flow."

Stream gauging

An important activity of the U.S. Geological Survey is the measurement, or gauging, of stream flow in the United States. In cooperation with states and municipalities this organization maintains over 6000 river-measurement stations on principal streams and their tributaries. The figures on discharge thus obtained are published by the Geological Survey in a series of *Water-Supply Papers*. Information on daily discharge and flood discharges is essential for planning the distribution

Figure 20.7   This stilling tower on Fish Creek, near Duarte, California, houses gauges for recording stream flow. (U.S. Geological Survey photograph.)

Figure 20.8 In gauging large rivers, the current meter is lowered on a cable by a power winch. Earphones, connected by wires to the meter, receive a series of clicks whose frequency indicates water velocity. (U.S. Geological Survey photograph.)

and development of surface waters as well as for design of flood-protection structures and for the prediction of floods as they progress down a river system.

A stream gauging station requires a device for measuring the height of the water surface, or *stage* of the stream. Simplest to install is a *staff gauge*, which is simply a graduated stick permanently attached to a post or bridge pier. This must be read directly by an observer whenever the stage

is to be recorded. More useful is an automatic-recording gauge, which is mounted in a *stilling tower* built beside the river bank (Figure 20.7). The tower is simply a hollow masonry shaft filled by water which enters through a pipe at the base. By means of a float connected by cable to a recording mechanism above, a continuous ink-line record of the stream stage is made on a graph paper attached to a slowly rotating drum.

To measure stream discharge, it is necessary to determine both the area of cross section of the stream and the mean velocity. This requires that a *current meter* (Figure 20.8) be lowered into the stream at closely spaced intervals so that the velocity can be read at a large number of points evenly distributed in a grid pattern through the stream's cross section (Figure 20.9). A bridge often serves as a convenient means of crossing over the stream; otherwise a cable car or small boat is used. The current meter has a set of revolving cups whose rate of turning is proportional to current velocity. The Price current meter, pictured in Figure 20.8, is in general use by the Geological Survey and will measure velocities from 0.2 to 20 ft (0.06 to 6 m) per second. As the velocities are being measured from point to point, a profile of the river bed is also made by sounding the depth. Thus, a profile is drawn and the cross-sectional area is measured from the profile. Mean velocity is computed by summing all individual velocity readings and dividing by the number of readings. Discharge can then be computed using the formula, $Q = AV$.

Stream flow and precipitation

By studying the records of stream discharge in relation to precipitation on a given watershed, the hydrologist has developed a set of basic principles applying to the variations in stream discharge with different lengths and intensities of storms and with different sizes of watersheds.

Figure 20.10 shows the water graph, or *hydrograph*, of Sugar Creek, Ohio, with a watershed area of 310 sq mi (805 sq km). Sugar Creek basin, a part of the much larger Muskingum River watershed, is outlined in Figure 20.11, a map showing

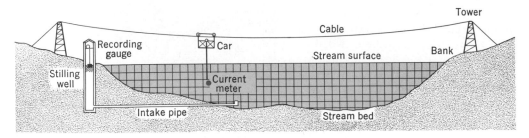

Figure 20.9 Idealized diagram of stream gauging installation.

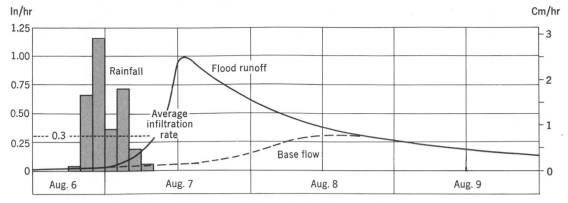

Figure 20.10 Four days of flow of Sugar Creek, Ohio. (After Hoyt and Langbein, *Floods*.)

by isohyetal lines the rainfall during the twelve-hour storm of August 6–7, 1935, for which the hydrograph was constructed. Over the area of Sugar Creek the average total rainfall was 6.3 in (16 cm) for the entire storm, but the total quantity discharged by Sugar Creek was only 3.0 in (7.5 cm). This means that 3.3 in (8.5 cm), or more than half of the rainfall, was retained on the watershed, having infiltrated to become part of the soil and ground water, or had evaporated.

Studying the rainfall and runoff graphs in Figure 20.10, we see that prior to the onset of the storm, Sugar Creek was carrying a small discharge. This was being supplied by the seepage of ground water into the channel and is termed *base flow*. After

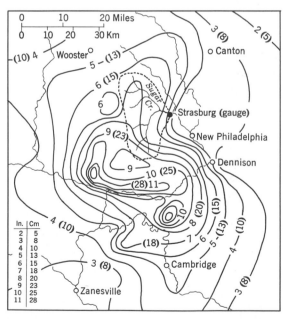

Figure 20.11 Isohyets of Sugar Creek Watershed, Ohio. Centimeter equivalents in parentheses. (After Hoyt and Langbein, *Floods*.)

the heavy rainfall began several hours elapsed before the stream gauge at the basin mouth began to show a rise in discharge. This time interval is known as the *lag* and indicated that the branching system of channels was acting as a temporary reservoir, receiving inflow more rapidly than it could be passed down the channel system to the stream gauge. The term *channel storage* is applied to runoff delayed in this manner during the early period of a storm.

The peak of flow in Sugar Creek was reached almost 24 hours after the rain began, or about 6 hours after the cessation of rainfall. Note also that the rate of decline in discharge was much slower than the rate of rise. In general, the larger a watershed, the longer is the lag of time between peak rainfall and peak discharge; the more gradual is the rate of decline of discharge after the peak has passed. Because much rainfall had entered the ground and had reached the water table, a slow but distinct rise is seen in the amount of discharge contributed by base flow.

Base flow and surface water flow

In regions of humid climates, where the water table is high and normally intersects the important stream channels, the hydrographs of larger streams will show clearly the effects of two sources of water: (*a*) *base flow* and (*b*) *surface water flow*. Figure 20.12 is a hydrograph of the Chattahoochee River, Georgia, a large river draining a watershed of some 3350 sq mi (8700 sq km), much of it in the humid southern Appalachian Mountains. The sharp, abrupt fluctuations in discharge are produced by overland flow following rain periods of one to three days duration. These are each similar to the hydrograph of Figure 20.10, except that they are here shown much compressed by the time scale.

After each rain period the discharge falls off rapidly, but if another storm occurs within a few

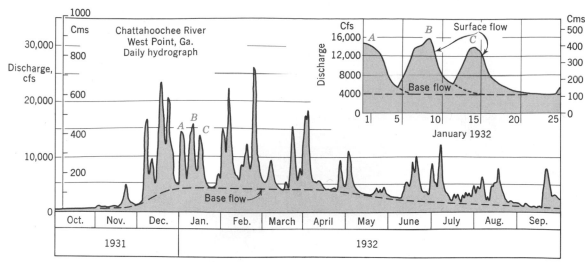

Figure 20.12   Flow peaks of the Chattahoochee River. (After E. E. Foster, *Rainfall and Runoff.*)

days, the discharge rises to another peak. The enlarged inset graph, showing details of the month of January, reveals how this effect occurs. Where a long period intervenes between storms, the discharge falls to a low value, the base flow, where it levels off. Throughout the year the base flow, which represents ground water inflow into the stream, undergoes a marked annual cycle. During the period of recharge (winter and early spring), water table levels are raised and the rate of inflow into streams is increased. For the Chattahoochee River, the rate of base flow during January, February, March, and April holds uniform at about 4000 cfs (110 cms). As the heavy evapotranspiration losses of spring reduce soil water, and therefore cut off the recharge of ground water by downward percolation (see Chapter 9), the base flow falls steadily. The decline continues through the summer, reaching by the end of October a low of about 1000 cfs (30 cms) supplied entirely from base flow.

Finally, examine the hydrograph of the Missouri River at Omaha, Nebraska, from October 1940, to September 1942 (Figure 20.13). This great river, draining 322,800 sq mi (840,000 sq km) of water-

shed, is a major tributary of the Mississippi River. Note that the discharge, ranging from 10,000 to 100,000 cfs (280 to 2800 cms), is many times greater than the discharges of smaller streams considered thus far. High rates of flow are chiefly from snowmelt, which occurs on the High Plains in spring and in the Rocky Mountain headwater areas in early summer. This event explains the sudden high discharges from April through June. During midwinter, when soil moisture is frozen and total precipitation small over the watershed as a whole, the discharge rises little above the base flow. Ground water recharge occurring in the spring raises summer levels of base flow to about 20,000 cfs (570 cms), or two to three times the winter base flow.

Floods

In our modern day of newspapers, movies, and television, everyone has seen enough pictures of river floods to have a good idea of the appearance of flood waters and the havoc wrought by their erosive power and by the silt and clay that they leave behind. Nevertheless, even the hydraulic engineer may not be fully satisfied that he can

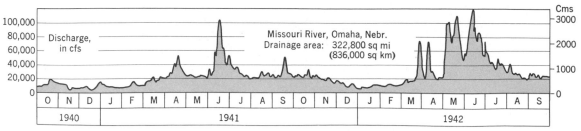

Figure 20.13   Discharge of the Missouri River. (After E. E. Foster, *Rainfall and Runoff.*)

exactly define the term *flood*. Perhaps it is enough to say that a condition of flood exists when the discharge of a river cannot be accommodated within the margins of its normal channel, so that the water spreads over adjoining ground upon which crops or forests are able to flourish.

Most larger streams of humid climates have a *floodplain*, a belt of low flat ground bordering the channel on one or both sides inundated by stream waters about once a year, at the season when abundant supplies of surface water combine with effects of a high water table and ample soil moisture to supply more runoff than can stay within the heavily scoured troughlike channel (Figure 20.14). Such annual inundation is considered a flood, even though its occurrence is expected and does not prevent the cultivation of crops after the flood has subsided, or does it interfere with the growth of dense forests which are widely distributed over low, marshy floodplains in all humid regions of the world. Still higher discharges of water, the rare and disastrous floods which may occur as infrequently as a decade or longer, inundate ground lying above the floodplain, principally affecting broad steplike expanses of ground known as *terraces* (Figure 20.15).

For practical purposes, the U.S. Weather Bureau, which provides a flood-warning service, designates a particular stage of gauge height at a given place as the *flood stage*, implying that the critical level has been reached above which overbank flooding may be expected to set in. Immediately at or below flood stage the river may be described as being in the *bank-full stage*, the flow being entirely within the limits of the heavily scoured channel.

Downstream progress of a flood wave

The rise of a river stage to its maximum height, or *crest*, followed by a gradual lowering of stage, is termed the *flood wave*. The flood wave is simply a large-sized rise and fall of river discharge of the type already analyzed in earlier paragraphs, and follows the same principles. Figure 20.16*A* shows the downstream progress of a flood on the Chattooga-Savannah river system. In the Chattooga River near Clayton, Georgia, the flood peak or crest was quickly reached—one day after the storm—and quickly subsided. On the Savannah River, 65 mi (105 km) downstream at Calhoun Falls, South Carolina, the peak flow occurred a day later, but the discharge was very much larger because of the larger area of watershed involved. Downstream another 95 mi (153 km), near Clyo, Georgia, the Savannah River crested five days after the initial storm with a discharge of over 60,000 cfs (1700 cms). This set of three hydrographs

Figure 20.14 The Wabash River in flood near Delphi, Indiana, February 1954. An ice dam clogs the river channel, while lines of trees mark the crest of the bordering natural levees. The floodplain itself is inundated on both sides of the channel and reaches to the base of the bluff, at left. (U.P.I. Telephoto.)

shows that (*a*) the time lag in occurrence of the crest increases downstream, (*b*) the entire period of rise and fall of flood wave becomes longer downstream, and (*c*) the discharge increases greatly downstream as watershed area increases.

Figure 20.16*B* is a somewhat different presentation of the same flood data, in that the discharge is given in terms of a common unit of area, the square mile, thus eliminating the effect of increase in discharge downstream and showing us only the shape or form of the flood crest.

Flood prediction

The U.S. Weather Bureau operates a River and Flood Forecasting Service through 85 selected offices located at strategic points along major river systems of the United States. Each office issues river and flood forecasts to the communities within

Figure 20.15 The city of Hartford was partly inundated by the Connecticut River flood of March 1936. The river channel is to the left, its banks marked by a line of trees. (Official Photograph, 8th Photo Section, A. C., U.S. Army.)

the associated district, which is laid out to cover one or more large watersheds. Flood warnings are publicized by every possible means. Close co-operation is maintained with such agencies as the American Red Cross, the U.S. Army Corps of Engineers, and the U.S. Coast Guard, in order to plan evacuation of threatened areas, and the re-moval or protection of vulnerable property.

Long and intensive study of stream flow data enables the U.S. Weather Bureau to prepare graphs of flood stages telling the likelihood of occurrence of given stages of high water for each month of the year. Figure 20.17 shows expectancy graphs for four selected stations. The meaning of the strange looking bar symbols is explained in the

key. The Mississippi River at Vicksburg illustrates a great river corresponding largely to spring floods so as to yield a simple annual cycle. The Colorado River at Austin, Texas, is chosen to illustrate a river draining largely semi-arid plains. Summer floods are produced directly by torrential rains from invading moist tropical air masses. Floods of the late summer and fall are often attributable to tropical storms (hurricanes) moving inland from the Gulf of Mexico. The Sacramento River at Red Bluff, California, has a winter flood season when rains are heavy, but a sharp dip to low stages in late summer, which is the very dry period for the California coastal belt. The flood expectancy graph for the Connecticut River at Hartford shows two

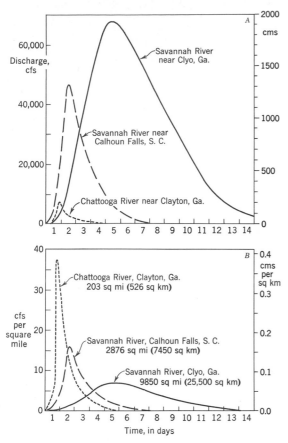

Figure 20.16 The downstream progress of a flood wave. (After Hoyt and Langbein, *Floods*.)

seasons of floods. The more reliable of the two is the early spring, when snowmelt is rapid over the mountainous New England terrain; the second, in the fall, when rare but heavy rainstorms, some of hurricane origin, bring exceptional high stages.

Flood control

In the face of repeated disastrous floods, vast sums of money have been spent on a wide variety of measures to reduce and control floods. The economic, social, and political aspects of flood control are beyond the scope of this book, but mention can be made of certain physical principles applied to the problem. Two basic forms of control are: (*a*) to detain and delay runoff by various means on the ground surfaces and in smaller tributaries of the watershed; (*b*) to modify the lower reaches of the river where floodplain inundation is expected.

The first form of control is aimed at treatment of watershed slopes, usually by reforestation or planting of other vegetative cover so as to increase the amount of infiltration and reduce the rate of overland flow. This type of treatment, together with construction of many small flood-storage dams

in the valley bottoms, may greatly reduce the flood crests and allow the discharge to pass into the main stream over a longer period of time.

Under the second type of flood control, designed to protect the floodplain areas directly, two quite different theories can be practiced. First, the building of *levees*, or *dikes*, parallel with the river channel on both sides can function to contain the overbank flow and prevent inundation of the adjacent floodplain (Figure 20.18). Such levees are broad embankments built of earth and must be designed with great care, not only to possess the physical resistance to water pressures, but must be high enough to contain the greatest floods; otherwise they will be breached rapidly by great gaps, termed *crevasses*, at the points where water spills over (Figure 20.19). Under the control of the Mississippi River Commission, which began in 1879, a vast system of levees was built along the Mississippi River in the expectation of containing all floods. Figure 20.18 shows such a levee during the flood of 1903, when it was necessary in Louisiana to add to the top of the levee by means of planks and sand-filled bags for a distance of 71 mi (114 km) to prevent overflow. Levees have been continuously improved and now total more than 2500 mi (4000 km) in length and in places are as high as 30 ft (10 m). Additional levees on the lower, or deltaic, alluvial plain form floodways, by means of which excessive discharges can be diverted in time of flood and passed directly to the sea.

The second theory, practiced in more recent years on the Mississippi River by the U.S. Army Corps of Engineers, is to shorten the river course by cutting channels directly across the great meander loops to provide a more direct river flow. Shortening has the effect of increasing the river slope, which in turn increases the mean velocity. Greater velocity enables a given flood discharge to be moved through a channel of smaller cross-sectional area; the flood stage is correspondingly reduced. Channel improvement has had a measurable effect in reducing flood crests along the lower Mississippi and the levees are thus not in such great danger of being overtopped. Certain parts of the floodplain are also set aside as temporary basins into which the river is to be diverted according to plan to reduce the flood crest. This planned flooding can make use of the least populated parts of the floodplain.

Ground water

The relation of soil water to ground water, and the distinction between zones of aeration and saturation were treated briefly in Chapter 9. We turn now to consider the distribution and move-

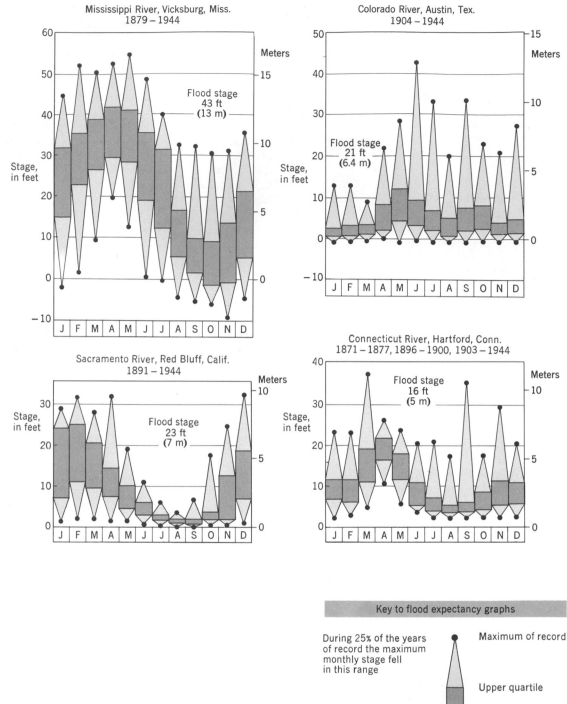

Figure 20.17 The highest water stage that occurred in each month is given in terms of percentages on these graphs of four rivers. (After U.S. Weather Bureau.)

Figure 20.18 This old photograph shows the artificial levee of the Mississippi River near Greenville, Mississippi, during the great flood of March 1903. A crevasse, or break, at the distant point, *x*, is discharging flood water into the lower flood plain on the left. (Mississippi River Commission photograph.)

Figure 20.19 This air view taken in April 1952, shows a break in the artificial levee adjacent to the Missouri River in western Iowa. Water is spilling from the high river level at right to the lower floodplain level at left. (Photograph by Forsythe, U.S. Department of Agriculture.)

ment of ground water and its geologic and economic aspects.

Table 20.1 summarizes the earth's water resources and shows how the total quantity of soil and ground water (subsurface water) compares with that held as surface water on the lands, as glacial ice, as atmospheric moisture, and as water in the world ocean. Note that the volume of ground water is vastly greater than that held in lakes and streams, but is itself less than one-third of the volume held as ice of icecaps and glaciers. Ground water is subdivided into that less than a half-mile (0.8 km) in depth, which is largely extractable for man's use, and that below a half-mile, much of which is not economically recoverable.

Pore spaces in the ground water zone

Ground water can saturate a great variety of geological materials ranging from relatively soft overburden, of both residual and transported types, to hard bedrock of any origin. The term

TABLE 20.1   DISTRIBUTION OF THE WORLD'S ESTIMATED WATER[a]

Location	Surface Area		Water Volume		Percent of Total
	Sq Mi	Sq Km	Cu Mi	Cu Km	
SURFACE WATER					
Fresh-water lakes	330,000	860,000	30,000	125,000	0.009
Saline lakes and inland seas	270,000	700,000	25,000	104,000	0.008
Average in stream channels	—	—	300	1,250	0.0001
SUBSURFACE WATER					
Soil moisture and intermediate-zone (vadose) water	50,000,000	130,000,000	16,000	67,000	0.005
Ground water within 0.5 mi (0.8 km) depth	50,000,000	130,000,000	1,000,000	4,200,000	0.31
Ground water, deep-lying	50,000,000	130,000,000	1,000,000	4,200,000	0.31
Total liquid water in land areas	50,600,000	132,000,000	2,070,000	8,630,000	0.635
ICECAPS AND GLACIERS	6,900,000	18,000,000	7,000,000	29,200,000	2.15
ATMOSPHERE	197,000,000	510,000,000	3,100	13,000	0.001
WORLD OCEAN	139,500,000	360,000,000	317,000,000	1,322,000,000	97.2
Totals (rounded)			326,000,000	1,360,000,000	100

[a] Data from Dr. Raymond L. Nace, U.S. Geological Survey, 1964.

porosity refers to the total volume of pore space present in a given volume of rock. The amount of water that can be held in storage in a rock is measured by its porosity. A knowledge of rocks enables us to understand something of the variation in porosity that might be expected in rocks. Among the sedimentary rocks, the coarse-grained clastic rocks, such as sandstone or conglomerate, can have large porosity. Similarly, any transported overburden consisting of sands and gravels laid down by streams or shore currents will have large porosity. Shale, because of its dense compaction, has relatively low porosity, but soft clay or mud, by contrast, has high porosity even though the pores are extremely tiny. Limestones may have large openings, such as caverns, produced by solution of the rock. Scoriaceous lavas commonly have great porosity.

Dense, massive rocks, such as the igneous and metamorphic types, have little or no pore space between the individual mineral crystals, but the presence of numerous fractures, such as joints and faults, offers a large number of interconnected openings in which water can be stored and through which it can move. In sedimentary strata the bedding planes may provide additional openings.

How far down into the earth does the ground water zone extend? No single depth can be stated in answer to this question, but it is certain from experience with deep wells that water becomes very scanty at depths of more than 2 mi (3.2 km). Furthermore, geologists have evidence that at depths greater than about 10 mi (16 km), rock is under such great pressure that it yields by slow flowage, tightly closing any natural openings in the rock and preventing any water from entering or remaining in the rock. We may call this the *region of rock flowage* and say that it limits the extent of the ground water body in depth.

Still another property besides porosity determines whether ground water will move through a rock mass. This is *permeability*, or the ease with which water may be forced through the rock. Obviously, pore spaces in a rock, even though large, may offer no free passage to water if each pore is sealed off from its neighbors by mineral matter. Thus the degree to which openings are interconnected is important in determining permeability.

Second, the size of the openings exerts a strong influence upon rate of flow. Coarse sands or gravels permit rapid flow and are therefore rated as highly permeable materials. The microscopic pores of a clay impede flow of water so effectively that clays and rocks containing much clay are commonly *impermeable* (not permeable), at least for practical purposes of obtaining useful flow of water from them. Although intrusive igneous rocks are highly impermeable where massive and not decomposed, they may actually constitute highly permeable bodies if broken by numerous, closely set systems of joints and fractures.

The water table

Ground water is extracted for man's use from wells dug or drilled to reach the ground water zone. In the ordinary shallow well, water rises to the same height as the ground water table (not including the capillary fringe). The upper limit of saturation is actually at the upper limit of the capillary fringe (Chapter 9).

If wells are numerous in an area, the position of the water table can be mapped in detail by plotting the water heights and noting the trends in elevation from one well to the other. When this is done it is usually seen that the water table is highest under the highest areas of surface, namely, hilltops and divides, but descends toward the valleys where it may appear at the surface close to streams, lakes, or marshes (Figure 20.20). The reason for such a configuration of the water table is that water percolating down through the zone of aeration tends to raise the water table, whereas seepage into streams, swamps, and lakes tends to draw off ground water and to lower its level.

Because ground water moves extremely slowly, a difference in water table level, or *head*, is built up and maintained between areas of high elevation and those of low elevation. In periods of excessive rainfall this head is increased by rise in water table over divide areas; in drought periods the water table slowly falls (Figure 20.20). Large fluctuations such as these are usually seen only in periods of excessive rain or drought lasting several years, but in humid, middle-latitude regions of strong seasonal contrasts, a distinct annual cycle of rise and fall of water table is present. This cycle is illus-

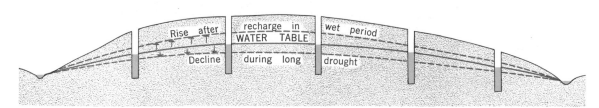

Figure 20.20   The water table conforms to surface topography.

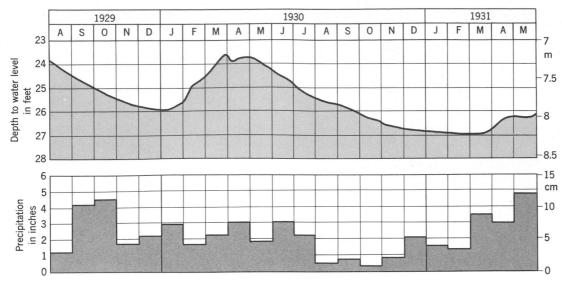

Figure 20.21   Water level in an observation well at Washington, D.C. (After Meinzer.)

trated in the graphical record of water table and precipitation by months based on the level of an observation well at Washington, D.C., extending from 1929 to 1931 (Figure 20.21). These were unusually dry years. Note that the water table dropped steadily in the latter part of 1929, even though considerable rain fell in September and October. This continued drop is explained by the fact that the soil belt and intermediate belt absorbed and held all of the water which infiltrated the surface, allowing none to get down to the water table. By late winter, however, considerable amounts of water were reaching the ground water zone, because the overlying belts in the zone of aeration had taken up all the water they could hold. Such a rise in water table illustrates the *recharge* of ground water supplies by percolation of surface water.

Aquifers, aquicludes, perched water tables, and artesian wells

Where rock layers lie nearly horizontal, or in gently inclined positions, the ground water relations may be quite different from the simple pattern illustrated in Figure 20.20 in which completely uniform geologic materials were assumed to exist. Suppose, for example, that the region is one of sedimentary strata with beds of sandstone alternating with beds of shale (Figure 20.22). Sandstone is commonly both porous and permeable, providing a large ground water storage reservoir through which water may move easily. Such a rock body is termed an *aquifer*. By contrast, a shale bed with low permeability virtually prevents flow of ground water and is called an *aquiclude*. In the particular case shown in Figure 20.22, a

thin bed of shale has effectively blocked the downward percolation of water to the main water table below, creating a *perched water table*, separated from the main water table by a zone of aeration. Where the perched water table meets the valley side a *seep* or *spring* (that is, a slow flow of water emerging from the ground) is formed.

Most natural springs are mere trickles of water, unseen and unnoticed under a cover of dense vegetation. A few springs, however, discharge enormous volumes of water where an unusually good aquifer, abundantly fed from a large source area, is exposed in a deep valley or canyon (Figure 20.23).

Where strata are inclined (dipping) a favorable situation may exist for development of an *artesian* spring or well, one in which the water flows upward toward the surface through its own pressure. In Figure 20.24 we see a highly diagrammatic representation of such conditions, the vertical exaggeration being very great merely to show the principle. The eroded edge of a sandstone aquifer is exposed to intake of water at a high position.

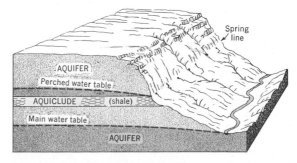

Figure 20.22   A perched water table.

Figure 20.23 Thousand Springs, Idaho, emerges from the north side of the Snake River Canyon, opposite the mouth of the Salmon River. The spring extends for 0.5 mi (0.8 km) and issues from a scoriaceous basalt layer with a nearly constant discharge of 500 cfs (14 cms). (Photograph by I. C. Russell, U.S. Geological Survey.)

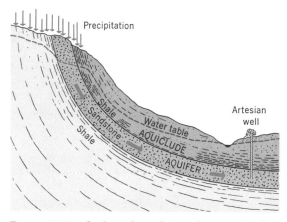

Figure 20.24 Geological conditions for artesian flow.

Water entering here passes deep underground to a position below the valley floor, at which point the water is under a strong pressure, or head, from the weight of the overlying water. This pressure is sufficient to force water up to the surface in a well drilled down through the impervious shale layer into the aquifer. Similar flow as an artesian spring may occur naturally if there are faults in the strata which permit water to seep upward through the shale layer.

Pumping, drawdown, and recharge of wells

Of increasing importance in economic geography is the effect on the water table of man's withdrawals of ground water. The drilling of vast numbers of wells, from which water is forced out in great volumes by powerful pumps, has profoundly altered nature's balance of ground water recharge and loss, which is a part of the hydrologic cycle. Increased urban populations and industrial developments require larger water supplies, needs

that cannot always be met from construction of new surface water reservoirs.

In agricultural lands of the semi-arid and desert climates, heavy dependence is placed upon irrigation water from pumped wells, especially since many of the major river systems have already been fully developed for irrigation from surface supplies. Wells can be drilled within the limits of a given agricultural or industrial property, and hence can provide immediate supplies of water without need to construct expensive canals or aqueducts. A few of the physical principles of water wells are treated here to aid the geography student in understanding the basis of complex economic and legal problems arising from ground water development.

Formerly the small well needed to supply domestic and livestock needs of a home or farmstead was actually dug by hand as a large cylindrical hole, lined with masonry where required. By contrast, the modern well put down to supply irrigation and industrial water is drilled by powerful machinery which may bore a hole 12 to 16 in (30 to 40 cm) or more in diameter to depths of 1000 ft (300 m) or more, although much smaller-scaled wells and well-boring machines suffice for domestic purposes. Drilled wells are sealed off by metal casings which exclude impure near-surface water and prevent clogging of the tube by caving of the walls. Near the lower end of the hole, where it enters the aquifer, the casing is perforated so as to admit the water through a considerable surface area. Rate of flow of a well or spring is stated in units of gallons or liters per minute or per day. The yields of single wells range from as low as a few gallons per day in a domestic well to many millions of gallons per day for large, deep industrial or irrigation wells.

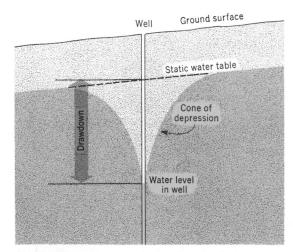

Figure 20.25 Drawdown and cone of depression in a pumped well.

In most wells, powerful pumps can easily bring water to the surface more rapidly than it can enter the well, so that the delivery of ground water is limited by the properties of the aquifer rather than by the mechanical equipment. Figure 20.25 shows the effects of rapid pumping. The rate at which water can enter the well depends on the permeability of the aquifer, which limits the rate of flow of water through the aquifer from the surrounding area. Flow of ground water is extremely slow, in any case, compared to flow of streams. It is estimated that ground water may move at a speed of 5 ft (1.5 m) per day through a formation in which wells of good yield are developed, that in exceptional cases of coarse gravels the velocity may reach 30 to 60 ft (10 to 20 m) per day. In dense clays and shales the rate may be immeasurably slow.

When rate of pumping of the well exceeds the rate at which water can enter, the level of water in the well drops and the surrounding water table is lowered in the shape of a conical surface, termed the *cone of depression*, the height of which is termed the *drawdown* (Figure 20.25). By producing a steeper gradient of the water table, the flow of ground water toward the well is also increased, so that the well will yield more water. This holds only for a limited amount of drawdown, beyond which the yield fails to increase. The cone of depression may extend as far out as 8 to 10 mi (13 to 16 km) or more from a well where very heavy pumping is continued. Where many wells are in operation, their intersecting cones produce a general lowering of the water table.

Depletion often greatly exceeds the rate at which the ground water of the area is recharged by percolation from rain or from the beds of streams. In an arid region, much of the ground water for irrigation is from wells driven into thick sands and gravels which are lowland deposits of transported overburden of a type termed *alluvium*. (These features are described in Chapter 21.) Recharge of such deposits depends on the seasonal flows of water from streams heading high in adjacent mountain ranges. Where such highly permeable materials exist, the extraction of ground water by pumping can greatly exceed the recharge by stream flow. Cones of depression deepen and widen; deeper wells and more powerful pumps are then required. Overdrafts of water accumulate and the result is exhaustion of a natural resource not renewable except by long lapses of time. In humid areas where annual rainfall is copious—from 30 to 50 in (75 to 125 cm) annually—natural recharge is by general percolation over the ground area surrounding the well. Here the prospects of achieving a balance of recharge and withdrawal are highly favorable through the control of pumping

and the return of waste waters or stream waters to the ground water table by means of other wells in which water flows down, rather than up.

Natural discharge of ground water

The subsurface phase of the hydrologic cycle is completed when the ground water emerges along lines or zones where the water table intersects the ground surface. Such places are the channels of streams and the floors of marshes and lakes. By slow seepage and spring flow the water must emerge in sufficient quantity to balance that which enters the ground water table by percolation through the zone of aeration.

Contrary to what the average person might predict, all of the ground water does not move directly from divides to the lines of seepage by flow close to the top of the water table. If such were the case, the lower parts of the ground water body would be stagnant. Certain geological phenomena, such as the cementation of rocks and the transfer of dissolved mineral matter from place to place, would not take place without some ground water flow, even though extremely slow.

Figure 20.26 shows the theoretical paths of flow of ground water as calculated by use of basic principles of the physics of fluids. Water follows paths curved concavely upward. Water entering the slope midway between divide and stream flows rather directly. Close to the divide point on the water table, however, the flow lines go almost straight down to great depths in the earth from which they recurve upward to points under the streams. Progress along these deep paths would be incredibly slow; that near the surface would be faster. The most rapid flow is encountered close to the line of discharge in the stream, where the arrows are shown to converge.

Of considerable interest in economic geography of coastal regions is the problem of the relation of fresh ground water to salt ground water, because wells put down close to the shore may encounter salt water, or may, through overdraft, cause salt

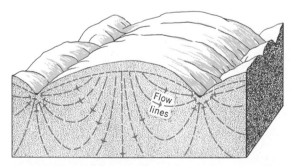

Figure 20.26 Theoretical paths of ground-water movement under divides and valleys. (After M. K. Hubbert.)

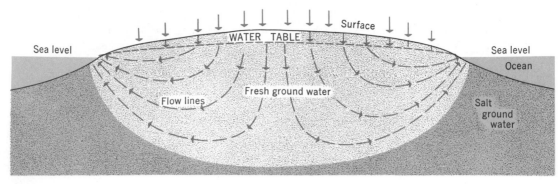

Figure 20.27 Fresh-water and salt-water relations in an island or peninsula. (After G. Parker.)

water to be drawn into the well and render it unfit for use. Figure 20.27 shows an idealized diagram through an island or a narrow peninsula. The body of fresh ground water takes the shape of gigantic lens with convex faces, except that the upper surface has only a broad curvature whereas the lower surface, in contact with the salt ground water, bulges deeply downward. Because fresh water is less dense than salt, we can think of this fresh water lens as floating upon the salt water, pushing it down much as the hull of an ocean liner pushes aside the surrounding water. The ratio of densities of fresh water to salt water is as 40 to 41. Hence, if the water table is, say, 10 feet above sea level, the bottom of the fresh water lens will be located 400 feet below sea level, or forty times as deep as the water table is high with respect to sea level.

The fresh ground water extends seaward some distance beyond the shoreline. Although the salt ground water is stagnant, the fresh water travels in the curved paths shown by arrows in Figure 20.27. If water is pumped excessively from wells close to the coast, the contact of salt with fresh ground water shifts landward, where it may intersect the wells and contaminate the fresh water.

When such a condition exists, the only cure is to cease pumping and allow the fresh water slowly to push the contact back toward its original seaward position, or to create a fresh water recharge barrier between the wells and the coastline. Resumed pumping must then be regulated to a lower rate.

REFERENCES FOR FURTHER STUDY

Foster, E. E. (1948), *Rainfall and runoff*, The Macmillan Co., New York, 487 pp.

Thomas, H. E. (1951), *The conservation of ground water*, McGraw-Hill Book Co., New York, 321 pp.

Leopold, L. and T. Maddock, Jr. (1954), *The flood control controversy*, The Ronald Press Co., New York, 255 pp.

Bennett, H. H. (1955), *Elements of soil conservation*, second edition, McGraw-Hill Book Co., New York, 358 pp.

Hoyt, W. G. and W. B. Langbein (1955), *Floods*, Princeton University Press, Princeton, N.J., 469 pp.

U.S. Dept. of Agriculture (1955), *Water; yearbook of agriculture 1955*, U.S. Govt. Printing Office, Washington, D.C., 751 pp.

Todd, D. K. (1959), *Ground water hydrology*, John Wiley and Sons, New York, 336 pp.

REVIEW QUESTIONS

1. What are the general characteristics of a natural drainage system? What is the shape of the drainage basin?

2. What branches of earth science deal with runoff and drainage systems? What governmental agencies take responsibility for various phases of this study? In what ways is a knowledge of principles of runoff of value to the student of geography?

3. Distinguish between overland flow and channel flow. To what extent is overland flow observable? How does flow in shoestring rills fit into this classification of types of flow?

4. Describe the development of overland flow, beginning with the onset of rainfall. What is interception? Surface detention? What equation relates overland flow to precipitation and infiltration?

5. How does the geologic norm of soil erosion differ from accelerated soil erosion? What is the role of splash erosion? What forms of land surface treatment bring about accelerated land erosion? Explain. How is the degree of ground slope a factor in intensity of soil erosion?

6. Describe the successively more severe forms of accelerated soil erosion. What happens to the eroded soil? Distinguish between *colluvium* and *alluvium*. In what ways is sedimentation harmful?

7. List the various geometrical elements of a stream channel, state how each is defined and measured. Explain the meaning of "percent of grade."

8. Describe the manner in which velocity varies throughout a stream, both across the channel and in vertical section. Why is the velocity not the same in all parts of the stream?

9. What is stream turbulence? What role does it play in stream transportation?

10. What is the mean velocity of a stream? State and explain in words the equation relating discharge to area and mean velocity.

11. Describe the methods and equipment used in stream gauging. What governmental agency is responsible for stream gauging in the United States? How is mean velocity actually measured and computed?

12. What is a hydrograph? Describe the manner in which runoff occurs following a rainstorm on a watershed. Why is there a time lag between rainfall and runoff? For larger drainage basins, how do these relationships differ?

13. What is base flow? What supplies base flow? What is channel storage and how does it affect the hydrograph of a large watershed?

14. Describe the annual changes in base flow and surface water contributions in a watershed typical of the humid southeastern United States.

15. How is a flood defined? How can the normal channel of a river be distinguished from the adjacent floodplain? What is the bank-full stage of a stream flow?

16. Describe the progress of a flood crest down a large river. How does discharge change with increasing distance downstream? How does the form of the flood wave change?

17. What government agency supplies river and flood forecasts? How are its activities organized and what is their function?

18. How do flood expectancies differ for the Mississippi River at Vicksburg, Mississippi, and the Colorado River at Austin, Texas? Explain.

19. What basic forms of flood control are used? What principles lie behind each? What are the advantages and disadvantages of constructing extensive levee systems? How can channels be modified to reduce flood crests?

20. How does the volume of ground water on earth compare with the average volume of water held in lakes and streams? With that held in glacial ice?

21. Distinguish between porosity and permeability of rocks. Cite varieties of rocks and transported overburden to illustrate extremes of porosity and permeability. What conditions exist in the region of rock flowage?

22. Describe the form of the water table under a hilly topography. What fluctuations may be expected in the water table? What is ground water recharge and how does it take place?

23. Explain how layered rocks can cause a perched water table. How can springs be so formed? What geologic conditions are required for artesian wells and springs?

24. How does pumping cause drawdown of water table? What is the cone of depression? How far may it extend from a well? How rapidly is ground water recharged from alluvial deposits of arid regions?

25. Where is ground water discharged from the ground to become runoff? Describe the paths of ground water flow from divide to stream.

26. Describe and explain the relation of a fresh ground water body to the salt ground water beneath an island. What physical principle governs the relation of water table elevation to depth of fresh water? How is salt water contamination of wells caused?

CHAPTER 21
Landforms Made by Streams

FAR from being mere systems for disposal of excess runoff, streams are important agents of land sculpture—the principal carriers of mineral matter from the lands to the oceans. Not only is this denudational role of interest to the geomorphologist, who studies the landforms produced, but also to the student of historical geology for whom the sedimentary strata of past geologic eras can be interpreted in many instances as the deposits brought by streams from ancient land masses.

Geologic work of streams

The geologic work of streams consists of three closely interrelated activities: *erosion, transportation,* and *deposition*. Erosion by a stream is the progressive removal of mineral material from the floor and sides of the channel, whether this be carved in bedrock, or in residual or transported overburden.

Transportation consists of movement of the eroded particles by dragging along the bed, by suspension in the body of the stream, or in solution. Deposition is the progressive accumulation of transported particles upon the stream bed and floodplain, or on the floor of a standing body of water into which the stream empties. Obviously, erosion cannot occur without some transportation taking place, and the transported particles must

eventually come to rest. Therefore, erosion, transportation, and deposition are simply three phases of a single activity.

Stream erosion

Streams erode in various ways, depending on the nature of the channel materials and the tools with which the current is armed. The force of the flowing water alone, exerting impact and a dragging action upon the bed, can erode poorly consolidated alluvial materials such as gravel, sand, silt, and clay, a process termed *hydraulic action*. Where rock particles carried by the swift current strike against bedrock channel walls, chips of rock are detached. The rolling of cobbles and boulders over the stream bed will further crush and grind smaller grains to produce an assortment of grain sizes. These processes of mechanical wear are combined under the term *corrasion*, or *abrasion*, which is the principal means of erosion in bedrock too strong to be affected by simple hydraulic action. Finally, the chemical processes of rock weathering—acid reactions and solution—are effective in removal or rock from the stream channel and may be designated as *corrosion*. Effects of corrosion are most marked in limestone, which is a hard rock not easily carved by abrasion, but yielding readily to the action of carbonic acid in solution in the stream water.

One interesting form produced by stream abrasion is the *pothole*, a cylindrical hole carved into the hard bedrock of a swiftly moving stream (Figure 21.1). Potholes range in diameter from a few inches to several feet; the larger ones may be many feet deep. Often a spherical or discus-shaped stone is found in the pothole and is apparently the tool, or *grinder*, with which the pothole was deepened. A spiraling flow of water in the pothole causes the grinder to be rotated at the base of the hole, thus boring gradually into the rock. Many other features of abrasion, such as plunge pools, chutes, and troughs lend variety to the rock channel of a swift mountain stream.

Stream transportation

The load of a stream is carried in three forms. Dissolved matter is transported invisibly in the

Figure 21.1 These potholes have been carved in granite in the channel of the James River, Henrico County, Virginia. (Photograph by C. K. Wentworth, U.S. Geological Survey.)

form of chemical ions. All streams carry some dissolved salts resulting from rock decomposition. Clay and silt are carried in *suspension*, that is, held up in the water by the upward elements of flow in turbulent eddies in the stream. This fraction of the transported matter is termed the *suspended load*. Sand, gravel, and still larger fragments move as *bed load* close to the channel floor by rolling or sliding and an occasional low leap.

The load carried by a stream varies enormously in the total quantity present and the size of the fragments, depending on the discharge and stage of the river. In flood, when velocities of 20 ft (6 m) per second or more are produced in large rivers, the water is turbid with suspended load. Boulders of great size may be moving over the stream bed, if the river gradient is steep. Frederick S. Dellenbaugh, a member of Major Powell's boat party which descended the Grand Canyon of the Colorado River in 1871 and 1872, wrote that as the men rested beside the river at night they could feel and hear the dull, thundering impacts of huge boulders rolled over and over on the channel bottom in the swift rapids.

The hydraulic action of flood waters is capable of excavating enormous quantities of unconsolidated materials in a short time (Figure 21.2). Not only is the channel often greatly deepened in flood, but the undermining of the banks causes huge masses of alluvium to slump into the river where the particles are quickly separated and become a part of the stream's load. This process, known as *bank caving*, is an important source of sediment during high river stages, and is associated with rapid sidewise shifts in channel position on the outsides of river bends.

Suspended loads of large rivers

The load carried by a large river is of considerable importance in planning for construction of large storage dams and in the construction of canal systems for irrigation. Sediment will be trapped in the reservoir behind a dam, eventually filling the entire basin and ending the useful life of the reservoir as a storage body. At the same time, depriving the river of its sediment in the lower course below the dam may cause serious upsets in river activity. Resulting deep scour of the bed and lowering of river level may upset the grades of irrigation systems. In designing for canal systems, the forms of artificial channels must be adjusted to the size and quantity of sediment carried by the water, otherwise obstruction by deposition or abnormal scour may follow.

The table at right gives comparative figures on the sediment loads of rivers in various stages.[1]

[1] Data from G. H. Matthes, "Paradoxes of the Mississippi," *Scientific American*, Vol. 184, pp. 19–23, 1951.

Figure 21.2 A river in flood eroded this huge trench at Cavendish, Vermont, in November 1927. An area 1 mi (1.6 km) wide and 3 mi (4.8 km) long, once occupied by eight farms, was cut away by the flood waters. Damage was great because the material consisted of sand and gravel which offered little resistance. (Photograph by Wide World Photos.)

Although information is scanty on quantity of sediment moved as bed load, the proportion of suspended load is generally very high. For the Mississippi, 90 per cent of the total load is carried in suspension. The table below shows that rivers draining semi-arid or arid lands (Missouri and Colorado) have very high suspended loads because

	Suspended Sediment Parts per Million	Fraction by Weight
Mississippi River		
Yearly average	500 to 600	1/1800 to 1/1660
Flood stage up to 2,000,000 cfs (56,600 cms)	2,600	1/400
Low stage 70,000 cfs (2,000 cms)	50 (water blue, clear)	1/2000
Missouri River		
Flood stage	20,000	1/50
Colorado River (before Hoover Dam)		
Flood stage 50,000 to 70,000 cfs (1400 to 2000 cms)	40,000	1/25
Yellow River, China		
Flood stage	Weight of solids may be greater than weight of water	

of the large expanses of barren soil from which sediment is easily swept by overland flow into the stream channels.

How channels change in flood

We tend to think of a river in flood as changing largely through increase in height of water surface, which causes channel overflow and inundation of the adjoining floodplain. Because of the turbidity of the water we cannot see the changes taking place on the stream bed, but these can be determined by sounding the river depth during stream-gauging measurements (Figure 21.3). At first the bed may be built up by large amounts of bed load supplied to the stream during the first phases of heavy runoff. This is soon reversed, however, and the bed is actively deepened by scour as stream stage rises. Thus, in the period of highest stage, the river bed is at its lowest elevation. When the discharge then starts to decline, the level of the stream's surface drops and the bed is built back up by the deposition of alluvium. In the example shown in Figure 21.3, about 10 ft (3 m) of thickness of alluvium was *reworked*, that is, moved about in the complete cycle of rising and falling stages.

Alternate deepening by scour and shallowing by deposition of load are responses to changes in the stream's ability to transport its load. The maximum quantity, or load, of debris that can be carried by a stream is a measure of the stream's *capacity*. Load is usually stated in terms of the weight of material moved through a given stream cross section in a given unit of time, commonly in units of tons per day. Total load includes both the bed load and the suspended load.

If a stream is flowing in a channel of hard bedrock, it may not be able to pick up enough alluvial material to supply its full capacity for bed load. Such conditions exist in streams occupying deep gorges and having steep gradients, so that

when flood occurs, the channel cannot be quickly deepened in response. In an alluvial river, however, where thick layers of silt, sand, gravel, and boulders underlie the channel, the rising river easily picks up and sets in motion all of the material that it is capable of moving. In other words, the increasing capacity of the stream for load is easily satisfied.

Capacity for load increases sharply with the stream's velocity, because the swifter the current the more intense is the turbulence and the stronger is the dragging force against the bed. Capacity to move bed load goes up about as the third to fourth power of the velocity. Thus, if a stream's velocity is doubled in flood, its ability to transport bed load is increased from eight to sixteen times. It is small wonder, then, that most of the conspicuous changes in the channel of a stream, such as sidewise shifting of the course, occur in flood stage, with very few important changes occurring in low water stages.

When the flood crest has passed and the discharge begins to decrease, the stream's capacity to transport load also declines. Therefore, some of the particles that are in motion must come to rest on the bed in the form of sand and gravel bars. First the largest boulders and cobblestones will cease to roll, then the pebbles and gravel, then the sand. Fine sand and silt carried in suspension can no longer be sustained, and settle to the bed. In this way the stream adjusts to its falling capacity. When restored to low stage the water may become quite clear, with only a few grains of sand rolling along the bed where the current threads are fastest.

Life history of a stream

Throughout its life history, a stream passes through a series of stages, each with certain definite characteristics (Figure 21.4). The initial stage

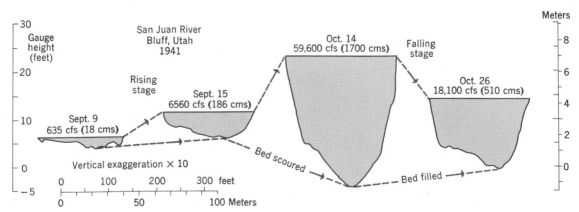

Figure 21.3   Changes in channel form of the San Juan River near Bluff, Utah, during the progress of a flood. (After Leopold and Maddock, U.S. Geological Survey.)

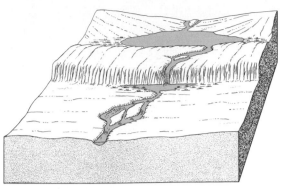

A. In the initial stage a stream has lakes, waterfalls, and rapids.

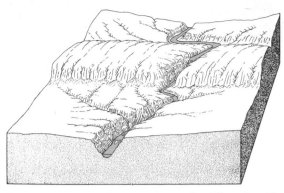

B. By middle youth the lakes are gone, but falls and rapids persist along the narrow incised gorge.

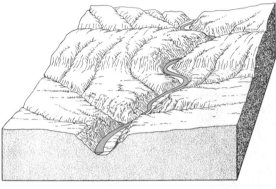

C. Early maturity brings a smoothly graded profile without rapids or falls, but with the beginnings of a flood plain.

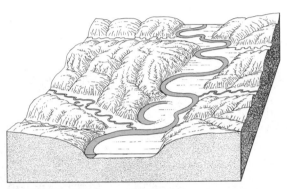

D. Approaching full maturity, the stream has a flood plain almost wide enough to accommodate its meanders.

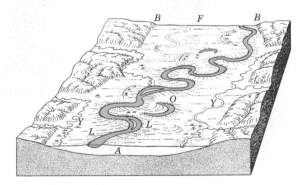

E. Full maturity is marked by a broad flood plain and freely developed meanders. *L* = levee; *O* = oxbow lake; *Y* = yazoo stream; *A* = alluvium; *B* = bluffs; *F* = flood plain.

Figure 21.4   Stages in the life history of a stream. (After E. Raisz.)

occurs as soon as a new land surface is created by uplift and dislocation of a portion of the earth's crust. It is assumed here for simplicity of discussion that the surface was formerly under ocean level and has now become exposed for the first time. The landscape is thus composed entirely of; initial landforms. Rain falling upon the land will produce overland flow. This must flow down the initial slopes, whatever their form. Water flow will be concentrated where slight depressions exist in the slopes, thus causing the development of stream channels, which are quickly deepened by erosive

action of the water and any loose rock particles it carries. Depressions will fill up with water, making lakes. Overflow at the lowest points on the rims of these lake basins will serve to make a connected system of drainage from higher to lower lakes. Thus the initial stream system comes into existence. It is characterized by falls, rapids, and lakes along its course (Figure 21.4*A*).

Once formed, the stream enters upon the stage of *youth*. Deepening of the channel is the principal activity of a young stream, whose capacity for load exceeds the load available to it. Lake outlets are cut through, draining the lakes and extending the stream across the old lake floors. Waterfalls are cut down at the lip until they are nothing more than rapids. A deepening *gorge* or *canyon* is perhaps the most striking landform associated with a young stream. The gorge is steep-walled and has a V-shaped cross section. The stream occupies all the bottom of the gorge. From the steep walls much weathered rock material is shed into the stream. Landslides occur frequently, large fallen masses sometimes temporarily damming the stream. Talus slopes of loose rock fragments may here and there extend down into the water. Because of rapid denudation of the steep valley walls, bedrock outcrops are conspicuous, locally forming bold cliffs (Figure 21.4*B*).

The geographic importance of a young river valley can be readily imagined. There is no room for roads or railroads between the stream and the

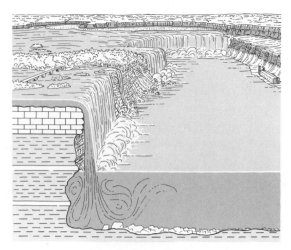

Figure 21.6 Niagara Falls is formed where the river passes over the eroded edge of a massive limestone layer. Continual undermining of weak shales at the base keeps the fall steep. (After G. K. Gilbert and E. Raisz.)

valley sides; hence road beds must be cut or blasted at great expense and hazard from the valley sides. Maintenance is expensive because of undercutting by the stream and the sliding and falling of rock, which can wipe out or damage the road bed. Yet a young gorge may afford the only passage through a mountain range. The Royal Gorge of the Arkansas River, in the Rocky Mountain Front Range of southern Colorado, is a striking example (Figure 21.5).

Another geographic consideration is that a young stream is not navigable, even though it might otherwise have a sufficient discharge.

The steep gradients of young streams, especially at waterfalls, sometimes make them important sources of hydroelectric power (Figures 21.6, 21.7). Most large young rivers, however, do not possess abrupt drops in gradient, and so it is necessary to build dams in order to create artificially the vertical drop necessary for turbine operation. An example is the Hoover Dam, behind which lies Lake Mead occupying the canyon of the young Colorado River.

As a stream progresses through the stage of youth it removes falls and rapids from its course, creating a smooth, even gradient. Deepening of the valley becomes greatly retarded, allowing the canyon or gorge walls to be worn down to more moderate slopes (Figure 21.8).

Stream equilibrium

The stage of *maturity* is reached when the stream has completed its phase of rapid downcutting and has prepared itself a smoothly graded course. It is now in a state of balance, or *equilibrium*, in which the average rate of supply of rock waste to the stream from all its tributaries and their

Figure 21.5 The Royal gorge of the Arkansas River in the Colorado Rockies illustrates the canyon of a young river with a steep gradient. Seen above is a suspension bridge, 1053 ft (321 m) above river level. (Photograph by Josef Muench.)

Figure 21.7   This air view of Victoria Falls of the Zambezi River shows that the river has excavated a long cleft in the bedrock, probably along a fault or other zone of weakness. (Photographer not known.)

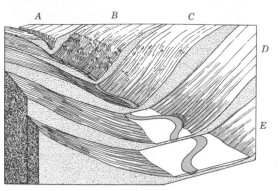

Figure 21.8   As a stream progresses from youth to maturity, its valley walls become more gentle in slope and the bedrock is covered by soil and weathered rock. (After W. M. Davis.)

slopes is equal to the average rate at which the stream can transport the load. In other words, the stream's capacity is satisfied by the load supplied.

The longitudinal profile, representing the stream channel from upper to lower end, is a profile of equilibrium (Figure 21.9). It may also be said that the stream is *graded*, which simply means that it possesses a profile of equilibrium.

It is important to understand that the balance between load and a stream's capacity exists only as an average condition over periods of many years. As already explained, streams scour their channels in flood and deposit load when in low stage. Thus in terms of conditions of the moment, a stream is rarely in equilibrium; but over long periods of time, the graded stream maintains its level by restoring those channel deposits temporarily removed by the excessive energy of flood flows.

Having attained this state of balance, the stream continues to cut sidewise on the outsides of banks. It cannot continue to cut down without destroying the equilibrium condition, but the lateral cutting does not materially affect the equilibrium (Figure 21.4C).

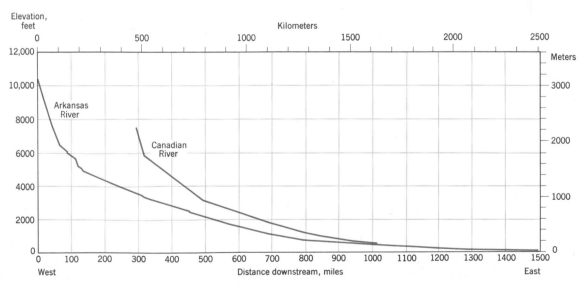

Figure 21.9   Longitudinal profiles of the Arkansas and Canadian rivers. The middle and lower parts of the profiles are for the most part smoothly graded, whereas the poorly graded upper parts reflect rock inequalities and glacial modifications within the Rocky Mountains. (From Gannett, U.S. Geological Survey.)

Floodplain development

Immediate evidence of the earliest stage of maturity is the beginning of development of a flat valley floor. During enlargement of a bend, the river channel shifts toward the outer part of the bend, leaving a strip of relatively flat land, or *floodplain*, on the inner side of the bend (Figure 21.10). The flood plain is built of bars composed largely of sand and gravel brought as bed load scoured from the outsides of bends immediately upstream. Innundation of the floodplain approximately yearly in frequency allows finer silt and clay to settle out over the surface, adding to the floodplain height and covering the coarser alluvium beneath.

As lateral cutting by the stream continues, floodplain strips grow wider and presently join to form more or less continuous belts along either side of the stream (Figure 21.11). The stream bends are now larger and more smoothly rounded. When the bends are developed into smooth, sinuous curves they are termed *meanders*. As valley development progresses the flood plain becomes wide enough to accommodate the meanders without cramping their form. The stream has then passed from *early maturity* to *full maturity* (Figure 21.4E).

The stage of early maturity of a stream is significant geographically. The floodplain, though narrow, will accommodate roads or railroads. The graded stream profile assures low, smooth grades for the roadbeds. Agriculture, impossible on the steep walls of a young valley, can be practiced on the narrow floodplain strips. With advanced maturity, the floodplain valley assumes more and more importance as a productive belt and is occupied by relatively greater numbers of persons than the upland areas between the valleys.[2] Furthermore, the absence of rapids in the stream channels

[2] We are speaking figuratively here, because the changes in stage of a stream's life cycle take tens or hundreds of thousands of years.

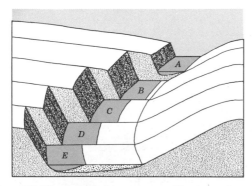

Figure 21.10   A graded stream cuts sidewise against the outside of a bend, leaving a flood-plain belt on the inside. (After W. M. Davis.)

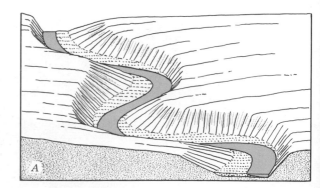

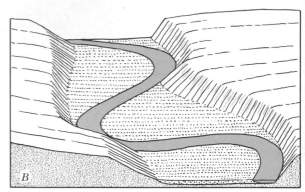

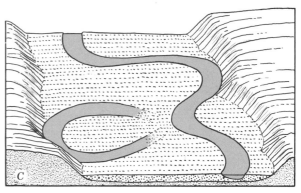

Figure 21.11   Widening of a valley by lateral cutting of a graded stream permits the free growth and cutoff of meander bends. (After W. M. Davis.)

permits navigation, although streams in the early mature stage require locks to assure navigability.

After a stream has reached full maturity, its principal activity is to widen the floodplain. Eventually the floodplain attains a width several times as great as the meander belt. By *meander belt* is meant the area included by two lines, each one drawn on the side of the meandering stream in such a way as to connect the outermost points of the bends. It is common practice for the hydraulic engineer to apply the term *alluvial river* to a freely meandering stream with a broad floodplain.

The ultimate goal to which the stream profile is being lowered is an imaginary inland extension of a sea-level surface. Below this level, which the

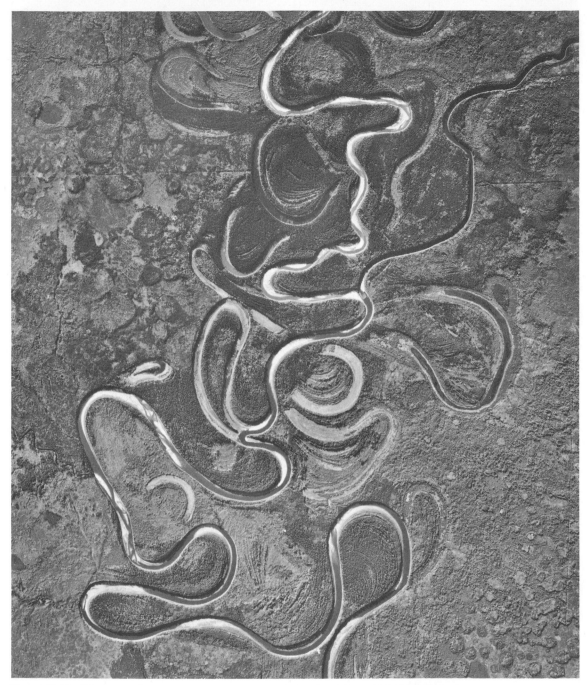

Figure 21.12 This vertical air photograph, taken from an altitude of about 20,000 ft (6100 m) shows meanders, cutoffs, oxbow lakes and swamps, and flood-plain of the Hay River, Alberta (lat. 58° 55′ N, long. 118° 10′ W). (Copyright photograph by Air Photo Division, Canadian Government Department of Energy, Mines and Resources.)

geologist calls *base level*, the valleys could not be deepened. The mouth of every stream that empties into the sea is at base level. Although theoretically the remainder of the stream might eventually reach base level, all alluvial rivers have a slight gradient, due to the fact that the lands within each drainage basin still stand well above sea level, and are shedding sediment which requires a sloping stream bed for transport.

Meanders of alluvial rivers

The floodplains of large alluvial streams have many special features of interest to the geographer (Figures 21.4*E* and 21.12). Meander bends grow as the stream undercuts the bank on the outside of the bend and deposits alluvium on the inside of the bend. These two sides of the bend are called the *undercut* and *slipoff slopes*, respectively. The bars of alluvium built on the slipoff slope are referred to as *point-bar deposits*.

The bends grow larger and larger until the channels meet, causing the intervening meander loop to be pinched off and abandoned. An occurrence of this type is called a *cutoff* (Figure 21.13). On a large river, such as the Mississippi, Missouri, or Arkansas, cutoffs are of considerable geographical importance. Where a state boundary is defined as the midline of a river, cutoffs cause portions of land within the cutoff bends to be transferred from one state to another, automatically altering

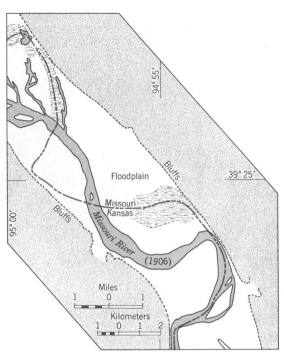

Figure 21.14 The Missouri-Kansas boundary was originally surveyed along the midline of the Missouri River, but the river has since shifted to a new course. (After U.S. Geological Survey.)

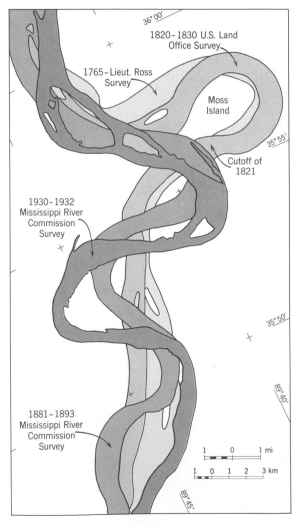

Figure 21.13 A series of four surveys of the Mississippi River shows considerable changes in the position of the channel and the form of the meander bends. Note that one meander cutoff has occurred (1821) and new bends are being formed. (After U.S. Army Corps of Engineers.)

the legal residences of persons who live on these lands and making them subject to laws and taxes of the other state. This difficulty can be overcome by fixing the boundary at a given time and maintaining it despite river changes. As a result, some maps show the old boundary meandering along a former river course quite different from the present course (Figure 21.14).

Along the Mississippi, towns grew up at many places along the steep undercut banks of meander bends. Here a river steamer could safely come close to the bank, dropping its gangplank directly upon the shore. If the meander bend was cut off the river took a new channel and the town immediately felt the effects of economic strangulation. Lakes formed by sealing off of the cutoff bend are called *oxbow lakes* (Figure 21.4E). Several oxbow lakes and marshes can be identified in Figure 21.12.

Natural levees

A floodplain, as the name implies, is normally inundated by flood stages. Unless protected by artificial levees (Chapter 20), a floodplain experiences inundation about once annually, at the season of highest runoff. Although the whole plain, from one valley wall, or *bluff*, to the other, is under water at such times (Figure 20.14), the water current is most rapid along the deep line of the river channel. Silt-bearing water, which spreads out and mingles with shallow flood waters on either side, quickly loses velocity, and much of the silt and mud settles out. Because the greatest amount of sediment settles out adjacent to the river channel there is built up by many such floods a belt of slightly higher ground, known as a *natural levee,* on both sides of the river (Figure 21.4E). The levee surface slopes gently downward away from the river to lower portions of the floodplain. Strangely enough, then, the highest ground on the floodplain is along the natural levees, immediately adjacent to the river. This narrow strip of ground may remain above water in all but the highest floods and is the safest place, aside from the floodplain bluffs themselves, of any on the floodplain.

Streams entering upon the floodplain cannot directly join the main river because of the levees; hence they flow down valley, parallel to the river, considerable distances until a point of entry can be found. Streams of this type are called *yazoo streams* (Figure 21.4E), after the Yazoo River of the Mississippi River floodplain. Heightening of the natural levees by artificial levees may prevent the river from overflowing its banks in most floods. Once the water breaks through, however, it is difficult or impossible to control and the whole floodplain between levee and bluffs is inundated (Chapter 20).

The extensive, flat lands of a broad floodplain are usually highly productive and densely populated agricultural regions. The natural levee slopes are intensively cultivated; ditches assist in the natural surface water drainage away from the river bank. Farther from the river, in the lower parts of the floodplain and in oxbows and other sections of abandoned channels, are swamps that support a dense forest of trees adapted to the wet environment. Because of natural and artificial levees, the river channel is often slowly built up to appreciably higher level than the floodplain, making these "bottom lands," as they are called, subject to repeated inundations and consequent heavy loss of life and property. Floodplains of many of the world's great rivers present serious problems of this type. In China, in 1887, the Hwang Ho inundated an area of 50,000 sq mi (130,000 sq km), causing the direct death of a million people and the indirect death of a still greater number through ensuing famine.

Braided streams and alluvial fans

When a stream is supplied with more rock waste than it can carry, the excess material is spread along the channel bottom. This is done by the stream to increase its gradient, which will in turn increase its velocity of flow and therefore increase its capacity to transport load. The process of building up the channel is termed *aggradation;* it is the opposite of *degradation,* the normal downcutting process so marked in young streams.

Aggradation gives a distinctive broad, shallow form to stream channels. The stream divides and subdivides into two, three, four, or more threads, which come together and redivide in a manner suggestive of the braided strands of rope. The term *braided stream* is, in fact, used to describe the pattern of an aggrading stream. The reason for braiding and constant shifting of the channels is that deposition of sand and gravel bars on the channel floor causes the stream to split into two or more channels which shift sideways toward lower adjacent ground. The stream thus forces itself out of its own channel. Aggrading streams are commonest in dry regions where stream flow is small, but where large quantities of rock waste are swept into stream valleys from relatively bare, unprotected valley slopes.

One very common landform built by braided, aggrading streams is the *alluvial fan,* a low cone of alluvial sands and gravels resembling in outline an open Japanese fan (Figure 21.15). The apex, or central point of the fan, lies at the mouth of a canyon, ravine, or gully and is built out upon an adjacent plain. Alluvial fans are of many sizes, ranging from tiny miniature fans a foot or two across, such as the ones seen alongside a roadcut, to huge fans many miles across.

Fans are built by young streams carrying heavy loads of coarse rock waste out from a mountain or upland region. Where the stream flows out upon the gentle slope of the plain, the current velocity is greatly reduced, thus forcing the stream to aggrade. The braided channel shifts constantly but, because it is firmly fixed in position at the canyon mouth, must sweep back and forth like the wagging

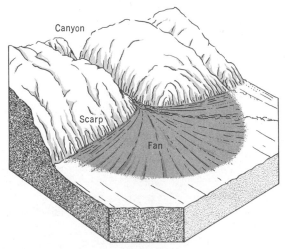

Figure 21.15 A simple alluvial fan.

Figure 21.16 A great alluvial fan in Death Valley, built of debris swept out of a large canyon. Observe the braided stream channels. (Copyrighted Spence Air Photos.)

tail of a gigantic dog. This fixed apex accounts for the semicircular form and the downward slope in all radial directions from the apex (Figure 21.16).

Fans are of considerable geographic importance. In many mountainous regions, populations and cities are concentrated on the gentler outer fringes of alluvial fans near the base of a mountain range. Water for irrigation is diverted from the streams that issue from the canyon mouths. Artesian water can be had from wells drilled into the permeable alluvial material of which the fan is made.

A striking example of fan utilization is found in the Los Angeles basin of southern California. Flash floods have been a serious menace to such communities as Burbank, Glendale, Montrose, and Pasadena, whose favored residential districts are on the higher, inner slopes of these fans. These *debris floods*, as they are called by engineers, are distinctive in that the sediment load is very great and may include large boulders as well as sand,

silt, and clay. The floods arise on the steep slopes of mountain watersheds, especially where brush fires have destroyed protective vegetation.

Terraces

If a stream aggrades its valley for a long time the alluvial deposits may reach a thickness of many tens of feet, as Diagram *A* of Figure 21.17 shows. Among several possible causes of aggradation is the onset of a more arid climate, which reduces stream discharge and requires streams to steepen their gradients by building up their channels. A steeper gradient enables a stream to increase its ability to move coarse bed load shed from slopes unprotected by vegetative cover. Perhaps the commonest cause of aggradation in fairly recent geologic time in North America has been the advance and wasting away of great ice sheets. (See Chapter 23). Melting water from the ice was heavily charged with rock debris, which caused virtually all streams near the ice front to fill their valleys with alluvium.

With return to normal conditions of reduced load, a stream will cut down through its alluvial deposit and eventually sweep most of it out of the valley. During this degradation a series of *alluvial terraces* is formed. A terrace is a relatively flat strip of ground bounded on one side by a steeply descending slope and on the other side by a steeply rising slope. A series of alluvial terraces resembles a flight of rather broad, low steps (Figure 21.18). As indicated in Diagrams *B* and *C* of Figure 21.17, terraces are made by the stream swinging from one side of the valley to the other as it slowly cuts down through the material. As each terrace is cut, the width of the next one above it is reduced.

All older terraces would be destroyed were it not that bedrock of the valley wall here and there projects through the alluvium and protects higher terraces. In Diagram *C*, the valley alluvium has been largely removed, but some *rock-defended terraces* remain on the valley sides, protected from stream attack by the rock which outcrops at the points labeled *R*. Notice that the scarps separating terraces are curved in broad arcs concave toward the valley. The curvature is easily explained as the result of cutting of the scarps by curved meander bends.

Terraces are of geographical importance similar to that of river floodplains. The relatively flat terrace surfaces are suitable for cultivation and make good sites for towns and cities, highways, and railroads. In all these utilizations terraces have one advantage over floodplains: their surfaces may

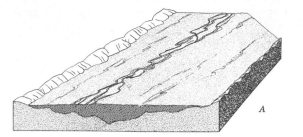

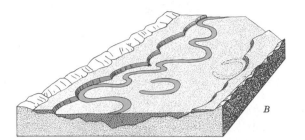

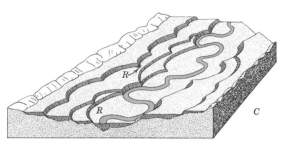

Figure 21.17 Alluvial terraces form when a graded stream slowly cuts away the alluvial fill in its valley.

Figure 21.18 Terraces of the Shoshone River near Cody, Wyoming, indicate former positions of the river flood plain. (Photograph by Frank J. Wright.)

Figure 21.19 This air view of the Kander delta in Switzerland shows a tongue of silt-ladened water being projected into Lake Thun. (Swissair photo.)

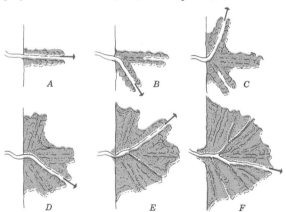

Figure 21.20 Stages in the formation of a simple delta. (After G. K. Gilbert.)

be well above the level of even the highest floods, whereas floodplains are normally subject to frequent inundation.

Deltas

The deposit of mud, silt, sand, or gravel made by a stream where it flows into a body of standing water is known as a *delta* (Figure 21.19). Deposition is caused by rapid reduction in velocity of the stream current as it pushes out into the standing water (Figure 21.20). The coarse particles settle out first; the fine clays continue out farthest and eventually come to rest in fairly deep water (Figure 21.21). Contact of fresh with salt water causes the finest clays to clot into larger aggregates which settle to the sea floor.

Deltas show a variety of shapes. The Nile delta, whose resemblance to the Greek letter "delta" suggested the name for this type of landform, has many *distributaries* which branch out in a radial arrangement (Figure 21.22A). Because of its broadly curving shoreline, causing it to resemble in outline an alluvial fan, this type may be described as an *arcuate* delta. The Mississippi River delta presents a very different sort of picture (Figure 21.22B). It is said to be of the *bird-foot* type because of the long, projecting fingers which grow far out into the water at the ends of each distributary. Where a river empties out upon a fairly straight shoreline along which wave attack is vigorous, the sediment brought out by the stream is spread along the shore in both directions from the river mouth, giving a pointed delta with curved sides. Because of its resemblance to a sharp tooth, this type is called a *cuspate* delta (Figure 21.22C). Where a river empties into a long, narrow estuary, the delta is confined to the shape of the estuary (Figure 21.22D). This type can be called an *estuarine* delta.

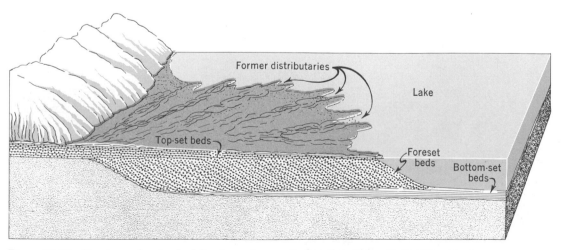

Figure 21.21 Structure of a simple delta shown in a vertical section. (After G. K. Gilbert.)

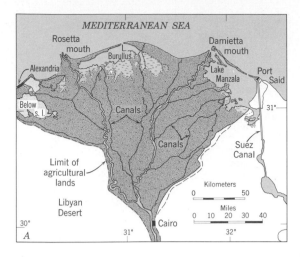

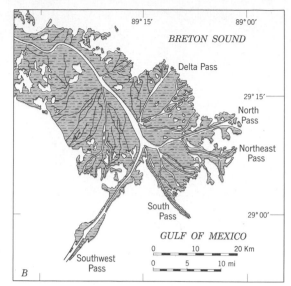

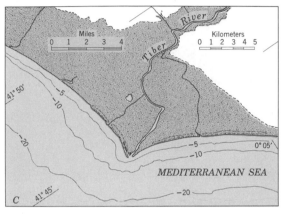

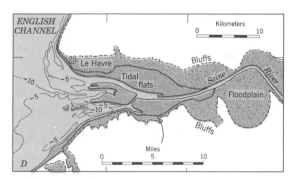

Figure 21.22   Deltas. *A*, The Nile delta has an arcuate shoreline and is triangular in plan. *B*, The Mississippi delta is of the branching, bird-foot type with long passes. *C*, The Tiber delta on the Italian coast is pointed, or cuspate, because of strong wave and current action. *D*, The Seine delta is filling in a narrow estuary.

Deltas of large rivers have been of great geographic importance from earliest historical times because their extensive flat areas support dense agricultural populations. Important coastal cities, linking ocean and river traffic, are often situated on or near deltas, as Alexandria on the Nile, Calcutta on the Ganges-Brahmaputra, Amsterdam and Rotterdam on the Rhine, Shanghai on the Yangtze, Marseilles on the Rhone, and New Orleans on the Mississippi, to mention but a few. Delta growth is often rapid, ranging from about 10 ft (3 m) per year for the Nile to 200 ft (60 m) per year for the Po and Mississippi rivers. Thus, some cities and towns that were at river mouths several hundred years ago are today several miles inland. An important engineering problem is to keep an open channel for ocean-going vessels which have to enter the delta distributaries to reach port. The ends of the Mississippi River delta distributaries, known as *passes*, have been extended by the con-

struction of jetties, between which the narrowed stream is forced to move faster, thereby scouring a deep channel (Figure 21.22*B*).

Rejuvenated streams and entrenched meanders

A mature stream, which has developed a graded profile with respect to a fixed sea level at its mouth, may experience a marked change if the land rises or the sea level falls. In either event, the base level of the stream is lowered and the stream is caused to begin rapid downcutting in order to reestablish its graded profile at a lower level. This process, termed *rejuvenation*, begins as a series of rapids at the stream's mouth, where the water passes from the former mouth down to the lowered sea level. The rapids quickly shift upstream, and soon the entire stream valley is being trenched to form a new, youthful valley.

If rejuvenation occurs when a stream has

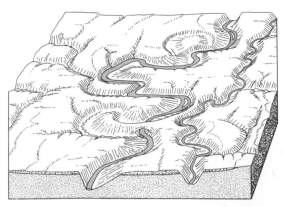

Figure 21.23 The winding valley of this stream resulted from entrenchment of the meandering river shown in Figure 21.4E. (After E. Raisz.)

reached maturity, the effect is to give a steep-walled inner gorge, on either side of which lies the former floodplain, now a flat terrace high above river level (Figures 21.23 and 21.24). Meanders which the river had formed on its floodplain have now become impressed into the bedrock and give the inner gorge a meandering pattern. These sinuous bends are termed *entrenched meanders* to distinguish them from the common floodplain meanders.

Figure 21.24 The Goose-Necks of the San Juan River in Utah are entrenched river meanders in horizontal sedimentary strata. (Spence Air Photos.)

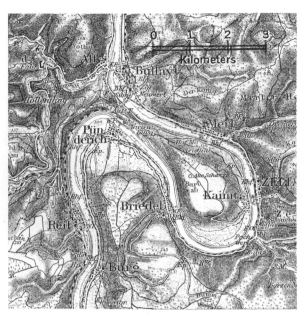

Figure 21.25 The Moselle River has a winding en-trenched-meandering gorge through the Eifel district of Western Germany. (Portion of German 1:100,000 topo-graphic map, 1890.)

Although entrenched meanders are not free to shift about as floodplain meanders, they can enlarge slowly so as to produce cutoffs. Cutoff of an entrenched meander leaves a high, round hill surrounded on three sides by the deep abandoned river channel and on the fourth by the shortened river course. As might be guessed these hills form ideal natural fortifications. Many European fortresses of the Middle Ages were built on such cutoff meander spurs. A good example is Verdun, near the Meuse River.

Under unusual circumstances, where the bed rock includes a strong, massive sandstone formation, meander cutoff leaves a *natural bridge*, formed by the narrow meander neck (Figure 21.23). One well-known example is Rainbow Bridge at Navajo Mountain, in southeastern Utah; other fine examples can be seen in Natural Bridges National Monument at White Canyon in San Juan County, Utah.

Entrenched meanders do not offer ideal locations for railroads and highways, but in a few instances they have been the best available choices for arteries of travel. This point is well illustrated by the Moselle River, whose winding entrenched meanders through the Ardennes mountain upland of Belgium and Western Germany have been utilized (Figure 21.25). Engineers have even cut tunnels through the narrow meander necks to shorten the distance of travel.

REFERENCES FOR FURTHER STUDY

Powell, J. W. (1875), *Exploration of the Colorado River of the West and its tributaries,* U.S. Govt. Printing Office, Washington, D.C. 291 pp.

Lobeck, A. K. (1939), *Geomorphology,* McGraw-Hill Book Co., 731 pp. See Chapters 5, 6, 7.

Davis, W. M. (1954), *Geographical essays,* Dover Publi-cations, New York, 777 pp. See pp. 514–580, 587–616.

Leopold, L. B., M. G. Wolman, and J. P. Miller (1964), *Fluvial processes in geomorphology,* W. H. Free-man, San Francisco, 522 pp.

Morisawa, M. (1968), *Streams: their dynamics and morphology,* McGraw-Hill Book Co., New York, 175 pp.

REVIEW QUESTIONS

1. What three geologic activities are carried on by a stream? Explain the ways in which stream erosion occurs. What is a pothole and how is it formed?

2. What are the modes of stream transportation? What sizes of particles are carried in each? When is bank caving of great importance?

3. How is the quantity of suspended sediment load of a stream stated? How does it vary in large rivers? Why?

4. Explain the changes in cross section of the channel of an alluvial river during a cycle of rising and falling discharge. Define stream capacity. What causes capacity to vary in a stream?

5. Describe the very first events in the formation of a new stream.

6. What features characterize the stage of youth of a stream? What is the geographical importance of young streams?

7. What event marks the attainment of the stage of maturity of streams? What is the profile of equilibrium? What is a graded stream? What is the form of the longitudinal profile of a graded stream?

8. Describe the floodplain and meanders of an alluvial river. What is the geographical importance of a flood-plain?

9. Explain the following terms: undercut and slipoff slopes, point-bar deposits, cutoffs, oxbow lakes, natural levees, yazoo streams.

10. What is the lower limit to which a stream can eventually degrade its channel? What stage of stream development would be found on a peneplain?

11. When does aggradation take place in a stream? Does aggradation change the stream gradient? Explain. What is a braided stream?

12. Describe and explain the development of an al-luvial fan. Where and why are alluvial fans of great importance to human beings?

13. Explain how stream terraces are formed. How did the advance and retreat of the continental ice sheet cause terraces to form in the northeastern United States?

14. Describe the stages in growth of a simple delta. What internal structures has a simple delta? How is the sediment sorted according to particle size?

15. Describe and explain the various forms commonly assumed by deltas of large rivers. Cite examples of the rates of growth of deltas.

16. How can a mature, graded river become rejuve-nated? How do entrenched meanders form?

CHAPTER 22
The Cycle of Landmass Denudation

THUS far, individual types of landforms produced by weathering, mass wasting, overland flow, and streams have been examined. Consider now the entire aspect of denudation of a large region under the combined attack of these agents. Imagine an area, such as a continent or a large portion of a continent, which is elevated by internal earth movements to provide a new landmass. This elevation will constitute the *initial* stage of a grand *cycle of landmass denudation*, in which the region passes through youth, maturity, and old age. For the sake of simplicity, it must be assumed that the elevation of the landmass occurs rapidly and that further crustal deformation then ceases, leaving the denudational agents a long period of uninterrupted activity.

A single, ideal cycle of landmass denudation will not cover all occurrences. There is a difference between landform development in humid climates and that in arid climates. It is consequently necessary to describe two cycles, one for each climate. Furthermore, some initial landmasses are relatively smooth surfaced, representing an even sea floor broadly uparched by epeirogenic crustal movement. Others are mountainous because of breaking and bending of the rock during orogeny. For our purposes, two combinations will be discussed: (1) a smooth-surfaced landmass in a region of humid climate; (2) a rugged, mountainous landscape in an arid climate.

Cycle of landmass denudation in a humid climate

A landmass formed by uparching of a relatively smooth sea floor would have gentle slopes dipping seaward from the high central area. A portion of this slope is shown in Block *A* of Figure 22.1; it is said to be in the *initial stage*. Overland flow upon the new surface would drain off in the most convenient downslope direction and would soon develop initial streams. In the manner already explained in connection with the life history of a stream, these would begin to trench youthful V-shaped valleys into the initial landmass. Marshes and lakes occupying shallow depressions in the initial surface would soon be drained.

Block *B* shows a stage of *early youth* in the cycle of landmass denudation. The *relief* of the area—that is, the difference in elevation between valley bottoms and divide summits—is now increasing rapidly because the streams are cutting down rapidly, whereas between the streams there remain relatively flat portions of the initial land surface (Figure 22.2). As the valleys deepen they also widen, because rock waste is swept down the valley sides into the streams. The unconsumed areas between valleys thus are reduced in proportionate area, while the steep valley slopes increase in extent. Small tributary valleys branch out from the larger streams, further cutting into the initial landmass.

Block *C* of Figure 22.1 shows remnants of the initial land surface, between which are well-developed valley systems, but the remnants shrink in area until the greater proportion of the region consists of steep valley slopes, a stage that may be termed *late youth*. Relief has been steadily increasing as the streams have been actively downcutting. Next, however, conditions show a marked change. When the larger streams become graded and begin to form their floodplains, the increase in relief is halted. The remaining flat remnants of the initial surface are finally consumed, and the valley slopes intersect in narrow divides.

When relief has reached the maximum, the stage of *maturity* is attained (Block *D*, Figure 22.1 and Figure 22.3). From this time on, the valley floors are lowered with extreme slowness, whereas the interstream divides are rapidly lowered. Thus the relief of the region decreases steadily. Slopes become progressively lower in angle (Block *E*, Figure 22.1). Sheet wash and gravity movements no longer are so active as in previous stages.

Base level and peneplain

After a period of time much longer than was required for maturity to be reached, the landscape is reduced to a low rolling surface and may be said to have entered the stage of *old age* (Block *F*, Figure 22.1). By this time most of the streams have extremely low gradients and extensive floodplains. The ultimate goal, which would be reached if infinite time were available for its accomplishment, would be the reduction of the land to a

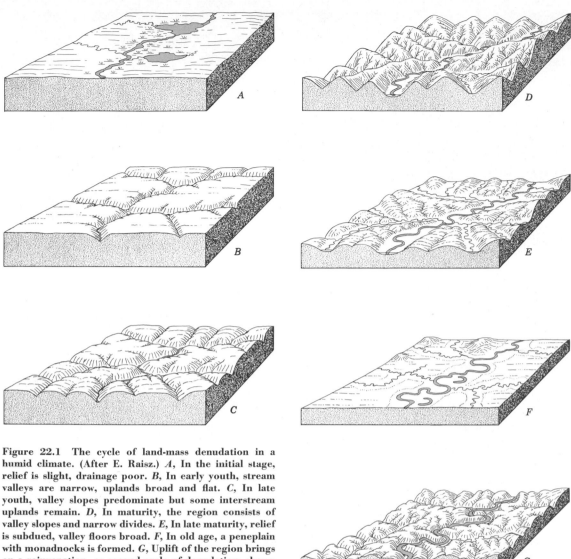

Figure 22.1 The cycle of land-mass denudation in a humid climate. (After E. Raisz.) *A*, In the initial stage, relief is slight, drainage poor. *B*, In early youth, stream valleys are narrow, uplands broad and flat. *C*, In late youth, valley slopes predominate but some interstream uplands remain. *D*, In maturity, the region consists of valley slopes and narrow divides. *E*, In late maturity, relief is subdued, valley floors broad. *F*, In old age, a peneplain with monadnocks is formed. *G*, Uplift of the region brings on a rejuvenation, or second cycle of denudation, shown here to have reached early maturity.

surface coinciding with sea level projected inland. To this imaginary surface is applied the word *base level,* for it is both a base toward which denudation is progressing and a water-level surface. Although the base level is attainable only in theory, a fairly close approach to it has been made in various parts of the world throughout eras of the geologic past. The word *peneplain* is given to a land surface of faint relief produced in the old-age stage of a cycle of denudation.

A peneplain is not perfectly flat but has gentle slopes. Because the streams are sluggish and the land slopes low, further denudation is extremely slow. It is not easy to set a particular figure for the number of years required for a region to pass from initial stage to old age, because this depends upon how high the landmass was elevated to begin with and how resistant the rocks are to weathering and erosion. Perhaps it would be safe to say that in known cases in the geologic record, several million years have been required to reduce a mountain mass to a peneplain.

Sometimes a region contains zones or patches of rock far more resistant to weathering and erosion than the rock of the region as a whole. As the cycle progresses through maturity into old age, these harder rocks are left standing in prominent hills or isolated mountains, which rise conspicuously above the surrounding peneplain (Figure 22.4). To a residual hill or mountain of this type is given the name *monadnock* , named for Mount Monadnock in southern New Hampshire.

Figure 22.2 The Grand Canyon of the Yellowstone River, viewed from over Inspiration Point, illustrates a youthful canyon carved into the initial surface of a lava plateau. (U.S. Army Air Service Photograph.)

Rejuvenation

Once formed, a peneplain is usually elevated again by crustal movement. This follows the principle, stated earlier, that the internal earth forces act spasmodically, and that periods of extreme stability of a particular part of the world are ended by uparching or severe deformation.

A peneplain that is badly folded or fractured during orogeny is quickly obliterated in the erosion cycle that follows. If, however, the region is merely elevated by epeirogenic movements of a few hundred or a few thousand feet, remains of the peneplain persist for a considerable period. In Block G of Figure 22.1 this occurrence is illustrated. Similarity between this and the young stage

illustrated in Block C is marked. The principal difference between the two is that the initial land surface in Block C is a former sea floor, whereas the "initial" surface in Block G is the uplifted peneplain. Drainage on the uplifted peneplain is already well established, so that the streams are merely rejuvenated and cut deep V-shaped valleys into their old shallow courses. When maturity of the second cycle is reached, the former peneplain is completely consumed, but its influence is seen generally in the accordant summits of hilltops over the region as a whole. There is no particular limit to the number of cycles a region can undergo. In some regions three or four previous cycles can be interpreted from a study of the topography.

Figure 22.3 This air view of a maturely dissected region shows a complex of stream channels and small drainage basins. (Spence Air Photos.)

Geographic aspects of the erosion cycle

The geographic importance of stage in the cycle of landmass denudation is very great. Regions in the initial stage of the cycle are relatively flat plains on which drainage is poor and marsh lands often extensive. Sandy beach deposits left by waves as the land surface emerged from beneath the sea usually produce infertile soils. An excellent example is the coastal plain region of Georgia and northern Florida. The great Okefenokee Swamp occupies a shallow depression in the former sea floor whereas long sandy strips of higher land parallel the coast. The porous sands permit plant nutrients to be leached out of the soil.

Not all regions in an initial stage have emerged from the sea; some, such as the High Plains of eastern Colorado, western Kansas, and northern Texas, were built by aggrading streams and, although remarkably flat, are well drained and do not have extensive marshes. The High Plains possess a high productivity in wheat, not only because the soil and climate are favorable but also because the flatness of the land permits enormous grain fields to be cultivated and harvested by machines.

Still other areas of initial land surface are formed by lava flows, poured out profusely to inundate the previous topography and produce a high, undulating lava plateau. This is the origin of the plateau into which the Yellowstone River

Figure 22.4 Stone Mountain, on the Piedmont upland near Atlanta, Georgia, is a striking monadnock about 1.5 mi (2.4 km) long and rising 650 ft (193 m) above the surrounding Piedmont peneplain surface. The rock is a light-gray granite, almost entirely free of joints, and has been rounded into a smooth dome by weathering processes. (Photograph by U.S. Army Air Service.)

has carved its gorge (Figure 22.2). The Snake River Plain of southern Idaho is another example. A region in youth of the cycle supports its population on the relatively flat areas between deep V-shaped valleys. Because these valleys are in a young stage they have no floodplains; hence, roads, railroads, cities, and farms are situated on the uplands. A mature region, on the other hand, has no flat uplands remaining, hence is not favorable to habitation, agriculture, or transportation. Many of the world's mountain regions are in the stage of maturity in the erosion cycle. Extremely great relief and steep slopes are the result of recent, high uplift of those portions of the earth's crust. A coastal region, on the other hand, elevated only 200 or 300 ft (60 to 90 m) above sea level, can never attain really mountainous relief because the streams can cut into the mass no more deeply than the mass itself rises above sea level. In some maturely dissected regions, the larger streams have already reached full maturity in their own individual life cycles and have sizable floodplains at the same time that the surrounding region is extremely rugged. Under such circumstances, human activity is concentrated in the valley floors.

Regions with a humid climate in late maturity or old age of the cycle are usually favorable to agriculture. Slopes are moderate or low and are well drained. Soils tend to be thick. Destructive soil erosion can be held in check. Roads and railroads cross the rolling surface without great difficulty or follow extensively developed floodplains.

Much of the Amazon Basin is a peneplain, but because of the heavy rainfall a dense forest vegetation renders the region passable only at the cost of great labor. Low, floodplain areas bordering the mature rivers are forbidding morasses.

Peneplains that have been uplifted and trenched by valleys in a new cycle are comparable geographically to regions in young stages of the cycle of landmass denudation.

Drainage networks

Much of the earth's land surface consists of landforms produced in the cycle of denudation in a humid climate, although little is in the initial and early youth stages of the cycle because of the short duration of those stages. Thus, well-developed drainage systems characterize vast areas of the continents. Even in deserts, where a somewhat different cycle is followed by the landmass as a whole, the mountains and plateaus are dominated by drainage systems quite like those of the cycle in humid climates. It is therefore important to devote some attention to the form development, or *morphology*, of stream networks and the associated systems of slopes and drainage basins by which they are fed.

If we place a sheet of tracing paper over a topographic map, the network of stream channels can be traced and examined, as in Figure 22.5. Only the larger streams are shown on topographic maps by actual lines, but most are clearly indicated by the sharp V-bends of contours. A larger-scaled map would probably show many more small fingertip branches than are drawn on Figure 22.5, but even a generalized map illustrates the fundamental principles. Major divides have also been drawn (dashed lines) to show the outlines of the larger watersheds.

Ground slopes converge upon the head of each channel, bringing overland flow in sufficient quantities during heavy rains to produce channel flow and keep the channel well scoured, free of soil and vegetation which might otherwise tend to

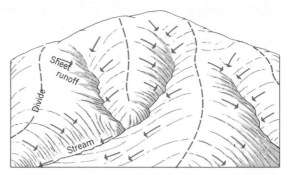

Figure 22.6 Overland flow from slopes in the headwater area of a stream system supplies water and rock debris to the smallest elements of the channel network.

obstruct and obliterate the channel (Figure 22.6). Over a long period of time, as the landmass cycle evolved through youth into maturity, the available ground has been apportioned to the individual small channels so that each receives the amount of runoff required to sustain it. Ground slopes and channel gradients have become mutually adjusted so that the rates of erosion are neither so rapid that a channel lengthens at the expense of its neighbor, nor is it so deficient in area of runoff that it becomes choked with debris of which it cannot dispose. Slow down-cutting of channels and slow lowering of slopes and divides continues, but with a general uniformity over the entire region.

Each finger-tip channel joins another, or enters a larger channel, in such a way that watersheds of increasing size are formed; exit channels are progressively larger in dimensions and discharge. Most stream junctions form acute angles, so that the discharge is carried in the general direction of the larger trunk streams in as direct a manner as possible and with as few channels as possible while, at the same time, providing each element of the stream network with an adequate area of surface runoff.

Playfair's law

Prior to about 1800 there was a widespread belief among naturalists that stream valleys and other landforms were the products of a great cataclysmic upheaval, or rending, of the earth's crust. It was supposed that streams later occupied the valleys produced in this manner, hence, that the valleys came first and were not the work of the streams themselves.

To John Playfair, an English geologist, is attributed the first clear and convincing statement of beliefs universally accepted today concerning streams and their valleys. In 1802, he published the following statement, now known as *Playfair's law:*

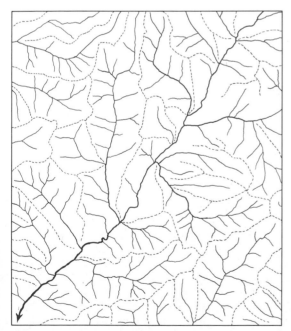

Figure 22.5 A drainage system consists of many small basins, each adjusted in size and shape to the magnitude of the stream it serves. Streams are shown by solid lines; divides by dashed lines.

Every river appears to consist of a main trunk, fed from a variety of branches, each running in a valley proportioned to its size, and all of them together forming a system of valleys connecting with one another, and having such a nice adjustment of their declivities that none of them join the principal valley on too high or too low a level; a circumstance which would be infinitely improbable if each of these valleys were not the work of the stream which flows in it.

The three main points of this law, restated in abstract form, are (1) valleys are proportioned in size to streams flowing in them; (2) stream junctions are accordant in level; (3) therefore, valleys are carved by the streams flowing in them, because both (1) and (2) would be "infinitely improbable" on the basis of chance alone. Exceptions occur in nature. Sometimes a tributary stream passes over a waterfall to reach the level of the master stream. Sometimes small streams occupy very large valleys. But for each of these exceptions, closer study reveals unusual or special conditions that have locally prevailed. Regions of homogeneous bedrock, which have been eroded by streams over a long period of time, invariably bear out the truth of Playfair's law.

Drainage density and texture of topography

If we study an area of badlands, the intricately eroded forms which develop in barren areas of soft clays in arid climates (Figure 22.7), we cannot fail to be impressed with the way that the landforms resemble miniature mountains. Innumerable tiny

Figure 22.7 This vertical air photograph of an area of one square mile in the Big Badlands of South Dakota illustrates ultrafine texture. (U.S. Dept. of Agriculture.)

channels carve tiny valleys to reproduce on a small scale the same great canyon and ridge forms seen in a rugged mountain range, such as the San Gabriel Range of California, or the Great Smoky Mountains of North Carolina. Evidently nature follows certain laws of geometry regardless of whether the smallest unit drainage basin is so small that one can stand astride it, or whether it is a full mile across. Because such similarity of geometry prevails in maturely eroded landmasses, it is necessary to have some means of describing and measuring the scale of magnitude of the forms.

If, for the drainage network map of Figure 22.5, we should measure the total length in miles of all channels, and divide this figure by the total area in square miles of the entire map or watershed, the *drainage density* is found:

$$\text{Drainage density} = \frac{\text{total length of streams (miles)}}{\text{area (square miles)}}$$

Suppose that a drainage density value of 12 is obtained; this is interpreted as meaning that there are twelve miles of channel for every square mile of land surface.[1]

Low drainage density, averaging three to four miles of channel per square mile, is typical of a region underlain by massive, hard, sandstone beds and under heavy forest cover. Such a region of low drainage density may be described as having *coarse texture*, since the individual elements of the topography are very large, or gross.

Medium drainage density, averaging 12 to 16, is typical of an area underlain by thin-bedded sandstones and thick shales, relatively easily eroded, but under a heavy deciduous forest cover. It is typical of large parts of the humid eastern United States where the stage of landmass erosion is mature. (See Figure 26.18.)

High drainage density, or *fine texture*, is developed in easily eroded, weak sedimentary strata where vegetative cover is sparse. Drainage density runs from 30 to 40 under such conditions. Much higher values of drainage density are found in badlands, where there may be from 200 to 500 or more miles of channel per square mile. Such topography would be described as *ultra-fine texture* (Figure 22.7).

What factors control drainage density? One highly important control is rock type. Hard, resistant rocks, such as intrusive granitic rock, gneiss, sandstone, and quartzite, tend to give low drainage density (coarse texture). This is because stream erosion is difficult and only a relatively large channel can maintain itself. In weak rocks, such as

---

[1]Metric units of length and area might also be used. To convert to kilometers per square kilometer multiply drainage density values by 0.62.

shales and clays, even a small watershed can supply enough runoff for channel erosion. A second factor is the relative ease of infiltration of precipitation into the ground surface and downward to the water table (see Chapter 9). Highly permeable materials, such as sand or gravel, tend to give low drainage density because infiltration is great and little water is available as surface runoff to maintain channels. Clays and shales, on the other hand, have a high proportion of surface runoff and this combines with their weakness to give high drainage density. A third major factor is the presence or absence of vegetative cover. A weak rock will have much lower drainage density in a humid climate, where a strong, dense, cover of forest or grass protects the underlying material, than in an arid region where no protective cover exists. For this reason, badlands are characteristic of arid climates, and the drainage density there tends to be markedly higher on all rock types, even the most resistant.

Cycle of landmass denudation in an arid climate

The general appearance of desert regions is strikingly different from that of humid regions, reflecting differences in both vegetation and landforms. It should be emphasized that rain falls in dry climates as well as in moist and that most landforms of desert regions are formed by running water. A particular locality in a dry desert may experience heavy rain only once in several years, but when it does fall, stream channels carry water and perform the same work as the constantly flowing streams of moist regions. Excess water runs off from the valley slopes into the streams, washing down rock particles into the channels, just as in the moister regions. We may even go further and say that, although running water is a rather rare phenomenon in dry deserts, it works with more spectacular effectiveness on the fewer occasions when it does act. This is explained by the meagerness of vegetation in dry deserts. The few small shrubs and herbs that survive offer little or no protection to soil or bedrock. Without a thick vegetative cover to protect the ground and hold back the swift downslope flow of water, excessive quantities of coarse rock debris are swept into the streams. A dry channel is transformed in a few minutes into a raging flood of muddy water, heavily charged with rock fragments.

From these statements it might be inferred that rainfall in desert regions is heavier and more violent than in moist regions. This is not true. In dry deserts, almost all rainfall is of the thunderstorm type, which is localized and intense, affecting only a small area directly under the storm. In humid middle-latitude climates there are even

more frequent and heavier thunderstorm downpours, but also many prolonged periods of steady, light rain during which the soil takes in moisture and becomes fully saturated. Furthermore, the air of humid regions tends to be moist, thus reducing evaporation from the ground. Under such conditions, vegetation can grow densely and maintain its protective hold upon the ground. Water tables are high and streams are fed in part by base flow. It is true that in the humid tropics most of the rainfall is of the violent thunderstorm type, but this occurs so frequently that dense vegetation can flourish. In a dry desert the periods between rains are so prolonged that only a few species of extremely hardy plants can grow.

Because desert rainfall is so localized, a stream flowing into an adjacent dry area will evaporate rapidly, leaving its load upon the stream bed. One of the most important generalizations made about streams in desert regions is that "the streams are shorter than the slopes." Instead of long, continuously flowing streams extending to the sea, streams of desert regions are often short and terminate in alluvial fans and shallow, dry lake floors where rock waste accumulates.

A cycle of landmass denudation in an arid region is illustrated in Figure 22.8. In this ideal cycle we imagine a mountainous region formed by folding or fracturing of the earth's crust and lying in an interior part of the continent (Block *A*). Relief is at the maximum in the initial stage and diminishes throughout successive stages. Numerous large depressions exist between mountain ranges. These do not fill up with water to form lakes, as in a moist climate, but remain dry because of excessive evaporation in the hot, dry climate. The flat central parts of such depressions provide the beds of temporary lakes and are known as *playas*. Playas lakes are shallow and fluctuate considerably in level, often disappearing entirely for long periods. Because they have no outlets, playa lakes contain salt water often more strongly saline than ocean water.

Throughout the erosion cycle, the intermontane depressions become filled with rock waste as alluvial fans are built out from the adjoining mountain masses (Figure 22.9). When the basins are filled with alluvium and the mountains masses are cut up into an intricate set of canyons, divides, and peaks, the region is said to be in the *mature stage* (Block *B*, Figure 22.8). As maturity progresses, the mountains are worn lower, at the same time shrinking in size as the alluvium of the fans encroaches progressively farther inward upon the mountain base.

When *old age* is reached the mountains are represented by small islandlike remnants of their former selves (Block *C*). Eventually even these

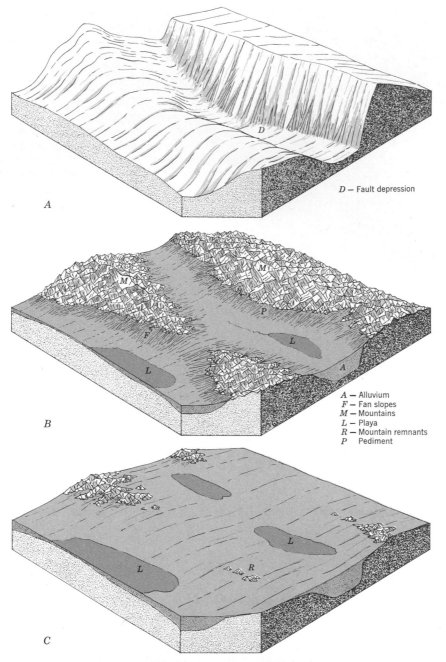

D — Fault depression

A — Alluvium
F — Fan slopes
M — Mountains
L — Playa
R — Mountain remnants
P   Pediment

Figure 22.8   The cycle of land-mass denudation in a mountainous desert. *A,* In the initial stage, relief made by crustal deformation is at the maximum. *B,* In the mature stage, the mountains are completely dissected and the basins are filled with alluvial fan material and playa deposits. *C,* In the old stage, relief is low and alluvial deposits have largely buried the eroded mountain masses, whose remnants project here and there as islandlike groups.

remnants, which may be compared to monadnocks on a peneplain, are eroded away and a vast plain remains. This surface is a type of peneplain, but it has not been developed with reference to sea level as a base because no streams drain out to the sea, and it may, therefore, even lie many hundreds of feet above sea level. It contains shallow depressions occupied by playas rather than by floodplains of meandering rivers. Wind action in a dry climate is effective in eroding shallow depressions and in making dunes of shifting sand.

Pediments

In most deserts, the sloping surfaces of boulders, gravel, and sand which extend from the abrupt base of steep mountain faces to the flat ground

Figure 22.9 This air view of Death Valley, California, shows a mature desert landscape comparable to that shown in Figure 22.8B. (Copyrighted Spence Air Photos.)

of the playas are alluvial fans, underlain by thick deposits of alluvium shed by the mountains. In some places, however, this alluvium, although outwardly taking the form alluvial fans, is nothing more than a veneer perhaps 10 to 20 ft (3 to 6 m) thick, overlying a smooth sloping floor of solid bedrock. To such a rock surface, fringing a desert mountain range or cliff line, the term *pediment* is applied. On the righthand cross section of Blocks *B* and *C*, Figure 22.8, pediment surfaces are shown in profile in a narrow zone between the thick basin alluvium and the rugged mountain masses.

Because the pediment gradient is approximately the same as that of the alluvial cover upon it, and because the alluvial cover is composed of the bedload deposits of graded streams issuing from canyons in the mountain front, it is reasonable to suppose that the pediment is carved by water erosion. The larger streams will be actively downcutting (youthful) in their upper reaches, deep within the mountain ranges. In their lower reaches, where the streams meet the flat basin floor, they are aggrading, or depositing. Therefore, in a narrow intermediate zone which lies close to the mouth of the canyon, the stream will be graded and can shift laterally to undercut the rock of adjacent mountain slopes, much as a saw blade might act

if turned on its side. This process, termed *lateral planation*, is a normal activity of any stream and is the same process by which floodplains are widened by alluvial rivers of humid climates.

Other processes of removal are believed to be important in carving pediments, the relative importance of each depending upon the location and type of bedrock. Between the larger canyons are smaller streams, issuing from steep-walled narrow ravines, while still smaller channels, which may be termed *rills*, drain the mountain slope and are important in removing weathered rock and wearing back the slope. These forms of running water must also adjust their gradients to conform to the gradients of the larger streams which they join, hence, can extend the pediment as a continuous rock floor with only minor undulations.

Once a pediment or alluvial fan is formed, flood flows from the canyons easily fill the shallow, braided channels to overflowing and spread as a more or less uniform layer, termed a *sheetflood*, over the entire graded slope. Some geologists consider that the erosive power of sheetfloods is an important agent in shaping pediments.

Because a pediment is normally veneered by alluvium while it is being formed, the only way to be certain that a pediment exists is to see the

Figure 22.10   Bedrock is widely exposed over this pediment surface at the foot of the Dragoon Mountains near Benson, Arizona. (Photograph by Douglas Johnson.)

bedrock widely exposed to view by later erosion. Thus, the pediment shown in Figure 22.10 has been cut into by stream action, exposing bedrock in numerous outcrops. The cause of stream rejuvenation in this case is not certain, but it may have been brought about by renewed faulting of the mountain block or by a climate change which upset the graded condition of the streams.

Because the old-age surface of the arid climate landmass cycle consists in part of pediments, the term *pediplain* has been introduced as equivalent to the term "peneplain" in the humid climate landmass cycle. A pediplain consists of alluvial fan and playa surfaces as well as pediments; it is thus partly erosional and partly depositional.

Geographical aspects of the arid cycle

The geographical aspects of mountainous deserts are well illustrated by the vast basin-and-range region which includes Nevada, southern California, western Utah, and the southern halves of Arizona and New Mexico. Here, excellent examples of various stages in the cycle can be observed as one travels through the desert on the main transcontinental highways, railroads, and airways. Sparseness of population goes hand in hand with sparseness of vegetation. In 1960, the state of Nevada, which lies wholly within this region, had only about 285,000 inhabitants in its 110,000 sq mi (286,000 sq km). Contrast this with Pennsylvania, whose 45,000 sq mi (117,000 sq km) contained over 11,000,000 persons. The vastness of waste land in this region can be partly appreciated when we realize that the explosion of the first test atom

bomb was kept a secret within a single intermontane basin in New Mexico.

To the traveler, this landscape of mountainous deserts seems to be composed of three distinctive elements: (1) intensely rugged and inhospitable mountains; (2) broad sloping pediments and alluvial fans scored with innumerable shallow dry stream channels, or *washes;* and (3) perfectly flat playa lake floors, covered with shallow water or white salt deposits. In the rugged mountains occur valuable mineral deposits, which are an outstanding source of economic wealth. The alluvial fan slopes are virtually worthless except in the few places where wells bring in a flow of water for the needs of isolated communities. The playas provide mineral wealth of a different sort. Various salts of calcium, sodium, and potassium are often present in sufficient quantities to exploit.

REFERENCES FOR FURTHER STUDY

Lobeck, A. K. (1939), *Geomorphology*, McGraw-Hill Book Co., New York, 731 pp. See Chapter 5.

Cotton, C. A. (1941), *Landscape*, Cambridge University Press, Cambridge, England, 301 pp. See Chapters 7, 14, 17, 18, 19, 20.

Cotton, C. A. (1942), *Climatic accidents in landscape-making*, Whitcombe and Tombs, Christchurch, N.Z. 354 pp. See Section I.

Davis, W. M. (1954), *Geographical essays*, Dover Publications, New York, 777 pp. See pp. 249–278, 296–322, 350–380, 381–412.

Thornbury, W. D. (1969), *Principles of geomorphology*, second edition, John Wiley and Sons, New York, 594 pp. See Chapters 5, 6, 8, 11.

1. Explain the concept of the cycle of landmass denudation. What possible combinations of initial relief and climate may be considered?

2. Describe the various stages in the landmass denudation cycle in a humid climate. Include discussion of the development of the drainage system, valley slopes and divides, and topographic relief. Which stage has greatest relief?

3. Describe the appearance of a peneplain. What is a monadnock? If elevation of the land causes a new cycle to begin, what happens to the peneplain?

4. Describe the appearance of a drainage network. How is it adapted to efficient discharge of runoff and sediment load?

5. State Playfair's law. What other view was held in Playfair's time? Does the law have any exceptions? Explain.

6. Define drainage density and give a series of typical values found in nature. Discuss the factors that control drainage density.

7. How does the role of running water in arid regions differ from that in humid climates? Contrast the forms of stream channels in humid and arid climates. Explain the statement that in deserts "streams are shorter than slopes."

8. Describe the stages of a landmass cycle in mountainous deserts. Compare this cycle, stage by stage, with the cycle of landmass denudation in a humid climate. Compare a peneplain with a pediplain.

9. Describe a pediment. How does it resemble an alluvial fan? How does it differ from an alluvial fan? What processes are considered to produce pediments?

10. Discuss the economic geographical aspects of mountainous deserts. What physical factors in the environment tend to limit human occupation of such areas?

CHAPTER 23
Landforms Made by Glaciers

MOST of us know ice only as a brittle, crystalline solid because we are accustomed to seeing it only in small quantities. Where a great thickness of ice exists, let us say 200 to 300 ft (60 to 90 m) or more, the ice at the bottom behaves as a plastic material and will slowly flow in such a way as to spread out the mass over a larger area, or to cause it to move downhill, as the case may be. This behavior characterizes *glaciers*, which may be

Figure 23.1 The Eklutna Glacier, Chugach Mountains, Alaska, seen from the air. A deeply crevassed ablation zone with a conspicuous medial moraine (foreground) contrasts with the smooth-surfaced firn zone (background). (Photograph by Steve McCutcheon, Alaska Pictorial Service.)

defined as any large natural accumulations of land ice affected by present or past motion.

Conditions requisite to the accumulation of glacial ice are simply that snowfall of the winter shall, on the average, exceed the amount of melting and evaporation of snow that occurs in summer. (The term *ablation* is used by glaciologists to include both evaporation and melting of snow and ice.) Thus, each year a layer of snow is added to what has already accumulated. As the snow compacts, by surface melting and refreezing, it turns into a granular ice, then is compressed by overlying layers into hard crystalline ice. When the ice becomes so thick that the lower layers become plastic, outward or downhill flow commences, and an active glacier has come into being.

At sufficiently high altitudes, whether in high or low latitudes, glaciers form both because air temperature is low and mountains receive heavy orographic precipitation (Chapter 6). Glaciers that form in high mountains are characteristically long and narrow because they occupy previously formed valleys and bring the plastic ice from small collecting grounds high upon the range down to lower elevations, and consequently warmer temperatures, where the ice disappears by ablation (Figure 23.1). Such *valley glaciers*, or *alpine glaciers*, are distinctive types.

In arctic and polar regions, prevailing temperatures are low enough that ice can accumulate over broad areas, wherever uplands exist to intercept heavy snowfall. As a result, areas of many thousands of square miles may become buried under gigantic plates of ice whose thickness may reach several thousand feet. The term *icecap* is usually applied to an ice plate limited to high mountain and plateau areas. During glacial periods an icecap spreads over surrounding lowlands, enveloping all landforms it encounters and ceasing its spread only when the rate of ablation at its outer edge balances the rate at which it is spreading. This extensive type of ice mass is called a *continental glacier* or *ice sheet*.

Figure 23.2 illustrates a number of features of alpine glaciers. The center illustration is of a simple glacier occupying a sloping valley between steep rock walls. Snow is collected at the upper

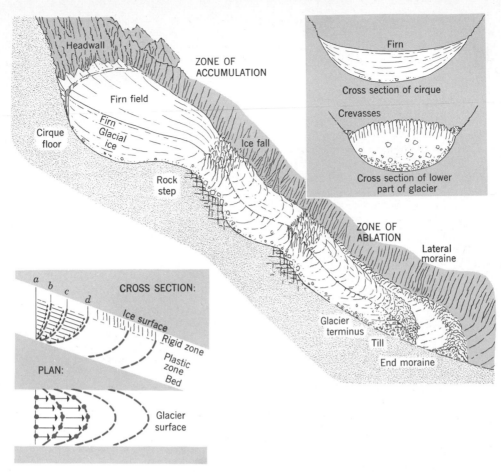

Figure 23.2 Structure and flowage of a simple alpine glacier.

end in a bowl-shaped depression, the *cirque*. The upper end constitutes the *zone of accumulation*. Layers of snow in the process of compaction and recrystallization constitute the *firn* (or *névé*). The smooth *firn field* is slightly concave up in profile (upper right). Flowage in the glacial ice beneath carries the excess ice out of the cirque, down valley. An abrupt steepening of the grade comprises a *rock step*, over which the rate of ice flow is accelerated and produces deep crevasses (gaping fractures) which form an *ice fall*. The lower part of the glacier lies in the *zone of ablation*. Here the rate of ice wastage is rapid and old ice is exposed at the glacier surface, which is extremely rough and deeply crevassed. The glacier *terminus*, or *snout* is heavily charged with rock debris. The lower part of the glacier is usually of upwardly convex cross-profile, the center being higher than the sides (upper right).

The uppermost layer of a glacier, perhaps 200 ft (60 m) in thickness is brittle and fractures readily into crevasses, whereas the ice beneath behaves as a plastic substance and moves by slow flowage (lower left). If one were to place a line of stakes across the glacier surface, the glacier flow would gradually deform the line into a parabolic curve, indicating that rate of movement is fastest in the center and diminishes toward the sides.

Altogether, a simple glacier forms a system that readily establishes a dynamic equilibrium in which the rate of accumulation at the upper end balances the rate of ablation at the lower end. Ice is transferred by flowage to maintain the glacier at approximately a given length and cross-sectional area. Equilibrium is easily upset by changes in the average annual rates of nourishment or wastage.

Rate of flowage of both alpine and continental glaciers is very slow indeed, amounting to a few inches per day for large ice sheets and for the more sluggish valley glaciers, up to several feet per day for an active valley glacier.

Glacial erosion

Most glacial ice is heavily charged with rock fragments, ranging from pulverized rock flour to huge angular boulders of fresh rock. This material is derived from the rock floor upon which the ice moves, or in alpine glaciers, from material that slides or falls from valley walls. Glaciers are capable of great erosive work, both by *abrasion*, erosion caused by ice-held rock fragments that scrape and grind against the bedrock, and by *plucking*, in which the moving ice lifts out blocks of bedrock that have been loosened by freezing of water in joint fractures. The process of plucking out of joint blocks is illustrated in Figure 23.2, at the rock steps.

The debris thus obtained must eventually be left stranded at the outer edge or lower end of a glacier when the ice is dissipated. Thus there are two glacial activities to consider, erosion and deposition. Both result in distinctive landforms, which in some cases are further differentiated according to the type of glacier, whether alpine or continental.

Landforms made by alpine glaciers

Landforms made by alpine glaciers can best be studied by a series of diagrams (Figure 23.3), in which a previously unglaciated mountainous

A. **Before glaciation sets in, the region has smoothly rounded divides and narrow, V-shaped stream valleys.**

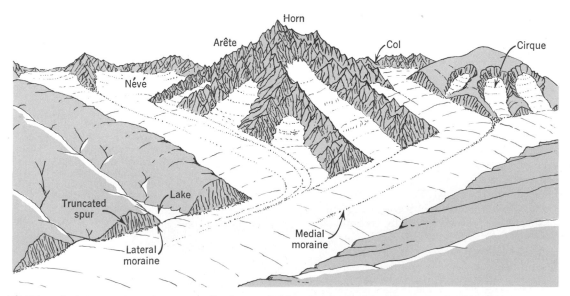

B. **After glaciation has been in progress for thousands of years new erosional forms are developed.**

Figure 23.3   Landforms produced by alpine glaciers. (After W. M. Davis and A. K. Lobeck.)

region is imagined to be attacked and modified by glaciers, after which the glaciers disappear and the remaining landforms are exposed to view.

Block *A* shows a mountainous region sculptured entirely by weathering, mass wasting, and streams. The mountains have a smooth, full-bodied appearance, with rather rounded divides. Although this is not always true, it is typical of the appearance of mountains in humid regions, for example, the Great Smoky Mountains of the southern Appalachians. Imagine now a climatic change in which the average annual temperature becomes several degrees lower and results in the accumulation of snow in the heads of most of the valleys high upon the mountain sides. An early stage of glaciation is shown at the right hand side of Block *B*, where snow is collecting and cirques are being carved by the outward motion of the ice and by intensive frost shattering of the rock near the masses of compacted snow.

In Block *B*, glaciers have filled the valleys and are integrated into a system of tributaries that feed a trunk glacier just as in a stream system. Glaciers are, of course, enormously thicker than streams, because the extremely slow rate of ice motion requires a great cross section if a glacier is to maintain a discharge equivalent to a swiftly flowing stream. Tributary glaciers join the main glacier with smooth, accordant junctions, but, as we shall see later, the bottoms of their channels are quite discordant in level.

Vigorous freezing and thawing of melt water from snows lodged in crevices high upon the walls of the cirque shatters the bare rock into angular fragments, which fall or creep down upon the snowfield and are incorporated into the glacier.

Figure 23.4   The Swiss Alps appear from the air as a sea of sharp arêtes and toothlike horns. In the foreground is a cirque. (Swissair Photo.)

Frost shattering also affects the rock walls against which the ice rests. The cirques thus grow steadily larger. Their rough, steep walls soon replace the smooth, rounded slopes of the original mountain mass. Where two cirque walls intersect from opposite sides, a jagged, knifelike ridge, called an *arête*, results. Where three or more cirques grow together, a sharp-pointed peak is formed by the intersection of the arêtes. The name *horn* is applied to such peaks in the Swiss Alps (Figure 23.4). One of the best known is the striking Matterhorn. Where the intersection of opposed cirques has

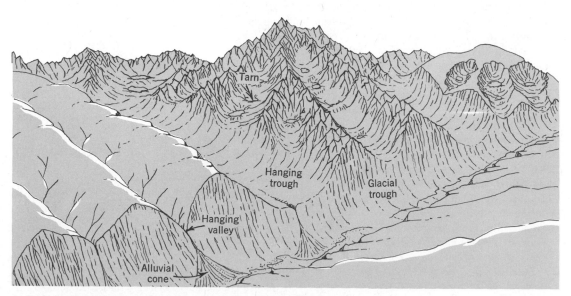

C. **With the disappearance of the ice a system of glacial troughs is exposed.**

Figure 23.5 This maturely dissected mountain mass presents a sea of cirques, arêtes, and horns. A lake occupies the cirque floor near center. San Juan Mountains, Colorado. (Photograph by Frank Jensen.)

and *rock steps.* Cirques and upper parts of troughs thus are occupied by numerous small lakes, called *tarns* (Figure 23.5). The major troughs frequently contain large, elongate *trough lakes,* sometimes referred to as *finger lakes.* Landslides are numerous because glaciation leaves oversteepened trough walls. In glaciated countries such as Switzerland and Norway slides are a major type of natural disaster, because most towns and cities lie in the trough floors where they are readily destroyed by mudflows and landslides (Chapter 19).

Debris may be carried by an alpine glacier within the ice, or it may be dragged along between the ice and the valley wall as a *lateral moraine* (Figures 23.2 and 23.3*B*). Where two ice streams join, this marginal debris is dragged along to form a *medial moraine,* riding upon the ice in midstream (Figure 23.1). At the terminus of a glacier debris accumulates in a heap known as a *terminal moraine,* or *end moraine.* This heap is usually in the form of a curved embankment lying across the valley floor and bending upvalley along each wall of the trough to merge with the lateral moraines (Figures 23.2 and 23.6). As the end of the glacier wastes back, scattered debris is left behind. Should irregularities in the rate of glacier wasting cause the front temporarily to stand still, another moraine ridge will be built. Successive halts in ice

been excessive, a pass or notch, called a *col,* is formed.

Glacier flow constantly deepens and widens its channel so that after the ice has finally disappeared there remains a deep, steep-walled *glacial trough,* characterized by a relatively straight or direct course and by the U-shape of its transverse profile (Block *C,* Figure 23.3). Tributary glaciers likewise carve U-shaped troughs, but they are smaller in cross section, with floors lying high above the floor level of the main trough, so are called *hanging troughs.* Streams, which later occupy the abandoned trough systems, form scenic waterfalls and cascades where they pass down from the lip of a hanging trough to the floor of the main trough. These streams quickly cut a small V-shaped notch in the trough bottom.

Valley spurs that formerly extended down to the main stream before glaciation occurred have been beveled off by ice abrasion and are termed *truncated spurs* (Block *B*). Under a glacier the bedrock is not always evenly excavated, so that the floors of troughs and cirques may contain *rock basins*

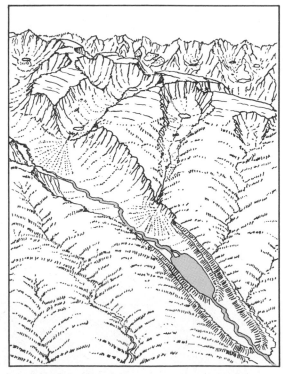

Figure 23.6 Moraines of a former valley glacier appear as looped embankments marking successive positions of the ice margins. (After W. M. Davis.)

Figure 23.7 Development of a glacial trough. (After E. Raisz.) *A*, During maximum glaciation, the U-shaped trough is filled by ice to the level of the small tributaries. *B*, After glaciation, the trough floor may be occupied by a stream and lakes. *C*, If the main stream is heavily loaded, it may fill the trough floor with alluvium. *D*, Should the glacial trough have been deepened below sea level, it will be occupied by an arm of the sea, or fiord.

retreat produce successive moraines, termed *recessional moraines.*

Glacial troughs and fiords

Many large glacial troughs now are nearly flat-floored because aggrading streams that issued from the receding ice front were heavily laden with rock fragments. Figure 23.7 shows a comparison between a trough with little or no fill and another with alluvial-filled bottom. The deposit of alluvium extending down valley from a melting glacier is the *valley train.*

When the floor of a trough open to the sea lies below sea level, the sea water will enter as the ice front recedes, thus producing a narrow estuary known as a *fiord* (Figure 23.8). Fiords may originate either by submergence of the coast or by glacial erosion to a depth below sea level. Most fiords are explained in the second way because ice is of such a density that, when floating, from three-fourths to nine-tenths of its mass lies below water level. Therefore, a glacier several hundred feet thick could erode to considerable depth below sea level before the buoyancy of the water reduced its erosive power where it entered the open water.

Fiords are observed to be opening up today along the Alaskan coast, where some glaciers are melting back rapidly and the fiord waters are being extended along the troughs. Fiords are found largely along mountainous coasts in latitudes 50° to 70° N and S. The explanation of this distribution lies in climate and is treated under the discussion of marine west coast climate (Chapter 11).

Geographical aspects of alpine glaciation

In general, the ruggedness of fully glaciated mountains such as the Alps, Pyrenees, Himalayas, or Sierra Nevada makes for sparseness of population and difficulty of access. Land above timber line is useless for any purpose except summer pastures and the extraction of such minerals as may lie in the rocks. Locally, recreational uses—mountain climbing and winter sports—are intensively developed.

Below timber line are rich forests. U-shaped glacial troughs provide broad, accessible strips of land at relatively low levels. These are utilized for town sites, for winter pasture, and as arteries of transportation. In the Italian Alps several great flat-floored glacial troughs extend from the heart

Figure 23.8 This Norwegian fiord has the steep rock walls of a deep glacial trough. (Photograph by Mittet and Co.)

of the Alps southward to the plain of northern Italy. These are important geographic controls because they provide smooth and easy access into the heart of the Alps and to the principal Alpine passes. The Brenner Pass lies at the head of a magnificent trough of this type, the Adige River valley.

The steep-walled troughs contain many waterfalls and rapids readily used for hydroelectric plants. Electrification of railroads and industry is highly advanced in Switzerland and Norway.

Ice sheets of the present

Two enormous accumulations of glacial ice are the Greenland and Antarctic icecaps. These may be imagined as huge plates of ice, several thousand feet thick in the central areas, resting upon landmasses of subcontinental size. The Greenland icecap has an area of 670,000 sq mi (1,740,000 sq km) and occupies about seven-eighths of the entire island of Greenland (Figure 23.9). Only a narrow, mountainous coastal strip of land is exposed.

The Antarctic icecap covers about 5,000,000 sq mi (13,000,000 sq km) and in places spreads out into the ocean to form floating *ice shelves* (Figure 23.10). One significant point of difference between these two icecaps is their position with reference to the poles. Whereas the antarctic ice rests almost squarely upon the south pole, the Greenland icecap is considerably offset from the north pole, with its center about at 75° N lat. This position illustrates a fundamental principle; that a large area of high land is essential to the accumulation of a great icecap. No land exists near the north pole; ice accumulation there is restricted to a thin layer of floating sea ice.

Contours drawn upon the surface of the Greenland icecap show that it is in the form of very broad, smooth domes. From a high point of about 10,000 ft (3000 m) elevation east of the center there is a gradual slope outwards in all directions. The rock floor of the icecap lies close to sea level under the central region, but is higher near the edges. Accumulating snows add layer upon layer of ice to the surface, while at great depth the plastic ice slowly flows outward toward the edges. At the outer edge of the sheet the ice thins down to a few hundreds of feet. Continual loss through ablation keeps the position of the ice margin relatively steady where it is bordered by a coastal belt of land. Elsewhere the ice extends in long tongues, called *outlet glaciers*, to reach the sea at the heads of fiords. From the floating glacier edge huge masses of ice break off and drift out to open sea with tidal currents. The breakup of the ice front is known as *calving* and is brought about by strains caused by the rise and fall of tide level as well as by the undercutting and melting at and below the water line. The calving of floating glacier fronts is an extremely rapid process compared to ablation of ice fronts on land. Consequently, ice-caps are limited in their seaward extent and rarely extend far into the ocean beyond the limits of bays and shallow continental shelves.

Ice thickness in Antarctica is even greater than that of Greenland. For example, on Marie Byrd Land a thickness of 14,000 ft (4250 m) was measured, the rock floor lying 8200 ft (2500 m) below sea level. On Victoria Land Plateau at an elevation of 8900 ft (2700 m) above sea level, the ice was found to be 13,000 ft (4000 m) thick. This means that the rock floor over certain parts of the Antarctic continent lies below sea level.

An important glacial feature of Antarctica is the presence of great plates of floating glacial ice, termed *ice shelves* (Figure 23.10). The largest of these is the Ross Ice Shelf with an area of about 200,000 sq mi (520,000 sq km) and a surface elevation averaging about 225 ft (70 m) above the sea. Ice shelves are fed by the icecap, but also accumulate new ice through the compaction of snow.

Ice sheets of the Pleistocene epoch

As if the vast ice sheets of Greenland and Antarctica do not seem fantastic enough, geologists have brought to light abundant and convincing evidence that much of North America and Europe and parts of northern Asia and southern South America were covered by enormous ice sheets in a period of time designated the Pleistocene epoch (see Chapter 18). Despite the fact that this ice age ended 10,000 to 15,000 years ago with the rapid wasting away of the ice sheets, the glacial epoch is the most recent major episode in geologic his-

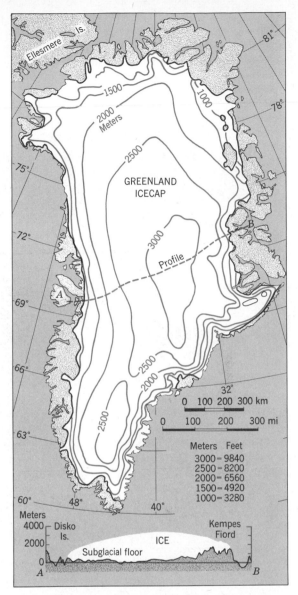

Figure 23.9 Generalized map of Greenland. (After R. F. Flint, *Glacial and Pleistocene Geology*.)

tory. Landforms made by the last ice advance and recession are very little modified by erosional agents.

Figures 23.11 and 23.12 show the extent to which North America and Europe were covered at the maximum known spread of the last advance of the ice. In the United States, all the land lying north of the Missouri and Ohio rivers was covered, as well as northern Pennsylvania and all of New York and New England. In Europe, the ice sheet centered upon the Baltic Sea, covering the Scandinavian countries and spreading as far south as central Germany. The British Isles were almost covered by an icecap that had several centers on highland areas and spread outward to coalesce with the Scandinavian ice sheet. The Alps at the same time were heavily inundated by enlarged alpine glaciers, fused into a single icecap. All high mountain areas of the world underwent greatly intensified alpine glaciation at the time of maximum ice-sheet advance. Today, only small remnant alpine glaciers exist as vestiges of these great valley glaciers. In less favorable mountain regions no glaciers remain.

Proof of the former great extent of ice sheets has been carefully accumulated since the middle of the nineteenth century when the great naturalist

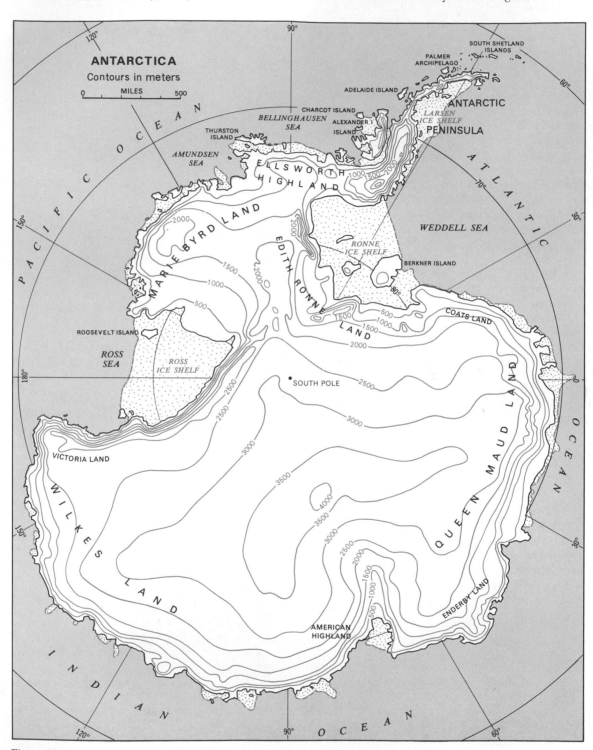

Figure 23.10   Antarctica. (John P. Tremblay, cartographer.)

Louis Agassiz first announced the bold theory. In general, the evidence of past glaciation lies in the recognition throughout North America and Europe of landforms identical with those now seen near the margins of the Greenland icecap and other glaciers. Although Agassiz's pronouncement was greeted with much skepticism a century ago, continental glaciation of the Pleistocene epoch is now universally accepted among scientists. Moreover, careful study of the deposits left by the ice has led to the knowledge that not one but four major advances and retreats occurred, spaced over a total period of about a million years. It is the deposits of the last advance, known as the *Wisconsin stage*, that form fresh and conspicuous landforms. For the most part, then, the discussion of glacial landforms will concern these most recent deposits.

Glacial stages

The four principal glacial stages of North America are matched by corresponding stages of Europe, as deciphered in the Alps. Together with the interglacial stages (periods of mild climate during which the ice sheets disappeared), these glacial stages in order of increasing age are:

North American Stages (Northcentral U.S.)	European Stages (Alps)
Wisconsin glacial	Würm glacial
Sangamon interglacial	Riss-Würm interglacial
Illinoian glacial	Riss glacial
Yarmouth interglacial	Mindel-Riss interglacial
Kansan glacial	Mindel glacial
Aftonian interglacial	Günz-Mindel interglacial
Nebraskan glacial	Günz glacial

Figure 23.13 shows the limit of southerly spread of ice of each glacial stage in the north-central United States. Where older limits were overridden by younger, the boundaries are conjectural. Notice that an area in Wisconsin (the *Driftless Area*) was surrounded by ice but never inundated.

The absolute age and duration in years of the Pleistocene epoch and its stages are extremely difficult to fix, even though the relative order of events is well established. The last ice disappeared 10,000 to 15,000 years ago from the north-central United States. The Wisconsin stage may have endured 60,000 years. The earliest onset of glacial advance (Nebraskan) may have occurred 300,000 to 600,000 years ago, although the beginning of the Pleistocene epoch of geologic time is commonly set by geologists as about one million years before present.

Cause of continental glaciation

It should not be inferred that the occurrence of an ice age as the last event of geologic history

Figure 23.11 Pleistocene ice sheets of North America at their maximum spread reached as far south as the present Ohio and Missouri rivers. (After R. F. Flint.)

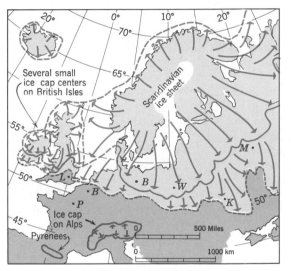

Figure 23.12 The Scandinavian ice sheet dominated northern Europe during the Pleistocene glaciations. Solid line shows limits of ice in the last glacial stage; dotted line on land shows maximum extent at any time. (After R. F. Flint.)

means that the earth as a planet is cooling off. There is excellent evidence in the form of rock deposits of similar periods of glaciation in the early part of the earth's geologic history. But, beyond the fact that glaciations are occasional events in geologic history, knowledge concerning the cause of glaciation is speculative. Obviously a prolonged

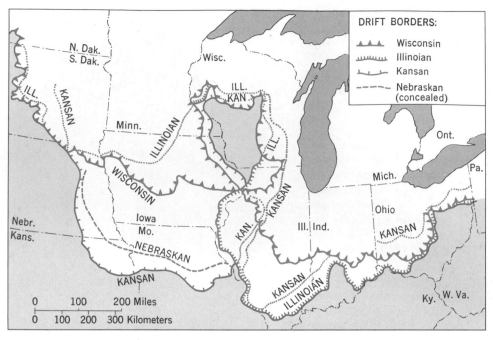

Figure 23.13 In each glacial stage, the ice sheet reached a different line of maximum advance. (After R. F. Flint, *Glacial and Pleistocene Geology.*)

period of colder climate with ample snowfall brings on the growth of ice sheets.

One possible contributing cause of glaciation is a decrease in the quantity of solar radiation intercepted by the earth. It has been reasoned that, should this quantity of energy have diminished somewhat at the beginning of the Pleistocene epoch, the average temperature of the earth's atmosphere would have dropped to a lower level. Providing that this change did not also diminish the quantity of snowfall, the reduced ablation of snowfields would increase the quantity of snow turned into glacial ice, with resultant growth and spread of icecaps. A second factor, possibly working in harmony with the first, is the known increase in elevation of large parts of the continents during the Pliocene and early Pleistocene epochs as a result of widespread mountain making (orogeny), as well as broad uparching of continental interiors (epeirogenic uplift). That mountains intercept large quantities of precipitation has already been explained (Chapter 6) in the discussion of orographic precipitation. The combined effects of reduced solar energy and increased altitude of continents would bring on colder climates with increased snowfall over favorable areas of the continents, such as the Laurentian Upland of eastern Canada and the Scandinavian peninsula. This composite theory of cause of glaciation may be called the *solar-topographic theory.*

An important and widely held theory attributes glaciation to a reduction of the carbon dioxide content of the atmosphere. The role of carbon dioxide in absorbing long-wave radiation and thus warming the atmosphere has been explained in Chapter 4. It is estimated that if the carbon dioxide content of the atmosphere, which is now 0.03 percent by volume, were reduced by half that amount, the earth's average surface temperature would drop about 7 F° (4 C°). Such a reduction in carbon dioxide content might be postulated in combination with increased continental altitude to bring on the growth of ice sheets.

Other theories invoke quite different mechanisms. It has been postulated that increased quantities of volcanic dust in the atmosphere might bring about a glacial period because more solar energy would be reflected back into space, permitting less to enter the lower atmosphere. Along with the reduced air temperature would be the increase in numbers of tiny dust particles to serve as nuclei for the condensation of moisture, thus favoring increased precipitation. Another group of theories propose shifts in the positions of the continents with respect to the poles, bringing various parts of the landmasses into favorable geographical positions for the growth of ice sheets. Still another theory requires that changes in oceanic currents, specifically the diversion or blocking of such warm currents as the Gulf Stream, would have brought colder climates to the subarctic regions. Variations in the earth's orbit, causing changes in the amounts of solar energy received by the earth, have also been considered as the cause of glacial periods.

Erosion by ice sheets

Like alpine glaciers, ice sheets are effective eroding agents. The slowly moving ice may scrape and grind away much solid bedrock, leaving behind smoothly rounded rock masses bearing countless minute abrasion marks. Scratches, or *striations*, trend in the general direction of ice movement (Figure 23.14), but variations in ice direction from time to time often result in intersecting lines. Certain kinds of rock were susceptible to deep grooving (Figure 23.15).

Where a strong, sharp-pointed piece of rock was held by the ice and dragged over the bedrock surface, there resulted a series of curved cracks fitted together along the line of ice movement. These *chatter marks*, and closely related *crescentic gouges*, whose curvature is the opposite, are good indicators of the direction of ice movement (Figure 23.14). Some very hard rocks have acquired highly polished surfaces from the rubbing of fine clay particles against the rock. The evidences of ice erosion described here are common throughout the northeastern United States. They may be seen on almost any exposed hard rock surface. Once understood, they may be recognized by any alert observer.

Commonly bearing the above abrasion marks is a type of conspicuous knob of solid bedrock that has been shaped by the moving ice (Figure 23.16). One side, that from which the ice was approaching, is characteristically smoothly rounded and shows a striated and grooved surface. This is termed the *stoss* side. The other, or *lee* side, where the ice plucked out angular joint blocks, is irregular, blocky, and steeper than the stoss side. The quaint term *roches moutonnées* has long been applied by glaciologists to such glaciated rock knobs.

Vastly more important than the minor abrasion forms are enormous excavations that the ice sheets made in some localities where the bedrock is of a weak variety and the ice current was accentuated by the presence of a valley paralleling the direction of ice flow. Under such conditions the ice sheet behaved much as a valley glacier, scooping out a deep, U-shaped trough. As fine examples may be cited the Finger Lakes of western New York State. Here a set of former stream valleys lay parallel to southward spread of the ice, which scooped out a series of deep troughs. Blocked at the north ends by glacial debris the basins now hold elongated lakes. Many hundreds of lake basins were created in a similar manner all over the glaciated portion of North America and Europe. Countless small lakes of Minnesota, Canada, and Finland occupy rock basins scooped out by ice action (Figure 23.17). Irregular debris deposits left by the ice are also important in causing lake basins.

Figure 23.14 Glacial striations and fracture marks, mostly crescentic gouges, cover the smoothly rounded surface of this rock knob. These marks were made by the East Twin Glacier, Alaska. The ice moved in a direction away from the photographer. (Photograph by Maynard Miller.)

Figure 23.15 Unusually smooth, deep glacial grooves were carved in limestone by the ice action on Kelley's Island near the south shore of Lake Erie. (Photograph by State of Ohio, Department of Industrial and Economic Development.)

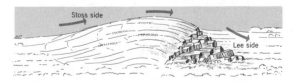

Figure 23.16 A glacially abraded rock knob. (From A. N. Strahler, *The Earth Sciences*, Harper and Row, New York.)

Figure 23.17 Seen from the air, this esker in the Canadian shield area appears as a narrow embankment crossing the terrain of glacially eroded lake basins. (Photograph by Canadian Department of Mines, Geological Survey.)

Deposits left by ice sheets

The term *glacial drift* has long been applied to include all varieties of rock debris deposited in close association with glaciers. Drift is of two major types: (1) *Stratified drift* consists of layers of sorted and stratified clays, silts, sands, or gravels deposited by melt-water streams or in bodies of water adjacent to the ice. (2) *Till* is a heterogeneous mixture of rock fragments ranging in size from clay to boulders and is deposited directly from the ice without water transport. Moraines of valley glaciers, previously described, are composed largely of till, whereas the valley train is composed of stratified drift.

Over those parts of the United States formerly covered by Pleistocene ice sheets, glacial drift averages from 20 ft (6 m) thick over mountainous terrain such as New England, to 50 ft (15 m) and more thick over the lowlands of the north-central United States. Over Iowa, drifts are from 150 to 200 ft (45 to 60 m) thick; over Illinois, it averages more than 100 ft (30 m) thick. Locally, where deep stream valleys existed prior to glacial advance, as in Ohio, drift may be several hundred feet deep.

In order to understand the form and composition of deposits left by ice sheets, it is desirable to consider the conditions prevailing at the time of existence of the ice, as shown in Figure 23.18. Block *A* shows a region partly covered by an ice sheet with a relatively stationary front edge. This condition occurs when the rate of ice ablation balances the amount of ice brought forward by spreading of the ice sheet. Any increase in ice movement would cause the ice to shove forward to cover more ground; an increase in the rate of wasting would cause the edge to recede and the

ice surface to become lowered. Although the Pleistocene ice fronts did advance and recede in many minor and major fluctuations, there were considerable periods when the front was essentially stable. This condition is represented in Block *A*.

The transportational work of an ice sheet may be likened to that of a great conveyor belt. Anything carried on the belt is dumped off at the end and if not constantly removed will pile up in increasing quantity. Rock fragments brought within the ice are deposited at the edge as the ice evaporates or melts. There is no possibility of return transportation.

Glacial till that accumulates at the immediate ice edge forms an irregular, rubbly heap known as a *terminal moraine*. After the ice has disappeared, as in Diagram *B*, the moraine appears as a belt of knobby hills interspersed with basinlike hollows, some of which hold small lakes. The term *knob and kettle topography* is often applied to morainal belts (Figure 23.19). Terminal moraines tend to form great curving patterns, the convex form of curvature being directed southward and indicating that the ice advanced as a series of great *lobes*, each with a curved front (Figure 23.20). Where two lobes came together, the moraines curved back and fused together into a single moraine pointed northward. This is termed an *interlobate moraine* (Figure 23.18, Block *B*). In its general recession accompanying disappearance, the ice front paused for some time along a number of lines, causing morainal belts similar to the terminal moraine belt to be formed. These belts, known as *recessional moraines* (Figures 23.18 and 23.20), run roughly parallel with the terminal moraine but are often thin and discontinuous.

Figure 23.18, Block *A*, shows a smooth, sloping plain lying in front of the ice margin. This is the *outwash plain*, formed of stratified drift left by braided streams issuing from the ice. Their deposits are in reality great alluvial fans upon which are spread layer upon layer of sands and gravels. The adjective *glaciofluvial* is often applied to stream-laid stratified drift.

Large streams issue from tunnels in the ice, particularly when the ice for many miles back from the front has become stagnant, without forward movement. Tunnels then develop throughout the ice mass, serving to carry off the melt water. After the ice has gone (Block *B*, Figure 23.18) the outwash plain remains in its original form, but may be bounded on the iceward side by a steep slope which is the mold of the ice against which the outwash was built. Such a slope is called an *ice-contact slope*. Farther back, behind the terminal moraine, the position of a former ice tunnel is marked by a long, sinuous ridge known as an *esker*. The esker is the deposit of sand and gravel for-

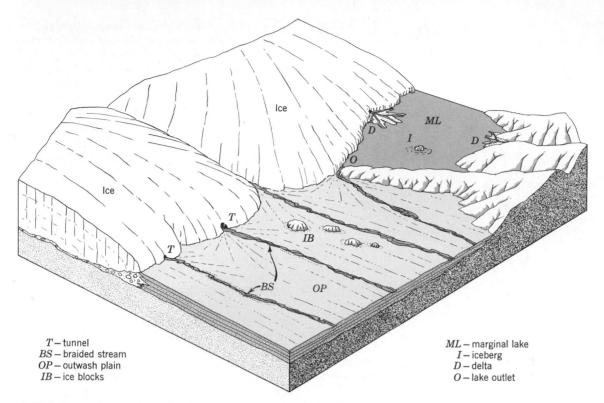

T – tunnel
BS – braided stream
OP – outwash plain
IB – ice blocks

ML – marginal lake
I – iceberg
D – delta
O – lake outlet

A. With the ice front stabilized and the ice in a wasting, stagnant condition, various depositional features are built by melt water.

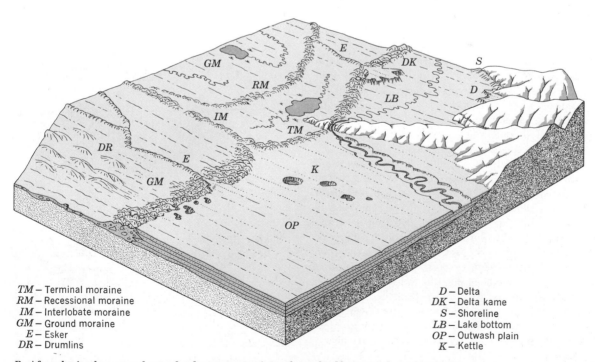

TM – Terminal moraine
RM – Recessional moraine
IM – Interlobate moraine
GM – Ground moraine
E – Esker
DR – Drumlins

D – Delta
DK – Delta kame
S – Shoreline
LB – Lake bottom
OP – Outwash plain
K – Kettle

B. After the ice has wasted completely away, a variety of new landforms made under the ice is exposed to view.

Figure 23.18 Marginal landforms of continental glaciers.

Figure 23.19 Rugged topography of small knobs and kettles characterizes this interlobate moraine northeast of Elkhart Lake, Sheboygan County, Wisconsin. (Photograph by W. C. Alden, U.S. Geological Survey.)

as *ground moraine.* This cover is often inconspicuous because it forms no prominent or recognizable topographic feature. Nevertheless, the ground moraine may be thick and may obscure or entirely bury the hills and valleys that existed before glaciation. Where thick and smoothly spread, the ground moraine forms an extensive, level till plain, but this condition is likely only in regions already fairly flat to start with. In more dissected regions, the preglacial valleys and hills retain their same general outlines despite glaciation.

Deposits built into standing water

Where the general land slope is toward the front of an ice sheet, a natural topographic basin is formed between the ice front and the rising ground. Valleys that may have opened out northward are blocked by ice. Under such conditions, *marginal glacial lakes* form along the ice front (Figure 23.18, Block *A*). These lakes overflow along the lowest available channel, which lies between the ice and the ground slope or over some low pass along a divide. Into marginal lakes streams of melt water from the ice build *glacial deltas,* similar in most respects to deltas formed by any stream flowing into a lake. Streams from the land likewise build deltas into the lake. When the ice has disap-

merly laid upon the floor of the ice tunnel. Inasmuch as ice formed the sides and roof of the tunnel, its disappearance left merely the stream-bed deposit, which now forms a ridge (Figure 23.21). Eskers are often many miles long; in parts of Maine a few are more than 100 mi (160 km) long. Some have branches just as streams do. Because the esker is made of highly porous sand and gravel, the rapid draining away of water from the crest may prevent the growth of trees along the top of some eskers, which look as if artificially cleared of forest (Figure 23.17).

Another curious glacial form is the *drumlin,* a smoothly rounded, oval hill resembling the bowl of an inverted teaspoon. It consists of glacial till (Figure 23.22). Drumlins invariably lie in a zone behind the terminal or recessional moraines. They commonly occur in groups or swarms, which may number in the hundreds. The long axis of each drumlin parallels the direction of ice movement, and the drumlins thus point toward the terminal moraines and serve as indicators of direction of ice movement. From a study of the composition and structure of drumlins, it has been generally agreed that they were formed under moving ice by a kind of plastering action in which layer upon layer of bouldery clay was spread upon the drumlin. This would have been possible only if the ice were so heavily choked with debris that the excess had to be left behind. Furthermore, some sort of knob or surface irregularity may have been required to start the plastering action and localize its occurrence.

Between the terminal, recessional, and interlobate moraines, the surface left by the ice is usually overspread by a cover of glacial till known

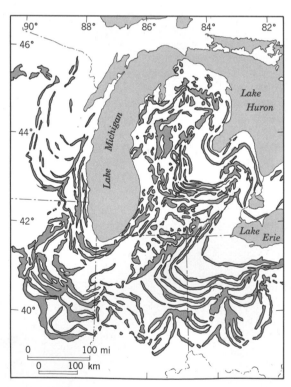

Figure 23.20 Moraine belts of the north-central United States have a festooned pattern left by ice lobes. (After R. F. Flint and others, *Glacial Map of North America,* 1945.)

peared the lake drains away, exposing the bottom upon which layers of fine clay and silt have been laid. These fine-grained strata, which have settled out from suspension in turbid lake waters, are called *glaciolacustrine* sediments and are a variety of stratified drift. The layers are commonly of banded appearance, with alternating dark and light layers, termed *varves*. Glacial lake plains are extremely flat, with meandering streams and extensive areas of marshland.

Deltas, built with a flat top at what was formerly the lake level, are now curiously isolated, flat-topped landforms known as *delta kames*. Delta and stream channel deposits built between a stagnant ice mass and the wall of a valley become *kame terraces*, whose steep scarps are ice-contact slopes (Figure 23.23). Kame terraces are difficult to distinguish from the uppermost member of a series of alluvial terraces, but most kames have undrained depressions or pits produced by the melting of enclosed ice blocks. Built of very well-washed and sorted sands and gravels, kames commonly show the steeply dipping foreset beds characteristic of deltas (Figure 23.24).

Geographical aspects of glacial landforms

Because much of Europe and North America was glaciated by the Pleistocene ice sheets, landforms associated with the ice are of fundamental geographical importance in influencing the activity of human beings. Agricultural influences of glaciation are both favorable and unfavorable, depending on preglacial topography and whether the ice eroded or deposited heavily. In hilly or mountainous regions, such as New England, the glacial till is thinly distributed and extremely stony. Cultivation is made difficult by countless boulders and stones in the soil. Along morainal belts the steep slopes, irregularity of topography, and abundance of boulders and stones in the till are unfavorable to cultivation. These same features, however, make morainal belts extremely desirable as suburban residential areas. A pleasing variety of hills, depressions, and small lakes makes ideal locations for large estates.

Extensive till plains, outwash plains, and lake plains, on the other hand, make for some of the most productive agricultural land in the world. In this class belong the prairie lands of Indiana, Illinois, Iowa, Nebraska, and Minnesota.

Glaciofluvial deposits are of great economic value. The sands and gravels of outwash plains, kames, and eskers provide the aggregate necessary for concrete and other building purposes (Figure 23.25). The purest sands may be used for molds needed for metal castings. Huge quantities of ground water are contained in glaciofluvial deposits. Where deep, preglacial valleys were filled

Figure 23.21 This esker has developed a cover of soil and vegetation, concealing the coarse gravel that lies within it. Dodge County, Wisconsin. (Photograph by W. C. Alden, U.S. Geological Survey.)

Figure 23.22 This small drumlin, located south of Sodus, New York, shows a tapered form from upper right to lower left, indicating that the ice moved in that direction (north to south). (Photograph by Ward's Natural Science Establishment, Inc., Rochester, N.Y.)

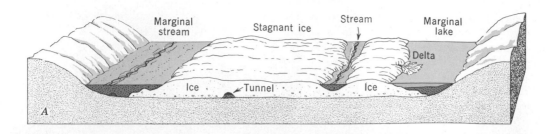

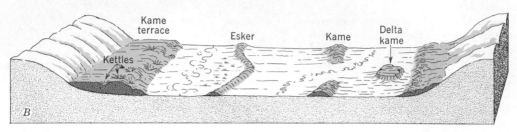

Figure 23.23   Kames may originate as stream or lake deposits laid between a stagnant ice mass and the valley sides. (After R. F. Flint.)

Figure 23.24   These cross-bedded, sorted sands were laid down in a glacial delta near North Haven, Connecticut. (Photograph by R. J. Lougee.)

Figure 23.25   Thick layers of outwash sands and gravels such as these on the north shore of Long Island are excavated in great quantities for use in highway and building construction. The dark layer at the top is a bed of glacial till, left by a glacial advance. Boulders in the foreground are glacial erratics that have rolled down from the till bed. (Photograph by A. K. Lobeck.)

with such materials, large quantities of water can be pumped from wells penetrating the deposit. In this way, many large cities and industrial plants in Ohio, Pennsylvania, and New York obtain their water supplies.

REFERENCES FOR FURTHER STUDY

Lobeck, A. K. (1939), *Geomorphology*, McGraw-Hill Book Co., New York, 731 pp. See Chapters 8 and 9.

Davis, W. M. (1954), *Geographical essays*, Dover Publications, New York, 777 pp. See pp. 617–634.

Flint, R. F. (1957), *Glacial and Pleistocene geology*, John Wiley and Sons, New York, 553 pp.

Hough, J. L. (1958), *Geology of the Great Lakes*, University of Illinois Press, Urbana, 313 pp.

Dyson, J. L. (1962), *The world of ice*, A. A. Knopf, New York, 292 pp.

Schultz, Gwen (1963), *Glaciers and the Ice Age*, Holt, Rinehart and Winston, New York, 128 pp.

Wright, H. E., and D. G. Frey, eds. (1965), *The Quaternary of the United States*, Princeton University Press, Princeton, N.J., 922 pp.

Thornbury, W. D. (1969), *Principles of geomorphology*, second edition, John Wiley and Sons, New York, 594 pp. See Chapters 14, 15, 16.

REVIEW QUESTIONS

1. What is a glacier? What conditions are necessary for the formation of glaciers? Define ablation. What is firn? What is the distinction between alpine and continental glaciers?

2. Describe a simple alpine glacier and explain how it operates. What is glacier equilibrium? How fast does a glacier flow?

3. How do glaciers erode their channels? Compare alpine glaciers with streams in regard to erosion and transportation activities.

4. How do alpine glaciers modify mountain topography? Describe and explain the following features: cirque, arête, horn, col, glacial trough, hanging trough, rock basin, rock step, tarn, finger lake.

5. What is the form of a glacial trough? How does this form differ from that of a normally eroded stream valley? Explain how a fiord is formed. Where are fiords found?

6. Describe the various kinds of deposits made by a valley glacier. What are the location and form of the following types of moraines: medial, lateral, terminal, and recessional? What is a valley train?

7. Where are the icecaps of the world today? About how thick is the ice in the central parts of these icecaps? Explain how an icecap is fed and how the ice moves. What are ice shelves? Outlet glaciers? What is calving?

8. Describe the extent of ice sheets of the Pleistocene epoch in North America and Europe. Describe the southern glacial limit in the United States. What great scientist was largely responsible for convincing the scientific world that the Pleistocene glaciation actually occurred?

9. Name in order the glacial and interglacial stages of North America. How long ago did the last glacial ice disappear from the United States?

10. Review the principal theories that attempt to explain the occurrence of glacial periods.

11. What erosional features on rock surfaces give evidence of the former presence of an ice sheet? How can the direction of ice movement be inferred?

12. How can the direction of ice movement be inferred from the shapes of a glaciated rock knobs (roches moutonnées)? How were the Finger Lakes of western New York State formed?

13. What is glacial drift? What is the distinction between stratified drift and till?

14. Describe and explain the various depositional forms associated with the margin of an ice sheet.

15. What kinds of moraines are left by ice sheets? What is a glacial outwash plain? Describe the surface topography of an outwash plain.

16. How is an esker formed? How long are eskers? Of what material are they composed?

17. What is a drumlin? Of what material is it composed? How is it formed? Where are most drumlins found in relation to the terminal moraine?

18. What is a till plain? How is it formed?

19. Explain how marginal glacial lakes are formed. What types of deposits are formed in them? What is a delta kame? A kame terrace? What are varves?

20. Discuss the geographical aspects of glacial landforms.

CHAPTER 24
Landforms Made by Waves and Currents

OCEAN waves work unceasingly to transform the shores of continents and islands. Waves travel across the deep ocean with only gradual loss of energy, but when shallow water is reached, the wave form changes radically and new water motions are developed. These motions take the form of powerful surges and currents capable of performing great erosive and transportational work.

Most shorelines have a rather smooth, sloping bottom extending out beneath the water level. As waves approach this shallow zone, wave velocity becomes less and wave crests become more closely spaced (Figure 24.1). Wave height and steepness increase rapidly until the crest rolls forward to make a *breaker* (Figure 24.2). After the breaker has collapsed, a foamy, turbulent sheet of water rides up the beach slope. This *swash*, or *uprush*, is a powerful surge causing a landward movement of sand and gravel on the beach. When the force of the swash has been spent against the slope of the beach, the return flow, or *backwash*, pours down the beach, but much disappears by infiltration into the permeable beach sand. Sands and gravels are swept seaward by the backwash.

Wave erosion

In times of relative calm or moderate winds, waves do little erosional work but tend instead to build beaches and sand bars out of sand and gravel. In times of storm, when enormous waves break

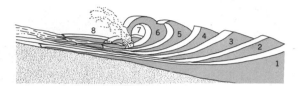

Figure 24.2 A breaking wave.

and throw tons of water against the shoreline, erosion is rapid (Figure 24.3). The violent uprush hurls cobbles and boulders against exposed bedrock along the shore, causing fragments to be broken free. A continual crushing and grinding action goes on as the stones are jostled together. The products of this breaking up are sorted according to size. Fine particles are carried out to sea, where they eventually come to rest in deep, quiet water to form silt and clay layers. Sands and gravels remain close to shore, forming beaches and bars.

Should the shoreline consist of hard rock, erosion will be slow and the storms of many years' time will make little visible change. Where soft materials, such as glacial moraines or outwash plains, form the shoreline, erosion is very rapid. The force of the water alone is sufficient to erode such deposits, and the sea cliff may recede many feet in a single storm (Figure 24.4). Some of the particular forms produced by wave erosion are described and explained in a later paragraph.

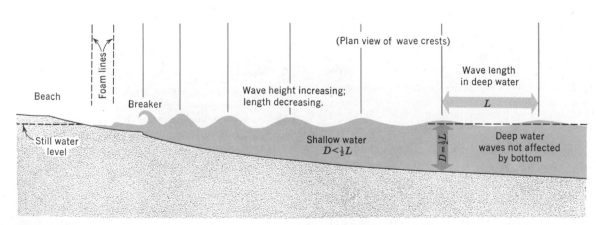

Figure 24.1 As waves enter shallow water the form changes until breaking occurs.

Figure 24.3 Tremendous forward thrust is evident in these storm waves breaking against a sea wall at Hastings, England. (Photographer not known.)

Shore and beach drifting

The unceasing alternate shifting of materials with swash and backwash of breaking waves results not only in movements of rock fragments in seaward and landward directions on the beach, but also in a sidewise movement known as *beach drifting* (Figure 24.5). Wave fronts usually approach the shore with a slight obliquity rather than directly head on. The swash is therefore directed obliquely up the beach, and the sand, gravel, and cobbles are consequently moved obliquely up the slope. After the water has spent its energy, the backwash flows down the slope of the beach, being controlled by the pull of gravity which urges it in the most direct downhill direc-

Figure 24.4 Storm waves breaking against a coast underlain by weak sand quickly undermined this shore home at Seabright, New Jersey. The barrier of wooden pilings (right) proved ineffective in preventing cutting back of the cliff. (Photograph by Douglas Johnson.)

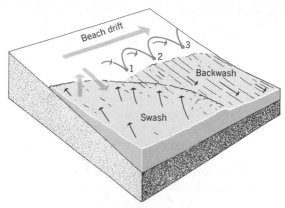

Figure 24.5 Swash and backwash.

tion. The particles are therefore dragged directly seaward and come to rest at a position to one side of the starting place. Because, on a particular day, wave fronts approach consistently from the same direction this movement is repeated many times. Individual rock particles thus travel a considerable distance along the shore. Multiplied many thousands of times to include the numberless particles of the beach, this form of mass transport is a major process in shoreline development.

Although the word *beach* is in common language use, it would be well at this point to define a beach as a deposit of sand, gravel, or cobbles formed inshore from the zone of breaking waves by the action of swash and backwash. If the sand is arriving at a particular section of the beach more rapidly than it can be carried away, the beach is widened and built shoreward, a change called *progradation*. If sand is leaving a section of beach more rapidly than it is being brought in, the beach is narrowed and the shoreline moves landward, a change called *retrogradation*.

A process related to beach drifting is *longshore drifting*. When waves approach a shoreline under the influence of strong winds, the water level is slightly raised near shore by a slow shoreward drift of water. There is thus an excess of water pushed

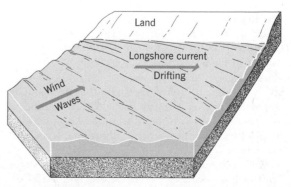

Figure 24.6 Longshore current drifting.

shoreward, which must escape. A *longshore current* is set up parallel to shore in a direction away from the wind (Figure 24.6). When wave and wind conditions are favorable, this current is capable of moving sand along the sea bottom in a direction parallel to the shore.

Both beach drifting and longshore drifting move particles in the same direction for a given set of onshore winds and oblique wave fronts and therefore supplement each other's influence in sediment transportation.

Tidal currents are still another type of water movement that causes the transportation of rock particles. The nature of these currents is discussed in a later paragraph. They are particularly effective in narrow bays and estuaries, but of little importance where a shoreline is fairly straight.

Wave refraction

The phenomenon of change in direction, or bending, of wave fronts as they approach the shore is known as *wave refraction*. Figure 24.7 shows a shoreline with bays and promontories. Successive positions of a wave are indicated by the lines numbered 1, 2, 3, etc. In deep water the wave fronts are parallel. As the shore is neared, the retarding influence of shallow water is felt first in the areas in front of the promontories. Shallowing of water reduces speed of wave travel at those places, but in the deeper water in front of the bays the retarding action has not yet occurred. Consequently, the wave front is bent, or *refracted*, in rough conformity with the shoreline. If the shoreline pattern consists of broad, open curves, the waves may break everywhere along the shore at the same time, but this is unusual. The wave ordinarily will break first upon the promontory and on the bay head last, as indicated in Figure 24.7.

Particularly important in understanding the development of embayed shorelines is the distribution of wave energy along the shore. On Figure 24.7, dashed lines (lettered *a, b, c, d*, etc.) divide the wave at position 1 into equal parts, which may be taken to include equal amounts of energy. Along the headlands the energy is concentrated into a short piece of shoreline; along the bays it is spread over a much greater length of shoreline. Consequently, the breaking waves act as powerful erosional agents on the promontories, but are relatively weak and ineffective at the bay heads. The important principle thus revealed is that headlands and promontories are rapidly eroded back, tending to produce a simple, straight shoreline as an ultimate form.

Wave refraction also occurs where oblique waves approach a perfectly straight shore (Figure 24.8). They are turned so as to break almost parallel with

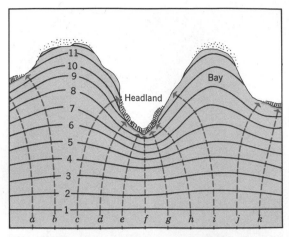

Figure 24.7 Wave refraction along an embayed coast.

the beach. Wave-refraction patterns can be studied from air photographs and may provide valuable information regarding the form of the bottom in the vicinity of the shoreline.

Development of sea cliffs

Where a steeply sloping land surface descends below the water the development of steep cliffs bordering the shoreline is especially favored. Such a condition may come about by a sinking of the land or a rise of ocean level, bringing the water line against what were formerly the steep slopes of mountains or hills.

Figure 24.9 illustrates the development of sea cliffs. Block *A* shows an early stage, termed the *nip stage*, in which wave attack has carved out a small cliff in the hard bedrock. At the base of the cliff is a small rock platform sloping seaward and lying just below water level. Disintegrated rock fragments are swept seaward because wave energy is excessive and will not permit sand and gravel to remain as a beach.

Block *B* shows a cliff developed to considerable height, because, as the cliff is cut back, the rise of land slope causes its height to increase. The

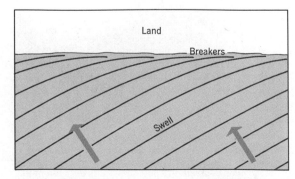

Figure 24.8 Wave refraction on a straight shoreline.

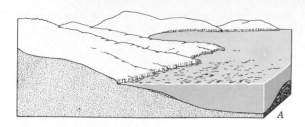

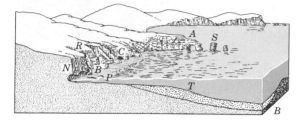

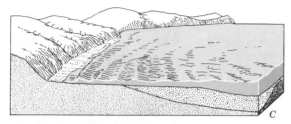

Figure 24.9 Development of sea cliffs. (After E. Raisz.) *A*. Breaking waves first cut out a small cliff, or nip. *B*. As the cliff is cut back it develops undercut notches, arches, stacks, and crevices. At the same time, an abrasion platform is cut and a terrace built. *A* = arch: *S* = stack: *C* = cave: *N* = notch: *P* = abrasion platform: *T* = shore face terrace: *B* = beach: and *R* = crevice. *C*. When equilibrium has been reached a broad platform is present and the cliff is no longer actively undermined.

waves have sought out points of comparative weakness in the rock, penetrating to form crevices and *sea caves*. Some more solid portions of rock project seaward. Where attached to the mainland they may form *sea arches;* where detached, they rise as *stacks* (Figure 24.10). At this stage, the cliff line has reached its greatest degree of irregularity. It is still being vigorously undercut, as evidenced by a wave-cut *notch* at the cliff base. The sloping rock-floored platform at the cliff base, known as the *abrasion platform* is now relatively wide. The inner edge is covered by water only at times of high tide or storm. A beach of sand or cobblestones may be present, but is transitory and may disappear during a single storm, to be built back very slowly. At the outer edge of the rock-abrasion platform, the slope is extended in a deposit of sand and larger rock particles known as the *shore-face terrace*, derived from ground-up rock that the waves have eroded from the cliff.

Streams that formerly flowed to the sea in valleys whose lower ends were at sea level may now be shortened and left as *hanging valleys*

(Figure 24.11), having been unable to deepen their valleys with sufficient rapidity to keep pace with the retreating sea cliff. If the rock material is of an unstable variety, large masses may slump or slide from the cliff (Figure 24.12).

A late stage in the development of sea cliffs, shown in Block *C* of Figure 24.9, may be considered the *mature stage*. The abrasion platform has been so greatly widened that all, or almost all, wave energy is expended in friction as the waves travel over the shallow platform, and in shifting sand across the beach. Consequently, wave attack upon the cliff base is greatly reduced. Weathering and rainbeat acting upon the cliff face wear it down to a lower slope. Irregularities such as sea caves and crevices disappear. The beach may now be broad and deep with little bedrock appearing at the surface.

The geographical influence of marine cliffs is felt in several ways. The coast may be inaccessible because of the high cliffs. In military operations this is especially significant, because a landing is hazardous and a beachhead difficult to expand where a sheer cliff parallels the beach. If the stream valleys are hanging, there are few points where access may be had to the coastal region. Along such coasts, which tend to be fairly straight, the only natural harbors are at the mouths of large streams that have been able to cut down to an accordant junction with the sea (Figures 24.10 and 24.11). This condition prevails along parts of the French channel coast.

A shore bordered by a marine cliff may be of limited value for summer beach use because of the inaccessibility of the shore from the land back of the cliff and because the beach may be thin or rocky even if present. At high tide, the abrasion platform is sometimes inundated to the cliff base, making the shore a dangerous place where bathers can be trapped by rising tide.

Sand bars and sand spits

Where the continued abrasion of a marine cliff has produced an ample supply of sand, or where the delta of a stream furnishes a sand supply, the sand is moved by shore drifting away from the source toward regions of less intense wave action in more sheltered locations. In the case of a straight shoreline indented by a deep bay, as shown in Figure 24.13, shore drifting extends the beach in a more or less direct line into the open water of the bay to produce a *sand spit*. Spits are usually curved toward the land at their extremities, and are described as *recurved spits*. Ultimately, a sand spit will join with another spit developed from the opposite side of the bay and will form a continuous sand bar, or *baymouth bar*, separating the bay from the open water body (Figure 24.14).

Figure 24.10 The chalk cliffs of Normandy, along the French channel coast, show stacks, arches, and sea caves. (Photographer not known.)

Figure 24.11 When a marine cliff is cut back rapidly, hanging valleys appear as notches in the cliff. The large stream at the right has been able to maintain an accordant junction with the sea level and provides a small harbor. (After W. M. Davis.)

Shore drift from an island commonly forms a *tombolo*, which is a sand bar connecting the island with the mainland (Figure 24.15). Drift of sand along a shore from opposing directions results in the building of a pointed sand-bar deposit known as a *cuspate bar* (Figure 24.16). Long-continued deposition, adding one beach upon another in a ribbed arrangement, produces a prominent *cuspate foreland* (Figure 24.17).

Tidal currents

Most marine coasts are to some degree influenced by the *tide*, or rhythmic rise and fall of sea level under the influence of changing attractive forces of moon and sun upon the rotating earth. Where tides are great, the effects of changing water level and the currents thus set in motion are of

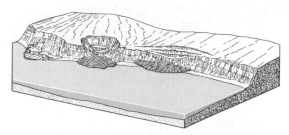

Figure 24.12 Coastal landslides in weak sedimentary strata are the result of oversteepening of a marine cliff by wave attack. (After W. M. Davis.)

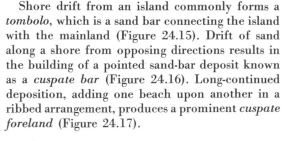

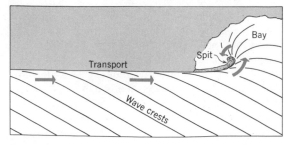

Figure 24.13 Shore drifting producing a sand spit in a bay. (From A. N. Strahler. *The Earth Sciences*, Harper and Row, New York.)

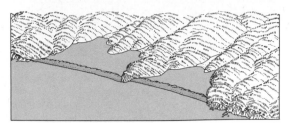

Figure 24.14 These baymouth bars have sealed off two bays. (After W. M. Davis.)

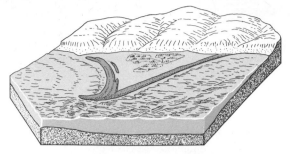

Figure 24.16 This cuspate bar, which has enclosed a triangular lagoon, receives drifted beach materials from both sides. (After E. Raisz.)

Figure 24.15 Two tombolos have connected this island with the mainland. (After W. M. Davis.)

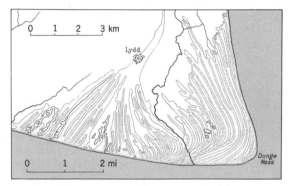

Figure 24.17 Dungeness Foreland, on the Dover Straits of southeastern England, is a large cuspate foreland with curving beach ridges.

major importance in shaping the coastal landforms.

Without going into the causes of ocean tides and their many variations, some of the important principles of tidal action can be understood by considering the common type of *tide curve* in which the cycle of rise and fall is *semidaily* (*semidiurnal*), taking approximately $12\frac{1}{2}$ hours to complete. If we make half-hourly observations of the position of water level against a measuring stick or tide staff attached to a pier or sea wall, we can plot the changes of water level and thus draw the tide curve. Figure 24.18 is a tide curve for Boston Harbor covering a day's time. The water reached its maximum height, or *high water*, at the 12-ft (3.7-m) mark on the tide staff, then fell to its minimum height, or *low water*, occurring about $6\frac{1}{4}$ hours later. A second high water occurred about $12\frac{1}{2}$ hours after the previous high water, completing a single semidaily tide cycle.

The *range of tide*, or difference between heights of successive high and low waters, is about 9 ft (2.7 m) for the example shown in Figure 24.18. The tide range and the form of the tide curve vary throughout the lunar month of about $29\frac{1}{2}$ days. Twice during the month the range is greater than average, comprising *spring tides*; one week later the range is less than average, comprising *neap tides*.

The importance of tides along a coast lies in the currents set up by the changing water level and by the fact that at high water the swash of breaking waves reaches high up on the beach, whereas at low water the waves act only upon the foot of the beach.

The rising tide sets in motion in bays and estuaries a landward current of water known as the *flood current*. When the tide begins to fall this flood current slackens. Flow ceases about the time when the water is midway between high and low water (*midtide*), a condition known as *slack water*. As the tide continues to fall, a seaward current, the *ebb current*, begins to flow and gains in strength to attain a maximum speed about at low tide. The relationships between tidal-current speed and the tide curve are shown in Figure 24.19, which is based on observations at the mouth of New York Harbor. Note that the ebb current is stronger than the flood current, a condition explained by the fact that the Hudson River contributes a considerable discharge of fresh water from the land which must escape to the sea. This stream discharge augments the ebb current but opposes the flood current.

Tidal current deposits

Ebb and flood currents generated by tides perform several important functions along a shoreline. First, the currents that flow into and out of bays through narrow inlets are very swift and can scour the inlet strongly to keep it open despite the tendency of shore drifting processes to close the inlet with sand. Second, the tidal currents carry much fine silt and clay in suspension, derived from streams which enter the bays or from bottom muds agitated by storm wave action. This fine sediment tends to become clotted into small aggregates (process of *flocculation*) where fresh water mixes with salt water. The sediment then settles to the floors of the bays and estuaries where it accumulates in layers and gradually fills the bays. Much organic matter is normally present in such sediment.

In time, tidal sediments fill the bays and produce *mud flats*, which are barren expanses of silt and clay exposed at low tide but covered at high tide (Figure 24.20). Next, there takes hold upon the mud flat a growth of salt-tolerant plants (such as the genus *Spartina*). The plant stems entrap more sediment and the flat is built up to approximately the level of high tide, becoming a *salt marsh* (Figure 24.21). Tidal currents maintain their flow through the salt marsh by means of a highly complex network of sinuous *tidal streams* in which the water alternately flows seaward and landward (Figure 24.22).

Salt marsh is of interest to geographers because it is land that can be drained and made agriculturally productive. The salt marsh is first cut off from the sea by construction of an embankment of earth, a *dike*, in which gates are installed to allow the fresh water drainage of the land to exit

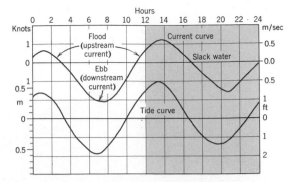

Figure 24.19   Ebb and flood currents at the entrance to New York Harbor. (After H. A. Marmer.)

in ebb flow but to prevent the ingress of salt water during flood flow. Gradually the salt water is excluded and soil water of the diked land becomes fresh. Such diked lands are intensively developed in Holland (*polders*) and southeast England (*fenlands*). Over many decades the surface of reclaimed salt marsh subsides because of compaction of the underlying peat layers and may come to lie well below mean sea level. The threat of flooding by salt water, when storm waves breach the dikes, hangs constantly over the inhabitants of such low areas. The reclamation of salt marsh by dike construction and drainage was practiced by New World settlers in New England and Nova Scotia.

Shoreline classification

Because shorelines display a wide variety of forms, it is desirable to have a classification to group them according to their origin and development. Five major classes of shorelines have been found to include most known types (Figure 24.23).

1. *Shorelines of submergence.* Wherever a sinking down of the earth's crust occurs near the border of a land area, or there is a rise of sea level, the new shoreline takes a position along what was approximately a former contour of the land surface. Below this level all the surface formerly exposed to the air is now submerged beneath the sea. In this way, the term *shoreline of submergence* is applicable.

It is possible to subdivide shorelines of submergence into subgroups, depending on the type of topography that existed before the submergence. Wherever a region was dissected by streams into a system of valleys and divides, submergence produces highly irregular, embayed shoreline termed a *ria shoreline.* Former valleys become deep embayments; former hilltops produce islands; former divides between valleys produce promontories or peninsulas. Variations in ria shoreline form depend on the stage of dissection and relief

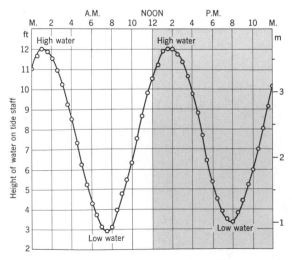

Figure 24.18   Height of water at Boston Harbor measured every half hour. (After H. A. Marmer.)

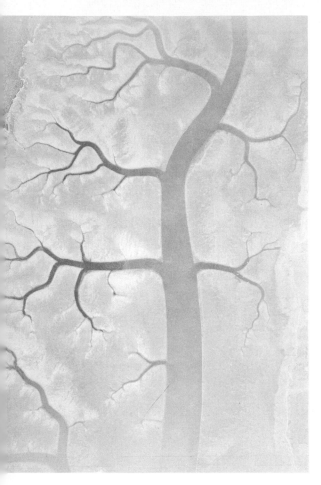

Figure 24.20 Viewed from the air at low tide these tidal mud flats near Yarmouth, Nova Scotia, show a well-adjusted branching system of tidal streams. The area shown is 1.5 mi (2.4 km) wide. (Royal Canadian Air Force Photograph.)

Figure 24.21 Salt marsh at low tide, Green Harbor, Boston Bay region, Massachusetts. Distant houses are on a higher sand ridge separating the salt marsh from the sea. (Photograph by Douglas Johnson.)

Figure 24.22 This broad tidal marsh along the east coast of Florida is laced with serpentine tidal channels. (Aerial photograph by Laurence Lowry.)

of the land immediately before submergence. Two possibilities, one a mountainous region (*1A*), the other a coastal plain of very low relief (*1B*), are illustrated in Figure 24.23.

Some coastal regions have been heavily eroded by valley glaciers, whose troughs were excavated below sea level. After the glaciers have disappeared, a *fiord shoreline* results (Figure 24.23, *1C*). (See also Chapter 23.) Such shorelines are distinctive because of the steepness of the fiord walls, the great depth of water, and the great inland extent of the fiords.

Other subtypes of shorelines of submergence result from the submergence of landscapes formed by continental glaciation (Figure 24.23, *1D*).

2. *Shorelines of emergence.* Wherever a rising of the earth's crust has occurred near the border of a continent, or the sea level has fallen, a *shoreline of emergence* is created. The water line takes a position against what was formerly a slope of the sea floor. Above the new shoreline lies a new coastal land belt which has emerged from the sea. Withdrawal of water to form extensive continental

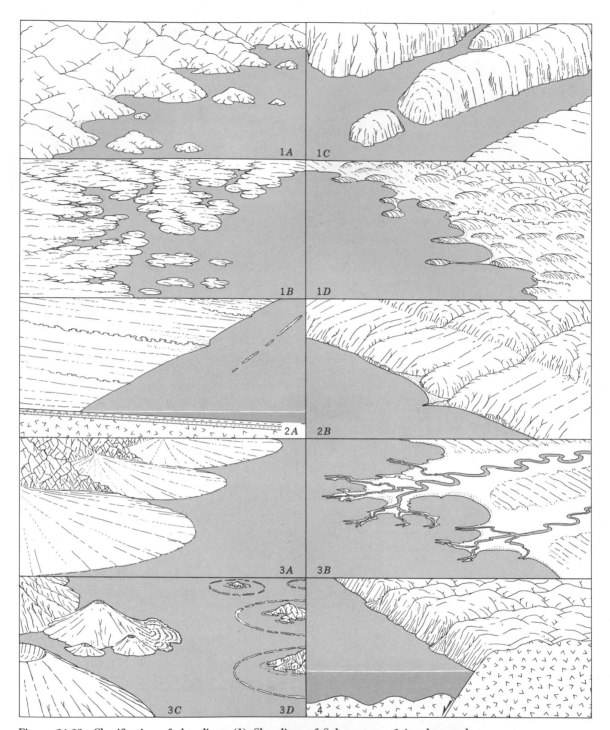

Figure 24.23 Classification of shorelines. (1) *Shorelines of Submergence:* 1A, submerged mountainous coast; 1B, submerged coastal plain, low relief; 1C, fiord coast; 1D, submerged glacial deposits (drumlins). (2) *Shorelines of Emergence:* 2A, coastal plain type, low relief; 2B, steeply sloping type, strong relief. (3) *Neutral Shorelines:* 3A, alluvial fan shoreline; 3B, delta shoreline; 3C, volcano shoreline; 3D, coral reef shoreline. (4) *Fault Shorelines.*

ice sheets is an effective cause of sea-level lowering. It is likely that shorelines of emergence were widely distributed during the time of maximum glacier advance in the Pleistocene epoch.

Most sea floors that have been submerged for a long period of geologic time near the margins of continents have been receiving layered deposits of muds, sands, and gravels derived from erosion of the lands and distributed by ocean currents. These continental shelves have a relatively smooth surface and gentle slope away from the continents. When a continental shelf is exposed by emergence it produces a low, smooth, gently sloping coastal plain, bounded by a simple, even shoreline, which may be termed a *coastal-plain shoreline* (Figure 24.23, *2A*).

Along some coastal belts, the submarine topography contains steep slopes. The shoreline of emergence here differs from that of the coastal-plain shoreline in that deep water lies close off-shore and the coastal belt may be relatively mountainous to within a short distance of the shore (Figure 24.23, *2B*). No simple name has been given to this subtype, but it may be designated a *steeply sloping shoreline of emergence*. Old wave-cut cliffs and benches at various levels above the sea indicate that emergence was spasmodic.

3. *Neutral shorelines.* Wherever a shoreline has been formed as a result of new materials being built out into the water, it is classified as a *neutral shoreline*. The word "neutral" implies that there need be no relative change between the level of the sea and the coastal region of the continent.

Several types of neutral shorelines are known. Each type is easily understood when referred to the agent responsible for building out the material into the water (Figure 24.23, *3A, 3B, 3C, 3D*). An *alluvial fan shoreline*, curved in outline, is built by braided streams building a fan in the manner already explained in Chapter 21. Very similar are shorelines formed by building of a glacial outwash plain into the sea where the ice front stood near the shore. A *delta shoreline* is built of material brought out by a stream system (Chapter 21). Where a volcanic eruption has occurred, the slope of a volcano or the edge of a lava flow may compose the shore, which is then classified as a *volcano shoreline*. A *coral-reef shoreline* is built by marine organisms in the shallow-water zones of tropical seas.

4. *Fault shorelines.* An unusual type of shoreline is produced by faulting of the earth's crust in such a way that there is a dropping down of a segment of the crust on the seaward side of the break and corresponding rising up of the landward side. Should the downdropped block subside below sea level, the shoreline will come to rest against the steeply sloping land surface which is the sur-

face of the plane of faulting, and can be termed a *fault shoreline* (Figure 24.23, *4*). A similar result would be obtained where a crustal mass bounded by fault planes arises from beneath the sea.

5. *Compound shorelines* are those that show the forms of two of the previous classes combined, for example, submergence followed by emergence, or vice versa.

Life history of a shoreline of submergence

A shoreline of submergence of the ria type passes through a series of developmental stages, as illustrated in Figure 24.24. In the initial stage, submergence has just occurred to modify the shoreline. The coast is deeply bayed, with long peninsulas or promontories. Numerous islands lie offshore.

In the stage of early youth, wave attack is vigorous upon the headlands and upon the seaward sides of islands. Wave refraction brings maximum attack to these points. Wave-cut cliffs form such minor features as sea caves, stacks, and arches. The term *cliffed headland* is applied to the beveled peninsulas. Rock abrasion platforms develop at the cliff base, but beaches are thin or absent.

In the stage of late youth, the cliffs have increased in height and have been cut back considerably. Some of the smaller islands have been entirely planed off by wave abrasion and the larger ones have lost considerable area. In this stage, large numbers of depositional forms begin to appear. They are built of sand and gravel by the processes of beach and longshore drifting, described earlier, and represent a series of sequential landforms built out of the waste products of wave abrasion against the cliffs.

Fronting the cliffed headlands are *headland beaches*. From these sources sands have drifted out across the bay mouths to produce spits extending into open water. These spits will normally be recurved, or bent toward the land. Continued growth may add new curved portions to the end of a recurved spit, forming a *compound recurved spit*. Should waves generated within the waters of the bay break upon the spit on its landward side, a secondary spit may be built by shore drifting, forming a complex spit.

As shoreline development continues the spits join to produce baymouth bars, which cut off the bays from the open ocean. A variety of other types of bars also forms along the coast. Tombolos connect islands with the mainland. Between an island and mainland may lie a harbor suitable for small craft. Sometimes a double tombolo is formed, enclosing a lagoon of quiet water between island and mainland (Figure 24.15). A *looped bar* may grow along the landward side of an island as a result of the drift of materials around the lee side

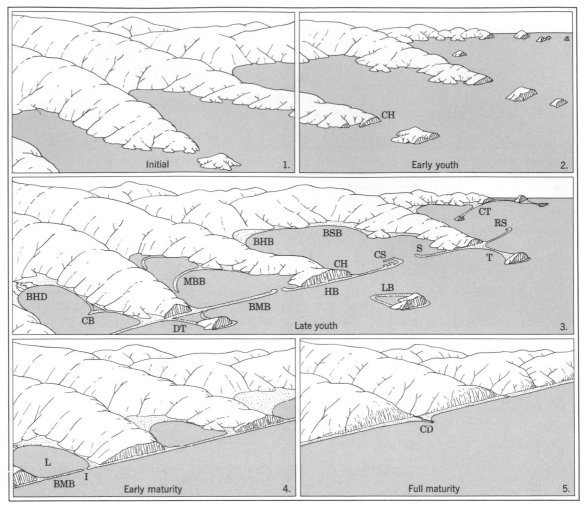

Figure 24.24   Development of the shoreline of submergence. T = tombolo; S = spit; RS = recurved spit: CS = complex spit; CT = complex tombolo; LB = looped bar; CH = cliffed headland; DT = double tombolo; HB = headland beach; BMB = baymouth bar; CB = cuspate bar; BHB = bayhead beach; BSB = bayside beach; BHD = bayhead delta; L = lagoon; I = inlet; CD = cuspate delta.

of the island from the cliffed portion on the seaward side.

Along the sides and ends of the bays are formed *bayside beaches* and *bayhead beaches*. These grow as a result of the drift of sand along the shore from the headlands. Because the bay heads are places of minimum wave attack, the sand tends to accumulate there. If the bays are long and narrow, curving bars will be built across the bay. If located in the middle portion of the bay, they are called *midbay bars;* if near the bay heads, *bayhead bars.* These bars are always smoothly curved, the concave part of the curve facing seaward and merging smoothly with the line of the bayside beaches. A cuspate bar, described in an earlier paragraph, sometimes forms along the side of a bay or on the outer shoreline (Figure 24.16).

At bay heads, deltas may be built into water, thus aiding in the process of filling the bays.

As the stage of youth draws to a close, the outlying islands are completely consumed and the cliffed headlands begin to form a fairly straight line. Baymouth bars carry the smooth line of the shore from headland to headland.

With the attainment of a simple, smooth shoreline the stage of early maturity is reached. This coast consists alternately of cliffed headlands and baymouth bars. The point of fundamental importance is that throughout the life cycle thus far described, a highly irregular shoreline has been replaced by a nearly straight shoreline. The bays now become filled in by delta materials supplied by streams and by deposition of silts as the tide rises and falls, and the tide-induced currents flow in and out through narrow passes in the baymouth bars. Some small bays lack any passes in the baymouth bars, but the permeability of the sand is sufficient to permit runoff to escape into the sea. Ultimately the bays are occupied by tide flats and salt marsh.

Throughout early maturity the shoreline continues to retreat landward. The sea cliffs are progressively cut back while the baymouth bars are pushed back to keep in a straight line. Eventually a position is reached where the shoreline coincides with the original line of the bay heads. The baymouth bars and all other depositional features except the outer beach disappear, and the cliff of bedrock extends along the entire shoreline, which is then said to have attained full maturity (Figure 24.24). No further major development occurs, aside from the continued landward retreat of the shoreline. This retrogression will, however, become a very slow process, as the increased relief of the land causes the cliffs to be heightened and to supply more detritus to the shore.

Geographical aspects of shorelines of submergence

The influence of shorelines on human activity is strong, especially that of shorelines of submergence. The deep embayments of the youthful shoreline make splendid natural harbors. Much of the shoreline of Scandinavia, France, and the British Isles is thus provided with harbor facilities. Consequently, these peoples have a strong tradition of fishing, ship-building, ocean commerce, and marine activity generally. Mountainous relief of ria and fiord coasts makes agriculture difficult or impossible, forcing the people to turn to the sea for a livelihood. Rich forests and cheap hydroelectric power have, however, stimulated lumbering and manufacturing. New England and the Maritime Provinces of Canada have a youthful shoreline of submergence with abundant good harbors. The influence of this environment has been to foster the same development of fishing, whaling, ocean commerce, shipbuilding, and manufacturing seen in the British Isles and Scandinavian countries.

A mature shoreline of submergence has few natural harbors but extensive sheer marine cliffs. Portions of the channel coast of France illustrate this development (Figure 24.10).

Development of barrier-island coasts

In contrast to ria and fiord coasts, with their bold relief and deeply embayed outlines, are coasts of low relief from which the land slopes gently beneath the sea. The coastal plain of the Atlantic and Gulf coasts of the United States presents a particularly fine example of such a gently-sloping surface. As explained in Chapter 26, this coastal plain is a belt of relatively young sedimentary strata, formerly accumulated beneath the sea as deposits on the continental shelf. Emergence as a result of repeated crustal uplifts of epeirogenic nature has characterized this coastal plain during the latter part of the Cenozoic era and into recent time. There exist various elevated marine features, such as wave-cut scarps and platforms, extensive sand beaches, and lagoonal deposits of tide-water origin, lying many miles inland and at elevations of many tens of feet.

Such evidence points toward the conclusion that the shoreline of the Atlantic and gulf coastal plain is one of emergence, but on closer examination we find that the lower portions of stream valleys along the entire coast are drowned by a rise of sea level and that tidal channels and tidal marshes extend in many places for miles inland in the valley bottoms. (See Figure 26.6.) Clearly the last event has been one of a rise of sea level, occasioned by the melting of the Pleistocene ice sheets, and has produced many characteristics of a shoreline of submergence.

Along with the rise of sea level there has developed along much of the Atlantic and Gulf coast shoreline a *barrier island*, which is a low ridge of sand built by waves and further increased in height by the growth of dunes shaped from beach sands by wind action. Behind the barrier island lies a *lagoon*, which is a broad expanse of shallow water, often several miles wide, and in places largely filled with tidal deposits. (See Figure 26.10.)

The manner in which barrier islands may have come into existence along gently sloping coasts of the world since the end of the last glacial stage is illustrated in Figure 24.25 which represents the Gulf coast of Texas. At the time when ice sheets were at their maximum extent over the continents, sea level was drawn down to perhaps as much as 330 ft (100 m) below present sea level. The shoreline then was many miles farther seaward than today and a broad sloping plain lay exposed. Streams draining the land were extended across this plain and carved deep trenches into it. Then, as the ice began to melt rapidly some 10,000 to 12,000 years ago, the sea level began its rise and the shoreline rapidly shifted landward. The forward part of the block diagram in Figure 24.25 shows conditions about 5000 years ago. A low barrier island was formed of beach sands derived from the shallow sea floor and from longshore drift. As the sea level rose, the waves continued to add material to the crest of the barrier, building it up to keep pace with rise of water level. Correspondingly, the lagoon widened and the inner

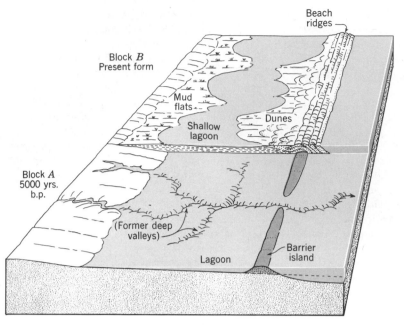

Figure 24.25 Upbuilding of a barrier island during post-glacial rise in sea level is an essential part of the history of the Texas Gulf Coast. (Based on data of H. N. Fisk, From A. N. Strahler, *The Earth Sciences*, Harper and Row, New York.)

shoreline encroached farther upon the gently sloping land surface. Today, as shown in the rear part of the block in Figure 24.25, the barrier island is a complex structure consisting of several wave-built beach ridges and of dunes which have widened the island on the landward side. The lagoon is partly filled by tidal muds.

Other examples of barrier-island coasts are found where glacial outwash plains were built out into the sea to produce a gently sloping plain. Post-glacial rise of sea level has partly submerged these outwash plains while at the same time a barrier beach has formed and produced a lagoon. A particularly striking example is the south shore of Long Island, New York, along which Fire Island separates Great South Bay, a lagoon over 5 mi (8 km) wide, from the Atlantic Ocean.

Tidal inlets and tidal deltas

A characteristic feature of most barrier islands—and of many baymouth bars as well—is the presence of gaps, known as *tidal inlets*, through which strong currents flow alternately seaward and landward as the tide rises and falls, building *tidal deltas*.

The spacing of tidal inlets in a barrier beach depends in part on the average range of tides along the coast, the spacing being closer where the range is greater. In heavy storms, the barrier may be breached by new inlets (Figure 24.26). Tidal currents will subsequently tend to keep a new inlet open, but it may be closed by shore drifting of

sand. Among these opposing activities a sort of balance is maintained, so that neither too many nor too few inlets exist.

Geographical aspects of barrier-island shorelines

Shallow water results in generally poor natural harbors along barrier-island shorelines. The lagoon itself may serve as a harbor if channels and dock areas are dredged to sufficient depths. Ships enter and leave through one of the passes in the barrier island, but artificial sea walls and jetties are required to confine the current and thereby keep sufficient channel depth. Frequently the major port cities are located where a large river empties into the lagoon. Drowning of the lower courses of large rivers provides tidal channels which may be dredged to accommodate large vessels and thus make seaports of cities many miles inland.

One of the finest examples of a barrier island and lagoon is along the Gulf coast of Texas. Here the island is unbroken for as much as 100 mi (160 km) at a stretch and passes are few. The lagoon is 5 to 10 mi (8 to 16 km) wide, indicating that the original slope of the sea bottom was very slight. Galveston is built upon the barrier island adjacent to an inlet connecting Galveston Bay with the sea. Most other Texas ports, however, are located on the mainland shore. Submergence has caused an embayed inner shoreline with extensive estuaries marking the mouths of the larger streams. Corpus Christi, Rockport, Texas City, Lavaca, and other

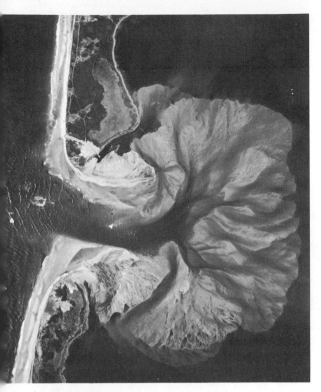

Figure 24.26 East Moriches Inlet was cut through Fire Island, a barrier island off the Long Island shoreline, during a severe storm in March 1931. This aerial photograph, taken a few days after the breach occurred, shows the underwater tidal delta being built out into the lagoon (right) by currents. The entire area shown is about 1 mi (1.6 km) long. North is to the right; the open Atlantic Ocean on the left. (U.S. Army Air Forces Photograph.)

Figure 24.27 The horizontal step seen in the mountain base is a wave-cut bench with associated gravel deposits. It represents a high stand of ancient Lake Bonneville, which occupied the Great Salt Lake basin during the glacial period. (Photograph by Hal Rumel. Utah Travel Council.)

ports are located along the shores of these embayments. (See Figure 26.10.)

Still another good illustration of the geographical aspects of a barrier-island shoreline is seen in the Atlantic coast of New Jersey, Delaware, Maryland, Virginia, and North Carolina. (See Figure 26.5.) Virtually the entire coast from Sandy Hook to Cape Lookout is bordered by a barrier beach. Extensive post-glacial submergence has caused vast embayments such as Chesapeake Bay and Delaware Bay. Great port cities, such as Baltimore, Wilmington, and Philadelphia are situated at the heads of these tidal estuaries. Along the New Jersey coast, however, the old inner shoreline of the lagoon is remarkably straight and shows vestiges of earlier wave cutting. The lagoon is largely filled now and forms a flat plain through which wander sinuous tidal creeks. The barrier beach makes a splendid resort and has been fully utilized with the building of such cities as Atlantic City and Asbury Park.

Elevated shorelines

A shoreline may, at any stage in its development, be raised above water level so as to become a land feature. At the same time, a new shoreline, which is a true shoreline of emergence, is produced at the new position of the water line. The raised shoreline, or *elevated shoreline*, is not a shoreline of emergence; in fact, it is not a true shoreline at all because it is no longer associated with wave and current action. Having once been elevated, it is attacked by weathering, mass wasting, and streams and will eventually be destroyed.

Elevated shorelines result either from crustal uplift along a coastal belt such as that of Alaska or California coast, which are subject to faulting and earthquakes, or from a falling of sea or lake level.

Lake Bonneville, the ancestor of the present Salt Lake in Utah, rose to a maximum level in the Pleistocene epoch when rainfall was greater and evaporation less than now. Excellent illustrations of elevated shorelines are to be found on the lower slopes of the mountain ranges in the Salt Lake region (Figure 24.27). From a study of these ancient wave-cut benches with their associated beaches, spits, and bars, it has been possible to reconstruct the history of the old lake and to make inferences as to climates of the past.

Where lake level falls steadily, or coastal regions rise steadily, the elevated shorelines become *strand lines* resembling natural contours of the land (Figure 24.28). Crustal rise of this continual and rather uniform nature has been occurring in the Baltic Sea area and along the Arctic coast of North America as a result of the recovery of the crust following its depression under the load of Pleistocene ice sheets.

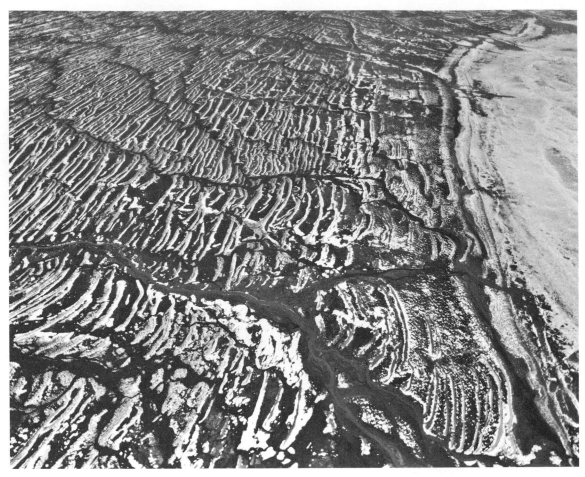

Figure 24.28 This great succession of elevated strand lines bordering the shore of Hudson Bay documents the almost-continuous post-glacial rise of the earth's crust which has followed the removal of ice load. (Photograph by Canadian Government Department of Energy, Mines, and Resources; Air Photo Division.)

Where elevated wave-cut platforms stand but a few feet above mean sea level, as they do on many islands of the Pacific (Figure 24.29), we have difficulty in deciding whether there has been a rise of land level or a sinking of sea level, inasmuch as the result would be the same in either event.

Extreme emergence of a coast may expose a wide belt of former sea floor (Figure 24.30). This new land is a coastal plain (Chapter 26) and is veneered with recently formed sedimentary layers.

Coral-reef shorelines

As a variety of neutral shoreline, coral-reef shorelines are unique in that the addition of new land is made by organisms: *corals*, which secrete lime to form their skeletons, and *algae*, plants that also make limy encrustations. Corals are colonial types of animals, that is, they occur in large colonies of individuals. As coral colonies die, new ones are built upon them, thus developing a coral limestone made up of the strongly cemented limy skeletons. Coral fragments torn free by wave attack and pulverized may be deposited to form beaches, spits, and bars, which later are cemented into a limestone.

Coral-reef shorelines occur in warm, tropical water between the limits 30° N and 25° S lat. Water temperatures above 68°F (20°C) are necessary for dense reef coral growth. Furthermore, reef corals live near the water surface, down to limiting depths of about 200 ft (60 m). Water must be free of suspended sediment and well aerated for vigorous coral growth; hence corals thrive in positions exposed to wave attack from the open sea. Because muddy water prevents coral growth, reefs are missing opposite the mouths of muddy streams. Coral reefs are remarkably flat on top (Figure 24.31) and have a level approximately equal to the upper one-third mark of the range of tide. Thus they are exposed at low tide and covered at high tide.

Three general types of coral reefs may be recog-

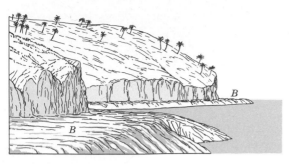

Figure 24.29   This wave-cut rock bench (*B*) on the shore of Tutuila in the Samoan Islands is about 10 ft (3 m) above mean sea level and may indicate a recent drop of sea level or a recent rise of the island. (After W. M. Davis.)

Figure 24.31   A fringing reef on the south coast of Java forms a broad bench between surf zone (left) and a white coral-sand beach. Inland is rainforest. (Photograph by Luchtvaart-Afdeeling, Bandung.)

nized: (1) fringing reefs, (2) barrier reefs, and (3) atolls. *Fringing reefs* are built as platforms attached to shore (Figure 24.32). They are widest in front of headlands where wave attack is strongest, and the corals receive clean water with abundant food supply. Fringing reefs are usually absent near the mouths and deltas of streams, where the water is muddy. This is a fact of great military importance where the problem is to find reef-free places for landing of troops and supplies. Fringing reefs may be from 0.25 to 1.5 mi (0.4 to 2.5 km) wide, depending on the length of time that the reef has been developing.

Barrier reefs lie out from shore and are separated from the mainland by a lagoon which may range from 0.5 to 10 mi (2.5 to 16 km) or more in width (Figure 24.33). The reef itself may be from 20 to 3000 ft (6 to 900 m) wide. The lagoon is shallow and flat-floored, usually 120 to 240 ft (35 to 75 m) deep. There are, however, many towerlike columns of coral in the lagoon. *Passes*, which occur at intervals in barrier reefs, are narrow gaps through which excess water from breaking

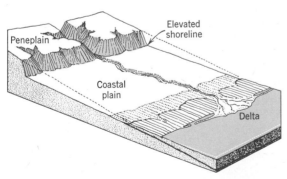

Figure 24.30   A coastal-plain strip 40 mi (65 km) wide and more than 1000 mi (1600 km) long indicates emergence along the east coast of peninsular India. The present shoreline is a young shoreline of emergence; the former shoreline, now of the elevated type, appears as a steep mountain front rising abruptly from a plain. (After W. M. Davis and S. W. Cushing.)

waves is returned from the lagoon to the open sea. They sometimes occur opposite deltas on the mainland shore, because of the inhibiting effect of mud on coral growth. Passes are of geographical and military importance because they provide the only means of entrance by ship into the lagoon.

Atolls are more or less circular coral reefs enclosing a lagoon, but without any land inside (Figure 24.34). In all other respects they are similar to barrier reefs. On large atolls, parts of the reef have been built up by wave action and wind to form low island chains, connected by the reef. A cross section of an atoll shows that the lagoon is flat-floored and shallow, and that the outer slopes are steep, often descending thousands of feet to great ocean depths.

Several plausible theories have been advanced for the origin of atolls and barrier reefs. To explain each one and discuss the advantages and disadvantages of each would take many pages. One interesting theory of origin, which has been popular since it was first outlined by a great scientist, Charles Darwin, in 1842, may be called the *subsidence theory* (Figure 24.35). He supposed that small islands, such as volcanoes, slowly subsided in a general downwarping of the earth's crust over parts of the ocean basin. Coral reefs, which were originally fringing reefs attached to the island shores, continued to build upward as the island subsided. Thus the area of the island shrank and a lagoon formed, creating a barrier reef. Finally the island sank out of sight, but the reef persisted, maintained at sea level by vigorous coral and algal growth.

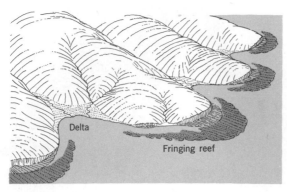

Figure 24.32 Fringing reefs are widest in front of headlands and may be absent near the mouths of streams. (After W. M. Davis.)

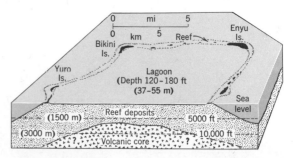

Figure 24.34 Bikini Atoll in the Pacific, scene of early atom bomb tests, is thought to consist of a great thickness of reef deposits resting on a sea mountain of volcanic rock. (After M. Dobrin, et al.)

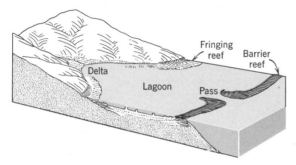

Figure 24.33 A barrier reef is separated from the mainland by a shallow lagoon. (After W. M. Davis.)

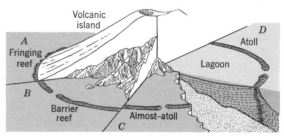

Figure 24.35 The subsidence theory of barrier-reef and atoll development is shown in four stages, beginning with a fringing reef attached to a volcanic island and ending with a circular reef. (After W. M. Davis.)

The geographical aspects of atoll islands are unique in some respects. First, there is no rock other than coral limestone, composed of calcium carbonate. This means that trees requiring other minerals, such as silica, cannot be cultivated without the aid of fertilizers or some outside source of rock from a larger island composed of volcanic or other igneous rock. The palm tree is native to atoll islands because it thrives on brackish water, and the seed, or palm nut, is distributed widely by floating from one island to another. Native inhabitants have cultivated the cocoanut palm to provide food, clothing, fibers, and building materials. Fresh water is scarce on small atoll islands because there is not enough surface area for the collection of rainfall, and the land is so low that a high water table of fresh water is not present to supply springs, streams, and wells. Rainfall must

be caught in open vessels or catchment basins and carefully conserved. Fish and other marine animals are an important part of the human diet on atoll islands. Calm waters of the lagoon make a good place for fishing and for beaching canoes. Coral islands of the western Pacific stand in continual danger of devastation by tropical cyclones (typhoons). Breaking waves wash over the low-lying ground, sweeping away palm trees and houses and drowning the inhabitants. There is no high ground for refuge. In the same way, great seismic sea waves of unpredictable occurrence may inundate atoll islands.

Mangrove coasts are of great extent in tropical and equatorial regions, where an abundance of fine sediment in suspension prevents coral reef growth. Mangrove as a form of natural vegetation is discussed in Chapter 16.

REFERENCES FOR FURTHER STUDY

Johnson, D. W. (1919), *Shore processes and shoreline development*, John Wiley and Sons, New York, 584 pp. Reprinted in 1965, Hafner Publ., New York.

Lobeck, A. K. (1939), *Geomorphology*, McGraw-Hill Book Co., New York, 731 pp. See Chapter X and XII.

Guilcher, A. (1958), *Coastal and submarine morphology*, John Wiley and Sons, New York, 274 pp.

King, C. A. M. (1960), *Beaches and coasts*, St. Martin's Press, New York, 403 pp.

Shephard, F. P. (1963), *Submarine geology*, second edition, Harper and Row, New York, 557 pp. See Chapters 6, 7, and 12.

Bascomb, W. (1964), *Waves and beaches*, Doubleday and Co., New York, 260 pp.

Thornbury, W. D. (1969), *Principles of geomorphology*, second edition, John Wiley and Sons, New York, 594 pp. See Chapter 17.

REVIEW QUESTIONS

1. Why do breakers form in shallow water? What do the terms *swash* and *backwash* mean? How are beach materials moved by these currents?

2. Under what conditions do waves do the greatest amount of shore erosion? When are beaches built?

3. Explain beach drifting and longshore drifting. What forms are built by these processes? What is a longshore current, and how is it caused?

4. Describe the phenomenon of wave refraction. Why does refraction occur? When waves are refracted along an embayed coast where is the energy of wave erosion concentrated? Where is it least? What important result does this have on the form of the shoreline?

5. Describe the development of sea cliffs, beginning with a newly submerged slope and continuing through to a mature sea cliff. Explain the following terms: nip, sea cave, sea arch, stack, wave-cut notch, abrasion platform, beach, shore-face terrace hanging valley.

6. What are sand spits? How and where do they form? What shape characterizes the end of a spit?

7. Explain the formation of baymouth bars, tombolos, and cuspate bars. How does a cuspate foreland differ from a cuspate bar?

8. Describe the rise and fall of water level in a simple semidaily tide cycle. What are spring and neap tides?

9. How are tidal currents produced? Define flood current, slack water, and ebb current.

10. What deposits of sediment are associated with tidal currents in bays and estuaries? Describe these deposits as to form and composition. How are salt marshes utilized by man?

11. How can shorelines be classified? Name the five principal classes.

12. What kinds of shorelines of submergence are recognized? Name and explain the subdivisions.

13. Describe the changing coastal landforms of a shoreline of submergence as it passes through its development cycle.

14. What is a barrier-island coast? How are barrier islands formed? How are tidal inlets related to barrier islands?

15. How have shorelines of submergence exerted an influence on human activities?

16. Discuss the geographical aspects of barrier-island shorelines. Compare harbor facilities of a barrier-island shoreline with those of a young shoreline of submergence.

17. Name several kinds of neutral shorelines.

18. What is meant by an elevated shoreline? How does it differ from a shoreline of emergence? Where may some excellent elevated shorelines be seen today?

19. How are coral reefs formed? Under what conditions do reef-building corals flourish?

20. What three general types of reefs are formed? Describe each type. Explain Darwin's subsidence theory of atolls.

CHAPTER 25
Landforms Made by Wind

WIND, the fourth of the agents of erosion thus far discussed, produces a variety of interesting sequential landforms, both erosional and depositional. In terms of total mass of material thereby removed or deposited, however, it ranks below mass wasting, running water, waves, and ice, except in certain desert regions especially favorable to its action. In humid regions, with ample soil moisture and dense vegetative cover, there are few evidences of the work of wind. These usually are coastal sand dunes. Elsewhere, vegetation holds the ground in place unless man has laid it bare.

Erosion by wind

Wind performs two kinds of erosional work. Loose particles lying upon the ground surface may be lifted into the air or rolled along the ground. This process is *deflation*. Where the wind drives sand and dust particles against an exposed rock or soil surface, causing it to be worn away by the impact of the particles, the process is *abrasion*. Abrasion requires cutting tools carried by the

wind; deflation is accomplished by air currents alone.

Deflation acts wherever the ground surface is thoroughly dried out and is littered with small, loose particles derived by rock weathering or previously deposited by running water, ice, or waves.

Thus, dry river courses, beaches, and areas of recently formed glacial deposits are highly susceptible to deflation. In dry climates, virtually the entire ground surface is subject to deflation because the soil or rock is everywhere bare. Wind is selective in its deflational action. The finest particles, those which constitute clay and silt, are lifted most easily and raised high into the air. Sand grains are moved only by moderately strong winds and tend to travel close to the ground. Gravel fragments and rounded pebbles up to 2 or 3 in (5 to 8 cm) in diameter may be rolled over flat ground by strong winds but do not travel far. They become easily lodged in hollows or between their fellows. Consequently, where a mixture of sizes of

Figure 25.1   This blowout hollow on the plains of Nebraska contains a remnant column of the original material, thus providing a natural yardstick for the depth of material removed by deflation. (Photograph by N. H. Darton, U.S. Geological Survey.)

particles is present on the ground, the finer sizes are removed; the coarser particles remain behind.

The principal landform produced by deflation is a shallow depression termed a *blowout,* or *deflation hollow.* This depression may be from a few yards to a mile or more in diameter, but is usually only a few feet deep. Blowouts form in plains regions in dry climates. Any small depression in the surface of the plain, particularly where the grass cover is broken through, may develop into a blowout. Rains fill the depression, creating a shallow pond or lake. As the water evaporates the mud bottom dries out and cracks, forming small scales or pellets of dried mud which are lifted out by the wind. In grazing lands, cattle may trample the margins of the depression into a mass of mud, breaking down the protective grass-root structure and facilitating removal when dry. Thus the depression is progressively enlarged (Figure 25.1). Blowouts are also found on rock surfaces where the rock is being disintegrated by weathering.

In the great deserts of southeastern California, Arizona, and New Mexico, the floors of intermontane basins are subject to deflation. The flat floors of the vast, shallow playas have in some places been reduced by deflation as much as several feet over areas of many square miles.

Where deflation has been active on a ground surface littered with loose fragments of a wide range of sizes, the pebbles that remain behind tend to accumulate until they cover the entire surface (Figure 25.2). By rolling or jostling about as the fine particles are blown away, the pebbles may become closely fitted together, forming a *desert pavement.* In North Africa such a pebble-covered surface is called a *reg.* The precipitation of calcium carbonate, gypsum, and other salts near the surface, as ground water is drawn to the surface and evaporated in dry weather, tends to cement the pebbles together, forming a highly effective protection against further deflation.

The sandblast action of wind against exposed rock surfaces is limited to the basal few feet of a cliff, hill, or other rock mass rising above a relatively flat plain, because sand grains do not rise high into the air. Wind abrasion produces pits, grooves, and hollows in the rock. Where a small rock mass projects above the plain it may be cut away at the base to make a *pedestal rock,* delicately balanced upon a thin stem. Most pedestal rocks, or *mushroom rocks,* are, however, produced by weathering processes.

Dust storms and sand storms

In dry seasons over plains regions, strong, turbulent winds lift great quantities of fine dust into the air, forming a dense, high cloud called a *dust storm.* The dust storm is generated where ground

Figure 25.2 This desert pavement of quartzite fragments was formed by action of both wind and water on the surface of an alluvial fan in the desert of southeastern California. Fragments range in size from 1 to 12 in (2.5 to 33 cm). Silt underlies the layer of stones. (Photograph by C. S. Denny, U.S. Geological Survey.)

surfaces have been stripped of protective vegetative cover by cultivation or grazing, or where they naturally carry no vegetation cover because of extreme aridity of the climate. A dust storm approaches as a great dark cloud extending from the ground surface to heights of several thousand feet (Figure 25.3). Within the dust cloud deep gloom or even total darkness prevails. visibility is cut to a few yards, and a fine choking dust penetrates everywhere.

It has been estimated that as much as 4000 tons of dust may be suspended in a cubic mile of air (875 metric tons per cubic kilometer). On this basis, a dust storm 300 mi (500 km) in diameter might be carrying more than 100 million tons (90 million metric tons) of dust—enough to make a

Figure 25.3 Front of an approaching dust storm, Coconino Plateau, Arizona. (Photograph by D. L. Babenroth.)

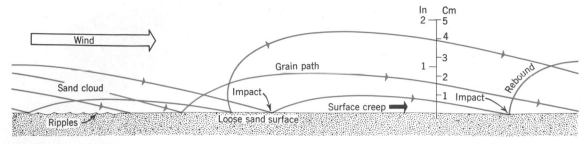

Figure 25.4 Sand particles travel in a series of long jumps. (After R. A. Bagnold.)

hill 100 ft (30 km) high and 2 mi (3 km) across the base.[1] A region that supplied the dust for thousands of such storms would thus lose a considerable mass over a span of thousands of years. Whether during the same period streams would remove more material from the same area is difficult to say, but in all probability they would remove much more.

Dust travels enormous distances in the air. That of individual dust storms is often traceable as far as 2500 mi (4000 km). Volcanoes erupt much extremely fine dust into the air. The renowned eruption of the volcano Krakatoa in Indonesia in the year 1883 cast out an enormous quantity of dust, some of which was caught by atmospheric circulation at high levels and carried around the entire earth. It is said that unusually brilliantly colored sunsets occurred in the British Isles in the years following 1883 as a result of the presence of the Krakatoa dust in the atmosphere. These were referred to as the "Chelsea sunsets," and were a favorite subject for paintings by English artists of the period.

The true desert *sandstorm* is a low cloud of moving sand that rises usually only a few inches

[1]A. K. Lobeck, *Geomorphology*, McGraw-Hill Book Co., New York. 1939, 731 pp., p. 380.

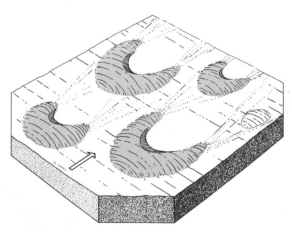

Figure 25.5 Barchans, or crescentic dunes. Arrow indicates wind direction.

and at most 6 ft (2 m) above the ground. It consists of sand particles driven by a strong wind. Those who have experienced sandstorms report that a man standing upright may have his head and shoulders entirely above the limits of the sand cloud. The reason why the sand does not rise higher is that the individual particles are engaged in a leaping motion, termed *saltation* (Figure 25.4). Grains describe a curved path of travel and strike the ground with considerable force but at a low angle. The impact causes the grain to rebound into the air. At the same time, the surface layer of sand grains creeps downwind as the result of the constant impact of the bouncing grains.

The erosional effect of blown sand is thus concentrated on surfaces exposed less than a foot or two (0.3 to 0.6 m) above the flat ground surface. Telephone poles on wind-swept sandy plains are quickly cut through at the base unless a protective metal sheathing or heap of large stones is placed around the base.

Sand dunes

A *dune* is any hill or accumulation of sand shaped by the wind. Dunes may be active, or *live*, when bare of vegetation and constantly changing form under wind currents. They may be inactive, or *fixed*, dunes, covered by vegetation that has taken root and serves to prevent further shifting of the sand.

Several common varieties of dunes are treated here. The *crescentic dune*, or *barchan* (also spelled barcan, barkhan, or barchane), is an isolated dune, which in plan view resembles a blunted crescent (Figure 25.5). The broadly rounded ends of the crescent point downwind and indicate the direction of dune motion and prevailing winds. On the windward side of the crest the sand slope is gentle, being the slope up which the sand grains move. On the lee side of the dune, within the crescent, is a steep curving dune slope, the *slip face*, which maintains an angle of about 35° from the horizontal (Figure 25.6). Sand grains fall or slide down the steep face after being blown free of the crest. When a strong wind is blowing, the flying sand makes a perceptible cloud at the crest. The term

Figure 25.6 Barchan dunes at Biggs, Oregon (Photograph by G. K. Gilbert, U.S. Geological Survey.)

smoking crest has been used for this feature. Crescentic dunes rest upon a flat, pebble-covered ground surface. The sand may originate as a drift in the lee of some obstacle, such as a small hill, rock, or clump of brush. Once a sufficient mass of sand has formed it begins to move downwind, taking the form of a crescent dune. Thus the dunes are commonly arranged in chains extending downwind from the source drifts.

Where sand is so abundant that it completely covers the ground, dunes take the form of wavelike ridges separated by troughlike furrows. The dunes are called *transverse dunes* because their crests trend at right angles to direction of wind (Figure 25.7). The entire area may be called a *sand sea*, for it resembles a storm-tossed sea suddenly frozen to immobility. The term *erg*, referring to any large expanse of dunes in the Sahara Desert, has been

Figure 25.7 This air photograph of a sand-dune field between Yuma, Arizona, and Calexico, California, shows a sand sea of transverse dunes in the background and a field of crescentic barchan dunes in the foreground. (Copyrighted Spence Air Photos.)

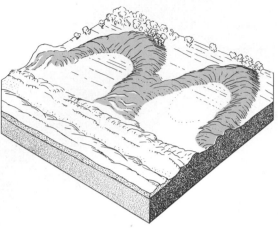

A. Coastal blowout dunes with saucerlike depressions.

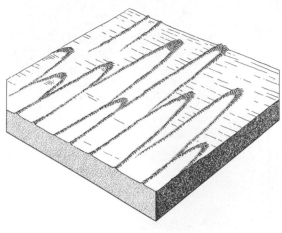

C. Parabolic dunes of hairpin form.

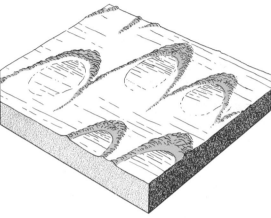

B. Parabolic blowout dunes on an arid plain.

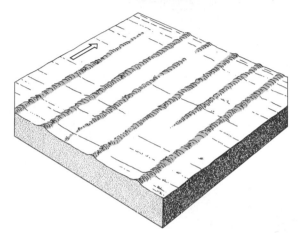

D. Longitudinal dune ridges on a desert plain.

Figure 25.8 Four types of dunes. Prevailing wind, shown by arrow, is the same for all diagrams.

Figure 25.9 At Beaufort Harbor, near Cape Lookout, North Carolina, coastal sand dunes are inundating a forest of live oak. (Photograph by J. A. Holmes, U.S. Geological Survey.)

adopted by geographers for this type of landscape. Individual sand ridges have sharp crests and are asymmetrical, the gentle slope being on the windward, the steep slope on the lee side. Deep depressions lie between the dune ridges. Sand seas require huge quantities of sand, often derived from weathering of a sandstone formation underlying the ground surface, or from adjacent alluvial plains. Still other transverse dune belts form adjacent to beaches which supply abundant sand and have strong onshore winds (Figure 25.10).

Another group of dunes belongs to a family in which the curve of the dune crest is bowed convexly downwind, the opposite of the curvature of crests in the barchan and transverse dunes. These may be described as *parabolic* in form. A common representative of this class, the *coastal blowout dune*, is formed adjacent to beaches, where large supplies of sand are available and are blown landward by prevailing winds (Figure 25.8*A*). A saucer-shaped depression is formed by deflation; the sand is heaped in a great curving ridge resembling a

Figure 25.10 The arrows on this photograph point to elongate blowout dunes of hairpin form, which once advanced from the beach and have since become stabilized by vegetation. Active transverse dunes are overriding the blowout dunes in a fresh wave, San Luis Obispo Bay, California. (Spence Air Photos.)

horseshoe in plan. On the landward side is a steep slip face which advances over the lower ground and buries forests, killing the trees (Figure 25.9). Coastal blowout dunes are well displayed along the southern and eastern shore of Lake Michigan, those of the southern shore being set aside for public use as the Indiana Dunes State Park.

In arid plains and plateaus, where vegetation is sparse and winds strong, groups of *parabolic blow-out dunes* develop to the lee of shallow deflation hollows (Figure 25.8*B*). Sand is caught by low bushes and accumulates in a broad, low ridge. These dunes have no steep slip faces, and may remain relatively immobile. In some cases, however, the dune ridge migrates downwind, drawing the parabola into a long, narrow form with parallel sides

(Figure 25.8*C*). This form resembles a hairpin in plan; hence has been named a *hairpin dune*, although it is a member of the parabolic family. Hairpin dunes stabilized by vegetative growth are seen in Figure 25.10.

Still another class of dunes is described as *longitudinal* because the dune ridges run parallel with the wind direction. On desert plains and plateaus, where sand supply is meager but winds are strong from one direction, *longitudinal dune ridges* are formed (Figure 25.8*D*). These are usually only a few feet high, but may be several miles long. In some areas, at least, the longitudinal dune is produced by extreme development of the hairpin dune, such that the parallel side ridges become the dominant form.

Figure 25.11 This oblique air photograph shows longitudinal sand-dune ridges reaching as far as the eye can see. Simpson Desert, southeast of Alice Springs, Australia. (Photograph by George Silk, *Life Magazine*, © Time, Inc.)

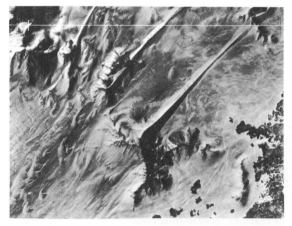

Figure 25.12 Longitudinal sand drifts appear on this air photograph as sharp-crested bladelike streamers of sand drawn out to the lee of a hill. Several barchan dunes are present in the lower left-hand corner. Width of area shown is about 0.75 mi (l.2 km). Air photomosaic of portion of Chao-Virú-Moche area, Peru. (Courtesy Ministerio de Fomento.)

Longitudinal dune ridges, oriented parallel with dominant winds, occupy vast areas of central Australia referred to as *sand-ridge deserts* (Figure 25.11). Ridges average 30 to 50 ft (10 to 15 m) in height, are spaced 0.25 to 1.5 mi (0.4 to 2.4 km) apart, and run in continuous length as much as 25 to 50 mi (40 to 80 km).

Also longitudinally oriented with respect to the wind, but not a true dune, is the *sand drift*, which is a long, tapering sharp-crested ridge of sand extending downwind from some topographic obstacle, such as a hill that might rise above a desert plain (Figure 25.12). Sand moving in saltation passes over or around the obstacle, lodging to the leeward and gradually building the drift until the zone of quiet air is filled.

Although the dunes described above are representative of types found in the United States, many other types have been described. In the vast deserts of North Africa, Arabia, and southern Iran are large, complex dune forms apparently not represented in the United States. One of these is the *seif dune*, or *sword dune*, which is a huge tapering sand ridge whose crestline rises and falls in alternate peaks and saddles and whose side slopes are indented by crescentic slip faces. Seif dunes may be a few hundred feet high and tens of miles long. Another Saharan type is the *star dune*, *pyramidal dune*, or *heaped dune*, a great hill of sand whose base resembles a many-pointed star in plan (Figure 25.13). Radial ridges of sand rise toward the dune center, culminating in sharp peaks as high as 300 ft (100 m) or more above the base. Star dunes seem to remain fixed in position for centuries and can serve as reliable landmarks for desert travelers.

Loess

In several parts of the world the ground is underlain by deposits of wind-transported silt, which has settled out from dust storms over many thousands of years. The material thus formed is known as *loess*. It generally has a uniform buff color and lacks any visible layering or other banding. Loess has a tendency to break, or *cleave*, along vertical cliffs wherever it is exposed by the cutting of a stream or by man (Figure 25.14). The cleavage is possibly produced by a slight shrinkage of the entire mass as it has compacted after being laid down.

Perhaps the greatest deposits of loess are in China, where thicknesses over 100 ft (30 m) are common and a maximum of 300 ft (90 m) has been measured. It covers many hundreds of square miles in northern China and appears to have been derived from the interior of Asia, out of which blow dry winter winds. Loess deposits are also important in central Europe, Argentina, and New Zealand, but not so extensive or thick as in China.

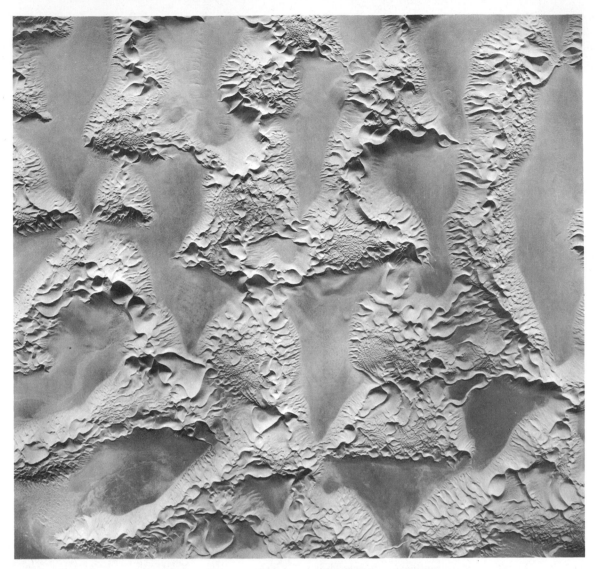

Figure 25.13   Seen from an altitude of 6 mi (10 km), these sand dunes of the Libyan desert appear as irregular patches which rise to star-shaped central peaks 300 to 600 ft (90 to 180 m) higher than the intervening flat ground. Width of the photograph represents about 7 mi (11 km). (Photograph by courtesy of Aero Service Corporation.)

In the United States, important loess deposits lie in the Mississippi River valley (Figure 25.15). Much of the prairie plains region of Illinois, Iowa, Missouri, Nebraska, and Kansas is underlain by a loess layer ranging in thickness from 3 to 100 ft (1 to 30 m). There are also extensive deposits along the lands bordering the lower Mississippi River floodplain on its east side, throughout Tennessee and Mississippi (Figure 25.14). Still other loess deposits are in northeast Washington and western Idaho. The American and European loess deposits are directly related to the continental glaciers of the Pleistocene epoch. At the time when the ice covered much of North America and Europe, it

is possible that a generally dry winter climate prevailed in the land bordering the ice sheets and that strong winds blew southward and eastward over the bare ground, picking up silt from the floodplains of braided streams which discharged the melt water from the ice. This dust settled upon the ground between streams, gradually building up to produce a smooth, level ground surface. The loess is particularly thick along the eastern sides of the valleys because of prevailing westerly winds, and is well exposed along the bluffs of most streams flowing through the region today (Figure 25.16).

The importance of loess in world agricultural resources cannot be easily overestimated. Loess

Figure 25.14 This perpendicular road cut in loess south of Vicksburg is typical of thick glacial loess accumulations on the eastern bluffs of the Mississippi River. (Photograph by Orlo Childs.)

plains and plateaus have developed rich, black soils especially suited to cultivation of grains. These are the prairie, chernozem, chestnut, and brown soils described in Chapter 14. The highly productive plains of southern Russia, the Argentine Pampa, and the rich grain region of north China are underlain by loess. In the United States, corn is extensively cultivated on the loess plains in those states, such as Iowa and Illinois, where rainfall is sufficient; wheat is grown farther west on loess plains of Kansas and Nebraska and in the Palouse region of eastern Washington.

Because loess forms vertical walls along valley sides and is able to resist sliding or flowage, but at the same time is easily dug into, it has been

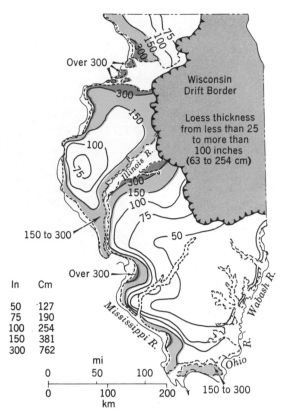

Figure 25.16 Thickness of loess in Illinois. (After R. F. Flint.)

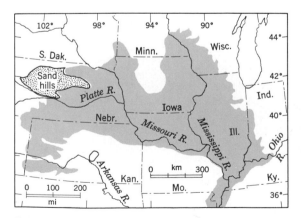

Figure 25.15 Loess deposits of the central United States (After E. T. Apfel.)

Figure 25.17 Road sunken deeply into loess, Shensi, China. (Photograph by Frederick G. Clapp, courtesy of The Geographical Review.)

widely used for cave dwellings both in China and in Central Europe. In China, old trails and roads in the loess have become deeply sunken into the ground as a result of the pulverization of the loess of the road bed and its removal by wind and water (Figure 25.17).

REFERENCES FOR FURTHER STUDY

Gautier, E. F. (1935), *Sahara: The great Desert*, translated by D. F. Mayhen, Columbia University Press, New York, 264 pp.

Lobeck, A. K. (1939), *Geomorphology*, McGraw-Hill Book Co., New York, 731 pp. See Chapter 11.

Bagnold, R. A. (1941), *The physics of blown sand and desert dunes*, Methuen and Co., London, 265 pp.

Cotton, C. A. (1942), *Climatic accidents in landscape-making*, Whitcombe and Tombs, Christchurch, N.Z., 354 pp. See Chapters 1 and 9.

Flint, R. F. (1957), *Glacial and Pleistocene geology*, John Wiley and Sons, New York, 533 pp. See Eolian features, pp. 176–194.

Thornbury, W. D. (1969), *Principles of geomorphology*, second edition, John Wiley and Sons, New York, 591 pp. See Chapter 12.

REVIEW QUESTIONS

1. Explain the processes of deflation and abrasion by wind.

2. What conditions favor deflation? What topographic forms are produced?

3. What is a desert pavement? How does it form? What is a reg?

4. What forms does wind abrasion produce?

5. How do dust storms originate? How much material might be carried in a single dust storm? How far does the dust travel?

6. How do sand grains travel in a sandstorm? How high do the grains rise? At what level is abrasion concentrated?

7. What are sand dunes? What is the distinction between live dunes and fixed dunes?

8. Describe a crescentic, or barchan, dune. In which direction does it move?

9. What are transverse dunes? What is a sand sea? What is the source of the sand? What places would be most favorable for the development of dune areas? What is an erg landscape?

10. Describe a coastal blowout dune, a parabolic blowout dune and a hairpin dune. How does each develop?

11. Describe longitudinal sand dunes and drifts. What do they indicate as to the direction of prevailing winds?

12. Where are seif dunes and star dunes found?

13. What is loess? How is it formed? What structure has loess?

14. Describe the distribution of loess throughout the world. What origin has the loess of northern China?

15. What states of the Mississippi-Missouri river region have loess deposits? Of what economic importance is loess?

CHAPTER 26
Coastal Plains, Horizontal Strata, Domes

THE foregoing chapters on landforms produced by weathering, mass wasting, streams, ice, waves, and wind have given little or no account of the manner in which variations in rock composition and structure are capable of exerting a powerful control upon the shapes and sizes of landforms. Instead, by assuming that all bedrock is of uniform composition throughout, it has been possible to describe the simple, ideal erosional landforms produced by each agent of denudation. There are, it is true, large land areas where bedrock is fairly uniform throughout, and it is in such areas that the denudational agents can produce the ideal forms. Elsewhere, sedimentary rocks are tilted, folded, domed, or faulted; metamorphic rocks are arranged in belted patterns; intrusive igneous rocks have solidified in a variety of bodies. It is with such structures and their topographic expression that these chapters deal.

Classification of landmasses

As illustrated in Figure 26.1, landmasses fall into several groups, distinguished according to the structure and composition of the bedrock comprising the mass.

A. **Undisturbed structures.**
1. **Coastal plains.** Recently emerged coastal belts underlain by sedimentary rock layers which lap over older rocks of the continents.
2. **Horizontal strata.** Sedimentary strata, essentially horizontal in attitude, which have been raised over a large area, but not otherwise seriously disturbed. Horizontal lava flows of great thickness and extent may be included in this group.

B. **Disturbed structures.**
3. **Domes and basins.** Circular or oval zones of uplift or depression causing sedimentary layers to be convexly or concavely bent.
4. **Folds.** Sedimentary strata that have been deformed by mountain-making crustal movements into long belts of wavelike folds. The folds may be broad and open or tightly compressed, depending on the degree of crustal compression.
5. **Fault blocks.** Crustal masses of any rock type or structure broken by faulting into sharply cut

blocks that have been displaced in relation to one another. Some tilting usually accompanies the faulting.
6. **Homogeneous crystallines.** Masses of intrusive igneous rock or metamorphic rock which are essentially uniform throughout as regards their resistance to weathering and erosion.
7. **Belted metamorphics.** Narrow zones of metamorphic rocks forming parallel mountain and valley belts (Figure 17.14).
8. **Complex structures.** Crustal masses that have suffered a combination of folding, faulting, or intrusion by igneous rocks so as to make a mass of irregular and complicated structures.

C. **Volcanoes and related forms.**
All types of rock masses resulting from the extrusion of molten rock. These include various types of volcanoes and lava flows.

There is a significant difference in the initial appearance of the undisturbed and the disturbed structures. The former have surfaces of low relief (plains or plateaus) before erosional modification sets in. The disturbed structures and volcanic forms, on the other hand, usually have bold, mountainous relief in the initial stage. Relief is greatest at the beginning of their life history, and the masses are ultimately reduced to surfaces of faint relief.

Each one of the structural types described above passes through an orderly series of erosional stages, patterned after the general cycle of landmass denudation already explained. For regions of horizontal strata or homogenous crystalline rocks, this cycle is very similar to the ideal general cycle because these rock masses are of uniform composition and structure in every direction horizontally. Folds, fault blocks, domes, and volcanoes, however, have life cycles quite different, not only from the ideal cycle, but also from one another, because the rock variations or initial shapes have a dominant control upon the denudational rates.

Coastal plains

Coastal plains pass through a series of stages illustrated in Figure 26.2. In Block *A* the region has recently emerged from beneath the sea, where

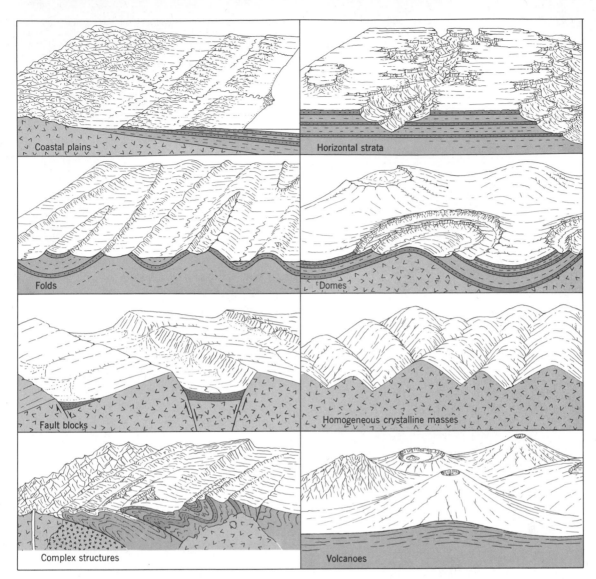

Figure 26.1 Land masses can be classified according to the groups illustrated here. Belted metamorphics are shown in Figure 17.14. Classification is based on variety and structure of rock.

it was formerly a shallow continental shelf accumulating successive layers of sediment brought from the land and distributed by currents. On the initial surface, streams flow directly seaward, down the slope of the new surface. A stream of this origin is a *consequent stream*, defined as any stream whose course is controlled by the initial slope of land surface. Consequent streams occur on many landforms, such as volcanoes, fault blocks, or beds of drained lakes. Streams that formerly drained the land surface inland from the coastal plain, but that now have become extended across it to reach the new shoreline, are called *extended consequent streams*. The term *oldland* is applied to the area

of older rock lying inland from the coastal plain. If the oldland had prominent hills upon its surface before the submergence which allowed sediments to be deposited, the hills may project through the coastal-plain layers as *mendips* (named after the Mendip Hills of southern England), or *inliers*.

In the mature stage of coastal-plain development a new series of streams and topographic features has developed (Block *B*). Where more easily eroded strata (usually clay or shale) are exposed, denudation is rapid, making *lowlands*. Between them rise broad low ridges or belts of hills comprising *cuestas*. The lowland lying between the oldland and the first cuesta is called the *inner*

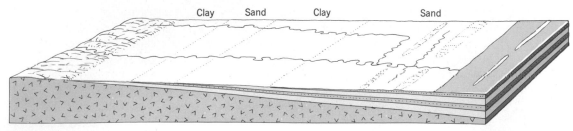

A. Initial stage; plain recently emerged.

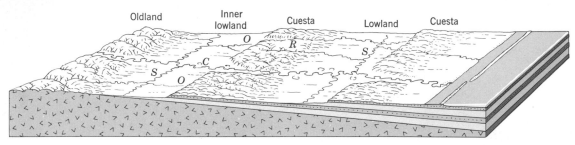

B. Mature stage; cuestas and lowlands developed. *S* = subsequent: *C* = consequent; *O* = obsequent; *R* = resequent.

C. Late mature or old stage; relief very low.

Figure 26.2   Development of a broad coastal plain. (After A. K. Lobeck.)

Figure 26.3   This sharply defined cuesta in the Paris basin of northern France has its steep face to the east (left), a very gentle slope westward from the crest. (Photograph by Douglas Johnson.)

lowland. Cuestas are commonly underlain by sand, sandstone, limestone, or chalk. They have a fairly steep slope on the landward side, or *inface*, because the edge of the eroded layer is exposed on this side. The seaward slope, or *backslope*, of the cuesta is gentle because it follows the top surface of the gently inclined harder layer. Where the resistant layer is very hard and is underlain by a weak layer, the cuesta face is often steep with occasional rock cliffs, as in the limestone cuesta near Rheims, France (Figure 26.3). More commonly, however, the cuesta is merely a belt of low hills.

Streams that develop along the trend of the lowlands, parallel with the shoreline, are of a class

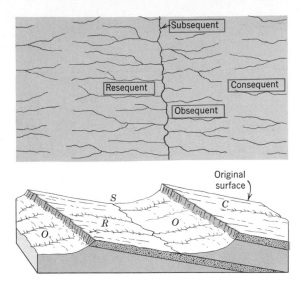

Figure 26.4 Four kinds of streams are associated with eroded dipping sedimentary layers. These form a trellis drainage pattern.

known as *subsequent* streams. They take their position along any belt or zone of weak rock and therefore follow closely the pattern of rock exposure. Subsequent streams occur in many regions and will be mentioned frequently in the discussion of folds, domes, and fault blocks.

Streams that flow down the inface of a cuesta to join the subsequent stream in a lowland are

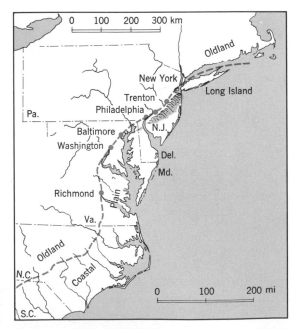

Figure 26.5 The coastal plain of the Atlantic seaboard states shows little cuesta development except in New Jersey. The inner limit of the coastal plain is marked by a series of fall-line cities. (After A. K. Lobeck.)

obsequent streams. Their direction of flow is opposite to the seaward slope of the strata, as well as opposite to the direction of flow of the consequent streams. Streams that came into existence on the backslopes of cuestas, but which were not originally present as consequent streams, are termed *resequent streams.* They flow in the same direction as the original consequent streams, but develop only as the backslope of the cuesta comes into being by removal of the weaker overlying layers (Figure 26.4).

The drainage lines on a maturely dissected coastal plain combine to form a *trellis* pattern (Figure 26.4), consisting of consequent, subsequent, obsequent, and resequent streams.

Diagram *C*, Figure 26.2, shows the coastal plain reduced to a peneplain in the stage of old age. The trellis drainage pattern persists, and the cuestas still show as faint hills rising above the nearly flat lowlands.

Coastal plains of the United States and England

Splendid examples of coastal plains are present along the Atlantic and Gulf coasts of the United States, in southeastern England, and in the Paris Basin region of north-central France.

The coastal plain of the United States is by far the largest of these, ranging in width from 100 to 300 mi (160 to 500 km) and extending for 2000 mi (3000 km) along the Atlantic and Gulf coasts. Strata are of Cretaceous and Tertiary age, the former being exposed nearest to the inner margin of the coastal plain because they lie directly upon the oldland rocks of Paleozoic and Precambrian age. The coastal plain starts at Long Island, which is a partly submerged cuesta, and widens rapidly southward so as to include much of New Jersey, Delaware, Maryland, and Virginia (Figure 26.5). Throughout this portion the coastal plain has but one cuesta, that which forms the Atlantic Highlands, Mt. Laurel, Pine Hills, and similar hill groups. The cuesta is underlain by a porous sand formation of Tertiary and Cretaceous age, which resists erosion by absorbing rainwater rapidly and thereby minimizing overland flow. The inner lowland is a continuous broad valley developed on a weak clay formation of Cretaceous age.

In Alabama and Mississippi the coastal plain is maturely dissected in all but the coastal area, which has recently emerged from the sea. Cuestas and lowlands run in belts roughly parallel with the coast (Figure 26.6). Hence the term *belted coastal plain* is applied to these regions. The cuestas tend to be underlain by sandy formations and support a growth of pines. Limestone forms fertile lowlands such as the Black Belt in Alabama.

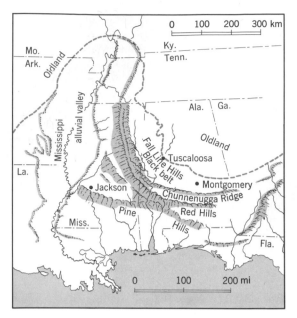

Figure 26.6 The Alabama-Mississippi coastal plain is belted by a series of sandy cuestas and shale or marl lowlands. (After A. K. Lobeck.)

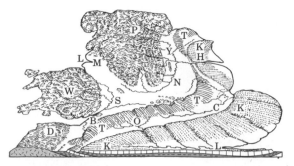

Figure 26.7 Southeastern England is a broadly curved mature coastal plain. L = London; K = Cretaceous chalk cuesta; C = Cambridge; O = Oxford; H = Humber River; Y = York; N = Nottingham; S = Severn River; B = Bristol; D = Dartmoor; W = Wales; M = Manchester; L = Liverpool; P = Pennine Range; T = Jurassic limestone cuesta. (After W. M. Davis.)

The principal steams cross the belted topography with little regard for lowlands and cuestas, suggesting that they originated as extended consequents. The shoreline of the Gulf Coast is one of fairly recent emergence, as explained in an earlier chapter.

The entire southeastern portion of England is a mature coastal plain (Figure 26.7). Two cuestas dominate the topography. The innermost is of Jurassic limestone and is locally named the Cotswold Hills. In England, the term *wold* is applied to a cuesta, *vale* to an intercuesta lowland. The outer or southeastern cuesta is of white chalk of Cretaceous age and includes the Chiltern Hills. Between cuestas is a lowland in which lie Oxford and Cambridge; hence it is called the Educational Lowland. An extensive inner lowland runs between the inner cuesta and the oldland rock masses of Cornwall, Wales, and the Pennine Range. In the inner lowland are the important cities of Bristol, Gloucester, Birmingham, Nottingham, Lincoln, and York, as well as extensive farm lands. This lowland is drained by a number of subsequent streams the Severn, Avon, Trent, and Ouse.

Geographical aspects of coastal plains

Broad coastal plains, such as those of the eastern United States and southeastern England, show intensive agricultural development because of the fertility and easy cultivation of broad lowlands. Although important seaport cities have developed on coastal plains there was not the same impelling necessity toward marine occupations that was induced by mountainous coastal belts bordered by shorelines of submergence.

Cuestas provide valuable forests, as in England and Europe and in the southern United States. Where excessively porous sands occur, as in the New Jersey coastal plain, pine and oak are supported.

Transportation tends to follow the lowlands and to connect the larger cities located there. For example, important roads and railroads connect New York with Trenton, Philadelphia, Baltimore and Washington, all of which are situated in an inner lowland. Cuesta topography, however, is rarely so rugged as to interfere seriously with the location of communication lines.

The seaward dip of sedimentary strata in a coastal plain provides a structure favorable to the development of artesian water wells. Water penetrates deeply into a sandy cuesta stratum, which is overlain by shales or clays impervious to the flow of underground waters. When a well is drilled into the sand formation considerably seaward of its surface exposure, water under hydraulic pressure reaches the surface (Figure 26.8). Artesian water in large quantities is available in many parts of the Atlantic and gulf coastal plains, although

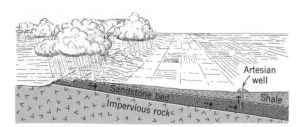

Figure 26.8 An artesian well requires a dipping sandstone layer. (After E. Raisz.)

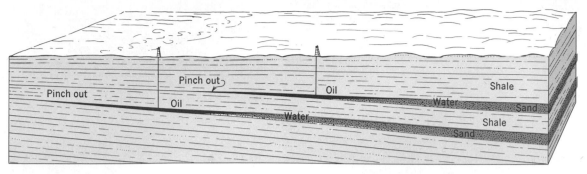

Figure 26.9   Oil pools can form in the fringes of sand formations which pinch out in the up-dip direction.

it is no longer sufficient to supply the demands of densely populated and industrialized localities.

The Gulf Coastal Plain of the United States contains petroleum and natural gas accumulations of enormous economic value. Oil occurs in *stratigraphic traps*, which are layers or lenses of permeable sand or sandstone capped by impermeable shales or clays. One kind of stratigraphic trap is the *pinch out*, illustrated in Figure 26.9. A sandstone formation in the coastal plain sequence of strata thins in the updip (landward) direction to a feather edge, where it disappears, whether through lack of deposition, or by later erosion that

preceded deposition of the next younger beds. Capped above by impermeable beds, the sandstone wedge forms a trap for oil migrating updip. Figure 26.10 show two curving bands of oil *pools* of this type in the Gulf Coast of Texas.

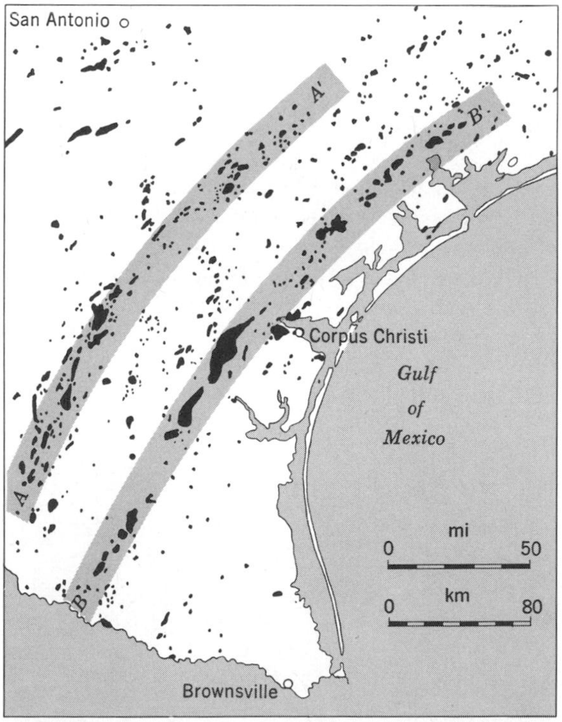

Another quite different type of oil pool common in coastal plains and other regions of thick sedimentary strata occurs on *salt domes*, or *salt plugs* (Figure 26.11). These strange, stalklike bodies of rock salt project upward through coastal plain strata. Apparently they were forced up by slow plastic flowage from thick salt formations lying in deep lower layers of the coastal plain. Surrounding strata are sharply bent up and faulted against the side of the salt plug, making traps for petroleum. Salt plugs commonly have a cap rock of limestone resting upon a plate of gypsum and anhydrite. Oil

Figure 26.10   Two zones of oil pools on up-dip pinchouts of sands of Eocene age (AA′) and Oligocene age (BB′). (After A. I. Levorsen.)

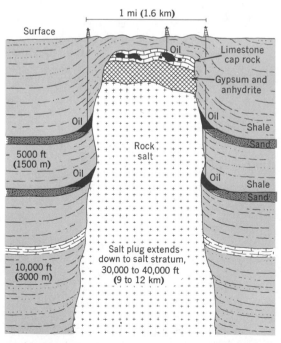

Figure 26.11   Idealized structure section of a salt dome.

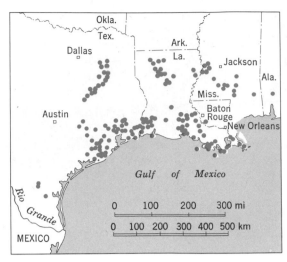

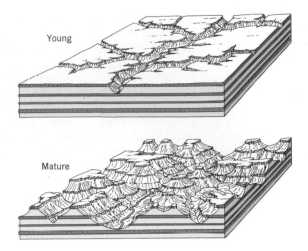

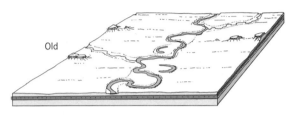

Figure 26.12 Distribution of salt domes of the Gulf Coast region is indicated by dots. (After K. K. Landes.)

Figure 26.13 Erosional development of horizontal sedimentary strata. The development of cliffs is accentuated here, as typical of an arid climate. (After E. Raisz.)

may collect in cavities in the limestone. Distribution of salt domes of the Gulf Coast is shown in Figure 26.12. The salt dome should not be confused with sedimentary domes discussed later in this chapter.

Other mineral deposits of economic importance in coastal plains include: *sulfur*, occuring in the coastal plain of Louisiana and Texas; *phosphate* beds, found in Florida; *lignite* (a low-grade, woody coal), used as a fuel in Alabama, Mississippi, and Texas; and *clays*, used in manufacture of pottery, tile, and brick in New Jersey and the Carolinas.

Horizontal strata

Considerable areas of the continental shields are covered by thick sequences of sedimentary-rock layers, which at one time in the geologic past were the bottom deposits of shallow inland seas or were stream deposits spread over vast alluvial plains. Strata of all three post-Cambrian geologic eras are represented. When uplifted with little disturbance other than a faint warping or minor faulting, these sedimentary strata pass through a series of stages of erosion such as those illustrated in the series of block diagrams of Figure 26.13.

In the initial stage the land is fairly smooth and is drained by consequent streams following the gentle slope of the surface. If the elevation of the initial surface is high, these streams soon cut deep canyons, leaving plateau surfaces between. Should the region have initially low elevation above sea level, the streams are prevented by the base level from cutting deeply, and hence strong relief can never develop in the region. Throughout the stage of youth, the region, whether of great or small relief, is dissected by streams whose valley network develops at the expense of the initial land surface.

When the initial land surface is entirely or almost entirely consumed and the region has reached its most rugged character, the stage of maturity has been reached. Throughout the remainder of the erosion cycle the relief diminishes and the slopes perhaps tend to become more gentle. In old age the region is reduced to a rolling plain upon which the larger streams have broad, flat floodplains. A few remnants of harder rock strata may remain as monadnocks.

The horizontal attitude of the rock layers gives rise to distinctive landforms where the layers are of alternately weak and resistant nature (Figure 26.14). The resistant layers, usually of sandstone and limestone (the latter particularly in arid cli-

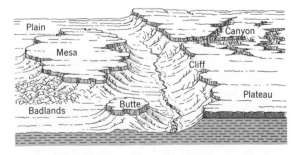

Figure 26.14 In arid climates, a distinctive set of landforms develops in flat-lying sedimentary formations.

mates), form *cliffs* or steep slopes. The weak layers, usually of shale, clay, or marl, are easily washed away from beneath the lower edges of the resistant layers, hence serving to accentuate the cliffs above them and form smoothly descending slopes at each cliff base. In dry climates, where vegetation is scant and the action of rainwash especially effective, sharply defined topographic forms develop. They comprise what may be described as *scarp-slope-shelf* topography, because the normal sequence of forms is a cliff, or scarp, at the base of which is a smooth slope. This in turn flattens out to make a shelf, terminated at the outer edge by the cliff of the next lower set of forms. In the walls of the great canyons of the Colorado Plateau region in Colorado, Utah, Arizona, and New Mexico, these forms are wonderfully displayed (Figure 26.15).

In plateau regions underlain by horizontal strata, the erosion processes tend to strip successive layers from the plateau surface. Cliffs, capped by hard rock layers, retreat as near-perpendicular surfaces because the weak clay or shale formations exposed at the cliff base are rapidly washed away by storm runoff and channel erosion. Thus undermined, the rock in the upper cliff face repeatedly breaks away along vertical fractures. Where a cliff has thus retreated a considerable distance from a canyon, there remains a broad, flat bench which is the exposed surface of the next layer below. The bench

so formed is termed an *esplanade* (Figure 26.15). Should the entire plateau surface be formed by the complete or almost complete removal of a rock series, leaving a plateau capped by a resistant layer, the plateau is termed a *stripped surface* or *stratum plain*.

In the later stages of erosion the landscape in an arid region has many *mesas* (Figure 25.14), tabletopped hills or mountains bordered on all sides by cliffs and representing the remnant of a formerly extensive layer of resistant rock. Often a mesa is capped by a lava flow, which is generally more resistant than the sedimentary rocks over which it once flowed. As a mesa is reduced in area by retreat of the cliffs that border it, it maintains its flat top and altitude. Before its complete consumption the final stage is a small, steep-sided hill or peak known as a *butte* (Figure 26.16).

Where extremely weak clays or shales, lacking a protective vegetative cover, are exposed to rainwash and gully erosion in dry regions, a very rugged topography resembling miniature mountains develops. Such areas are termed *badlands* (Figures 26.14 and 26.17).

In humid climates the elements described above, namely, scarp-slope-shelf topography, stripped surfaces, mesas, and buttes, are present only in greatly subdued aspect. This is due to the thick cover of vegetation that protects the ground from

Figure 26.15 This panoramic drawing by the noted geologist-artist, W. H. Holmes, published in 1882, shows the Grand Canyon at the mouth of the Toroweap. In this part of the canyon, rarely seen by tourists, a broad bench called The Esplanade is well developed. (From Dutton, *Atlas to accompany Monograph II*, U.S. Geological Survey.)

Figure 26.16   This early photograph shows a butte of horizontal red sandstones capped by a gypsum layer, near Cambria, Wyoming. (Photograph by N. H. Darton, U.S. Geological Survey.)

Figure 26.17   Badlands, such as these in the Petrified Forest National Monument, Arizona, are like miniature mountain topography on bare clay formations. (Photograph by B. Mears, Jr.)

rapid rainwash and permits a layer of soil and residual overburden to cloak the bedrock. Nevertheless, occasional lines of cliffs do form, and mesalike hills and mountains are developed. Vast areas in Pennsylvania, New York, Ohio, West Virginia, Kentucky, Tennessee, and Alabama consist of maturely dissected horizontal strata. Much of this land is mountainous and heavily forested (Figure 26.18).

In a maturely dissected region of horizontal strata the stream system forms a *dendritic drainage pattern*, in which the smaller streams show no predominant directional orientation or control (Figure 22.5). This pattern has been likened to the branching of an apple tree. Usually the larger trunk streams show a rough parallelism, because they were the original consequent streams which followed a perceptible slope of the region as a whole. The smaller streams, whose pattern is dendritic, are said to be *insequent* in origin. An insequent stream is controlled in its growth by very minor inequalities of rock resistance and slope which are not themselves systematically oriented. On the whole, therefore, the direction of branching is very much a matter of chance.

Figure 26.18 Seen from the air, the maturely dissected Allegheny Plateau of West Virginia appears largely forested. Relief of 700 to 800 ft (210 to 240 m) is here developed on shales of Devonian age. (Photograph by J. L. Rich, Courtesy of the *Geographical Review*.)

In addition to regions underlain by sedimentary rocks, the regions of horizontal strata may be made to include thick accumulations of lava flows. In some parts of the world, such as the Columbia Plateau region of eastern Washington and Oregon, or the Deccan Plateau of western India, the vast outpourings of highly fluid basalt lavas now cover thousands of square miles and are several thousand feet thick. Interbedded with the lavas are lake and stream deposits of sands, gravels, and clays, as well as much volcanic ash and dust. Consequently, the structure exhibits alternately weak and resistant layers in which erosion produces landforms very similar to those of sedimentary strata.

Geographical aspects of horizontal strata

Generalizations cannot readily be made about the utilization of areas underlain by horizontal strata because of the great variations in surface relief that exist. On initial and old surfaces, where the topography is plainlike, agriculture is widely developed. On the high plains of western Kansas and Nebraska, eastern Wyoming and Colorado, New Mexico, and Texas, wheat farming is the predominant activity. Here the plain is in its initial or very early stage of erosion. Despite elevations of 3000 to 5000 ft (900 to 1500 m), the plain is trenched only by a few major through-flowing streams.

Some regions of horizontal strata in the interior United States, including much of Illinois, Indiana,

Ohio, Missouri, Montana, Kansas and Iowa, are maturely dissected but a mantle of glacial drift has reduced relief so that slopes are gentle and are highly cultivated. Where relief is strong, as in the mountain areas of the Alleghenies or the Cumberland Mountains, cultivation is limited to a few small tracts, such as the floodplain belts of larger streams, despite the favorable humid climate (Figure 26.18). In the canyon lands of the Colorado Plateau an extremely low population density exists, not only because the high relief and aridity do not favor agriculture, but also because human access is virtually impossible across the network of sheer-walled canyons.

As with coastal plains, regions of horizontal strata have only those minerals and rocks of economic value that are associated with sedimentary rocks (or lavas). Building stone, such as the Bedford limestone in Indiana or the Berea sandstone in Ohio, is a valuable product. Limestone may be quarried for use in manufacture of portland cement or as flux in iron smelting. Some important deposits of lead, zinc, and iron ores occur in sedimentary rocks. For example, the lead and zinc mines of the Tristate district (Missouri, Kansas, Oklahoma) are in horizontal limestones.

Perhaps the greatest mineral resources occurring in sedimentary strata are coal and petroleum. Where the strata are undisturbed, coal is of the *bituminous*, or soft, variety and lies in horizontal layers from a few inches to several feet thick and hundreds of square miles in extent. Where the relief of the region is great, coal seams outcrop along the valley walls into which mine openings termed *drifts* can be tunneled to obtain the coal. This is common practice in the bituminous field of western Pennsylvania, eastern Ohio, and West Virginia. Where the coal seams are not exposed at the surface they must be reached by vertical shafts, as in the Illinois coal fields. Anywhere that a rich coal seam is found outcropping in a hillside or lies near the surface it can be reached by removing the overlying of rock and soil with powerful earth-moving machines. This process, termed *strip mining*, leaves great gashes and adjacent embankments of rock debris.

Petroleum occurs within permeable sandstone layers, in which the oil is trapped by overlying impermeable shales. Because the strata are not perfectly horizontal, but are affected by minor warpings and faults as well as by changes in thickness and character of the sandstone layers, there are many structures in which petroleum is localized into pools. Stratigraphic traps, similar in principle to those of coastal plains, form important pools. Traps also result from faulting, which offsets the edges of the strata (Figure 27.16).

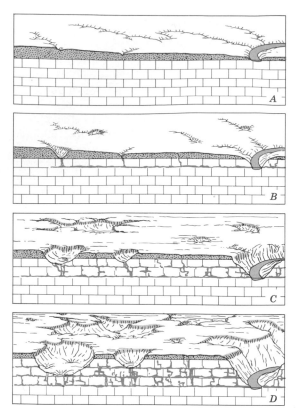

Figure 26.19 As a cavern system develops, surface stream flow is diverted to underground flow. (After A. K. Lobeck.)

Figure 26.20 Outcrops of horizontal limestone strata show in the walls of this deep sinkhole on the Kaibab Plateau of northern Arizona. Because of high elevation, 8500 ft (2500 m), the climate here is cool and humid, favoring limestone solution. (Photograph by A. N. Strahler.)

Limestone caverns

Limestone, composed of calcium and magnesium carbonate, is readily dissolved by weak solutions of carbonic acid present in soil and ground water. Ground waters carry away the dissolved carbonates, leaving elaborate systems of subterranean caverns. Most persons are familiar with the names of famous caverns, such as Mammoth Cave, Carlsbad Caverns, Luray Caverns, or Cave of the Winds, and many Americans have visited one or more of these famous tourist attractions. Although caverns may develop in folded, faulted, and steeply dipping limestone layers, most caverns occur in areas of flat-lying strata, so that it is appropriate to take up the subject at this point.

The development of a cavern system is illustrated in Figure 26.19. According to one widely held theory of cavern formation, it is necessary that a thick limestone formation be present and that the region be an elevated one with the valleys of the larger rivers deeply cut so as to provide topographic relief. At first (Stage A) the limestone is solid and is covered by a sandstone layer. The water drains off by means of surface streams. As the large river cuts more deeply, water percolating downward toward the lowered water table flows through joint cracks in the limestone to emerge at the river bank (Stage B). The passageways in the limestone become larger and deeper as time goes on (Stage C). By now, all surface water pours down into cavities known as *sinkholes* (Figure 26.20). These are usually filled at the surface by broken rock and soil through which the water filters. Occasionally they are enlarged by collapse so as to make a gaping hole. In Stage D the sinkholes are enlarged into small valleys and the cavern system is fully developed, with many passageways and shafts. On the lowest level run underground rivers, which emerge at the banks of the main river. Abandoned caves at higher levels may be dry except for the localized dripping of water.

A second theory of cavern development holds that the removal of carbonate rock takes place below the water table. As water slowly moves in the paths shown in Figure 20.26, the dissolved solids are brought to streams and carried off. This

A. Entrance to caves.

C. A drip curtain.

B. A stalagmite.

D. A travertine terrace.

Figure 26.21   The Jenolan Caves of New South Wales, Australia, have remarkably fine displays of cavern deposits. (Courtesy N. S. W. Government Printer.)

explanation, termed the *phreatic-water theory*, accounts for many cavern features not explained by the first theory. Once formed, the caverns come to lie above the water table as the adjacent stream valleys are carved to greater depths.

Cave deposits (or "formations," as they are usually called by tourist guides in our American caverns) are encrustations of calcium carbonate or gypsum formed after the cavern is fully enlarged and is no longer the site of rapid solution. The slow drip of water from the ceilings is accompanied by evaporation so that a small amount of calcium carbonate is left behind. Many beautiful forms result. Some of these are pictured in Figure 26.21. Spikelike forms hanging from the ceilings are *stalactites* (*D*); blunt columns growing up from the floor are *stalagmites* (*B*). Where the water drips from a long joint crack on the ceiling, a *drip curtain* results (*C*). Stalactites and stalagmites join to form *columns;* and rows of columns formed under a large joint crack may coalesce to make a *wall*. Among the blocks of stone that litter the cavern floor, pools of water overflow from one to another, making *travertine terraces* (*D*).

The geographical importance of caves is felt in several ways. Throughout man's early development, caves were an important habitation. Now we find the skeletal remains of these people, together with their implements and cave drawings, preserved through the centuries in caves in many parts of the world. Today, with increasing destructiveness of weapons of warfare, caverns are achieving importance as possible sites for storage of valuable materials, as living quarters, and as factories for important types of production.

Caverns have provided some valuable deposits of *guano*, the excrement of birds or bats, which is rich in nitrates and is used in the manufacture of fertilizers and explosives. Bat guano was taken from Mammoth Cave for making gunpowder during the war of 1812. Much more recently a valuable guano deposit in a limestone cavern in the wall of the Grand Canyon of the Colorado River has been mined and lifted to the canyon rim by cable car.

Karst landscapes

Where limestone solution has been especially active there results a landscape with many unique landforms. This is especially true along the Dalmatian coastal area of Yugoslavia, where the topography is termed *karst*. The term may be applied to the topography of any limestone area where sinkholes are numerous and small surface streams nonexistent. Four stages of development of a karst landscape are shown in Figure 26.22. Exposed limestone surfaces are deeply grooved and fluted

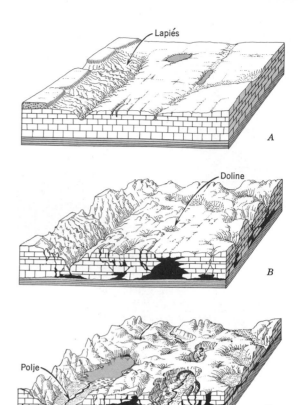

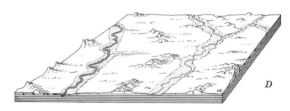

Figure 26.22   Stages of evolution of a karst landscape show increased relief and cavern development followed by decreasing relief and the removal of the limestone mass. (Drawn by E. Raisz.)

into *lapiés* (*A*). Deep, steep-walled funnel-like sinkholes, called *dolines* (*B*), are numerous. In places, these have coalesced to make open, flat-floored valleys termed *poljes* (*C*). Here surface streams flow and the soil may be suitable for agriculture.

Some other regions of karst or karstlike topography are the Mammoth Cave region of Kentucky, the Causses region of France, the Yucatan Peninsula, and parts of Cuba and Puerto Rico.

Domes and basins

Sedimentary layers in many places show a warping into broad domelike or basinlike structures, which may range from 100 to 200 mi (160 to 325 km) in diameter, but in which the strata are

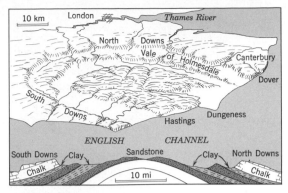

Figure 26.23 The Weald region of southeastern England is a broad dome of sedimentary strata from which the top has been removed. (After A. K. Lobeck.)

nowhere inclined more than 1° or 2° from the horizontal. Such domes and basins tend to form concentric cuestas when dissected. An example is the Weald Uplift of southeastern England (Figure 26.23). As the central part of the dome is eroded away, older rock layers are exposed. As each layer is cut through, it is eroded back from the center. Thus the cuestas have their steep inface toward the center of the dome and retreat away from the center. An example of a broad, shallow basin structure is the Paris Basin. The cuestas have their steep infaces outward, and as erosion progresses the cuestas retreat toward the center of the basin.

Domes may also be steep-sided and high, differing from the broad, low type in that the strata on the flanks dip outward at angles up to 25° or more. Thus the dome forms a conspicuous hill or mountain and, when maturely dissected, may constitute a truly rugged mass of peaks. To distinguish this type of dome it will be referred to as a *mountainous dome*. Basins of comparable development are not known because, although the strata may be strongly domed up by the intrusion of molten rock under great pressure, there is no geologic mechanism by which the strata can be sharply punched down.

The erosion of a mountainous dome is shown in the diagrams of Figure 26.24. In the stage of early youth a series of consequent streams drains outward in a *radial* drainage pattern. These streams entrench the flanks of the dome and quickly expose the underlying layers. As erosion progresses, the resistant strata begin to stand out

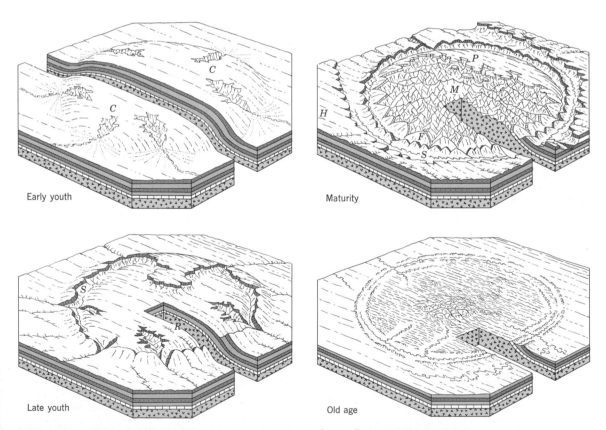

Figure 26.24 Stages in the development of a mountainous dome. C = consequent stream; S = subsequent stream; F = flatiron; P = plateau in center of dome; M = mountains of crystalline rock; H = horizontal strata surrounding dome; R = resequent stream.

Figure 26.25   In this air photo, the Virgin River is seen cutting across a hogback of steeply dipping strata on the flank of the Virgin anticline in southwestern Utah. (Photograph by Frank Jensen.)

as sharp-crested ridges, or *hogbacks*, encircling the dome (Figure 26.25). Hogbacks are very striking landforms; they owe their development to the rapid removal of weaker shale or clay beds on either side (Figure 26.26). A concentric arrangement of alternate hogback ridges and valleys develops on the mature dome. Streams occupying the weak rock valleys are subsequent in origin, forming a concentric, or *annular*, drainage pattern (Figure 26.27). Tributaries that flow down from the in-tervening hogback ridges are resequent and obsequent streams, the former flowing outward, the latter inward.

As dome erosion progresses, older and deeper rocks are exposed in the center. If geologic conditions are favorable, erosion may reveal in the center a core of intrusive igneous rock representing material that was forced up to produce the dome. In this event the igneous rock is younger than the sedimentary rock of the dome. One variety of

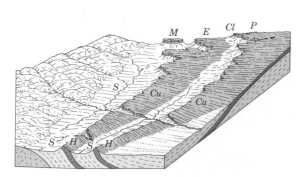

Figure 26.26 Hogbacks may gradually merge into cuestas, the cuestas into plateaus and esplanades, if the dip of the strata becomes less from one place to another. *S* = subsequent stream; *H* = hogback ridge; *Cu* = cuesta; *M* = mesa; *E* = esplanade; *Cl* = cliff; *P* = plateau. (After W. M. Davis.)

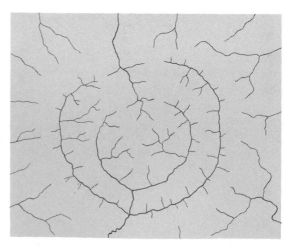

Figure 26.27   The drainage pattern on a maturely eroded dome combines annular and radial elements. It resembles a trellis pattern bent into a circle.

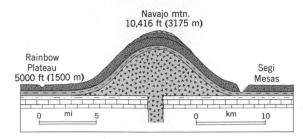

Figure 26.28 Navajo Mountain, Utah, is thought to be a laccolith whose igneous rock core is still covered by sedimentary strata. (After H. E. Gregory.)

intrusive dome is the *laccolith*, a convex but flat-bottomed mass of igneous rock which came up between strata and spread out, forcing the overlying layers to rise (Figure 26.28). In other domes the central core is of ancient rock, much older than even the sedimentary layers, and represents the rock upon which those sediments were deposited. It shows through in the dome core because the strata are not thick enough to cover it when the dome is fully dissected.

The last sedimentary rock layer to be stripped from the central core of crystalline rock clings to the sides of the core in triangular patches known

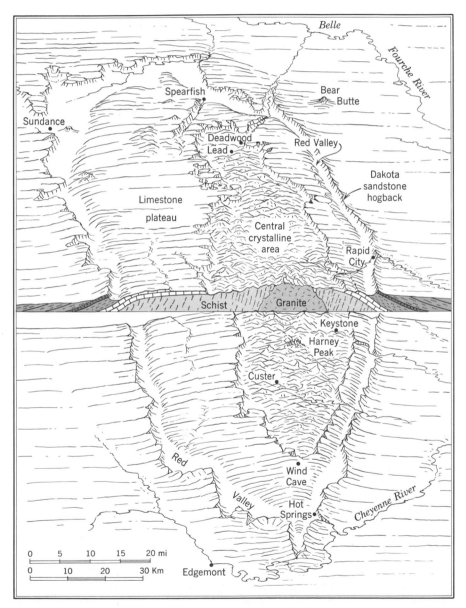

Figure 26.29 The Black Hills consist of a broad, flat-topped dome, deeply eroded to expose a core of crystalline rock.

as *flatirons*. The flatirons cap the ends of mountain spurs and are separated by V-shaped canyons (Figure 26.24).

In the old-age stage, a mountainous dome has been reduced to a peneplain on which the hogback ridges are represented by faint rows of hills. In the central core, a few monadnock masses may rise conspicuously above the peneplain level.

Geographical aspects of domes

For broad, low domes and basins the human-geographic and economic features are essentially those of coastal plains or regions of horizontal strata. Mountainous domes, however, have some unique features. These are illustrated by the great Black Hills dome of western South Dakota and eastern Wyoming (Figure 26.29).

The annular subsequent valleys that encircle the dome are splendid locations for railroads and highways. Thus it is natural that towns and cities should grow in these valleys. In the Black Hills dome, one annular valley in particular, the Red Valley, is continuously developed around the entire dome and has been termed the Race Track because of its shape. It is underlain by a weak shale which is easily washed away. In the Red Valley lie such towns as Rapid City, Spearfish, and Sturgis. On the outer side of the Red Valley is a high, sharp hogback of Dakota sandstone, known simply as the Hogback Ridge. It rises some 400 to 500 ft (120 or 150 m) above the level of the Red Valley. Farther out toward the margins of the dome the strata are less steeply inclined and form a series of cuestas. Artesian water is obtained from wells drilled in the surrounding plain.

The eastern central part of the Black Hills consists of a mountainous core of intrusive and metamorphic rocks. These mountains are richly forested, whereas the intervening valleys are beautiful open parks. Thus the region is attractive as a summer resort area. Harney Park, elevation 7242 ft (2207 m), is highest of the peaks of the core. In the northern part of the central core, in the vicinity of Lead and Deadwood, are valuable ore deposits. At Lead is the fabulous Homestake Mine, one of the world's richest gold-producing mines. In the southern part of the central crystalline area, at Pennington, is the Etta Mine, known widely for its enormous pegmatite crystals of spodumene, a source of lithium. These occurrences illustrate the principle that the interior cores of mature domes may be favorable places for mineral deposits.

The western central part of the Black Hills consists of a limestone plateau deeply dissected by streams. The original dome has a flattened summit. The limestone plateau represents one of the last remaining sedimentary rock layers to be stripped from the core of the dome.

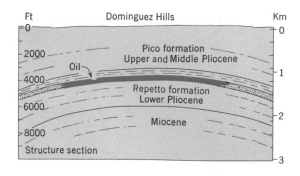

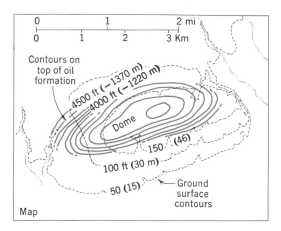

Figure 26.30 The Dominguez Hills, a low dome in an early stage of erosion, has beneath it a valuable oil pool. (After H. W. Hoots and U.S. Geological Survey.)

The subject of economic importance of domes cannot be passed by without mention of the important accumulations of oil and gas in many domes of sedimentary strata. Although not illustrated in the Black Hills themselves, many domes of the Rocky Mountain region have been important petroleum producers, for example, the Rock Springs Dome and Teapot Dome in Wyoming. Oil tends to accumulate in the domed sandstone layers which are overlain by impervious shales. An example of a very low dome in the initial stage of development, which is a valuable producer of oil, is the Dominguez Hills dome in the Los Angeles Basin (Figure 26.30).

REFERENCES FOR FURTHER STUDY

Dutton, C. E. (1882), *Tertiary history of the Grand Canyon district*, U.S. Geological Survey, Monograph 2, 264 pp.

Johnson, D. W. (1921), *Battlefields of the World War*, Amer. Geog. Soc. Research Series, No. 3, 684 pp.

Fenneman, N. M. (1931), *Physiography of the western United States*, McGraw-Hill Book Co., New York, 534 pp.

Fenneman, N. M. (1938), *Physiography of the eastern*

United States, McGraw-Hill Book Co., New York, 691 pp.

Lobeck, A. K. (1939), *Geomorphology*, McGraw-Hill Book Co., New York, 731 pp. See Chapters 13, 14, and 15.

Thornbury, W. D. (1965), *Regional geomorphology of the United States*, John Wiley and Sons, New York, 609 pp.

Hunt, C. B. (1967), *Physiography of the United States*, W. H. Freeman and Co., San Francisco, 480 pp.

REVIEW QUESTIONS

1. What groups of geologic structures make up the bedrock landmasses of the earth? Name and describe briefly each variety.

2. What is the basic difference in surface relief of the undisturbed and disturbed structural groups of landmasses in the initial stages of their respective erosion cycles?

3. How is a coastal plain formed? By what rock types is it underlain? What is a consequent stream? Where and how does it form on a new coastal plain?

4. What is the oldland? What is a mendip?

5. Describe the topography of a mature coastal plain. Name the component parts as they appear along a profile line beginning on the oldland and extending to the sea.

6. What is a subsequent stream? An obsequent stream? A resequent stream? Show by a sketch map how these stream types are related in a trellis drainage pattern.

7. Describe the coastal plain of the Atlantic and gulf coast states. What is notable about the drainage pattern of the Gulf coastal plain? What kind of shoreline is associated with the Gulf coastal plain?

8. Briefly state the salient features of the coastal plain topography of southeastern England.

9. How are agriculture and transportation influenced by coastal-plain topography? Explain how artesian wells can function on a coastal plain.

10. What are some of the mineral products of coastal plains? Which of these is most important in the United States?

11. Describe the stages of erosional development of a region of horizontal strata. What determines the maximum relief that can occur?

12. What special features develop during erosion of horizontal layers of shale and sandstone in an arid climate? Explain the following: mesa, butte, esplanade, stripped surface (stratum plain), badlands.

13. What is a dendritic drainage pattern? Of what kind of streams is it composed? How does a dendritic drainage pattern develop?

14. How can horizontal lava flows result in topography similar to that on horizontal sedimentary strata? Where are great lava plateaus found today?

15. How does the topography of a region of horizontal strata in the various stages of the erosion cycle influence agriculture and human settlements? What are the principal economic mineral products of regions of horizontal strata?

16. In what kind of rock are caverns formed? Why? What are sinkholes? Describe stalactites, stalagmites, drip curtains, and columns. What is travertine? What is guano?

17. What is karst? From what region does this term originate? Describe lapiés, dolines, and poljes. Name some regions of karst topography.

18. Describe the various kinds of dome and basin structures that occur in regions underlain by sedimentary rock formations. What variations may be found in the sizes of these structures and the steepness of dip of the strata?

19. Outline the systematic changes in topography that occur as a mountainous dome is eroded through the stages of youth, maturity, and old age. What type of drainage pattern is formed? Of what kinds of streams is this pattern composed? Compare this pattern with that of a coastal plain.

20. What is a hogback? How is it formed? What are flatirons?

21. Describe a laccolith. How does it differ from a batholith?

22. Describe briefly the important structural and topographical features of the Black Hills. In what ways are these features related to the economic products of the Black Hills?

23. How can domes act as traps for petroleum? Give examples of domes that have become valuable oil producers.

CHAPTER 27
Folds, Faults, and Fault Blocks

REGIONS of sedimentary strata that have been compressed into sets of parallel, wavelike folds pass through a series of stages illustrated in the diagrams of Figure 27.1. Short consequent streams drain the flanks of the folds to join major consequent streams that follow the axes of the troughs.

In geological terminology, a downfold is known as a *syncline;* an upfold, an *anticline* (Block *A*). It may help in remembering these terms to know that the root word *clino* means "lean," as, for example, in the word *incline.* The prefix *syn* means "together." Hence, a syncline is a structure in which strata dip toward the center line of the trough. *Anti*, meaning against or opposite, implies that in an anticline the layers dip away from the center line. In the initial stages of the development of folds, anticlines are identical with the mountains, or ridges; synclines, with the valleys.

Block *A* (Figure 27.1) shows erosion of anticlines occurring during the last stages of folding. Synclines are being filled with alluvial fan materials swept down from the adjacent anticlines. After folding has ceased the upper layers of soft, unconsolidated rock are removed until, as shown in Block *B*, a hard, well-cemented sandstone layer is exposed, reflecting the full amplitude of the folding. Coinciding with synclines are *synclinal valleys;* with anticlines, *anticlinal mountains.*

Streams that drain the flanks of the anticlines quickly cut deep ravines, exposing the underlying layers. The breaching spreads rapidly to the crest of the anticline, where a long, narrow valley is opened out along the summit (Block *B*). This valley is occupied by a subsequent stream excavating a belt of weak rock and is known as an *anticlinal valley*, because it lies upon the center line of the anticline. As this valley grows in length, depth, and breadth it replaces the original anticlinal mountain. A reversal of topography thus occurs. The synclinal valley, which originally contained the major stream, is now shrunken between the growing anticlinal valleys on either side. Moreover, the anticlinal valleys are the more rapidly deepened because of the core of weak rock exposed to attack, so that eventually the syncline becomes a mountain ridge, termed a *synclinal mountain* (Block *C*). This change of landform might well bring to mind the words of the prophet, Isaiah,

"Every valley shall be exalted, and every mountain and hill shall be made low." At this stage, which is that of maturity of the folds, the original topography has been completely reversed. The drainage pattern is a trellis type, similar in most respects to the trellis pattern of a mature coastal plain, but different in that the major subsequent streams are more closely spaced and the obsequent and resequent tributaries are shorter (Figure 27.2).

Here and there in a belt of folds, a principal stream crosses several folds at nearly right angles, passing through the sharply defined ridges by narrow *watergaps.* These streams are likely to have existed previously to the folding and to have maintained themselves as the folds were formed. The term *antecedent* has been applied to such streams. Some are illustrated in Block *B* of Figure 18.12, crossing the series of folds produced in the Appalachian revolution of Permian time.

As the dissection of the folded strata progresses, there is a continuous change in the form and position of the various types of ridges and valleys. Following the reversal of topography shown in Block *C*, Figure 27.1, the synclinal ridges will be completely removed by erosion. Meanwhile, new ridges are appearing in the centers of the anticlinal valleys. These form as a result of the uncovering of still older resistant strata which were folded along with the rest, but which previously lay below the general level of the land surface. The new ridges, which may be thought of as second-generation *anticlinal ridges*, grow in height as the weak rock is stripped from both sides. In time the anticlinal ridges, like the original anticlines of the initial land surface, are breached by streams and finally are transformed into anticlinal valleys. Thus an inversion of topography is again accomplished. The hard sandstone layers stand as narrow, sharp ridges separated by long, parallel valleys. Where the strata in a ridge dip in one direction only, representing one flank of an anticline or syncline, the ridge is termed a *homoclinal ridge* (Block *C*). Likewise, a valley of weak shales or limestones in which the layers all dip in one direction is termed a *homoclinal valley.*

In summary, mature topography developed on alternately resistant and weak sedimentary strata may have three types of ridges, or mountains: anticlinal, synclinal, and homoclinal; and three types of valleys: anticlinal, synclinal, and homo-clinal.

Ultimately the belt of folds is reduced to a peneplain (Figure 27.1, Block *D*). Even here, the ridges rise as rows of low hills and the streams maintain their trellis drainage pattern.

Zigzag ridges and plunging folds

The folds illustrated in Figure 27.1 are continuous and even-crested, hence they produce ridges that are approximately parallel in trend and continue for great distances. In some fold regions, however, the folds are not continuous and level-crested but instead have crests that rise or descend from place to place. When maturely dissected, such folds give rise to a zigzag line of ridges (Figure 27.3).

The topographic form of a syncline whose trough descends, or *plunges*, differs from that of an anticline which plunges in the same direction, when both are maturely dissected. Figure 27.3 compares the two forms. The plunging syncline is represented by a ridge with a slightly concave summit but steeply descending cliffs on the end and sides. Along the direction of plunge of the

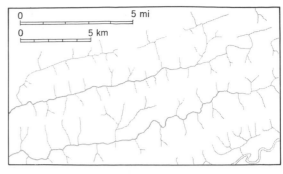

Figure 27.2 A trellis drainage pattern on folds. This resembles Figure 26.4, but opposing sets of tributaries are shorter and of more nearly equal length than on cuesta topography.

fold center line, or *axis*, this ridge develops an increasing concavity, then separates into two diverging homoclinal ridges. The plunging anticline is represented by a ridge that points in a direction opposite to the plunging synclinal mountain. The end is smoothly rounded and descends gradually to the level of the valley in the direction of plunge. In the opposite direction the mountain splits into

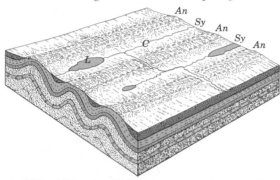

A. While folding is still in progress, erosion cuts down the anticlines; alluvium fills the synclines, keeping relief low. *An* = anticline; *Sy* = syncline; *C* = consequent stream; *L* = lake.

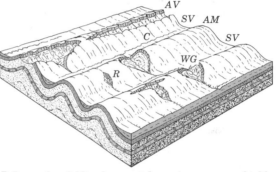

B. Long after folding has ceased, erosion exposes a highly resistant layer of sandstone or quartzite. *AV* = anticlinal; *SV* = synclinal valley; *C* = consequent stream; *WG* = water-gap; *R* = resequent stream.

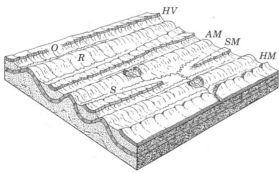

C. Continued erosion partly removes the resistant formation but reveals another below it. *AM* = anticlinal mountain; *SM* = synclinal mountain; *HM* = homoclinal ridge; *HV* = homoclinal valley: *O* = obsequent stream; *S* = subsequent stream.

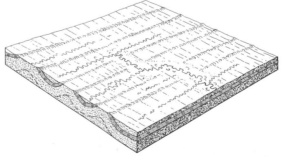

D. Peneplanation reduces the fold belt to low relief, but the hard-rock ridges still show.

Figure 27.1 Stages in the erosional development of folded strata.

two homoclinal ridges. Enclosed is a valley bounded by steep cliffs (Figure 27.3). The end of this valley, where the cliff line swings around, is often termed an *anticlinal cove*. In comparing the two forms it is apparent that the cliff slope faces outward on a plunging syncline, but faces inward in a plunging anticline.

Geographical aspects of fold regions

Some of the human-geographical and economic aspects of maturely dissected fold regions are illustrated by the Appalachians of south-central and eastern Pennsylvania (Figure 27.4). The ridges, of resistant sandstones and conglomerates, rise boldly to heights of 500 to 2000 ft (150 to 600 m) above broad lowlands underlain by weak shales and limestones. Major highways run in the valleys, crossing from one valley to another through the watergaps of streams that have cut through the ridges. Important cities may be situated near the watergaps of major streams. An example is Harrisburg, where the Susquehanna River issues from a series of watergaps cut in Blue Mountain, Second Mountain, and Peters Mountain (Figure 27.4). Where no watergaps are conveniently located, the roads must climb in long, steep grades over the ridge crests. The ridges are heavily forested; the valleys are rich agricultural belts. Fire towers are situated upon the ridge crests and command splendid views of distant ridges.

In various fold regions of the world, for example, in the Pennsylvania Appalachians, an important resource is *anthracite*, or hard coal (Figure 27.5). This occurs in strata that have been folded and squeezed. Pressure has converted the coal from bituminous into anthracite. Because of extensive erosion, all coal has been removed except that which lies in the central parts of synclines. The coal seams dip steeply; workings penetrate deeply to reach the coal that lies in the bottoms of the synclines. Seams near the surface are worked by strip mining.

Broad, gentle anticlinal folds may form impor-

tant traps for the accumulation of petroleum. The principle is the same as in low domes of sedimentary strata (Figure 26.30). Oil migrates in permeable sandstone beds to the anticlinal crest, where it is trapped by an impervious cap rock of shale. Many of the oil and gas pools of western Pennsylvania, where petroleum production first succeeded, are on low anticlines.

Pressures which created folds also changed shales into slates of considerable commercial value. The slate quarries of Pennsylvania occur in rows paralleling the base of a great sandstone ridge. Limestone, if of satisfactory quality for the manufacture of portland cement, is quarried along narrow belts where a steeply dipping bed appears at the surface.

Not all fold regions contain as extensive agricultural valleys as the Pennsylvania Appalachians. In parts of Maryland, West Virginia, and Virginia the ridges are predominant and form a great rugged mountain belt difficult to cross and thinly populated.

Other fold regions somewhat like the Appalachians are the Ouachita Mountains of Arkansas and the Jura Mountains of the Swiss and French Alps region (Figure 27.6). The Jura Mountains consist almost entirely of anticlinal limestone ridges. Good illustrations of fold belts likewise occur in North Africa, principally in Tunisia and Algeria, and in the Union of South Africa, not far north of Capetown.

Faults and fault blocks

A *fault* is a break in the brittle surficial rocks of the earth's crust as a result of unequal stresses. Faulting is accompanied by a slippage or displacement along the plane of breakage. Faults are often of great horizontal extent, so that the *fault line* can be traced along the ground for many miles,

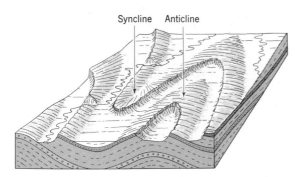

Figure 27.3   Folds with crests that plunge downward give zigzag ridges when maturely eroded. (After E. Raisz.)

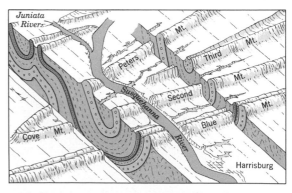

Figure 27.4   A great synclinal fold involving three resistant quartzite-conglomerate formations and thick intervening shales has been eroded to form bold ridges through which the Susquehanna River has cut a series of watergaps. (After A. K. Lobeck.)

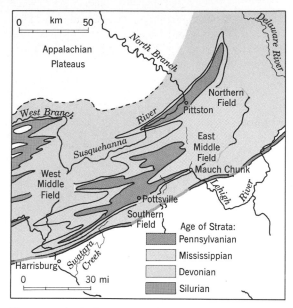

Figure 27.5 Anthracite coal basins of central Pennsylvania correspond with areas of Pennsylvania strata, downfolded into long synclinal troughs.

Age of Strata:
Pennsylvanian
Mississippian
Devonian
Silurian

0 km 50

Appalachian Plateaus

North Branch

Delaware River

West Branch

Susquehanna River

Pittston

Northern Field

East Middle Field

Mauch Chunk

West Middle Field

Lehigh River

Pottsville

Southern Field

Harrisburg

Swatara Creek

0   30 mi

sometimes even 100 mi (160 km) or more. Little is known of what happens to faults at depth, but in all probability most extend down for at least several thousands of feet.

Faulting occurs in sudden slippage movements which generate *earthquakes*, the wavelike ground tremors that start in the zone of maximum movement. A particular fault movement may result in a slippage of as little as an inch (2.5 cm) or as much as 25 or 50 ft (8 or 15 m). Successive

movements may occur many years apart, even many tens of hundreds of years apart, but aggregate total displacements of hundreds or thousands of feet. In some places clearly recognizable sedimentary rock layers are offset on opposite sides of a fault and the amount of displacement can be accurately measured.

According to the nature and relative direction of the displacement, several types of faults can be recognized (Figure 27.7). A *normal fault* has a steep or nearly vertical *fault plane*. Movement is predominantly in a vertical direction, so that one side is raised or *upthrown* relative to the other, which is *downthrown*. A normal fault results in a steep, straight *fault scarp*, whose height is an approximate measure of the vertical element of displacement (Figure 27.8). Fault scarps range in height from a few feet to a few thousand feet. Their length is measurable in miles; often they attain lengths of 100 to 200 mi (160 to 320 km). Normal faulting is an expression of tension in the earth's outer crust. It is an evident geometric observation that sliding upon an inclined surface of the type indicated in Figure 27.7 *A* must result in a spreading apart of points situated on opposite sides of the fault.

In a *reverse fault* the inclination of the fault plane is such that one side rides up over the other and a crustal shortening occurs (Figure 27.7*B*). Reverse faults produce fault scarps similar to those of normal faults, but the possibility of landsliding is greater because an overhanging scarp tends to be formed.

A *strike-slip fault* is unique in that the movement is predominantly in a horizontal direction

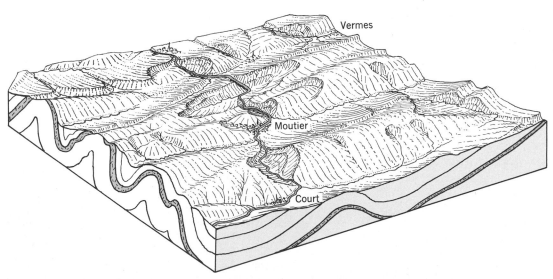

Figure 27.6 The Jura Mountains of France and Switzerland, a foothill range lying northwest of the Alps, are formed of a series of folded limestones. Almost all the mountains are anticlinal. Streams have cut valleys through these ridges to make deep watergaps. (Drawn by E. Raisz.)

Vermes

Moutier

Court

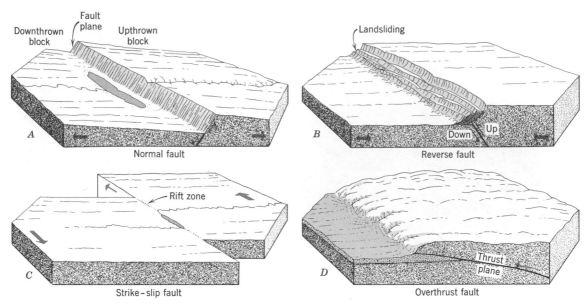

Figure 27.7 Four types of faults and their topographic expression.

(Figure 27.7C). Hence no scarp results, or a very low one at most. Instead, only a thin line is traceable across the surface. Streams sometimes turn and follow the fault line for a short distance. Sometimes a narrow trench, or *rift*, marks the fault line (Figure 27.17).

Figure 27.8 Formation of this fresh fault scarp in alluvial materials accompanied the Hebgen Lake earthquake of August 17, 1959, in Gallatin County, Montana. Displacement was about 19 ft (6 m) at the maximum point. Vehicle stands on the upthrown side of the fault. (Photograph by J. G. Stacy, U.S. Geological Survey.)

A *low-angle overthrust fault* (Figure 27.7D) likewise involves predominantly horizontal movement, but the fault plane is in a horizontal position and one slice of rock rides up over the adjacent ground surface. A thrust slice may be a few hundred or thousand feet thick but up to 25 or 50 mi (40 to 80 km) wide. Overthrusting of this type is generally associated with strong crustal compression in which intense folding also occurs. The scarp produced by low-angle overthrusting is not straight or smooth, as in normal and reverse faults; instead in may be irregular in plan.

Normal faults are not always simple, clean breaks. The fault may be split in such a manner that the end of one fault is overlapped by the end of another (Figure 27.9). Between the overlapping ends a sloping ramp, or *fault splinter*, is formed. Paradoxically, it is thus possible to go from the downthrown to the upthrown block without crossing any fault!

Normal faults may occur as a series of parallel, closely set fractures between which the rock is broken into thin slices (Figure 27.9). At the surface a series of steplike levels occurs, giving rise to the term *step faulting* for this multiple-fault arrangement.

Closely related to normal faulting is *monoclinal flexing* (Figure 27.10), in which sedimentary layers are sharply bent between the upthrown and downthrown sides instead of being fractured. A monocline passes through a series of stages of erosion quite similar to an anticline, except that only half the anticline is represented.

Faults rarely are isolated features. More often they occur in multiple arrangements, commonly

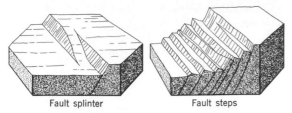

Figure 27.9   Normal faults.

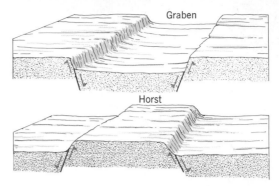

Figure 27.11   Graben and horst.

as a parallel series of faults. This gives rise to a grain or pattern of rock structure and topography. A narrow block dropped down between two normal faults is a *graben* (Figure 27.11). A narrow block elevated between two normal faults is a *horst*. Grabens make conspicuous topographic trenches, with straight, parallel walls. Horsts make blocklike plateaus or mountains, often with a fairly flat top, but steep, straight sides.

Erosional development of a fault scarp

Some of the various stages and forms attained by normal faults throughout the erosional period that follows their formation are illustrated in Figure 27.12. At the rear part of the block is the original fault scarp, produced directly by crustal movement, and therefore belonging to the group of initial landforms. Stage *A* is a *young fault scarp* in which the scarp is still straight and smooth, with only a few stream-cut canyons and some talus cones and alluvial fans built along the scarp base.

Stage *B* is a *mature fault scarp*. Weathering and erosion have caused the scarp to retreat from the original line and to become highly irregular.

In Stage *C* the scarp results entirely through erosion and is designated a *fault-line scarp*. In Stage *D* the region has been reduced to a peneplain and the scarp obliterated entirely.

In Stage *E* further erosion has revealed still another fault-line scarp. In appears because the weak shale formation has been stripped from both sides of the fault, revealing a basement of resistant igneous rocks upon which the shale was once deposited (Figure 27.13).

Fault-block mountains

In regions where normal faulting is on a grand scale, with displacements up to several thousands of feet, huge mountain masses are produced. In general, these faulted mountain blocks can be classed as *tilted* and *lifted* (Figure 27.14). A tilted block has one steep face, the fault scarp, and one gently sloping side. The initial divide lies near the top of the fault scarp and is hence situated over to one side of the block. A lifted block, which is a type of horst, is bound by steep slopes on both sides.

Figure 27.15 shows stages in the life history of a tilted fault block. Most numerous good examples of fault-block ranges are in the desert of the western United States. It should not be supposed,

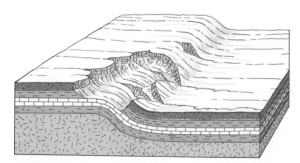

Figure 27.10   A monocline.

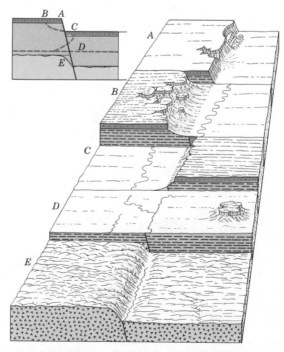

Figure 27.12   Erosional stages of a fault line.

In youth, the fault block is asymmetrical and has generally even sides, despite the presence of numerous small stream valleys that have been developing as the fault block was elevated.

In maturity (Figure 27.15), the range is dissected into a great number of divides, spurs, and peaks separated by deep canyons. The main crest line of the range is now pushed back to a more nearly central position and the simple blocklike aspect of the mountain range has disappeared. Along the base of the fault scarp, between canyon mouths, remain some parts of the original fault scarp. These make *triangular facets*, aligned nicely along the base line (Figure 27.15). Fans are more extensive than in early youth. The adjoining basins are filled higher with alluvium. In a humid climate, a rolling landscape of moderate relief develops in late maturity, whereas in old age the range is reduced to a subdued peneplain surface.

Geographical aspects of faults and block mountains

Faults are of economic and human-geographic significance for both geologic and topographic reasons. Fault planes are usually zones along which the rock has been pulverized, or at least considerably fractured. This has the effect of permitting ore-forming chemical solutions to rise along fault planes. Many important ore deposits lie in fault planes or in rocks that faults have broken across.

Another related phenomenon is the easy rise of underground water along fault planes. Springs, both cold and hot, are commonly situated along fault lines. They occur along the bases of young

Figure 27.13 Fault-line scarp, MacDonald Lake, near Great Slave Lake, Northwest Territories, Canada. (Royal Canadian Air Force Photograph.)

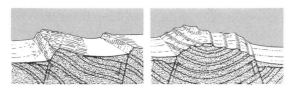

Figure 27.14 Fault block mountains may be of tilted type (left) or lifted type (right). (After W. M. Davis.)

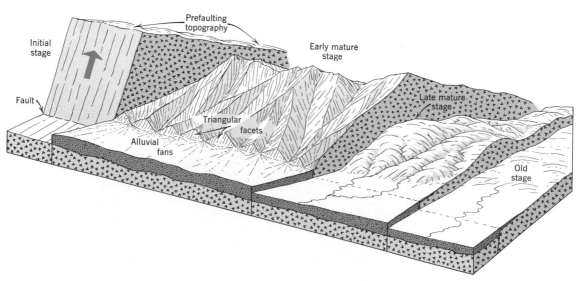

Figure 27.15 Stages in the erosion of a tilted fault block. (After W. M. Davis.)

fault-block mountains, as, for example, Arrowhead Springs along the base of the San Bernardino Range and Palm Springs along the foot of the San Jacinto Mountains, both in southern California.

Petroleum, too, finds its way along fault planes where the rocks have been rendered permeable by crushing, or it becomes trapped in porous beds that have been faulted against impervious shale beds (Figure 27.16). Some of the most intensive searches for oil center about areas of faulted sedimentary strata because of the great production that has been achieved from pools of this type.

Fault scarps and fault-line scarps may form imposing topographic barriers across which it is difficult to build roads and railroads. The great Hurricane Ledge of southern Utah is a feature of this type, in places a steep wall 2500 ft (760 m) high.

Active faults, those on which repeated fault movements are in progress, may impose serious threat of earthquake damage to cities, bridges, aqueducts, and other structures located on or near the fault lines. The California coastal belt is faced with such problems. Through the San Francisco Bay area passes the famed San Andreas fault (Figure 27.17). The devastating earthquake of 1906 resulted from slippage along this fault, which is of the strike-slip type. The San Andreas fault, 600 mi (965 km) long, extends southward and is in places represented by a rift valley, elsewhere by a normal fault bordering the San Bernardino mountain block. Where the Los Angeles aqueduct crosses this fault line and is particularly vulnerable to breakage through fault movement, special structures have been prepared to assure rapid repair in the event of damage. Vulnerability of the urban communities in the San Francisco and Los Angeles areas to earthquake shock is especially great be-

Figure 27.17 The valley followed by the road is the San Andreas Rift, California, marking a strike-slip type of fault along which earthquake-generating movements occur from time to time. In the foreground is Grant Lake. The view is southeast toward Palo Prieto Pass. (Spence Air Photos.)

cause the alluvial materials of the plains shake with great violence in an earthquake which might otherwise cause little disturbance to solid bedrock.

Grabens may be of such size as to form broad lowlands. An illustration is the Rhine graben of Western Germany. Here a belt of rich agricultural land, 20 mi (32 km) wide and 150 mi (240 km) long, lies between the Vosges and Black Forest ranges, both of which are block mountains faulted up in contrast with the downdropped Rhine graben block.

REFERENCES FOR FURTHER STUDY

Fenneman, N. M. (1931), *Physiography of the western United States*, McGraw-Hill Book Co., New York, 534 pp.

Johnson, Douglas (1931), *Stream sculpture on the Atlantic slope*, Columbia University Press, New York, 142 pp.

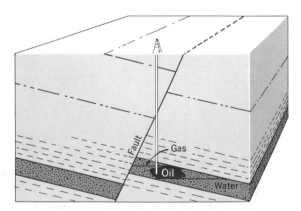

Figure 27.16 An oil pool has accumulated in the permeable sandstone beds and is prevented from escaping by the impermeable shales faulted against the edge of the sandstone layer.

Fenneman, N. M. (1938), *Physiography of the eastern United States*, McGraw-Hill Book Co., New York, 691 pp.

Lobeck, A. K. (1939), *Geomorphology*, McGraw-Hill Book Co., New York, 731 pp., See Chapters 16 and 17.

Davis, W. M. (1954), *Geographical essays*, Dover Publications, New York, 777 pp. See pp. 413–484, 725–772.

Thornbury, W. D. (1965), *Regional geomorphology of the United States*, John Wiley and Sons, New York, 609 pp.

Hunt, C. B. (1967), *Physiography of the United States*, W. H. Freeman and Co., San Francisco, 480 pp.

REVIEW QUESTIONS

1. What is an anticline? A syncline? How are these features related in a series of folds?

2. What is an anticlinal valley? A synclinal valley? How does reversal of topography normally occur in the process of erosion of a fold region?

3. How are watergaps formed in a fold region? Of what geographic importance are watergaps? What is an antecedent stream? How might antecedent streams develop in a fold region?

4. What type of drainage pattern is developed in a maturely dissected region of folds? Of what types of streams is this pattern composed?

5. How do homoclinal ridges and valleys differ from anticlinal and synclinal ridges and valleys? Show all these types by simple cross-sectional diagrams.

6. What effect does a downplunge of fold axes have upon the forms of ridges and valleys? How can a plunging synclinal mountain be distinguished from a plunging anticlinal mountain?

7. Discuss the geographical aspects of fold regions of rugged ridge-and-valley topography. What economic mineral products are important in fold regions?

8. What is a fault? What is a fault line? Explain how earthquakes are related to faults.

9. Describe a normal fault, and explain how it differs from a reverse fault. What kind of topographic feature is produced by these types of faulting? Explain how fault splinters and fault steps are associated with normal faults.

10. What is a strike-slip fault? How does it differ in topographic expression from a normal fault? What is a rift? Describe the San Andreas fault of California. Is this an active fault line at the present time?

11. Describe a low-angle overthrust fault. With what kind of crustal deformation is it associated? What topographic expression does an overthrust fault have?

12. How does monoclinal flexing differ from faulting? Are the two structures basically related?

13. Distinguish between a graben and a horst. Briefly describe the Rhine graben region as an illustration of graben and horst forms.

14. Why is it necessary to distinguish between a fault scarp and a fault-line scarp? Which form comes first? What varieties of fault-line scarps are recognized?

15. What kinds of block mountains can be formed? Compare the stages of youth and maturity of a large, tilted fault-block mountain. What are the triangular facets? How do they differ from flatirons?

16. In what ways do faults influence the occurrence of ground water and springs? How is petroleum concentrated by fault structures? Why are fault zones favorable for occurrences of ore minerals?

CHAPTER 28
Crystalline Masses and Volcanic Forms

THE term *crystalline rock* is useful in referring collectively to intrusive igneous rock and the metamorphic rocks, such as schists and gneisses. It is of value in understanding regional landform assemblages to recognize the range of landmass types possible under the general class of crystalline rocks, even though distinct subgroups cannot always be separated.

Homogeneous crystallines

Intrusive igneous rocks, such as granites (Figure 17.4), generally occur in enormous batholiths. One batholith in Idaho is exposed over an area of 16,000 sq mi (40,000 sq km), a region almost as large as New Hampshire and Vermont combined. The rock seems to extend down many thousands of feet, and for practical purposes may be considered bottomless. A smaller body of igneous rock less than 40 sq mi (100 sq km) in surface extent is termed a *stock*.

Batholiths and stocks do not reach the surface of the earth when formed, hence produce no initial landforms and have no initial stage in the erosional development cycle. They appear only after prolonged erosion has stripped away the older, overlying rock when the landmass is in a mature or old stage of denudation (Figure 28.1). An additional illustration of the process of exposure of deep-seated rocks was given in the discussion of the manner in which the central core of a dome becomes exposed (Chapter 26).

Topography developed on batholiths varies somewhat according to texture and composition of the rock and whether or not the mass has been faulted. Where the rock is quite uniform and free of strong faults, it is eroded into a maze of canyons and ravines which follow no predominant trend (Figure 28.2). The drainage pattern is dendritic and is composed of insequent streams just as for horizontal sedimentary strata. In fact, the two patterns may be virtually indistinguishable.

Where faulting has occurred, making a series of intersecting zones of crushed and weakened rock, the drainage follows the fault lines and forms a rectangular pattern (Figure 28.3). The streams are of subsequent type because they developed in the zones of weakness.

Certain areas of metamorphic rocks, such as gneisses and schists, also develop a dendritic drainage pattern of insequent streams because the vari-

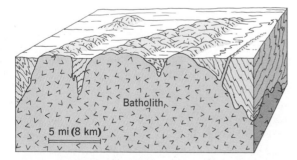

Figure 28.1 Deep-seated igneous rocks appear at the surface only after long-continued erosion has removed thousands of feet of overlying rocks. (After Longwell, Knopf, and Flint.)

Figure 28.2 This dendritic drainage pattern is developed on the maturely dissected Idaho batholith.

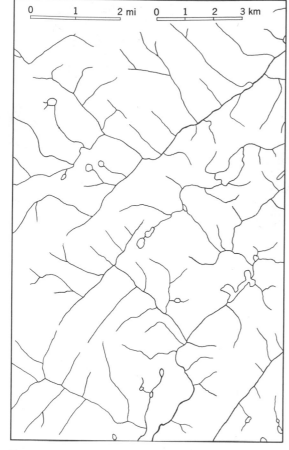

Figure 28.3   A rectangular drainage pattern.

ations in rock texture and composition seem to have little influence upon valley development. The topography of such areas may be identical with that of batholiths. Therefore it is convenient to use the term *homogeneous crystallines* to include both the igneous intrusive and metamorphic rock masses.

Belted metamorphics

Regions of metamorphic rock normally show a strong grain in the topography. Ridges tend to be elongate in one direction and to be separated by long, roughly parallel valleys (see Figure 17.14). Neither ridges nor valleys have the sharpness of folded sedimentary strata, but the drainage pattern is clearly of trellis or rectangular form. Regions of this type may be classed as *belted metamorphics* because the topography is a reflection of different rates of denudation of parallel belts of metamorphic rocks, such as schist, slate, quartzite, and marble. Marble tends to form distinctive valleys; slate and schist make belts of medium to strong relief; quartzite usually stands out boldly and may produce conspicuous narrow hogback ridges. Furthermore, most metamorphic rocks have been broken by reverse and overthrust faults which run parallel with the different belts and often separate one rock type from another. Subsequent stream valleys occupying these fault lines help bring out the grain of the topography. Much of New England, particularly the Taconic and Green Mountains, illustrates these principles well. The larger valleys trend north and south and are underlain by marble. These are flanked by ridges of gneiss, schist, slate, or quartzite. The highlands of the Hudson and of northern New Jersey continue this belted pattern southward where it joins the Blue Ridge. Near Harpers Ferry, Maryland, quartzite ridges rise prominently above broad valley belts of schist.

Areas of complex structure

Some parts of the earth's crust, particularly the continental shields, have undergone several periods of folding, faulting, intrusion, and volcanism. As each event occurred, new rocks or new structures were added to the mass, so that it may appear today as a region of *complex structure* (Figure 28.4). Because of the variety of rock types and structures, the landforms show a variety of forms.

Figure 28.5 shows by a series of diagrams the cycle of erosion in a geologically complex region. In the initial stage, mountains result from faulting, folding, doming, and volcanic activity. In the mature stage, a highly irregular drainage pattern develops. The rugged topography consists of fault and fault-line scarps, hogbacks, old volcanoes, and other landform types. Intrusive and sedimentary rocks of various ages and shapes are exposed. In the old-age stage, the region is reduced to a peneplain, but the harder rock masses stand out as monadnocks.

Geographical aspects of batholiths, metamorphic belts, and complex areas

Regions of intrusive igneous rock, metamorphic rock, and complex structure are often rich in mineral wealth. Where igneous intrusion occurred repeatedly, metallic ores were deposited. Examples from the Rocky Mountains are the copper, silver, gold, and lead of Butte, Montana; the silver and lead of the Coeur d'Alene District in Idaho; and the lead, zinc, and silver of Leadville, Colorado.

Metamorphic rocks, such as slates, quartzites, marbles, and schists, are not likely to contain metallic ores of importance, unless intrusive rocks have penetrated them. They do, however, have economic value for the slate or marble which can be quarried from them. Vermont has these rocks, along with intrusive granites which are quarried as well.

Among the foreign examples of valuable metallic mineral deposits occurring in igneous, meta-

Figure 28.4 Complexly folded and faulted strata of Precambrian age, northern Rockies of Glacier National Park, Montana. (Photograph by Chapman, U.S. Geological Survey.)

morphic, and complex rocks might be cited the tin deposits of the Katanga district, Republic of the Congo.

Much of the batholithic, metamorphic, and complex areas of the world are mountainous, hence, heavily forested and thinly populated. Examples are the Salmon River Mountains, underlain by the great Idaho batholith; the Great Smoky Mountains of North Carolina, Georgia, and Tennessee underlain by intrusive and metamorphic rocks in generally complex arrangement; or the Taconic and Green mountain ridges of belted metamorphics in Vermont and Massachusetts. Lumber is thus an important resource of these regions, in addition to their mineral wealth. At the same time, transportation is often difficult, and in the absence of agricultural land the population is thinly distributed if not actually absent from large areas.

Other extensive regions of intrusive or metamorphic rocks are of low relief—the continental shields—having been reduced to peneplains and only slightly dissected in the present erosion cycle. Much of eastern Canada is of this type of topog-

raphy, as are parts of Sweden and Finland. Not only are they similar in rock conditions, but these regions because of glaciation have countless lakes, and because of similar climate support a needleleaf evergreen forest.

In the United States, the outstanding example of a peneplain on intrusive and metamorphic rocks is the Piedmont Upland of Virginia, the Carolinas, and Georgia. A rolling landscape of monotonously uniform hill-top level extends in a vast belt between the Blue Ridge Mountains on the west and the Coastal Plain on the east. Above this surface rise a few monadnocks. On the Piedmont, soils are thick, slopes are low enough to permit agriculture, and transportation is easy between most points.

Volcanoes and associated landforms

Volcanoes are built by the eruption of molten rock and heated gases under pressure from a relatively small pipe, or *vent*, leading from a magma reservoir at depth. Both explosive and quiet types of eruption occur, the forms built differing for the two types.

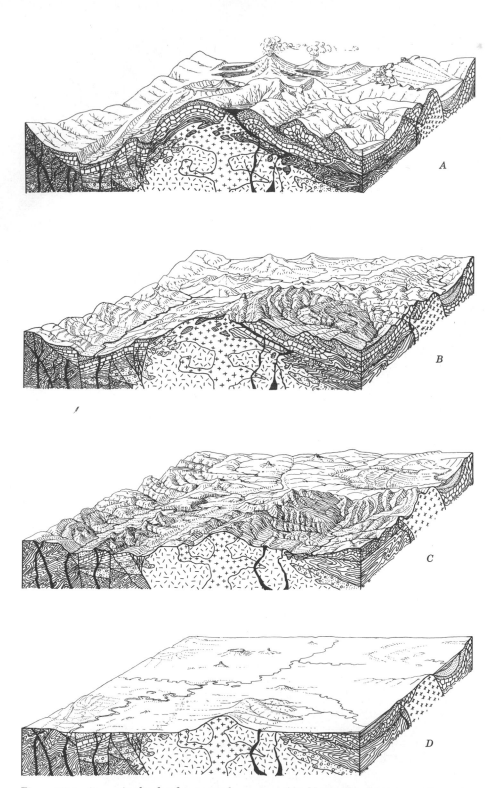

Figure 28.5 Stages in the development of a region of highly complex structure. (Drawn by E. Raisz.)

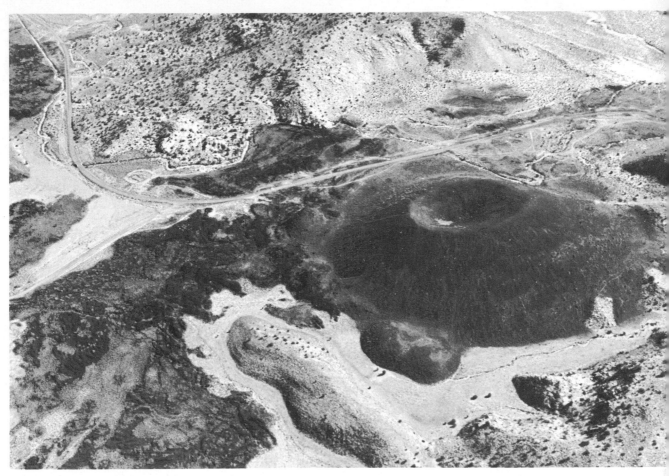

Figure 28.6   Seen in this oblique air view, a fresh cinder cone and its associated basaltic lava flow (left) have partially blocked a valley. Dixie State Park, about 17 mi (27 km) northwest of St. George, Utah. (Photograph by Frank Jensen.)

Volcanoes of explosive eruption are *cinder cones* and *composite cones;* those formed by relatively quiet outflow of lava are *lava domes.* Quiet eruption of lava, if issuing from extensive cracks, or *fissures,* in sufficient quantities may make great plains or plateaus of lava, classified with horizontal strata (Chapter 26).

Cinder cones

Smallest of the volcanoes are the cinder cones, built entirely of pieces of solidified lava thrown from a central vent. They form where a high proportion of gas in the molten rock causes it to froth into a bubbly mass and to be ejected from a vent with great violence. The froth breaks up into small fragments which solidify as they are ejected and fall as solid particles near the vent (Figure 28.6). The fragments resemble clinkers and ash taken from a coal furnace. Large pieces up to several tons in weight are *volcanic bombs;* they may be somewhat plastic when ejected. Smaller pieces, a fraction of an inch up to an inch or two

(1 to 5 cm) in size, are *cinders;* these make up the bulk of the cinder cone. Still finer particles are termed *ash* and *volcanic dust.* The ash falls like snow upon the ground within a few miles of the eruption (see Figure 28.9). Finer dust is carried by winds to distant regions and may settle out only after years of drifting in the atmosphere.

Figure 28.7   A cinder cone with its lava flows has dammed a valley, making a lake. Farther down valley, in the distance, another lava mass has made a second dam. (After W. M. Davis.)

Cinder cones rarely grow to more than 500 or 1000 ft (150 to 300 m) in height. Growth is rapid. Monte Nuovo, near Naples, Italy, grew to a height of 400 ft (120 m) in the first week of its existence. Paricutin, in Mexico, started as a cinder cone and reached a height of 1000 ft (300 m) in the first three months. The angle of slope of a recently formed cinder cone ranges between 26° and 30°. So loose is the material that it absorbs heavy rain without permitting surface runoff. Erosion is thus delayed until weathering produces a soil which fills the interstices.

Cinder cones normally have proportionately large central craters (Figure 28.6). The rim is often much higher on one side than the other, as the prevailing wind blows the finer cinders and ash to one side of the vent.

Lava flows sometimes issue from the same vent as a cinder cone. They may burst apart the side of the cone but more commonly do not alter its form. Cinder cones may erupt in almost any conceivable topographic location, on ridges, on slopes, and in valleys (Figure 28.7). Cinder cones usually occur in groups, often many dozens in an area of a few tens of square miles. They sometimes show an alignment parallel with fault lines in the underlying rock.

Composite volcanoes

Most of the world's great volcanoes are composite cones. They are built of layers of cinder and ash alternating with layers of lava, and for this reason have been called *strato-volcanoes* by some writers. The steep-sided form is governed by the angle at which the cinder and ash stands, whereas the lava layers provide strength and bulk to the volcano (Figure 28.8). Among the outstanding examples of recently formed composite volca-

Figure 28.8 Looking into the crater of an active stratovolcano, Mt. Spurr, el. 11,070 ft (3535 m). West side of Cook Inlet, near Anchorage, Alaska. (Steve McCutcheon, Alaska Pictorial Service)

A. This distant view of Sakurajima shows the great cauliflower cloud of volcanic gases and condensed steam.

C. Reaching the sea, the hot lava makes clouds of steam.

B. A blocky lava flow is advancing slowly over a ground surface littered with volcanic bombs and ash.

D. Volcanic ash has buried this village.

Figure 28.9   Sakurajima, a Japanese volcano, erupted violently in 1914. These pictures show various scenes from the eruption. (Photograph by T. Nakasa.)

noes are Fujiyama in Japan, Mayon in the Philippines, Mt. Hood in Oregon, and Shishaldin in the Aleutians. Other famous ones, less perfectly formed, are Vesuvius, Etna, and Stromboli in Italy and Sicily. Heights of several thousand feet and slopes of 20° to 30° are characteristic.

Many composite volcanoes lie in a great belt, the *circum-Pacific ring*, extending from the Andes in South America, through the Cascades and the Aleutians, into Japan; thence south into the East Indies and New Zealand. (See Figure 18.13.) There is also an important Mediterranean group, mentioned above, which includes active volcanoes of Italy and Sicily. Otherwise, Europe has no active volcanoes.

The eruption of large composite volcanoes is usually accompanied by explosive issue of steam, cinders, bombs, and ash, and by lava flows (Figure 28.9). The crater may change form rapidly, both from demolition of the upper part and from new accumulation.

Calderas

One of the most catastrophic of natural phenomena is a volcanic explosion so violent as to destroy the entire central portion of the volcano. There remains only a great central depression, termed a *caldera*. Whether the upper part of the volcano is largely blown outward in fragments or subsides into the ground beneath the volcano is not certain. Although calderas have been formed in historic time, conditions near the volcano do not permit observation of the process. Vast quantities of ash and dust are emitted and fill the atmosphere for many hundreds of square miles around.

Krakatoa, a volcanic island in Indonesia, exploded in 1883, leaving a great caldera. It is estimated that 18 cu mi (75 cu km) of rock disappeared during the explosion. Great seismic sea waves, or *tsunamis*, generated by the explosion killed many thousands of persons living on low coastal areas of Java and Sumatra. Another historic explosion was that of Katmai, on the Alaskan Peninsula, in 1912. A caldera more than 2 mi (3 km) wide and 2000 to 3700 ft (600 to 1100 m) deep was produced at this time. The explosion was heard at Juneau, 750 mi (1200 km) distant, while at Kodiak, 100 mi (160 km) away, the ash formed a layer 10 in (25 cm) deep.[1]

A classic example of a caldera produced in prehistoric time is Crater Lake, Oregon (Figure 28.10). Mt. Mazama, the former volcano, is estimated to have risen 4000 ft (1200 m) higher than the present rim. Valleys previously cut by streams and glaciers into the flanks of Mt. Mazama were beheaded by the explosive subsidence of the central portion and now form distinctive notches in the rim. Wizard Island, a recent volcano with associated flows, has since grown in the floor of the caldera.

Erosion cycle of volcanoes

Figure 28.11 shows successive stages in the erosion of volcanoes, lava flows, and a caldera. In

[1]A. K. Lobeck, *Geomorphology*, McGraw-Hill Book Co., New York, 1939, p. 685.

the first block are active volcanoes in the process of building. These are in their initial stage. Lava flows issuing from the volcanoes have spread down into a stream valley, following the downward grade of the valley and forming a lake behind the lava dam.

In the next block some changes have taken place, the most conspicuous of which is the destruction of the largest volcano to produce a caldera. A lake occupies the caldera, and a small cone has been built inside. One of the other volcanoes, formed earlier, has become extinct. It has been dissected by streams, losing the initial form, and may be said to be in a stage of late youth. Smaller, neighboring volcanoes are still active, and the contrast in form is marked. The examples of large, beautifully formed volcanoes cited above are all in their initial or very young stage.

The drainage pattern of streams upon a volcanic cone is of necessity *radial* in pattern. Because these streams take their positions upon a slope of an initial land surface they are of consequent origin. It is often possible to recognize volcanoes from a drainage map alone (Figure 28.12) because of the perfection of the radial pattern. Where a well-formed crater exists, small streams flow from the crater rim toward the bottom of the crater. Here the water is absorbed in the porous layers of ash

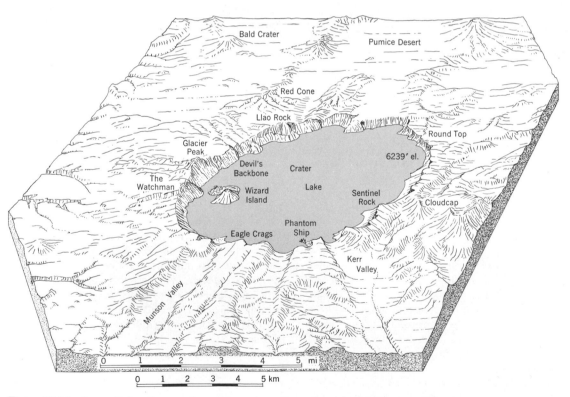

Figure 28.10 Crater Lake, Oregon, is an outstanding illustration of a caldera, now holding a lake. A great composite volcano which existed here was destroyed in prehistoric time by violent explosion. (After E. Raisz.)

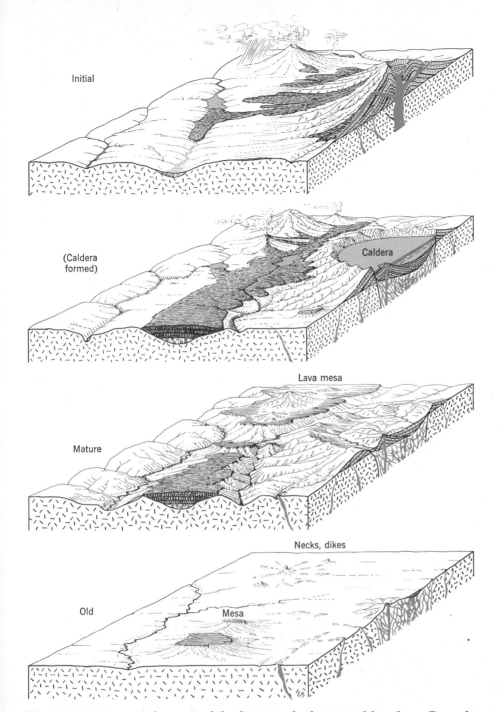

Figure 28.11 Stages in the erosional development of volcanoes and lava flows. (Drawn by E. Raisz.)

within the cone, or is conducted outward by means of a single gap in the crater rim. This inward drainage, described as *centripetal*, often adds to the certainty of interpreting a volcano from a drainage pattern.

In the third block, Figure 28.11, all volcanoes are extinct and have been eroded into the stage of maturity. The caldera lake has been drained and

the rim worn to a low, circular ridge. The lava flows which formerly flowed down stream valleys have been able to resist erosion far better than the rock of the surrounding area and have come to stand as mesas high above the general level of the region.

An example of a maturely dissected volcano is Mt. Shasta in the Cascade Range (Figure 28.13).

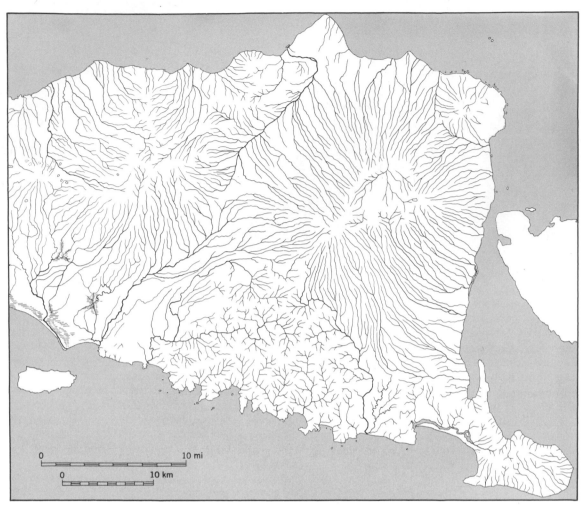

Figure 28.12 Radial drainage patterns of volcanoes in the East Indies. (After Verbeck and Fennema.)

Figure 28.13 Mt. Shasta in the Cascade Range is a maturely dissected volcano. The bulge on the left-hand slope is a more recent subsidiary cone, Shastina. (Infrared photograph by Eliot Blackwelder.)

Figure 28.14 Shiprock, New Mexico, is a volcanic neck. Radiating from it are dikes. (Spence Air Photos.)

A smaller subsidiary cone of more recent date, named Shastina, is attached to the side of the mountain.

Figure 28.11 shows the old-age stage of erosion of volcanoes. There remains now only a small sharp peak, or *volcanic neck*, representing the solidified lava in the pipe, or neck, of the volcano. Radiating from this are wall-like *dikes*, formed of lava, which previously filled fractures around the base of the volcano. Perhaps the finest illustration of a volcanic neck with radial dikes is Ship Rock, New Mexico (Figure 28.14). Because the central neck and radial dikes extend to great depths in the rock below the base of the volcano, they may persist as landforms long after the cone and its associated flows have been removed.

Lava domes or shield volcanoes

A very important type of volcano, differing greatly in form from those already discussed, is the *lava dome* or *shield volcano*. The best examples are from the Hawaiian Islands, which consist entirely of lava domes (Figure 28.15).

Lava domes are characterized by gently rising, smooth slopes which tend to flatten near the top, producing a broad-topped volcano. The Hawaiian domes range to elevations up to 13,000 ft (4,000 m) above sea level, but including the basal portion lying below sea level they are more than twice that high. In width they range from 10 to 50 mi (16 to 80 km) at sea level and up to 100 mi (160 km) wide at the submerged base.

Lava domes, as the name implies, are built by repeated outpourings of lava. Explosive behavior and emission of fragments are not important, as they are for cinder cones and composite cones. The lava, which in the Hawaiian lava domes is of a dark basaltic type, is highly fluid and travels far down the low slopes, which do not usually exceed 4° or 5°.

Instead of the explosion crater, lava domes have a wide, steep-sided *central depression*, or *sink*, which may be 2 mi (3.2 km) or more wide and several hundred feet deep. These large depressions are a type of caldera produced by subsidence accompanying the removal of molten lava from beneath. Molten basalt is actually seen in the floors of deep *pit craters*, steep-walled depressions 0.25 to 0.5 mi (0.4 to 0.8 km) wide or smaller, which occur on the floor of the sink or elsewhere over the surface of the lava dome (Figure 28.16). Most lava flows issue from cracks, or *fissures*, on the sides of the volcano.

Lava domes of the Hawaiian islands are in various stages of erosion (Figure 28.15). Active volcanoes such as Kilauea and Mauna Loa are in the initial stage and have smooth slopes. Others, such as East Maui, are partly dissected by deep canyons but still possess sizable parts of the original surface. Still others, such as West Maui, are fully dissected. Rising from the sea are some steep-walled stacks representing the last vestiges of old domes; and there exist submarine banks some 250 ft (75 m) below sea level, representing the final stage in destruction.

Geographical aspects of volcanoes

Volcanic eruptions count among the earth's great natural disasters. Wholesale loss of life and destruction of towns and cities are frequent in the history

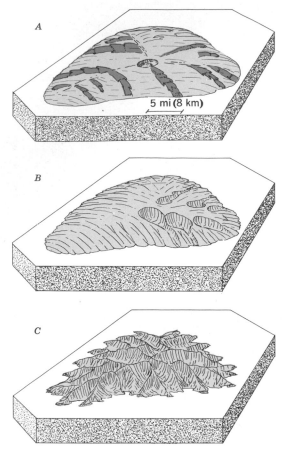

Figure 28.15 Lava domes in various stages of erosion make up the Hawaiian Islands. (Data from Stearns and Macdonald.) *A.* **Initial dome with central depression and fresh flows issuing from radial fissure lines.** *B,* **Young stage with deeply eroded valley heads.** *C,* **Mature stage with steep slopes and great relief.**

Figure 28.16 Halemaumau, a pit crater on Mauna Ioa, with a fire fountain of molten lava on its floor, 1952. (National Park Service, U.S. Dept. of the Interior.)

of peoples who live near active volcanoes. Loss occurs principally from sweeping clouds of incandescent gases that descend the volcano slopes like great avalanches; from lava flows whose relentless advance engulfs whole cities; from the descent of showers of ash, cinders, and bombs; from violent earthquakes associated with the volcanic activity; and from mudflows of volcanic ash saturated by heavy rain. For habitations along lowlying coasts there is the additional peril of great seismic sea waves, generated by submarine earth faults. These do not necessarily accompany volcanic activity and may occur without warning in the ocean basins which are bordered by belts of active mountain-making.

The surfaces of volcanoes and lava flows remain barren and sterile for long periods after their formation. Certain types of lava surfaces are extremely rough and difficult to traverse; the Spaniards who encountered such terrain in the southwestern United States named it *malpais* (bad ground). Most volcanic rocks in time produce highly fertile soils that are extensively cultivated.

Volcanic ash may have a remarkably beneficial effect upon productivity of soil where the ash fall is relatively light. The eruption of Sunset Crater, near Flagstaff, Arizona, about 800 A.D., spread a layer of sandy volcanic ash over the barren reddish soil of the surrounding region and caused it to become highly productive because of the moisture-conserving effect of the ash, which acted as a mulch in the semi-arid climate. Because Hopi Indian corn grows well in sand, this development attracted Indians, who settled the area thickly. As the ash was gradually washed off of the slopes by heavy summer rains or blown into thick dunes by wind, the fertility declined and after about 200 years of occupation the region was abandoned to its previous state.

Young and mature volcanoes possess most of the geographical attributes of rugged mountains of any sort. Steep slopes prevent extensive agriculture, although providing valuable timber resources. Thus the San Francisco Mountains, a group of maturely dissected volcanoes in northern Arizona, are clothed in what is perhaps the finest known western yellow pine forest (ponderosa pine). A lumber industry centered about the towns of Flagstaff and Williams has flourished for many years.

As scenic features of great beauty, attracting a heavy tourist trade, few landforms outrank volcanoes. National parks have been made of Mt. Rainier, Mt. Lassen, and Crater Lake in the Cascade Range. Mt. Vesuvius and Fujiyama also attract many visitors.

Mineral resources, particularly the metallic ores, are conspicuously lacking in volcanoes and lava flows, unless later geologic events have resulted in the injection or diffusion of ore minerals into the volcanic rocks. The gas-bubble cavities in some ancient lavas have become filled with copper or other ores. The famed *kimberlite* rock of South Africa, source of diamonds, is the pipe of an ancient volcano.

As a source of crushed rock for concrete aggregate or railroad ballast, and other engineering purposes, lava rock is often extensively used. Thus the ancient lava layers that make up the Watchung ridges of northern New Jersey have in places been virtually leveled in quarrying operations continued over several decades.

REFERENCES FOR FURTHER STUDY

Tyrell, G. W. (1931), *Volcanoes*, Butterworth and Co., London, 252 pp.

Lobeck, A. K. (1939), *Geomorphology*, McGraw-Hill Book Co., New York, 731 pp. See Chapters 18 and 19.

Williams, H. (1941), *Crater Lake, the story of its origin*, University of California Press, Berkeley and Los Angeles, 97 pp.

Cotton, C. A. (1944), *Volcanoes as landscape formers*, Whitcombe and Tombs, Christchurch, N.Z., 416 pp.

Jaggar, T. A. (1945), *Volcanoes declare war*, Paradise of the Pacific, Honolulu, 166 pp.

Bullard, F. M., (1962), *Volcanoes in history, in theory, in eruption*, University of Texas Press, Austin, 441 pp.

REVIEW QUESTIONS

1. What is a batholith? A stock? How large are these features? Of what kind of rock is a batholith composed? What kind of topography does it produce in a mature stage?

2. What is the meaning of a rectangular drainage pattern in a region underlain by homogeneous crystalline rock of a batholith?

3. What kind of topography develops in regions of belted metamorphic rocks such as gneiss, schist, slate, marble, and quartzite? Which of these rock types would form valleys? Which would form narrow ridges; which broad ridges?

4. What are some of the varieties of rocks and structure that might be found in a region of complex structure? From the standpoint of geologic time would a complex region be ancient or comparatively recent in development? Would it be a likely region for mineral deposits of economic value?

5. What types of volcanoes are recognized? How do they differ?

6. Describe a cinder cone. Of what material is it formed? What size is normally attained by cinder cones? How fast do they form? How are groups of cinder cones arranged in plan?

7. How is a composite type of volcano formed? Name several famous large composite volcanoes. Along what belts are most of the world's active, or recently active, large volcanoes located?

8. What is a caldera? What volcanic activity occurred at Krakatoa (1883) and at Katmai (1912)? Describe Crater Lake as an illustration of a caldera.

9. Describe the stages of youth, maturity, and old age in the erosion of a large volcano. What form of drainage pattern is typically present on volcanoes?

10. Describe a volcanic neck with radial dikes. Give an example.

11. Discuss the form and development of large lava domes, or shield volcanoes, as exemplified by the Hawaiian Islands. What feature of a lava dome corresponds to the crater of a composite volcano?

12. Discuss the geographical aspects of volcanic landforms. In what way are soils and vegetation affected by volcanic forms?

APPENDIX *I*

Soil Orders of the United States

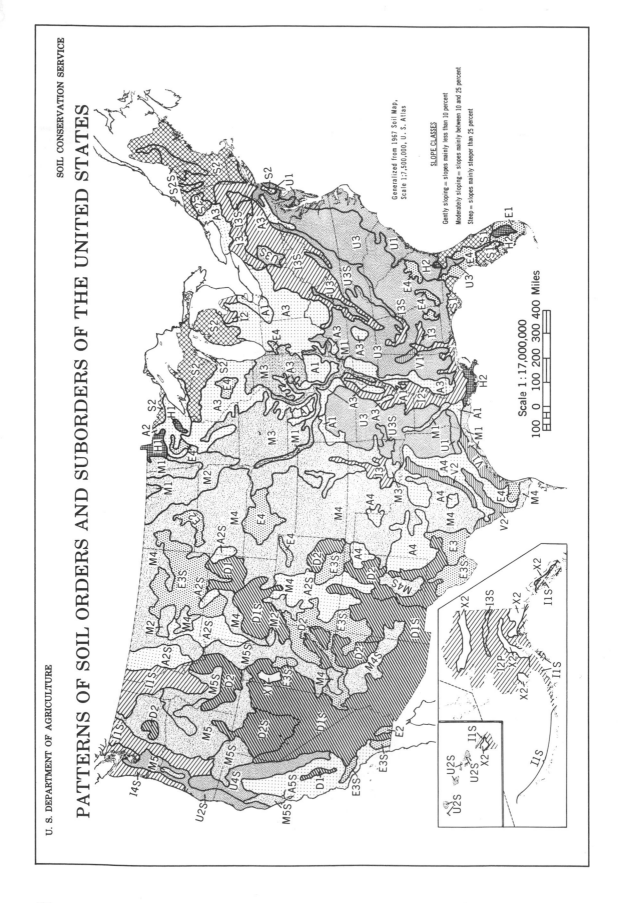

PATTERNS OF SOIL ORDERS AND SUBORDERS OF THE UNITED STATES

Generalized from 1967 Soil Map,
Scale 1:7,500,000, U. S. Atlas

SLOPE CLASSES

Gently sloping = slopes mainly less than 10 percent
Moderately sloping = slopes mainly between 10 and 25 percent
Steep= slopes mainly steeper than 25 percent

Scale 1:17,000,000

100 0 100 200 300 400 Miles

LEGEND

Only the dominant orders and suborders are shown. Each delineation has many inclusions of other kinds of soil. General definitions for the orders and suborders follow. For complete definitions see Soil Survey Staff, Soil Classification, A Comprehensive System, 7th Approximation, Soil Conservation Service, U. S. Department of Agriculture, 1960 (for sale by U. S. Government Printing Office) and the March 1967 supplement (available from Soil Conservation Service, U. S. Department of Agriculture). Approximate equivalents in the modified 1938 soil classification system are indicated for each suborder.

 ALFISOLS . . . Soils with gray to brown surface horizons, medium to high base supply, and subsurface horizons of clay accumulation; usually moist but may be dry during warm season

A1 AQUALFS (seasonally saturated with water) gently sloping; general crops if drained, pasture and woodland if undrained (Some Low–Humic Gley soils and Planosols)

A2 BORALFS (cool or cold) gently sloping; mostly woodland, pasture, and some small grain (Gray Wooded soils)

A2S BORALFS steep; mostly woodland

A3 UDALFS (temperate or warm, and moist) gently or moderately sloping; mostly farmed, corn, soybeans, small grain, and pasture (Gray–Brown Podzolic soils)

A4 USTALFS (warm and intermittently dry for long periods) gently or moderately sloping; range, small grain, and irrigated crops (Some Reddish Chestnut and Red–Yellow Podzolic soils)

A4S XERALFS (warm and continuously dry in summer for long periods, moist in winter) gently sloping to steep; mostly range, small grain, and irrigated crops (Noncalcic Brown soils)

 **ARIDISOLS** . . . Soils with pedogenic horizons, low in organic matter, and dry more than 6 months of the year in all horizons

D1 ARGIDS (with horizon of clay accumulation) gently or moderately sloping; mostly range, some irrigated crops (Some Desert, Reddish Desert, Reddish- Brown, and Brown soils and associated Solonetz soils)

D1S ARGIDS gently sloping to steep

D2 ORTHIDS (without horizon of clay accumulation) gently or moderately sloping; mostly range and some irrigated crops (Some Desert, Reddish Desert, Sierozem, and Brown soils, and some Calcisols and Solonchak soils)

D2S ORTHIDS gently sloping to steep

ENTISOLS . . . Soils without pedogenic horizons

E1 AQUENTS (seasonally saturated with water) gently sloping; some grazing

E2 ORTHENTS (loamy or clayey textures) deep to hard rock; gently to moderately sloping; range or irrigated farming (Regosols)

E3 ORTHENTS shallow to hard rock; gently to moderately sloping; mostly range (Lithosols)

E3S ORTHENTS shallow to hard rock; steep; mostly range

E4 PSAMMENTS (sand or loamy sand textures) gently to moderately sloping; mostly range in dry climates, woodland or cropland in humid climates (Regosols)

HISTOSOLS . . . Organic soils

H1 FIBRISTS (fibrous or woody peats, largely undecomposed) mostly wooded or idle (Peats)

H2 SAPRISTS (decomposed mucks) truck crops if drained, idle if undrained (Mucks)

 INCEPTISOLS . . . Soils that are usually moist, with pedogenic horizons of alteration of parent materials but not of accumulation

I1S ANDEPTS (with amorphous clay or vitric volcanic ash and pumice) gently sloping to steep; mostly woodland; in Hawaii mostly sugar cane, pineapple, and range (Ando soils, some Tundra soils)

I2 AQUEPTS (seasonally saturated with water) gently sloping; if undrained, mostly row crops, corn, soybeans, and cotton; if undrained, mostly woodland or pasture (Some Low–Humic Gley soils and Alluvial soils)

I2P AQUEPTS (with continuous or sporadic permafrost) gently sloping to steep; woodland or idle (Tundra soils)

I3 OCHREPTS (with thin or light-colored surface horizons and little organic matter) gently to moderately sloping; mostly pasture, small grain, and hay (Sols Bruns Acides and some Alluvial soils)

I3S OCHREPTS gently sloping to steep; woodland, pasture, small grains

I4S UMBREPTS (with thick dark-colored surface horizons rich in organic matter) moderately sloping to steep; mostly woodland (Some Regosols)

MOLLISOLS . . . Soils with nearly black, organic-rich surface horizons and high base supply

M1 AQUOLLS (seasonally saturated with water) gently sloping; mostly drained and farmed (Humic Gley soils)

M2 BOROLLS (cool or cold) gently or moderately sloping, some steep slopes in Utah; mostly small grain in North Central States, range and woodland in Western States (Some Chernozems)

M3 UDOLLS (temperate or warm, and moist) gently or moderately sloping; mostly corn, soybeans, and small grains (Some Brunizems)

M4 USTOLLS (intermittently dry for long periods during summer) gently to moderately sloping; mostly wheat and range in western part, wheat and corn or sorghum in eastern part, some irrigated crops (Chestnut soils and some Chernozems and Brown soils)

M4S USTOLLS moderately sloping to steep; mostly range or woodland

M5 XEROLLS (continuously dry in summer for long periods, moist in winter) gently to moderately sloping; mostly wheat, range, and irrigated crops (Some Brunizems, Chestnut, and Brown soils)

M5S XEROLLS moderately sloping to steep; mostly range

SPODOSOLS . . . Soils with accumulations of amorphous materials in subsurface horizons

S1 AQUODS (seasonally saturated with water) gently sloping; mostly range or woodland; where drained in Florida, citrus and special crops (Ground–Water Podzols)

S2 ORTHODS (with subsurface accumulations of iron, aluminum, and organic matter) gently or moderately sloping; woodland, pasture, small grains, special crops (Podzols, Brown Podzolic soils)

S2S ORTHODS steep; mostly woodland

ULTISOLS . . . Soils that are usually moist with horizon of clay accumulation and a low base supply

U1 AQUULTS (seasonally saturated with water) gently sloping; woodland and pasture if undrained, feed and truck crops if drained (Some Low–Humic Gley soils)

U2S HUMULTS (with high or very high organic–matter content) moderately sloping to steep; woodland and pasture if steep, sugar cane and pineapple in Hawaii, truck and seed crops in Western States (Some Reddish–Brown Lateritic soils)

U3 UDULTS (with low organic– matter content; temperate or warm, and moist) gently to moderately sloping; woodland, pasture, feed crops, tobacco, and cotton (Red–Yellow Podzolic soils, some Reddish–Brown Lateritic soils)

U3S UDULTS moderately sloping to steep; woodland, pasture

U4S XERULTS (with low to moderate organic–matter content, continuously dry for long periods in summer) range and woodland (Some Reddish–Brown Lateritic soils)

VERTISOLS . . . Soils with high content of swelling clays and wide deep cracks at some season

V1 UDERTS (cracks open for only short periods, less than 3 months in a year) gently sloping; cotton, corn, pasture, and some rice (Some Grumusols)

V2 USTERTS (cracks open and close twice a year and remain open more than 3 months); general crops, range, and some irrigated crops (Some Grumusols)

 AREAS with little soil . . .

X1 Salt flats

X2 Rockland, ice fields

NOMENCLATURE

The nomenclature is systematic. Names of soil orders end in sol (L. solum, soil), e. g., ALFISOL, AQUALF, and contain a formative element used as the final syllable in names of taxa in suborders, great groups, and sub-groups.

Names of suborders consist of two syllables, e. g., AQUALF. Formative elements in the legend for this map and their connotations are as follows:

and — Modified from Ando soils; soils from vitreous parent materials

aqu — L. aqua, water; soils that are wet for long periods

arg — Modified from L. argilla, clay; soils with a horizon of clay accumulation

bor — Gr. boreas, northern; cool

fibr — L. fibra, fiber; least decomposed

hum — L. humus, earth; presence of organic matter

ochr — Gr. base of ochros, pale; soils with little organic matter

orth — Gr. orthos, true; the common or typical

psamm — Gr. psammos, sand; sandy soils

sapr — Gr. sapros, rotten; most decomposed

ud — L. udus, humid; of humid climates

umbr — L. umbra, shade; dark colors reflecting much organic matter

ust — L. ustus, burnt; of dry climates with summer rains

xer — Gr. xeros, dry; of dry climates with winter rains

APPENDIX II
Topographic Map Reading

Topographic maps

SEVERAL methods have been used to show accurately the configuration of the land surface on topographic maps: *plastic shading, altitude tints, hachures,* and *contours* (Figure II.1). The first three processes give a strong visual effect of three dimensions so that even untutored persons can grasp the essential character of the landscape features without preliminary explanation. But, as compensation for ease of understanding, such methods of showing relief are inadequate because they do not tell the reader the elevation above sea level of all points on the map, or how steep the slopes are. The method of topographic contours, however, gives this information and makes the most useful type of topographic map.

Plastic shading

Maps using plastic shading to show relief look very much like photographs taken down upon a plaster relief model of the land surface illuminated from directly above or from an oblique angle (Figure II.1). They may also be likened to air photographs taken when the sun's rays are striking the earth at a fairly low angle (Figure II.2). The effect of relief is produced by gray or brown tones applied according to one of two principles. In the *oblique illumination* method, light rays are imagined as coming from a point in the northwestern sky somewhere intermediate between the horizon and zenith. Thus all slopes facing southwest receive the heaviest shades and are darkest where the slopes are steepest (Figure II.3).

Altitude tints

In its simplest form, altitude tinting consists of assigning a certain color, or a certain depth of tone of a color, to all areas on the map lying within a specified range of elevation. Schoolroom wall maps commonly show low areas by deep green, intermediate ranges of elevations by successive shades of buff and light brown, and high mountain elevations in darker shades of brown or red. The method is effective for small-scale maps that are viewed at some distance.

Hachures

Hachures are tiny, short lines arranged in such a way as to look as if someone had placed thousands of match sticks side by side into roughly parallel rows. Each hachure line lies along the direction of the steepest slope and represents the direction that would be taken by water flowing down the surface.

A precise system of hachures, adapted to representation of detailed topographic features on accurate, large-scale maps, was invented by a Saxon army officer, Major J. G. Lehmann (1765–1811), and was widely used on military maps of European countries throughout the nineteenth century. In the Lehmann system, steepness of slope is indicated by thickness of the hachure line.

Because hachures do not tell the elevation of surface points, it is necessary to print numerous elevation figures on hilltops, road intersections, towns, and other strategic locations. These numbers are known as *spot heights.* Without them a hachure map would be of little practical value.

Contour lines

A *contour line* may be defined as an imaginary line on the ground, every point of which is at the same altitude, or elevation, above sea level. Contour lines on a map are simply the graphic representations of ground contours, drawn for each of a series of specified elevations such as 10, 20, 30, 40, or 50 feet or meters above sea level or any other chosen base, known as a *datum plane.* The resulting line pattern not only gives a visual impression of topography to the experienced student of maps but also supplies accurate information about true elevations and slopes.

In order to clarify the contour principle, various commonplace things can be used for illustration. Imagine, for example, a small island, as shown in Figure II.5. the shoreline is a natural contour line because it is a line connecting all points having zero elevation. Suppose that sea level could be made to rise exactly ten feet (or that the island could be made to sink exactly ten feet); the water would come to rest along the line labeled "10."

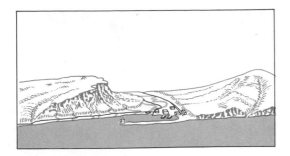

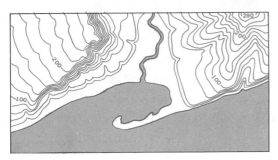

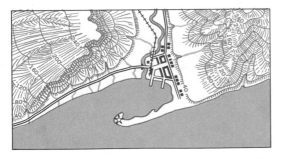

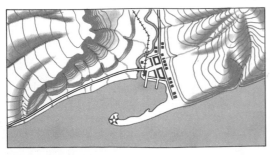

Figure AII.1 Various ways in which relief can be shown are, from top to bottom: (1) perspective diagram or terrain sketch, (2) hachures, (3) contours, (4) hachures and contours combined, and (5) plastic shading and contours combined. (Drawn by E. Raisz.)

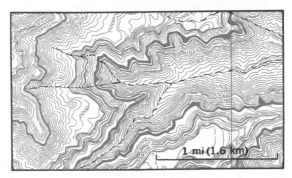

Figure AII.2 A vertical air photograph (above) is a kind of topographic map, showing all relief details but lacking elevation information. The contour topographic map (below) covers the same area and shows a small side canyon of the Grand Canyon. North is toward the bottom of these maps to give the proper effect of relief in the photograph. (U.S. Forest Service and U.S. Geological Survey.)

This would be the 10-foot contour because it connects all points on the island that are exactly ten feet higher than the original shoreline. By successive rises in water level, each exactly ten feet more than the last, the positions of the remaining contours would be fixed.

Contour interval and slope

Contour interval is the vertical distance separating successive contours. The interval remains constant over the entire map, except in special cases where two or more intervals are used on the map sheet.

Because the vertical contour interval is fixed, horizontal spacing of contours on a given map varies with changes in land slope. The general rule is: close crowding of contour lines represents a steep slope; wide spacing represents a gentle slope. Figure II.6 shows a small island, one side of which is a steep, clifflike slope. From the summit point *B* to the cliff base at *A* the contours are crossed within a short horizontal distance and therefore appear closely spaced on the map. From *B* to the shore at *C* the same total vertical descent is made, but because the slope is gentle, the horizontal

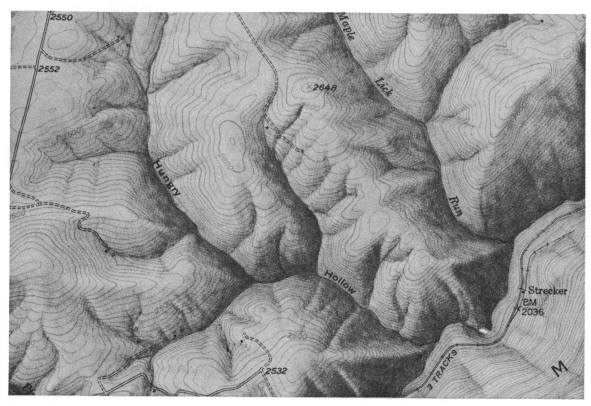

Figure AII.3 Plastic shading combined with contours greatly enhances the visual effect of relief. (Portion of U.S. Geological Survey, Kitzmiller, Md.–W. Va., topographic quadrangle. scale 1:24,000.)

distance is much greater. Hence, the contours between *B* and *C* are widely spaced on the map.

Selection of contour interval depends both on relief of the land and on scale of the map. Topographic maps showing regions of strong relief require a large interval, such as 50, 100, or 200 ft (10, 25, or 50 m); regions of moderate relief, intervals such as 10 or 25 ft (2 or 5 m).

Because by far the greatest parts of the earth's land surfaces are sculptured by streams flowing in valleys, special note should be made of how contours behave when crossing a stream valley. Figure II.7 is a small contour sketch map illustrating some stream valleys. Notice that each contour is bent into a V whose apex lies on the stream and points in an upstream direction.

Determining elevations by means of contours

Figure 11.7 can be used to illustrate the determination of elevations. Point *B* is easy to determine because it lies exactly on the 1300-foot contour. Point *C* requires interpolation. Because it lies midway between the 1100- and 1200-foot lines, its elevation is likewise the midvalue of the vertical interval, or 1150 feet. Point *D* lies about

one-fifth of the distance from the 1000 to the 1100-foot contours. Inasmuch as one-fifth of the contour interval is 20 feet, point *D* has an approximate elevation of 1020 feet. For the last two points only a guess has been made as to the true elevation, but if the ground is not too irregular the error will probably be small. Determination of the summit elevation, point *A*, involves still more uncertainty. It is certain that the summit point is more than 1700 feet and less than 1800 because the 1700-foot contour is the highest one shown. Because a sizable area is included within the 1700-foot contour, it may be supposed that the summit rises appreciably higher than 1700. A guess would place the true elevation at about 1750 feet.

On many topographic maps the elevation of hilltops, road intersections, bridges, lakes, etc., is printed on the map to the nearest foot or meter. These spot heights do away with the need for estimating elevations at key points.

Federal agencies, such as the U.S. Coast and Geodetic Survey and the U.S. Geological Survey, in mapping an area carefully determine elevation and position of convenient reference points. These are known as *bench marks*. On the map they are

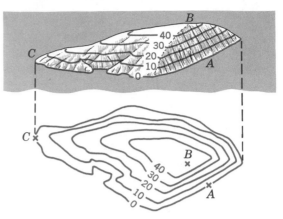

Figure AII.6   On the steep side of this island the contours appear more closely spaced.

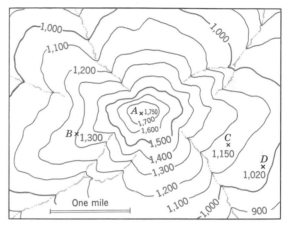

Figure AII.7   Stream valleys produce V-shaped indentations of the contours.

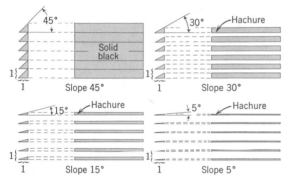

Figure AII.4   A portion of the Metz sheet is shown above to correct scale. This map is one of the French 1:80,000 topographic series using black hachures and spot heights. The Lehmann system of hachuring, shown below, varies the thickness of hachure line according to ground slope.

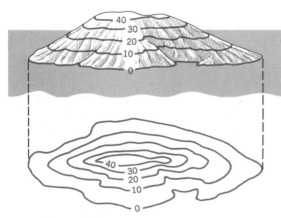

Figure AII.5   Contours on a small island.

designated by the letters B.M., together with the elevation stated to the nearest foot.

Depression contours

A special type of contour is used where the land surface has basinlike hollows, or *closed depressions*, which would make small lakes if they could be filled with water. This is the *hachured contour*, or *depression contour*. Figure II.8 is a sketch of a depression in a gently sloping plain. Below it is the corresponding contour map. Hachured contours have the same elevations and contour intervals as regular contours on the same map.

Topographic profiles

In order to get a better idea of the nature of the relief, *topographic profiles* are sometimes drawn. These are lines that show the rise and fall of the land surface along a selected line crossing the map. Figure II.9 illustrates the construction of a profile. A line, *XY*, is lightly drawn across

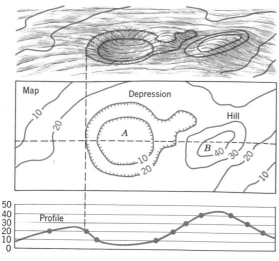

Figure AII.8 Contours which close in a circular manner show either closed depressions or hills.

the map at the desired location. A piece of paper, ruled with horizontal lines, is placed so that its top edge lies along the line *XY*.

Each horizontal line represents a contour level and is so numbered along the left-hand side. Starting at the left, a perpendicular is dropped from the point *a* where the map contour intersects the profile line, *XY*, down to the corresponding horizontal level. A point *a′* is marked on the horizontal line. Next, the procedure is repeated for the 400-foot contour at point *b*, and so on, until all points have been plotted. A smooth line is then drawn through all the points, completing the profile. Where contours are widely spaced, some judgment is required in drawing of the profile.

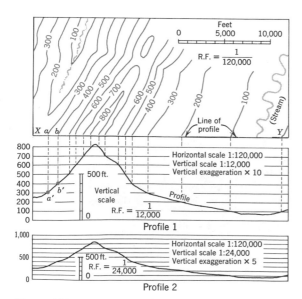

Figure AII.9 A topographic profile can be constructed from a contour map along any desired line.

Figure II.11 shows two profiles, both of which are drawn along the same line *XY*. The difference is one of degree of exaggeration of the vertical scale. In this illustration, horizontal map scale is 1 inch to 10,000 feet, or 1:120,000, whereas the vertical scale of the upper profile is 1 inch to 1000 feet, or 1:12,000. The vertical scale is thus ten times as large as the horizontal map scale, and the profile is said to have a *vertical exaggeration* of ten times. In the lower profile the horizontal scale remains the same, of course, but the vertical scale is 1 inch to 2000 feet, or 1:24,000. The vertical exaggeration is therefore five times. Some degree of vertical exaggeration is usually needed to bring out the nature of the topography. A *natural scale* profile, one in which the vertical and horizontal scales are the same, would give a profile with such tiny fluctuations as to be not only difficult to read but also difficult to draw or reproduce.

Map scale

Distance between-points shown on a map depends on the scale of the map—the ratio between map distances and the actual ground distances which the map represents (Chapter 1). Given a fractional scale (representative fraction, or R.F.) on which to draw his map, the cartographer must convert this fraction to units of measurement, for he must use a ruler scaled in inches or centimeters to measure lengths of line on the map to represent ground distances in miles or kilometers.

For example, if the map scale is given as 1/63,360, or 1:63,360, as explained in Chapter 1, this fractional scale may be interpreted as "one inch on the map represents one mile on the ground." A map scale of 1:100,000 could be read as "one centimeter represents one kilometer."

Most maps of small areas carry a fractional scale printed on the map margin. Conversion to equivalent ratios of inches to miles or centimeters to kilometers is left to the reader to compute. For practical map use, however, a *graphic scale* is printed on the map margin. This is a length of line divided off into numbered segments (Figure II.10). The units are in conventional terms of measurement, such as feet, yards, and miles, or meters and kilometers. To use the graphic scale, a piece of paper with a straight edge is held along the line to be measured on the map and the distance marked on the edge of the paper. The paper is then placed along the graphic scale and the length of the line read directly.

Large-scale and small-scale maps

The relative size of two different scales is determined according to which fraction is the larger and which the smaller. For example, a scale of 1:10,000 is twice as large as a scale of 1:20,000.

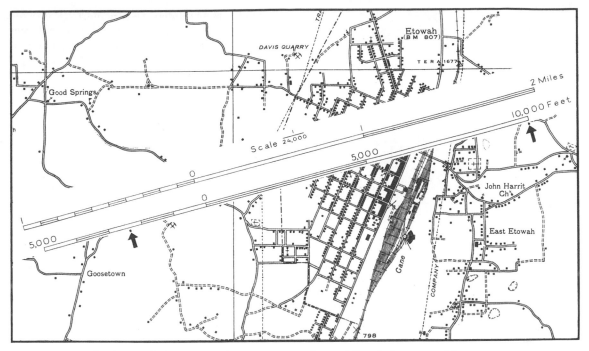

Figure AII.10   The distance between two points on a map can be read directly from a graphic scale. (Portion of U.S. Geological Survey map.)

Maps with scales ranging from 1:600,000 down to 1:100,000,000 or smaller are known as *small-scale maps*. Those of scale 1:600,000 to 1:75,000 are *medium-scale* maps; those of scale greater than 1:75,000 are *large-scale* maps.

For representing details of the earth's surface configuration, large-scale maps are needed and the area of land surface shown by an individual map sheet must necessarily be small. A topographic sheet 10 by 20 inches, on a scale of 1,63,360 (1 inch to 1 mile) would, of course, include an area 10 by 20 miles, or 200 square miles. Of the common sets of topographic maps published by national governments for general distribution, most fall within the scale range of 1:20,000 to 1:250,000.

Relation between scales and areas

Assuming that two maps, each on a different scale, have the same dimensions, what is the relation between the ground areas shown by each? In Figure II.11 are shown three maps, each having the same dimensions, but representing scales of 1:20,000, 1:10,000 and 1:5,000, respectively, from left to right. Although map B is on twice the scale of map A, it shows a ground area only one-fourth as great. Map C is on four times as large a scale as map A, yet it covers a ground area only one-sixteenth as much. From this example can be deduced the following rule: the ground area that is represented by a map of given outside dimen-

sions varies inversely with the square of the change in scale. Thus, if the scale is reduced to one-third its original value, the area that can be shown on a map of fixed dimensions increases to nine times the original value.

Map orientation and declination of the compass

It is a well-known convention to draw large-scale topographic maps in such a way that north is in a direction toward the top of the map, south to the bottom of the map, and east and west to the right- and left-hand sides, respectively.

The geographic north pole, to which all meridians converge, forms a reference point for the *true north*, or *geographic north*, direction. There is, however, another place, the *magnetic north pole*, to which magnetic compasses point (Figure II.12). The magnetic north pole is located in the

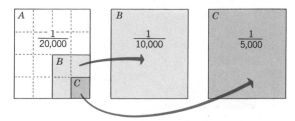

Figure AII.11   Area shown on a map decreases as scale increases.

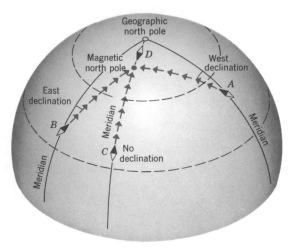

Figure AII.12 Whether declination is east or west depends on the observer's global position with respect to the magnetic and geographic north poles. (From A. N. Strahler, *The Earth Sciences*, Harper and Row, Inc., N.Y.)

Northwest Territories of Canada on Prince of Wales Island about lat. 73° N, long. 100° W. Most published large-scale maps have printed on the margin two arrows stemming from a common

point. One arrow designates true north, the other one *magnetic north*. The angular distance between the two directions is known as the *magnetic declination*.

Magnetic declination varies greatly in different parts of the world, depending principally on one's position relative to the geographic and magnetic poles. Lines on a map drawn through all places having the same compass declination are known as *isogonic lines* (Figure II.13). The line of zero declination (*agonic line*) runs through the eastern United States. Anywhere along this line the compass points to true geographic north, and no adjustment is required.

Magnetic declination changes appreciably with the passage of years. The amount of annual change in declination is usually stated on the margin of a good map.

Bearings and azimuths

In using maps it is frequently necessary to state the direction followed by a road or stream, or to describe the direction that can be taken to locate a particular object with respect to some known reference point. In air and marine navigation the direction from one point to another must be stated.

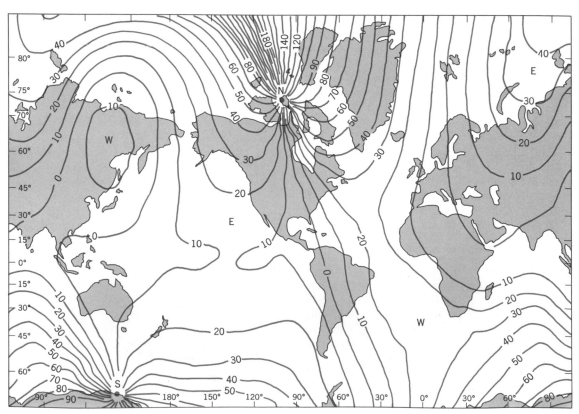

Figure AII.13 On this world isogonic map, declination is given in intervals of ten degrees. (Based on data of U.S. Navy Oceanographic Office. From A. N. Strahler, *The Earth Sciences*, Harper and Row, Inc., N.Y.)

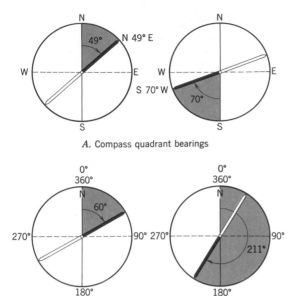

A. Compass quadrant bearings

B. Azimuths

Figure AII.14 **Directions expressed as bearings or azimuths.**

For these purposes the angle that the given line makes with a north-south line is measured. The unit of angular measurement most common in map work is the *degree*, 360 of which comprise a complete circle, but other systems of angular measurement, such as the *mil* (of which there are 6400 in a complete circle), are sometimes preferable for special applications.

Two systems of stating directions with respect to north can be used (Figure II.14). (*a*) *Compass quadrant bearings* are angles measured eastward or westward of either north or south, whichever happens to be the closer. Examples are shown in Figure II.14*A*. The direction from a given point to some object on the map is thus written as "N 49° E" or "S 70° W." All bearings range between 0° and 90°. Compass bearings may be magnetic bearings, related to magnetic north, or true bearings, related to geographic north. Unless specifically stated otherwise, a bearing should be assumed to be a true bearing.

(*b*) *Azimuths* are used by military services and in air and marine navigation generally. As shown in Figure II.14*B*, all azimuths are read in a clockwise direction from north and therefore range between 0° and 360°. Azimuths are usually measured from either magnetic north or true north, referred to as *magnetic azimuth* and *true azimuth*, respectively.

Topographic quadrangles

Any system whereby points on the earth's surface are located with reference to a previously determined set of intersecting lines can be called a *coordinate system*. The system of parallels and meridians used throughout the world, and the designation of any point on the earth's surface in terms of latitude and longitude are treated in Chapter 1. The system of *geographic coordinates* constitutes the commonest of the coordinate systems in general use. Two other systems are the *military grid* and the *township grid of U.S. Land Office Survey*.

Most published sets of large-scale topographic maps use the geographic grid to determine the position and size of individual map sheets in a series. A single map sheet, or *quadrangle*, is bounded on the right- and left-hand margins by meridians, and on the top and bottom by parallels, which are a specified number of minutes or degrees apart. Thus, individual sheets may be fitted together to form unified groups.

Seven standard scales comprise the National Topographic Map Series:

Series	R.F.	Unit Equivalents
7.5 minute	1:24.000	1 inch to 2000 feet
7.5 minute	1:31,680	1 inch to one-half mile
15 minute	1:62,500	1 inch to about 1 mile
Alaska	1:63,360	1 inch to 1 mile
30 minute	1:125,000	1 inch to about 2 miles
1:250,000	1:250,000	1 inch to about 4 miles
1:1,000,000	1:1,000,000	1 inch to about 16 miles

Figures II.15 and II.16 compare the coverages and sizes of standard quadrangles on several of these series.

The United States Land Office Survey

Topographic maps of the central and western United States show the civil divisions of land according to the United States Land Office Survey.

In 1785, Congress authorized a survey of the territory lying north and west of the Ohio River. To avoid the irregular and unsystematic type of land subdivision that had grown up in seaboard states during colonial times, Congress specified that the new lands should be divided into six-mile squares, now called *congressional townships,* and that the grid of townships should be based upon a carefully surveyed east-west base line, designated the "geographer's line." Meridians and parallels laid off at six-mile intervals from the base line were to form the boundaries of the townships. This general plan, believed to have been proposed by Thomas Jefferson, was subsequently carried out to cover the balance of the central and western states.

The *principal meridians* and *base lines*, from which rows of townships were laid off, are shown in Figure II.17. Principal meridians run north or

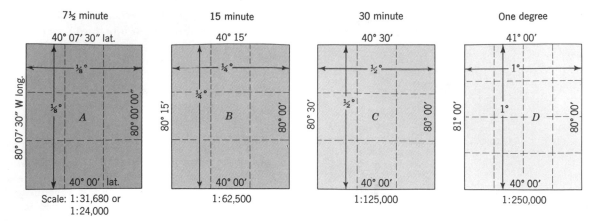

Figure AII.15 Large-scale maps of the U.S. Geological Survey are bounded by parallels and meridians to form quadrangles. The four quadrangles shown here represent the scales and areas commonly used in the United States, exclusive of Alaska.

south, or both, from selected points whose latitude and longitude were calculated by astronomical means. Some 32 principal meridians have been surveyed. Westward from the Ohio-Pennsylvania boundary these are numbered from 1 through 6, after which they are designated by names.

Through the initial point selected for starting the principal meridian, an east-west base line was run, corresponding to a parallel of latitude through that point. North and south from the base line, horizontal tiers of townships were laid off and numbered accordingly. Vertical rows of townships, called *ranges*, were laid off to the right and left of the principal meridians, and were numbered accordingly (Figure II.18).

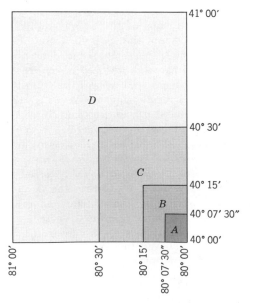

Figure AII.16 If the four quadrangles of Figure AII.15 were reduced to the same scale, their areas would compare as shown here.

The area governed by one principal meridian and its base line is restricted to a particular section of country, usually about as large as one or two states. Where two systems of townships meet, they do not correspond, because each system was built up independently of the others.

Because the range lines, on eastern and western boundaries of townships, are meridians converging slightly as they are extended northward, the width of townships is progressively diminished in a northward direction. In order to avoid a considerable reduction in township widths in the more northerly tiers, new base-lines, known as *standard parallels*, are surveyed for every four tiers of townships. They are designated 1st, 2nd, 3rd Standard Parallel N, etc. (Figure II.18). The ranges will be found to offset at the standard parallels, and in consequence, roads which follow range lines make an offset or jog when crossing standard parallels.

Subdivisions of the township are square-mile *sections*, of which there are 36 to the township. These are usually numbered as illustrated in Figure II.19. Each section may be subdivided into halves, quarters, and half-quarters, or even smaller units. These divisions, together with the number of acres contained in each, are illustrated in Figure II.20.

Topographic map symbols

Large-scale topographic maps of the U.S. Geological Survey use a standard set of symbols to show many kinds of features that cannot be represented to true scale. Reproduced on the front end paper of this book is the set of symbols used by the U.S. Geological Survey on its most recent series of large-scale topographic maps, together with a representative example of a recent map on the scale of 1:24,000.

In general, it is conventional to show relief

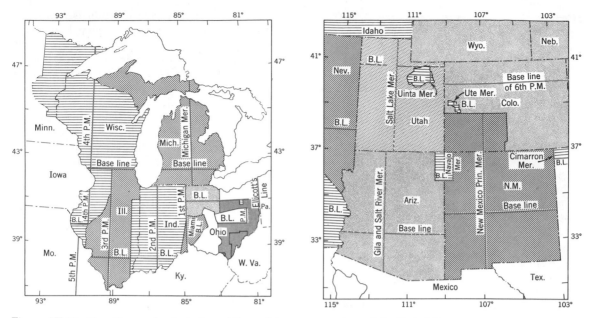

Figure AII.17 Base lines and principal meridians of these two portions of the United States are representative of the system used by the U.S. Land Office. (After U.S. Dept. Interior, General Land Office Map of the United States, 1937.)

features in brown, hydrographic (water) features in blue, vegetation in green, and cultural (man-made) features in black or red.

Geologic maps and structure sections

The *areal geologic map* shows by means of colors or patterns the surface distribution of each rock unit with special emphasis upon the lines of contact of rocks unlike in variety or age. Faults are shown as lines. Strikes and dips of strata are added by special symbols.

Figure II.21 is a simple geologic map of the same area shown by a perspective diagram in

	R. 6 W.				
6	5	4	3	2	1
7	8	9	10	11	12
18	17	16	15	14	13
19	20	21	22	23	24
30	29	28	27	26	25
31	32	33	34	35	36

T. 2 S.

Figure AII.19 A township is divided into 36 sections, each one a square mile.

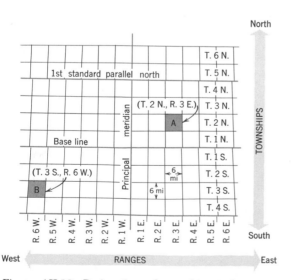

Figure AII.18 Designation of townships and ranges.

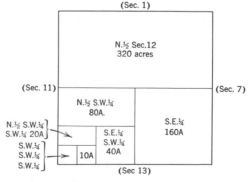

Figure AII.20 A section may be subdivided into many units. (After Willis E. Johnson.)

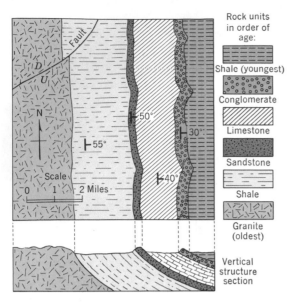

Rock units
in order of
age:

Shale (youngest)

Conglomerate

Limestone

Sandstone

Shale

Granite
(oldest)

Vertical
structure
section

Figure AII.21 A geologic map shows the surface distribution of rocks and structures. The structure section shows rocks at depth.

Figure 17.11. If map reproduction is limited to black and white, patterns are applied to differentiate the rock units. Shorthand letter combinations may be added as a code to set apart formations of different ages. Small T-shaped symbols, seen on the map, tell strike-and-dip. The long bar gives direction of strike; the short bar which abuts it at right angles show direction of dip. Amount of dip in degrees is given by a figure beside the symbol. A small fault, cutting across the northwest corner of the map, is shown by a solid line. The letters D and U indicate which side slipped downward, which upward.

To show the internal geologic structure of an area the geologist resorts to the *structure section,* an imaginary vertical slice through the rocks. A structure section is illustrated in the lower part of Figure II.21. The upper line of the section is a topographic profile.

Index

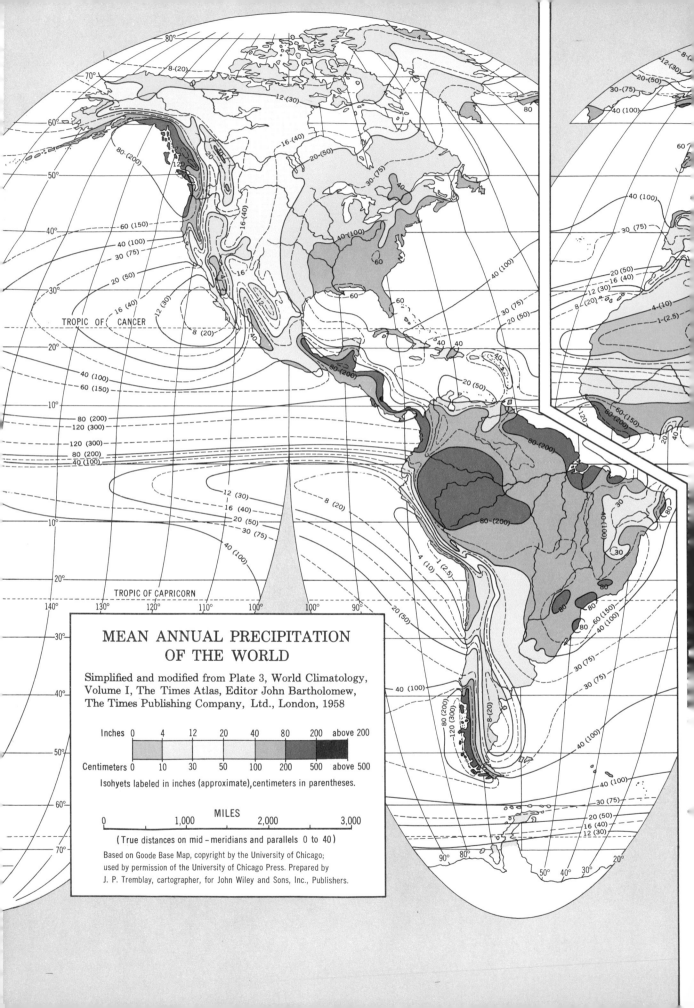

MEAN ANNUAL PRECIPITATION
OF THE WORLD

Simplified and modified from Plate 3, World Climatology,
Volume I, The Times Atlas, Editor John Bartholomew,
The Times Publishing Company, Ltd., London, 1958

Inches 0   4   12   20   40   80   200   above 200

Centimeters 0   10   30   50   100   200   500   above 500

Isohyets labeled in inches (approximate), centimeters in parentheses.

MILES

0   1,000   2,000   3,000

(True distances on mid–meridians and parallels 0 to 40)

Based on Goode Base Map, copyright by the University of Chicago;
used by permission of the University of Chicago Press. Prepared by
J. P. Tremblay, cartographer, for John Wiley and Sons, Inc., Publishers.

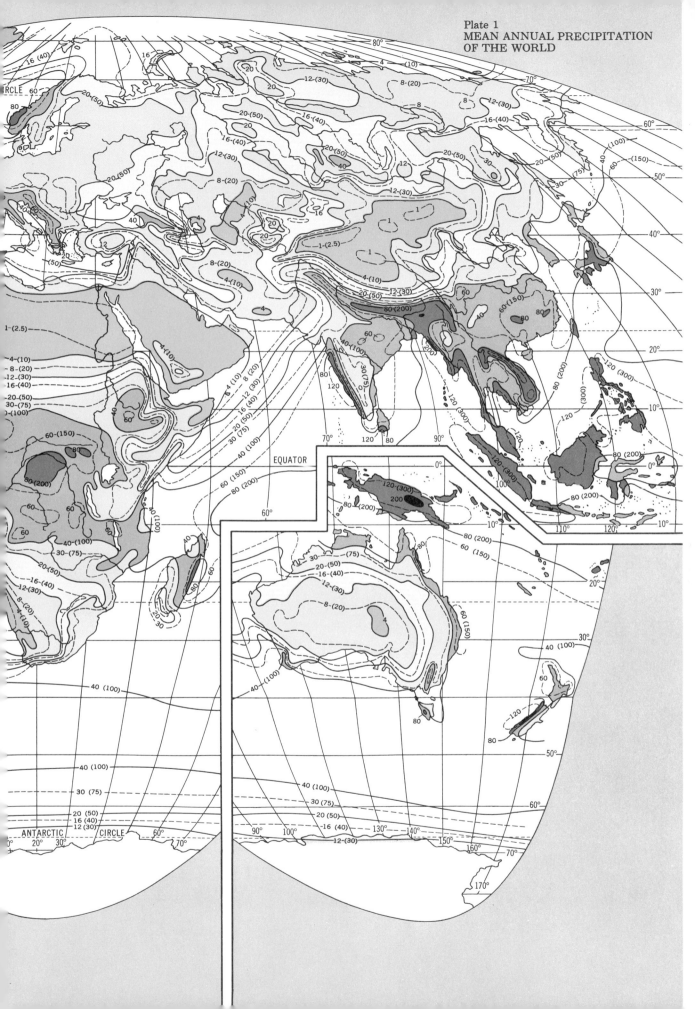

Plate 1
MEAN ANNUAL PRECIPITATION
OF THE WORLD

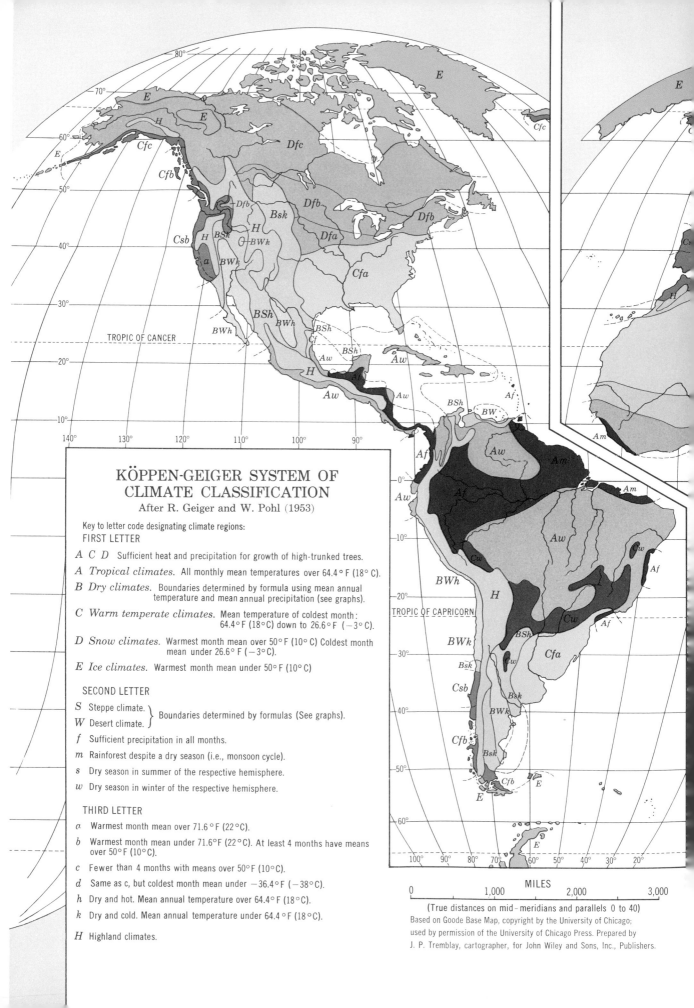

KÖPPEN-GEIGER SYSTEM OF CLIMATE CLASSIFICATION

After R. Geiger and W. Pohl (1953)

Key to letter code designating climate regions:

FIRST LETTER

A C D Sufficient heat and precipitation for growth of high-trunked trees.

A *Tropical climates.* All monthly mean temperatures over 64.4°F (18°C).

B *Dry climates.* Boundaries determined by formula using mean annual temperature and mean annual precipitation (see graphs).

C *Warm temperate climates.* Mean temperature of coldest month: 64.4°F (18°C) down to 26.6°F (−3°C).

D *Snow climates.* Warmest month mean over 50°F (10°C) Coldest month mean under 26.6°F (−3°C).

E *Ice climates.* Warmest month mean under 50°F (10°C)

SECOND LETTER

S Steppe climate. ⎫
W Desert climate. ⎬ Boundaries determined by formulas (See graphs).

f Sufficient precipitation in all months.

m Rainforest despite a dry season (i.e., monsoon cycle).

s Dry season in summer of the respective hemisphere.

w Dry season in winter of the respective hemisphere.

THIRD LETTER

a Warmest month mean over 71.6°F (22°C).

b Warmest month mean under 71.6°F (22°C). At least 4 months have means over 50°F (10°C).

c Fewer than 4 months with means over 50°F (10°C).

d Same as c, but coldest month mean under −36.4°F (−38°C).

h Dry and hot. Mean annual temperature over 64.4°F (18°C).

k Dry and cold. Mean annual temperature under 64.4°F (18°C).

H Highland climates.

MILES

0   1,000   2,000   3,000

(True distances on mid-meridians and parallels 0 to 40)

Based on Goode Base Map, copyright by the University of Chicago; used by permission of the University of Chicago Press. Prepared by J. P. Tremblay, cartographer, for John Wiley and Sons, Inc., Publishers.

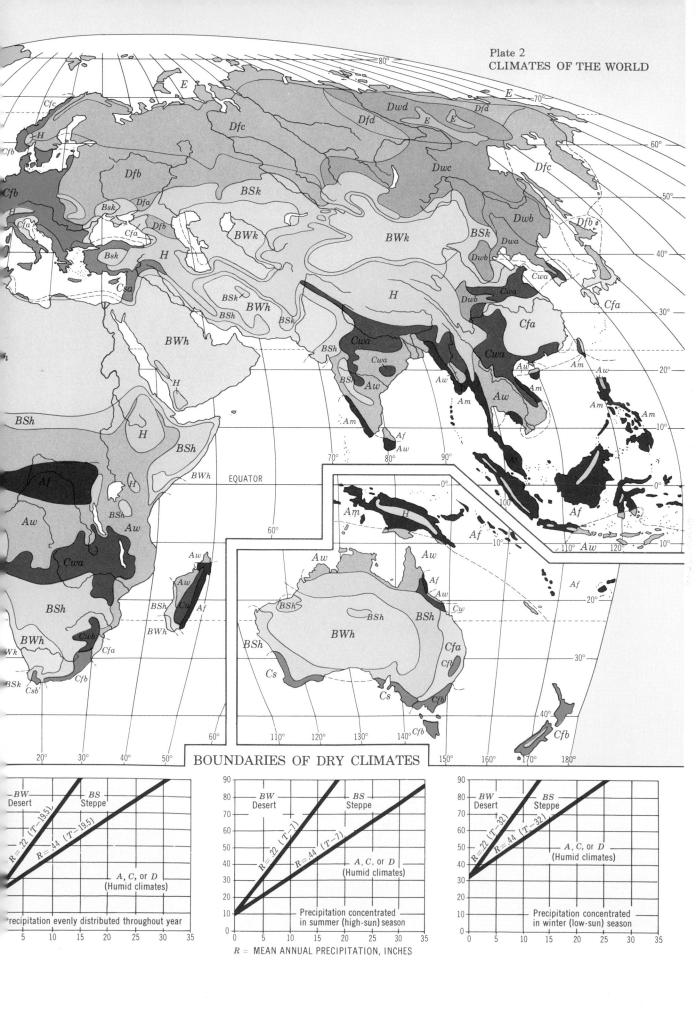

Plate 2
CLIMATES OF THE WORLD

BOUNDARIES OF DRY CLIMATES

$R = 22 (T-19.5)$ $R = 44 (T-19.5)$

BW Desert   BS Steppe

A, C, or D (Humid climates)

Precipitation evenly distributed throughout year

$R = 22 (T-7)$ $R = 44 (T-7)$

BW Desert   BS Steppe

A, C, or D (Humid climates)

Precipitation concentrated in summer (high-sun) season

$R = 22 (T-32)$ $R = 44 (T-32)$

BW Desert   BS Steppe

A, C, or D (Humid climates)

Precipitation concentrated in winter (low-sun) season

R = MEAN ANNUAL PRECIPITATION, INCHES

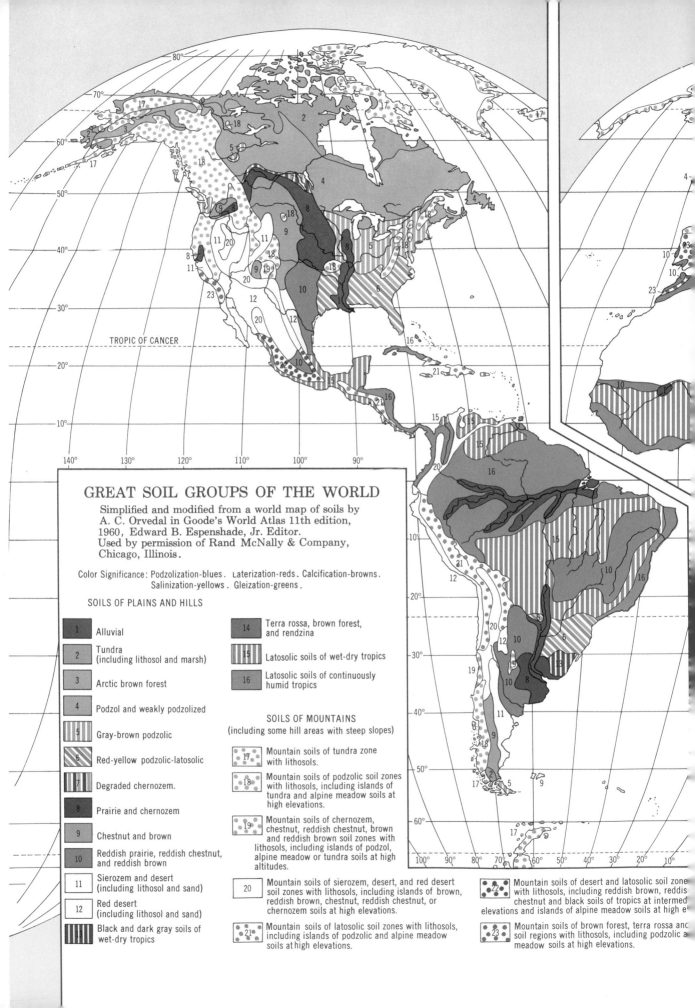

GREAT SOIL GROUPS OF THE WORLD

Simplified and modified from a world map of soils by
A. C. Orvedal in Goode's World Atlas 11th edition,
1960, Edward B. Espenshade, Jr. Editor.
Used by permission of Rand McNally & Company,
Chicago, Illinois.

Color Significance: Podzolization-blues. Laterization-reds. Calcification-browns.
Salinization-yellows. Gleization-greens.

SOILS OF PLAINS AND HILLS

1   Alluvial

2   Tundra
(including lithosol and marsh)

3   Arctic brown forest

4   Podzol and weakly podzolized

5   Gray-brown podzolic

6   Red-yellow podzolic-latosolic

7   Degraded chernozem.

8   Prairie and chernozem

9   Chestnut and brown

10   Reddish prairie, reddish chestnut,
and reddish brown

11   Sierozem and desert
(including lithosol and sand)

12   Red desert
(including lithosol and sand)

Black and dark gray soils of
wet-dry tropics

14   Terra rossa, brown forest,
and rendzina

15   Latosolic soils of wet-dry tropics

16   Latosolic soils of continuously
humid tropics

SOILS OF MOUNTAINS
(including some hill areas with steep slopes)

17   Mountain soils of tundra zone
with lithosols.

18   Mountain soils of podzolic soil zones
with lithosols, including islands of
tundra and alpine meadow soils at
high elevations.

19   Mountain soils of chernozem,
chestnut, reddish chestnut, brown
and reddish brown soil zones with
lithosols, including islands of podzol,
alpine meadow or tundra soils at high
altitudes.

20   Mountain soils of sierozem, desert, and red desert
soil zones with lithosols, including islands of brown,
reddish brown, chestnut, reddish chestnut, or
chernozem soils at high elevations.

21   Mountain soils of latosolic soil zones with lithosols,
including islands of podzolic and alpine meadow
soils at high elevations.

22   Mountain soils of desert and latosolic soil zone
with lithosols, including reddish brown, reddis
chestnut and black soils of tropics at intermed
elevations and islands of alpine meadow soils at high e

23   Mountain soils of brown forest, terra rossa and
soil regions with lithosols, including podzolic a
meadow soils at high elevations.

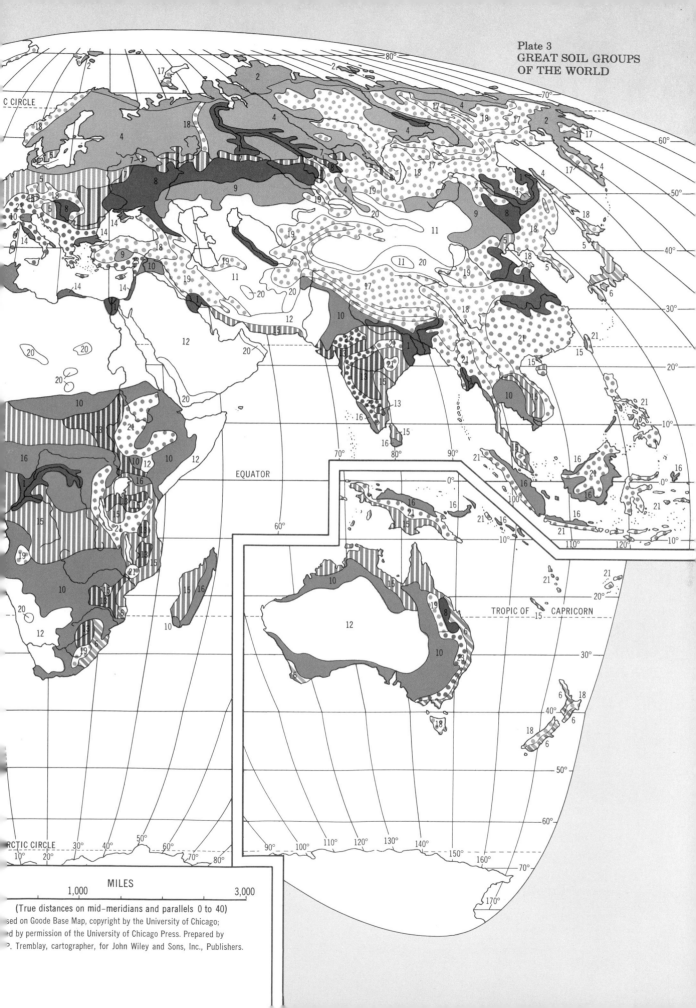

Plate 3
GREAT SOIL GROUPS
OF THE WORLD

EQUATOR

TROPIC OF CAPRICORN

MILES

1,000   3,000

(True distances on mid-meridians and parallels 0 to 40)
sed on Goode Base Map, copyright by the University of Chicago;
d by permission of the University of Chicago Press. Prepared by
. Tremblay, cartographer, for John Wiley and Sons, Inc., Publishers.

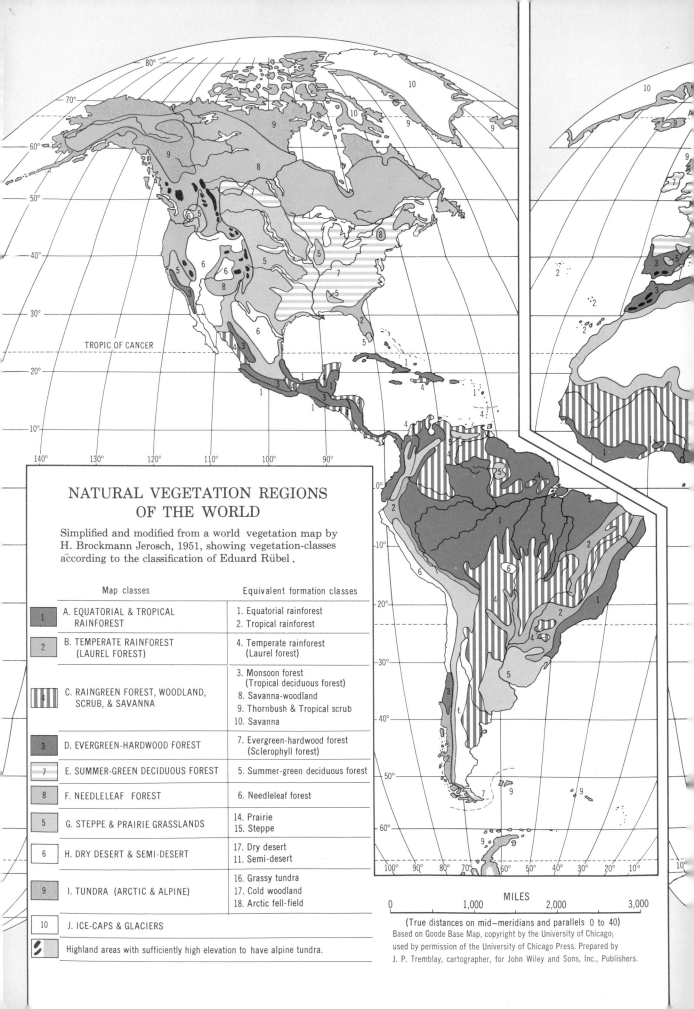

NATURAL VEGETATION REGIONS
OF THE WORLD

Simplified and modified from a world vegetation map by
H. Brockmann Jerosch, 1951, showing vegetation-classes
according to the classification of Eduard Rübel.

	Map classes	Equivalent formation classes
1	A. EQUATORIAL & TROPICAL RAINFOREST	1. Equatorial rainforest 2. Tropical rainforest
2	B. TEMPERATE RAINFOREST (LAUREL FOREST)	4. Temperate rainforest (Laurel forest)
4	C. RAINGREEN FOREST, WOODLAND, SCRUB, & SAVANNA	3. Monsoon forest (Tropical deciduous forest) 8. Savanna-woodland 9. Thornbush & Tropical scrub 10. Savanna
3	D. EVERGREEN-HARDWOOD FOREST	7. Evergreen-hardwood forest (Sclerophyll forest)
7	E. SUMMER-GREEN DECIDUOUS FOREST	5. Summer-green deciduous forest
8	F. NEEDLELEAF FOREST	6. Needleleaf forest
5	G. STEPPE & PRAIRIE GRASSLANDS	14. Prairie 15. Steppe
6	H. DRY DESERT & SEMI-DESERT	17. Dry desert 11. Semi-desert
9	I. TUNDRA (ARCTIC & ALPINE)	16. Grassy tundra 17. Cold woodland 18. Arctic fell-field
10	J. ICE-CAPS & GLACIERS	
	Highland areas with sufficiently high elevation to have alpine tundra.	

TROPIC OF CANCER

MILES

0   1,000   2,000   3,000

(True distances on mid—meridians and parallels 0 to 40)
Based on Goode Base Map, copyright by the University of Chicago;
used by permission of the University of Chicago Press. Prepared by
J. P. Tremblay, cartographer, for John Wiley and Sons, Inc., Publishers.

Plate 4
NATURAL VEGETATION REGIONS
OF THE WORLD

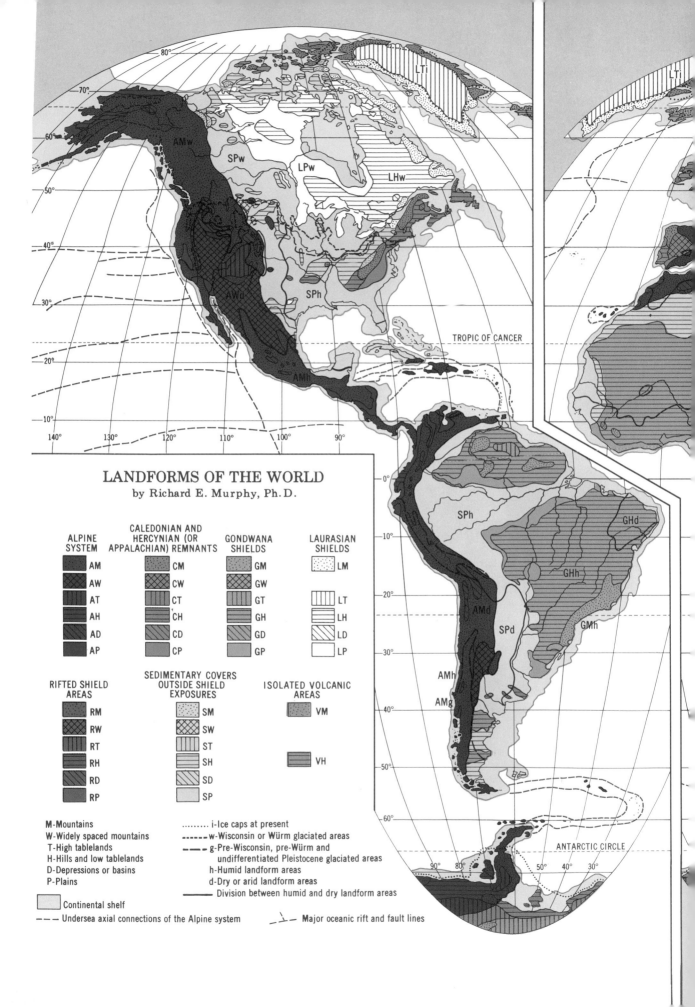

LANDFORMS OF THE WORLD
by Richard E. Murphy, Ph.D.

ALPINE SYSTEM	CALEDONIAN AND HERCYNIAN (OR APPALACHIAN) REMNANTS	GONDWANA SHIELDS	LAURASIAN SHIELDS
AM	CM	GM	LM
AW	CW	GW	GW
AT	CT	GT	LT
AH	CH	GH	LH
AD	CD	GD	LD
AP	CP	GP	LP

RIFTED SHIELD AREAS	SEDIMENTARY COVERS OUTSIDE SHIELD EXPOSURES	ISOLATED VOLCANIC AREAS
RM	SM	VM
RW	SW	
RT	ST	
RH	SH	VH
RD	SD	
RP	SP	

M-Mountains
W-Widely spaced mountains
T-High tablelands
H-Hills and low tablelands
D-Depressions or basins
P-Plains

.... i-Ice caps at present
------ w-Wisconsin or Würm glaciated areas
— — g-Pre-Wisconsin, pre-Würm and undifferentiated Pleistocene glaciated areas
h-Humid landform areas
d-Dry or arid landform areas
——— Division between humid and dry landform areas

Continental shelf

- - - Undersea axial connections of the Alpine system   Major oceanic rift and fault lines

TROPIC OF CANCER

ANTARCTIC CIRCLE

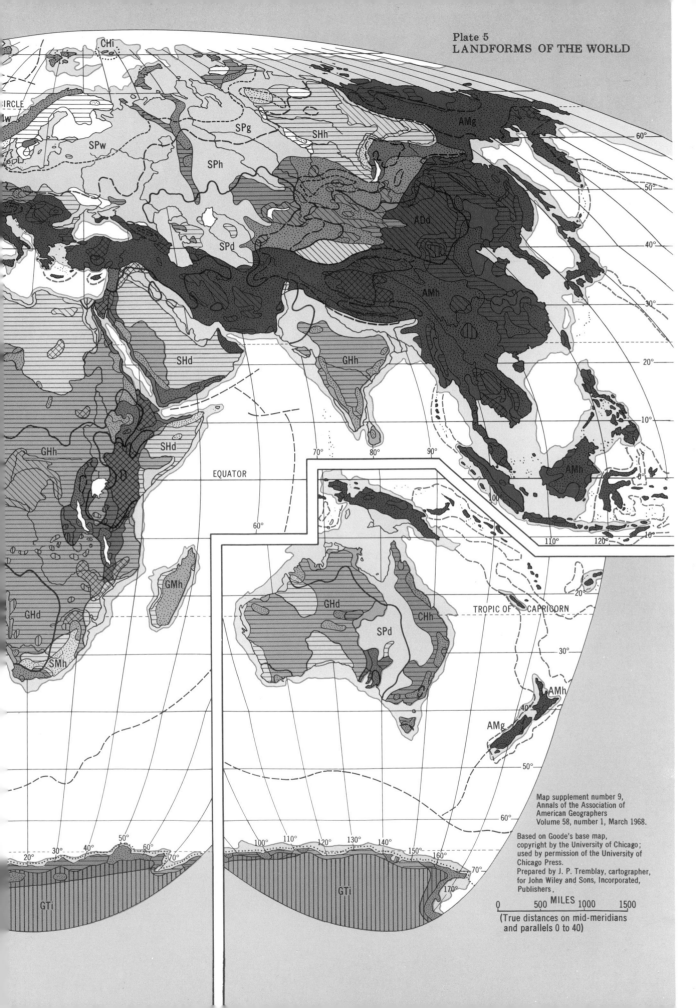

Plate 5
LANDFORMS OF THE WORLD

CHi

IRCLE
w

SPw

SPg

SHh

AMg

SPh

ADa

SPd

AMh

SHd

GHh

SMh

GHh

SHd

EQUATOR

70°

80°

90°

60°

100

110°

120°

GMh

TROPIC OF CAPRICORN

GHd

SPd

CHh

GHd

SMh

AMh

AMg

Map supplement number 9,
Annals of the Association of
American Geographers
Volume 58, number 1, March 1968.

Based on Goode's base map,
copyright by the University of Chicago;
used by permission of the University of
Chicago Press.
Prepared by J. P. Tremblay, cartographer,
for John Wiley and Sons, Incorporated,
Publishers.

100° 110° 120° 130° 140° 150° 160°

20° 30° 40° 50° 60° 70°

GTi

GTi

170°

0   500   MILES 1000   1500

(True distances on mid-meridians
and parallels 0 to 40)

60°

50°

40°

30°

20°

10°

0°

10°

20°

30°

40°

50°

60°

70°

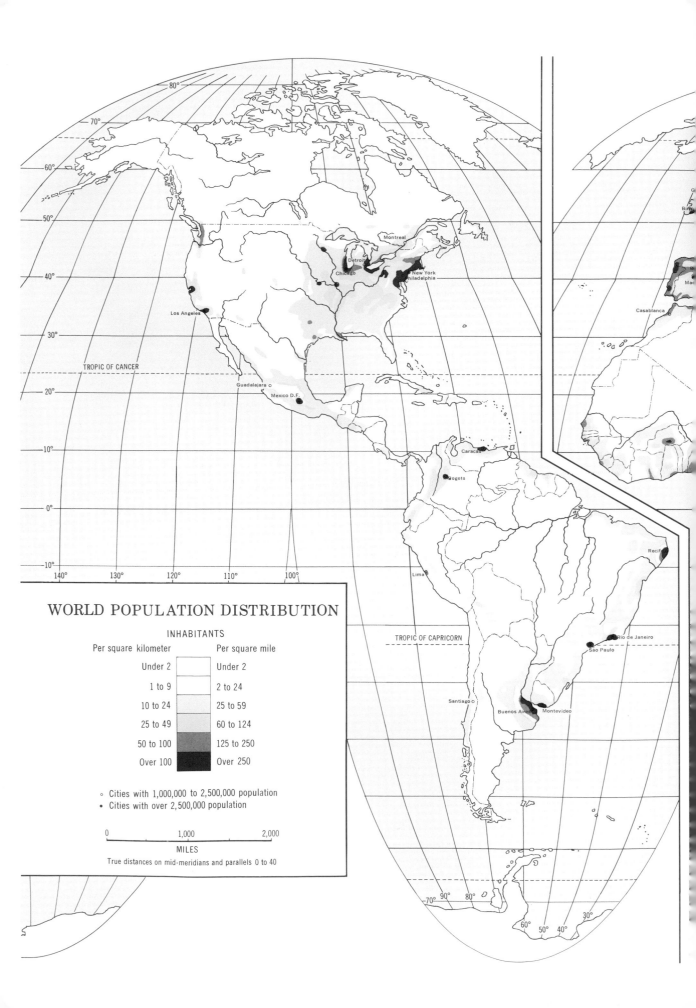

WORLD POPULATION DISTRIBUTION

INHABITANTS

Per square kilometer	Per square mile
Under 2	Under 2
1 to 9	2 to 24
10 to 24	25 to 59
25 to 49	60 to 124
50 to 100	125 to 250
Over 100	Over 250

○ Cities with 1,000,000 to 2,500,000 population
● Cities with over 2,500,000 population

0   1,000   2,000
MILES
True distances on mid-meridians and parallels 0 to 40

TROPIC OF CANCER

TROPIC OF CAPRICORN

Montreal
Detroit
Chicago
New York
Philadelphia
Los Angeles
Guadalajara
Mexico D.F.
Caracas
Bogota
Lima
Recife
Rio de Janeiro
Sao Paulo
Santiago
Buenos Aires
Montevideo
Casablanca
Madrid

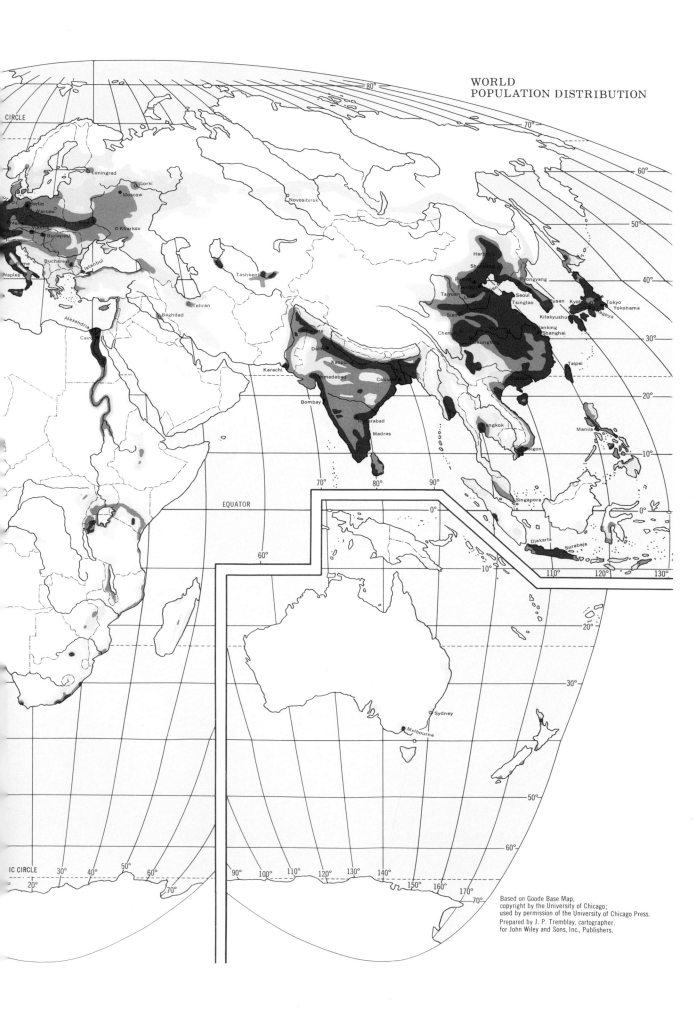

WORLD
POPULATION DISTRIBUTION

CIRCLE

80°

70°

60°

50°

Leningrad

Gorki

Moscow

Berlin
Warsaw

Vienna
Budapest

Kharkov

Novosibirsk

Harbin

Shenyang

Rome
Bucharest

Istanbul

Naples

Tashkent

Peiping
Tientsin
Taiyuan

Pyongyang

Seoul
Tsingtao

Pusan
Kyoto
Tokyo
Yokohama
Nagoya

Alexandria

Baghdad

Tehran

Sian

Chengtu

Nanking
Shanghai

Kitakyushu

Cairo

Lahore

Chungking

Wuhan

Taipei

Karachi

Delhi

Kanpur

Calcutta

Bombay

Ahmedabad

Hyderabad

Madras

Bangkok

Saigon

Manila

EQUATOR

Singapore

Djakarta
Surabaja

Sydney

Melbourne

IC CIRCLE

Based on Goode Base Map,
copyright by the University of Chicago;
used by permission of the University of Chicago Press.
Prepared by J. P. Tremblay, cartographer,
for John Wiley and Sons, Inc., Publishers.

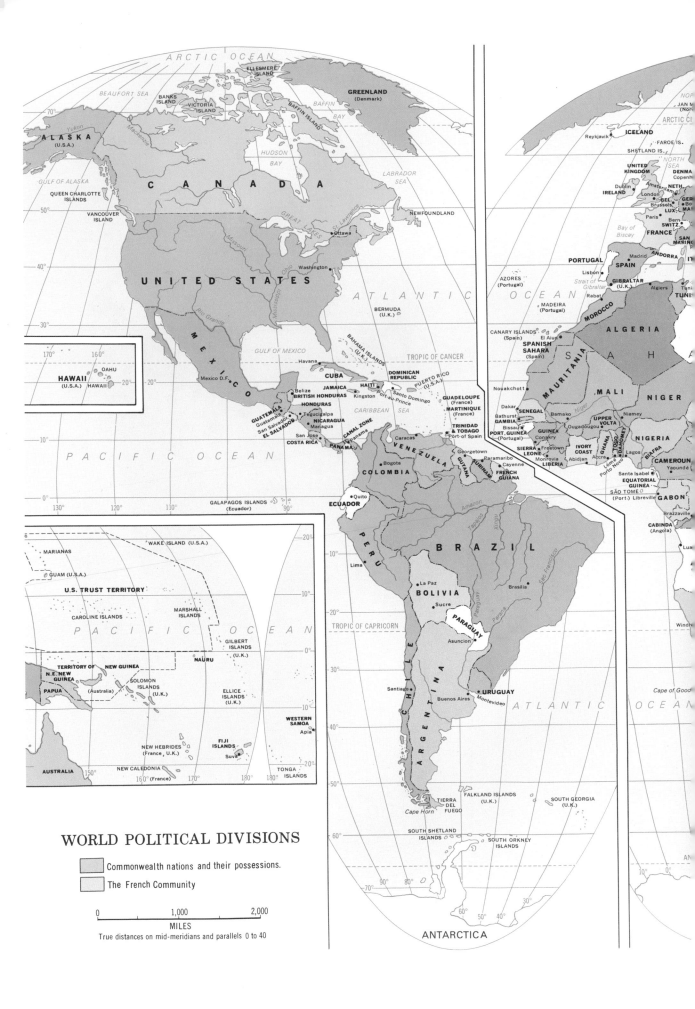

WORLD POLITICAL DIVISIONS

Commonwealth nations and their possessions.

The French Community

0   1,000   2,000

MILES

True distances on mid-meridians and parallels 0 to 40

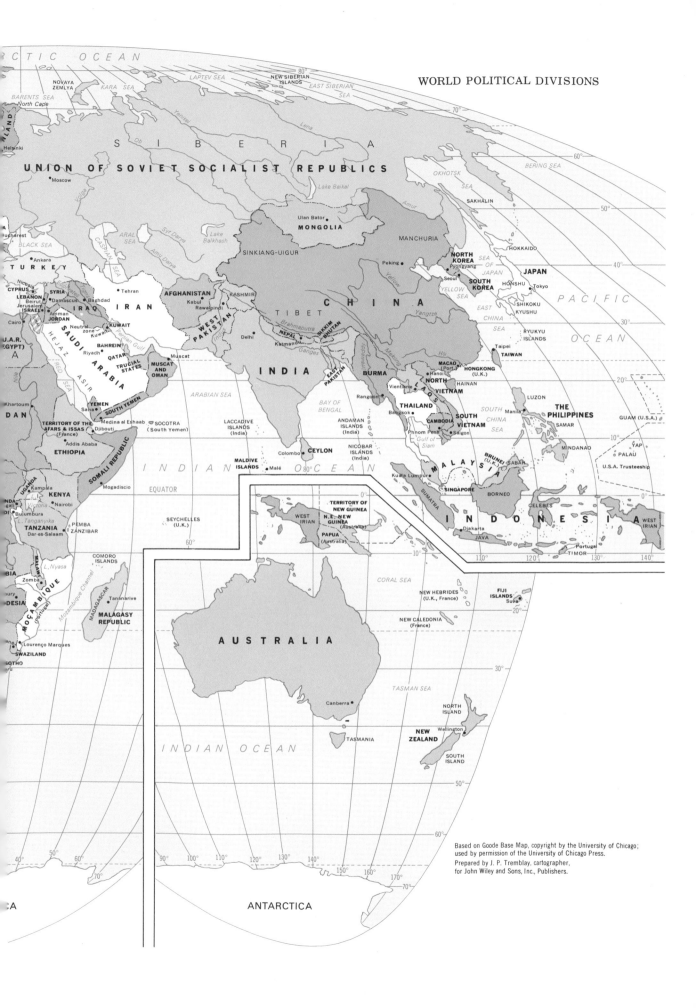

ARCTIC OCEAN

LAPTEV SEA
EAST SIBERIAN SEA
NEW SIBERIAN ISLANDS
NOVAYA ZEMLYA
KARA SEA
BARENTS SEA
North Cape
LAND
Helsinki

SIBERIA

Ob
Yenisei
Lena

BERING SEA

UNION OF SOVIET SOCIALIST REPUBLICS
• Moscow
Volga
Amur
SAKHALIN
OKHOTSK SEA
HOKKAIDO

ARAL SEA
Syr Darya
Lake Balkhash
Amu Darya
CASPIAN SEA
Ulan Bator •
MONGOLIA
MANCHURIA
Peking •
NORTH KOREA
Pyongyang •
SEA OF JAPAN
HONSHU
JAPAN
Tokyo •

ucharest
BLACK SEA
Ankara
TURKEY
Nicosia
CYPRUS
LEBANON Beirut
Damascus •
SYRIA
Baghdad •
IRAQ
Tehran •
AFGHANISTAN
Kabul •
Rawalpindi •
KASHMIR
SINKIANG-UIGUR
TIBET
CHINA
Yellow
Yangtze
Seoul •
SOUTH KOREA
SHIKOKU
KYUSHU
YELLOW SEA
EAST CHINA SEA
RYUKYU ISLANDS
Taipei •
TAIWAN

PACIFIC OCEAN

Jerusalem
ISRAEL
Amman
JORDAN
Cairo •
U.A.R. (EGYPT)
A
Neutral zone
KUWAIT
Kuwait •
BAHREIN
QATAR
Persian Gulf
TRUCIAL STATES
MUSCAT AND OMAN
Muscat •
WEST PAKISTAN
Delhi •
NEPAL
Katmandu •
SIKKIM
BHUTAN
Brahmaputra
Ganges
Indus
EAST PAKISTAN
BURMA
Irrawaddy
Mekong
Hsi
Hanoi •
LAOS
Vientiane •
NORTH VIETNAM
HAINAN
MACAO (Port.)
HONGKONG (U.K.)

SAUDI ARABIA
HEJAZ
Riyadh •
ASIR
IRAN

Khartoum •
DAN
YEMEN
Sana •
SOUTH YEMEN
Medina al Eshaab
Djibouti •
TERRITORY OF THE AFARS & ISSAS (France)
Addis Ababa •
ETHIOPIA
SOCOTRA (South Yemen)
ARABIAN SEA
LACCADIVE ISLANDS (India)
BAY OF BENGAL
Rangoon •
ANDAMAN ISLANDS (India)
THAILAND
Bangkok •
CAMBODIA
Phnom Penh •
Saigon •
SOUTH VIETNAM
Gulf of Siam
SOUTH CHINA SEA
Manila •
LUZON
SAMAR
THE PHILIPPINES
MINDANAO
GUAM (U.S.A.)
YAP
PALAU
U.S.A. Trusteeship

UGANDA
Kampala •
SOMALI REPUBLIC
Mogadiscio •
NDA
gali
Bujumbura
KENYA
Nairobi •
PEMBA
ZANZIBAR
L. Tanganyika
TANZANIA
Dar-es-Salaam
Colombo •
CEYLON
Malé •
MALDIVE ISLANDS
NICOBAR ISLANDS (India)
INDIAN OCEAN
EQUATOR
SEYCHELLES (U.K.)
TERRITORY OF NEW GUINEA
WEST IRIAN
N.E. NEW GUINEA (Australia)
PAPUA (Australia)
Kuala Lumpur •
MALAYSIA
SINGAPORE
SUMATRA
BORNEO
BRUNEI (U.K.)
SABAH
CELEBES
INDONESIA
WEST IRIAN

BIA
L. Nyasa
Zomba •
MOÇAMBIQUE (Portugal)
bury
DESIA
Lourenço Marques •
SWAZILAND
SOTHO
ru
COMORO ISLANDS
Mozambique Channel
MADAGASCAR
Tananarive •
MALAGASY REPUBLIC
Djakarta •
JAVA
Portugal
TIMOR

CORAL SEA
NEW HEBRIDES (U.K., France)
FIJI ISLANDS
Suva •
NEW CALEDONIA (France)

AUSTRALIA

Canberra •

TASMAN SEA

NORTH ISLAND
Wellington •
NEW ZEALAND
SOUTH ISLAND

INDIAN OCEAN

TASMANIA

Based on Goode Base Map, copyright by the University of Chicago; used by permission of the University of Chicago Press.

Prepared by J. P. Tremblay, cartographer, for John Wiley and Sons, Inc., Publishers.

CA

ANTARCTICA